Mechanisms of Cytotoxicity by NK Cells

Academic Press Rapid Manuscript Reproduction

Based on The Third International Workshop on
NK Cells held at Meadow Brook Hall, Oakland
University, Rochester, Michigan in May 1984

Mechanisms of Cytotoxicity by NK Cells

Edited by

Ronald B. Herberman
Biological Therapeutics Branch
Biological Response Modifiers Program
National Cancer Institute
National Institutes of Health
Frederick, Maryland

Denis M. Callewaert
Department of Chemistry
Oakland University
Rochester, Michigan

1985

ACADEMIC PRESS, INC.
(Harcourt Brace Jovanovich, Publishers)
Orlando San Diego New York London
Toronto Montreal Sydney Tokyo

ACADEMIC PRESS, INC.
Orlando, Florida 32887

United Kingdom Edition published by
ACADEMIC PRESS INC. (LONDON) LTD.
24–28 Oval Road, London NW1 7DX

LIBRARY OF CONGRESS CATALOG CARD NUMBER: 85-70441

ISBN 0-12-341370-2

PRINTED IN THE UNITED STATES OF AMERICA

85 86 87 88 9 8 7 6 5 4 3 2 1

Contents

II Characterization of NK Target Structures and Recognition Sites on NK Cells

III Mechanisms of Postrecognition Interactions between NK Cells and Targets

V Cultured Cells with NK-Like Activity: Models for Analysis of NK Activity

Contributors

Numbers in parentheses indicate the pages on which the authors' contributions begin.

E. W. Ades (231), *Department of Immunology Research, Lilly Research Laboratories, Indianapolis, Indiana 46268*

Paola Allavena (617), *Biological Therapeutics Branch, Biological Response Modifiers Program, DCT, NCI-Frederick Cancer Research Facility, Frederick, Maryland 21701*

Sumio Arai (489), *Department of Laboratory Medicine, Kumamoto University Medical School, Kumamoto, Japan*

Emira Ayroldi (421), *Institute of Pharmacology, University of Perugia, Perugia, Italy*

Theodore F. Beals (195), *Department of Pathology, University of Michigan Veterans Administration Medical Center, Ann Arbor, Michigan 48109*

J. G. Bekesi (117), *Department of Neoplastic Diseases, Mount Sinai Medical School, New York, New York 10029*

Philip J. Bierman (465), *Experimental Immunology Laboratory, Department of Internal Medicine, Veterans Administration Medical Center and the University of Nebraska Medical Center, Omaha, Nebraska 68101*

Isaac Blanca (335), *Biological Response Modifiers Program, National Cancer Institute, Frederick Cancer Research Facility, Frederick, Maryland 21701*

Robert Blumenthal (305), *Laboratory of Theoretical Biology, NCI, National Institutes of Health, Bethesda, Maryland 20205*

Reinder L. H. Bolhuis (543, 631), *Rotterdam Radio-Therepeutic Institute, Rotterdam and Radiobiological Institute TNO, Rijswijk, The Netherlands*

Benjamin Bonavida (323, 477), *Department of Microbiology and Immunology, UCLA School of Medicine, Los Angeles, California 91701*

Claudio Bordignon (93), *Department of Pathology, State University of New York at Buffalo, Buffalo, New York 14214*

Colin G. Brooks (245, 263, 645), *Basic Immunology, Fred Hutchinson Cancer Research Center, Seattle, Washington 98104*

Michael J. Brunda (397), *Department of Experimental and Applied Biology, Roche Research Center, Hoffmann-La Roche Inc., Nutley, New Jersey 07110*

Gordon F. Burns (589), *Lions Clinical Cancer Research Laboratory, Clinical Research Unit of the Walter and Eliza Hall Institute of Medical Research, Melbourne, Victoria, Australia*

Denis M. Callewaert (17, 381), *Department of Chemistry, Oakland University, Rochester, Michigan 48063*

Kathleen Carine (213), *UC San Diego Cancer Center, University of California, San Diego, La Jolla, California 92093*

Olli Carpén (185), *Department of Pathology, University of Helsinki, Helsinki, Finland*

William H. Chambers (351), *Division of Oncology, Department of Medicine, Vanderbilt University, Nashville, Tennessee 37203*

Allan R. Chen (287), *Division of Immunology, Duke University Medical Center, Durham, North Carolina 27706*

M. Culwell (231), *Department of Immunology Research, Lilly Research Laboratories, Indianapolis, Indiana 46268*

John P. Daley (93), *Department of Pathology, State University of New York at Buffalo, Buffalo, New York 14214*

Viveca Davatelis (397), *Department of Experimental and Applied Biology, Roche Research Center, Hoffmann-La Roche Inc., Nutley, New Jersey 07110*

Richard L. Deem (155), *Geriatric Research, Education and Clinical Center, Wadsworth VA Medical Center, Los Angeles, California and Departments of Medicine, and Microbiology and Immunology, University of California, Los Angeles School of Medicine, Los Angeles, California 90024*

Gino Degliantoni (29), *The Wistar Institute for Anatomy and Biology, Philadelphia, Pennsylvania 19104*

Gunther Dennert (173), *Comprehensive Cancer Center, University of Southern California, School of Medicine, Los Angeles, California 90033*

Patricia A. Fitzgerald (125), *Laboratory of Herpesvirus Infections, Memorial Sloan-Kettering Institute for Cancer Research, New York, New York*

James T. Forbes (351), *Division of Oncology, Department of Medicine, Vanderbilt University, Nashville, Tennessee 37203*

Guido Forni (433), *Istituto di Microbiologia Universitat, degli Studi di Torino, Turin, Italy*

Luigi Frati (433), *Istituto di Patologia Generale Universitat, degli Studi ''La Sapienze'', Rome, Italy*

Yoshiaki Fujimiya (521), *Departments of Pathology and Microbiology, University of Southern California School of Medicine, Los Angeles, California 90007*

Antonio Giampietri (421), *Institute of Pharmacology, University of Perugia, Perugia, Italy*

Ronald H. Goldfarb (137, 205, 263), *Department of Immunology and Infectious Disease, Central Research Division, Pfizer, Inc., Groton, Connecticut*

R. J. van de Griend (631), *Rotterdam Radio-Therepeutic Institute, Rotterdam and Radiobiological Institute TNO, Rijswijk, The Netherlands*

Maryanna Henkart (305), *Immunology Branch, NCI, National Institutes of Health, Bethesda, Maryland 20205*

Pierre Henkart (305), *Immunology Branch, NCI, National Institutes of Health, Bethesda, Maryland 20205*

Ronald B. Herberman (1, 335, 421, 433, 617), *Biological Therapeutics, Biological Response Modifiers Program, DCT, National Cancer Institute, Frederick Cancer Research Facility, Frederick, Maryland 21701*

J. P. Hester (507), *Department of Hematology, The University of Texas System Cancer Center, M.D. Anderson Hospital and Tumor Institute, Houston, Texas*

A. Hinson (231), *Department of Immunology Research, Lilly Research Laboratories, Indianapolis, Indiana 46268*

John C. Hiserodt (195), *Department of Pathology, University of Michigan, Veterans Administration Medical Center, Ann Arbor, Michigan 48109*

Dorothy Hudig (71, 213), *Department of Microbiology, School of Medicine, University of Nevada, Reno, Nevada 89507*

Margot M. Ip (253), *Grace Cancer Drug Center, Roswell Park Memorial Institute, Buffalo, New York*

Donald R. Johnson (361), *Department of Pathology and Laboratory Medicine, Department of Microbiology and Eppley Institute for Research on Cancer and Allied Diseases, University of Nebraska Medical Center, Omaha, Nebraska, 68105*

Joseph Kaplan (61, 103), *Department of Pediatrics, Wayne State University School of Medicine, Detroit, Michigan 48201*

Klas Kärre (81), *Department of Tumor Biology, Karolinska Institute, S-104 01 Stockholm, Sweden*

H. David Kay (263, 465), *Experimental Immunology Laboratory, Department of Internal Medicine, Veterans Administration Medical Center and the University of Nebraska Medical Center, Omaha, Nebraska 68105*

M. J. Keating (507), *Department of Hematology, The University of Texas System Cancer Center, M.D. Anderson Hospital and Tumor Institute, Houston, Texas*

Tatsuo Kiyohara (223), *Department of Microbiology and Immunology, Queen's University, Kingston, Ontario, Canada*

Lynell W. Klassen (465), *Experimental Immunology Laboratory, Department of Internal Medicine, Veterans Administration Medical Center, and the University of Nebraska Medical Center, Omaha, Nebraska 68105*

Eva Klein (443), *Department of Tumor Biology, Karolinska Institutet, S-104 01 Stockholm, Sweden*

Michiko Kobayashi (29), *The Wistar Institute for Anatomy and Biology, Philadelphia, Pennsylvania 19104*

Hillel S. Koren (253, 287, 373, 607), *Department of Microbiology and Immunology, Duke University Medical Center, Durham, North Carolina 27706*

Katsuo Kumagai (489), *Department of Microbiology, Tohoku University, School of Dentistry, Sendai, Japan*

Abraham Kupfer (173), *Department of Biology, University of California, San Diego, La Jolla, California 92093*

Lewis L. Lanier (563), *Becton Dickinson Monoclonal Center, Inc., Mountain View, California 94043*

Edmund C. Lattime (409), *Immunobiology Program, Memorial Sloan-Kettering Cancer Center, New York, New York*

Robert J. Lauzon (223), *Department of Microbiology and Immunology, Queen's University, Kingston, Ontario, Canada*

Kam H. Leung[1] (253), *Grace Cancer Drug Center, Roswell Park Memorial Institute, Buffalo, New York*

T. Juhani Linna (49), *Central Research and Development Department, Glenolden Laboratory, E.I. du Pont de Nemours and Company, Glenolden, Pennsylvania*

Lucille London (29), *The Wistar Institute for Anatomy and Biology, Philadelphia, Pennsylvania 19104*

Carlos Lopez (125), *Laboratory of Herpesvirus Infections, Memorial Sloan-Kettering Institute for Cancer Research, New York, New York*

E. Lotzová (507), *Laboratory of Immunogenetics, Department of General Surgery, The University of Texas System Cancer Center, M.D. Anderson Hospital and Tumor Institute, Houston, Texas*

Stuart Macphail (409), *Immunobiology Program, Memorial Sloan-Kettering Cancer Center, New York, New York*

Norma H. Mahle (381), *Department of Chemistry, Oakland University, Rochester, Michigan 48063*

E. J. Margalioth (117), *Department of Obstetrics and Gynecology, Hadassh University Hospital, Jerusalem, Israel*

Llewellyn Mason (617), *Biological Therapeutics Branch, Biological Response Modifiers Program, DCT, NCI-Frederick Cancer Research Facility, Frederick, Maryland 21701*

Maria-Grazia Masucci (443), *Department of Tumor Biology, Karolinska Institutet, S-104 01 Stockholm, Sweden*

Graziella Migliorati (421), *Institute of Pharmacology, University of Perugia, Perugia, Italy*

Paul Millard (305), *Immunology Branch, NCI, National Institutes of Health, Bethesda, Maryland 20205*

Lory Minning (71, 213), *UC Cancer Center, University of California, San Diego, La Jolla, California 92093*

Madhavan P. N. Nair (453), *Departments of Pediatrics and Epidemiology, The University of Michigan, Ann Arbor, Michigan 48109*

Ichiro Nakamura (93), *Department of Pathology, State University of New York at Buffalo, Buffalo, New York 14214*

Walter Newman (655), *Department of Molecular Immunology, Ortho Pharmaceutical Corporation, Raritan, New Jersey 08869*

Thomas Oeltmann (351), *Division of Oncology, Department of Medicine, Vanderbilt University, Nashville, Tennessee 37203*

[1]Present address: Central Research and Development Department, Glenolden Laboratory, E. I. du Pont de Nemours and Company, Glenolden, Pennsylvania.

John R. Ortaldo (335, 617), *Biological Response Modifiers Program, DCT, National Cancer Institute–Frederick Cancer Research Facility, Frederick, Maryland 21701*

Mahadev R. Patel (49), *Central Research and Development Department, Glenolden Laboratory, E.I. du Pont de Nemours and Company, Glenolden, Pennsylvania*

Paul K. Pattengale (521), *Departments of Pathology and Microbiology, University of Southern California, School of Medicine, Los Angeles, California 90007*

Graham Pawelec (655), *Immunology Laboratory, Medizinische Klinik, D-7400 Tübingen, West Germany*

Peter Pereira (499), *Department of Biological Structure, University of Washington, Seattle, Washington 98105*

Bice Perussia (29), *The Wistar Institute for Anatomy and Biology, Philadelphia, Pennsylvania 19104*

Howard T. Petrie (465), *Experimental Immunology Laboratory, Department of Internal Medicine, Veterans Administration Medical Center and the University of Nebraska Medical Center, Omaha, Nebraska 68101*

Joseph H. Phillips (563), *Becton Dickinson Monoclonal Center, Inc., Mountain View, California 94043*

Mario Piccoli (433), *Istituto di Patologia Generale Universitá, degli Studi "La Sapienze", Rome, Italy*

Sylvia B. Pollack (39, 499), *Department of Biological Structure, University of Washington, Seattle, Washington 98105*

Hugh F. Pross (529), *Departments of Radiation Oncology and Microbiology and Immunology, Queen's University at Kingston, Ontario, Canada*

Oscar Ramos (443), *Department of Tumor Biology, Karolinska Institutet, S-104 01 Stockholm, Sweden*

Doug Redelman (213), *Department of Microbiology, School of Medicine, University of Nevada, Reno, Navada 89507*

Craig Reynolds (305), *Biological Therapeutics Branch, NCI, Frederick Cancer Research Facility, Frederick, Maryland 21701*

Carlo Riccardi (421), *Institute of Pharmacology, University of Perugia, Perugia, Italy*

Claire P. Robles (499), *Department of Biological Structure, University of Washington, Seattle, Washington 98105*

John C. Roder (223), *Department of Microbiology and Immunology, Queen's University, Kingston, Ontario, Canada*

Robert C. Roozemond (477), *Laboratory of Histology and Cell Biology, University of Amsterdam, Amsterdam, The Netherlands*

Eero Saksela (185), *Department of Pathology, University of Helsinki, Finland*

Angela Santoni (433), *Istituto di Patologia Generale Universitá, degli Studi "La Sapienze", Rome, Italy*

C. A. Savary (507), *Laboratory of Immunogenetics, Department of General Surgery, The University of Texas System Cancer Center, M.D. Anderson Hospital and Tumor Institute, Houston, Texas*

Thomas E. Schindler[2] (125), *Laboratory of Herpesvirus Infections, Memorial Sloan-Kettering Institute for Cancer Research, New York, New York*

M. Schlesinger (117), *Hubert H. Humphrey Center for Experimental Medicine and Cancer Research, The Hebrew University, Hadassh Medical School, Jerusalem, Israel*

Stanley A. Schwartz (453), *Departments of Pediatrics and Epidemiology, The University of Michigan, Ann Arbor, Michigan 48109*

William Seaman (245), *Department of Medicine, VA Medical Center and University of California, San Francisco, California, 94143*

J. S. Singer (173), *Department of Biology, University of California, San Diego, La Jolla, California 92093*

Osias Stutman (409), *Immunobiology Program, Memorial Sloan-Kettering Cancer Center, New York, New York*

Ryuji Suzuki (489), *Department of Microbiology, Tohoku University, School of Dentistry, Sendai, Japan*

Satsuki Suzuki (489), *Department of Microbiology, Tohoku University, School of Dentistry, Sendai, Japan*

Stephan Targan (155), *Geriatric Research, Education and Clinical Center, Wadsworth VA Medical Center, Los Angeles, California and Departments of Medicine, and Microbiology and Immunology, University of California, Los Angeles School of Medicine, Los Angeles, California 90024*

Tuomo Timonen (599), *Department of Pathology, University of Helsinki, Helsinki, Finland*

William W. Travis[3] (607), *Division of Immunology, Duke University Medical Center, Durham, North Carolina 27706*

Tony Triglia (589), *Lions Clinical Cancer Research Laboratory, Clinical Research Unit of the Walter and Eliza Hall Institute of Medical Research, Melbourne, Victoria, Australia*

Giorgio Trinchieri (29), *The Wistar Institute for Anatomy and Biology, Philadelphia, Pennsylvania 19104*

Ismo Virtanen (185), *Department of Pathology, University of Helsinki, Finland*

Elizabeth A. Wayner (263), *Basic Immunology, Fred Hutchinson Cancer Research Center, Seattle, Washington 98104*

Monica Weitzen (477), *Department of Microbiology and Immunology, UCLA School of Medicine, Los Angeles, California 21701*

[2]Present address: Xytronyx, Inc., 10 West 35th Street, Chicago, Illinois 60616.

[3]Present address: Department of Microbiology and Immunology, Baylor College of Medicine, Houston, Texas 77030.

[4]Present address: The Walter and Eliza Hall Institute of Medical Research, Melbourne, Australia.

Jerome A. Werkmeister (529, 589), *Departments of Radiation and Oncology and Microbiology and Immunology, Queen's University at Kingston, Ontario, Canada and Clinical Research Unit of the Walter and Eliza Hall Institute of Medical Research, Melbourne, Victoria, Australia*

Peter Wernet (655), *Immunology Laboratory, Medizinsche Klinik, D-7400 Tübingen, West Germany*

Stanley M. Wilbur (323), *Department of Microbiology and Immunology, UCLA School of Medicine, Los Angeles, California 21701*

Janet Woodcock (245), *Department of Medicine, VA Medical Center and University of California, San Francisco, California 94143*

Pascale Wortley (499), *Department of Biological Structure, University of Washington, Seattle, Washington 98105*

Susan C. Wright (323), *Department of Microbiology and Immunology, UCLA School of Medicine, Los Angeles, California 21701*

Preface

About a dozen years have elapsed since the first reports appeared of cytotoxic reactivity of lymphocytes from normal donors against tumor cell lines. While the phenomenon of natural cytotoxicity was at first considered by many immunologists to be an *in vitro* artifact and for several years was only studied by a few investigators, recent rapid advances in our understanding of these cells, coupled with mounting evidence for multiple *in vivo* functions, have catapulted work on NK cells into a major aspect of modern immunological research. In addition to their likely role in antitumor defenses, NK cells appear to play an important role in natural resistance to some viral and other microbial infections, in rejection of bone marrow transplants, and in regulation of other immune responses, including antibody production.

Increasing suggestions for the *in vivo* importance of NK cells, along with rapid advances in the ability to purify and characterize NK cells, have stimulated the interest of increasing numbers of immunologists as well as scientists from other disciplines who, in turn, have applied new techniques to further enhance our understanding of these cells and their functions. There have also been parallel advances in our insight in other natural effector cells, which also may make important contributions to natural immunity against disease. The availability of highly purified populations of effector cells, of monoclonal antibodies reactive with them, and of homogenous recombinant cytokines that affect their reactivity has led to particularly impressive progress in our understanding of the mechanism(s) by which NK cells lyse susceptible target cells and of the mechanisms of the regulation of NK activity.

Attempts to help the scientific community stay abreast of the latest developments in this field have led to the publication of several review articles on different aspects of natural cell-mediated immunity, and up-to-date detailed summaries of findings from leading investigators in the field have been compiled in two major volumes: ''Natural Cell-Mediated Immunity against Tumors'' (R. B. Herberman, editor; Academic Press Inc., 1980) and ''NK Cells and Other Natural Effector Cells'' (R. B. Herberman, editor; Academic Press, Inc., 1982). The length of the latter work (over 1500 pages) and the continued rapid increase in the number of relevant publications in peer-reviewed journals attest to the fact that work in this field has become so extensive that it can no longer be adequately reported in breadth as well as depth. This volume is based on an international workshop on these topics, devoted to an extensive summary of studies related to some important and rapidly advancing aspects of the field.

Major immunology meetings are devoting additional sessions to NK cells and other natural effector mechanisms, and yet the increasing number and diversity of presentations on both basic and clinical aspects of natural immunity have made it more and more difficult to keep pace with all of the developments in this area. For scientists primarily interested in NK cells, it has also become more difficult to interact in an environment suitable for extensive exchange of information and discussion. For these reasons, the First International Workshop devoted exclusively to fundamental research on NK cells, held in North Carolina in the summer of 1982, was organized by R. B. Herberman and H. Koren. The following summer, in conjunction with the International Immunology Congress in Kyoto, Japan, another international workshop on NK cells was organized (by T. Hoshino). The success of these meetings, coupled with continued rapid advances in our understanding of NK cells, prompted the organization (by D. Callewaert, R. B. Herberman, J. Kaplan, H. Koren, and H. Pross) of a Third International Workshop which was held at Meadow Brook Hall on the grounds of Oakland University, Rochester, Michigan, in the spring of 1984. This three-day workshop, attended by fifty leading investigators, focused on recent studies that augment our understanding of the basic mechanisms involved in the function and regulation of NK cells. The limited number of participants, the residential setting, the focus of the meeting, and the format of the sessions contributed to lively and intense discussions that highlighted major advances as well as points of continued contention.

The expenses of the workshop, including transportation for a number of scientists, was made possible by a conference grant (no. CA/A1 37226) from the National Institutes of Health as well as by gifts and grants from the Oakland University Research Fund; the Oakland University Alumni Association; Meadow Brook Hall, a cultural enterprise of Oakland University; Advance Medical and Research Laboratories, Inc.; Boehringer Manneheim Biochemicals; E. I. DuPont De Nemours and Company; Hoffmann-La Roche, Inc., The New England Nuclear Company; Pfizer Central Research; The Monsanto Company; and The Upjohn Company.

The organizers and participants at the Meadow Brook Hall Workshop agreed that it would be important to have a rapid and concise publication of the key findings which were presented. However, rather than merely prepare a report of the workshop or of the workshop proceedings, a decision was made to use the workshop and the topics discussed as a focus for this volume since they represent some of the most important progress in research related to natural effector cells. Formal presentations at the workshop have not been included as such. Rather, selected participants were asked to provide more detailed accounts of their most significant findings than was possible at the workshop and to discuss the broader implications of their results and relate them to information published by other investigators. In some cases, several investigators were asked to combine related findings so as to provide the reader with the newest information in the briefest fashion possible. In addition, where appropriate, scientists who did not participate in the workshop were invited to contribute manuscripts to provide greater depth of coverage of relevant topics.

The book is divided into five major sections corresponding to the workshop sessions. Each section begins with an introduction in which the session chairman summarizes the presentations and discussions which took place during that session. Tape recordings were provided to the chairmen to insure the accuracy and comprehensiveness of these summaries. We believe that these carefully prepared summaries relate both the highlights of the sessions and the flavor of the extensive discussions.

This volume should impart to the reader a clear view of the present status of knowledge of the mechanisms responsible for the cytotoxic effects of NK cells and the regulation of their activity. It should therefore prove invaluable to researchers studying basic or applied aspects of natural immunity and to investigators concerned with cytotoxic mechanisms or regulation of other types of effector cells. It should also be of benefit to the broader immunological community as well as to clinicians interested in keeping abreast of the rapid advance being made in this exciting and important field.

The special attention given the assemblage and organization of this information will provide the reader with a rapid overview of the status of and prospects for this dynamic field of modern immunobiology. By printing the text directly from the authors' typewritten manuscripts production time has been cut to only several months after manuscript submission, thus ensuring publication of current material.

DENIS M. CALLEWAERT
RONALD B. HERBERMAN

INTRODUCTION

Ronald B. Herberman

Biological Response Modifiers Program
National Cancer Institute
Frederick, Maryland

It has become increasingly clear over the last several years that there is a variety of different effector mechanisms in addition to T cells. For example, natural killer (NK) cells certainly have arrived on the scene as one type of effector cell that needs to be considered. One of the diffficulties that investigators in the field are facing now is to agree on what an NK cell is and particularly how to distinguish clearly an NK cell from other effector cells. On the one hand, there have been a number of observations over the last couple of years that have made it simpler to define an NK cell, however, other observations have clouded the issue. For that reason, I think it is essential to reconsider the definition of NK cells.

An international committee of investigators with expertise related to various aspects of natural cell-mediated cytotoxicity are currently considering this issue, in conjunction with representatives from the Nomenclature Committee of the Reticuloendothelial Society. They will soon offer a consensus definition for general use by the scientific community. In this context, I offer the following suggestions:

> NK cells are lymphoid cells with azurophilic cytoplasmic granules, which have spontaneous cytotoxic reactivity against tumor cells, and also against virus-infected cells and some normal cells. They have several salient characteristics that set them apart from other cells:

ISBN 0-12-341370-2

- They lack some characteristic features of cytotoxic T cells (CTL). For example, mouse NK cells lack Lyt 2, and human NK cells do not express antigens detected by OKT3 and other related monoclonal antibodies.
- Similarly, NK cells also lack some characteristic cell-surface markers of monocytes or macrophages. For example, human NK cells lack reactivity with the MO2 monoclonal antibody. It appears that adherence properties do not provide a clear distinction from macrophages, since at least some NK cells, particularly activated NK cells, can adhere to plastic or nylon wool. In fact, because of this, it is becoming a challenge to prove that the spontaneous cytotoxicity associated with monocytes or macrophages is really due to such effector cells and not to a few percent of contaminating adherent NK cells (e.g. 1).
- NK cells are closely associated with large granular lymphocytes (LGL). There is now strong evidence that mouse (2,3) as well as human (4) and rat (5) NK cell activity is attributable almost exclusively to cells with the LGL morphology.
- As a major point of distinction from typical CTL activity, NK activity is not restricted by the major histocompatability complex.

Another point that needs to be considered is the clear heterogeneity of NK cells and how to designate such heterogeneity. In considering the various proposals that have been made in this regard over the last several years, perhaps the most straightforward approach would be to define subsets of NK cells in terms of the particular target cell that is being worked with. For example, one could indicate NK(K562) when one is measuring NK against K562, and similarly NK(YAC-1) or NK(HSV-fibroblasts). This would be a way to resolve the question of the relationship of natural cytotoxic (NC) cells to NK cells, i.e. to designate them in terms of the target that is being used, e.g. NK(WEHI 164).

As mentioned above, it is becoming increasingly clear that mouse NK cells as well as NK cells of other species are closely associated with LGL. However, in studies with NK cells in mouse spleens, it has remained quite difficult to obtain highly purified populations of LGL. In contrast, in recent studies by Wiltrout et al. (6), highly purified preparations of mouse LGL have been isolated from the liver after treatment of mice with various immunomodulators. Rather unexpectedly, it has been found that most of the nonadherent infiltrating cells from the liver are LGL. The immunomodulators stimulate about a 20-fold increase in the number of LGL in the liver, thereby providing substantial numbers of cells with this morphology. The augmented NK activity in the liver of these mice has been shown to be attributable to the LGL and not to the other types of infiltrating cells (6). In addition to strengthening the association between NK activity and LGL, these findings also point to a potentially broader role for NK cells than previously envisioned. The ability of NK cells to extensively infiltrate hepatic and pulmonary tissues, and other sites of inflammation (7), indicate that NK cells may play an important role in host resistance at the tissue level as well as in the circulation.

The recent evidence I referred to above, which has led to considerable controversy about the definition of NK activity, has come from studies with cells maintained in culture, especially in the presence of interleukin 2. In a variety of situations, cultured mouse or human lymphoid cells have been found to develop potent cytotoxic reactivity against NK-susceptible targets like K562 or YAC-1. Because of such reactivity, the cultured effector cells have been presumed to be NK cells or at least NK-like. However, I don't think it is satisfactory to consider an NK cell as any cell which kills YAC-1 or K562. A diverse array of cells, including Lyt 2^+ mouse thymocytes or T cells (8,9) and clones of CTL (10,11), and monocytes (12), has been reported to lyse these target cells. It, therefore, seems necessary to distinguish between NK cells which have developed or been maintained in culture and effector cells which have developed in culture from other cell types which are unrelated to NK cells. The clearest situation would seem to be the maintenance of cytotoxic reactivity in culture by highly purified LGL, particularly those positively selected for an NK cell-associated marker like B73.1 or Leu 11 (e.g. 13). However, one must also consider the possibility that T cells (14) other cells lacking the phenotype of NK cells might under some circumstances be able to differentiate in culture into what could be considered to be true NK cells. How might one determine whether such cultured effector cells were NK cells? Analysis of the cell surface markers might be helpful, particularly the markers that are most selectively characteristic of NK cells. For example, in the human the development of B73.1 or Leu 11 markers on the cultured effector cells might be taken as good evidence for a relationship to NK cells. However, two more central features might be considered to be essential characteristics to define a cultured effector cell as an NK cell:

a. Expression of NK recognition receptors for target cells.

Unfortunately, the recognition receptors on NK cells have not yet been defined. However, studies with rat LGL leukemia cell lines with strong NK activity have been found to lack expression of the beta chain of the T cell receptor and to have no detectable rearrangement of this gene (15), indicating that the NK receptor is probably distinct from the T cell receptor. The finding of rearranged T cell receptor genes and/or expression of T cell receptors in some clones of cytotoxic effector cells (16, 17) probably points to a derivation of such cells from CTL or other T cells rather than indicates the nature of the NK receptor.

In the absence of a clear definition of the NK receptor, the detailed specificity characteristics would seem to be a useful parameter. This, to my mind, would seem to require not only that the effector cells would kill an NK-susceptible target but also that there would be no reactivity against targets which are resistant to lysis by NK cells, even after their activation by interferon or IL2. For example, the high levels of cytotoxicity by cultured Lyt 2^+ mouse effectors against P815 target cells (8, 9, 11) which are quite resistant to lysis by NK cells, would seem to point to their distinctness from NK cells.

b. Cytotoxicity mediated by the same mechanism responsible for NK activity.

If the cultured effector cells were found to lyse their targets by a mechanism distinct from that utilized by NK cells, then it would seem that the effector cells could not be considered NK cells. However, the converse finding of the culutred effectors having the same lytic mechanism as NK cells might not be sufficient to conclude identity with NK cells, since CTL or other cell types might in fact share the same lytic pathway with NK cells.

This latter issue, the mechanism of killing by NK cells, is the central theme of this volume. I have been quite impressed over the last year as to how much progress has been made in our understanding of the mechanism of lysis by NK cells. Although research on NK cells has been relatively recent compared to that of CTL, I think we are in a position now to know at least as much about the mechanism of killing by NK cells. Figure 1 shows the model which John Ortaldo, Craig Reynolds, and I have developed, based on a considerable amount of data which has been accumulated, not only from our laboratory but also from many others. It is rather

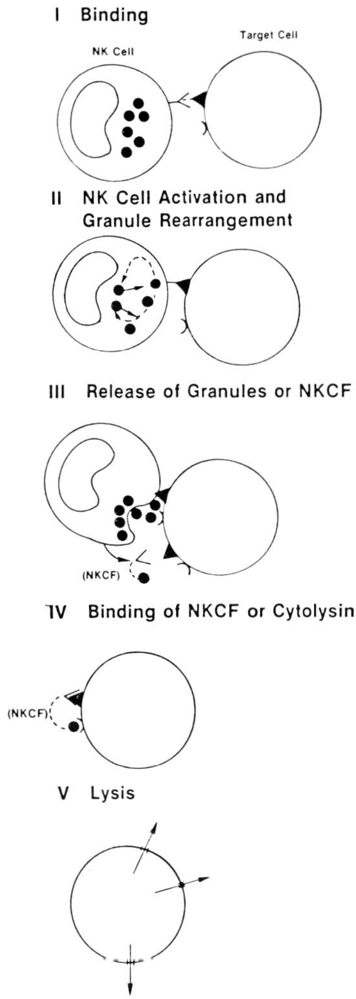

Fig. 1 Model for phases in interaction between NK cells and NK-susceptible target cells.

clear that the NK cell first has to bind to the target cell, through an NK receptor which recognizes particular structures on the target cell. The second phase appears to involve the triggering or activation of intracellular processes in the NK cells probably affecting the cytoplasmic granules and leading to the rearrangement of the granules and other cytoplasmic organelles towards the site of binding with the target cell. Several years ago, Saksela, et al. (18) presented evidence for such rearrangement after binding to target cells, and recently Dennert, et al. (19) have obtained evidence along the same lines. Based on the isolation of the LGL granules and the demonstration of their potent and rapid cytotoxic activity (20), it seems likely that the next step in the process is the release of some granules or their contents from the effector cell, at or in close proximity to the junction with the target. Then some molecules of the granule cytolysin would bind to the target cell, presumably at sites separate from those recognized by the NK receptor. One might expect that most of the target selectivity for lysis is based on the NK receptors, and the finding that LGL granules have less specificity than NK cells (20) is consistent with this expectation. The soluble NK cytotoxic factor (NKCF), which Susan Wright and Ben Bonavida originally described (21), might be an alternative but less efficient mechanism of lysis or it might in some way be related to the granule cytolysin. The observation that antibodies produced against LGL granules will neutralize lysis by NKCF as well as granule-mediated cytotoxicity (20) certainly supports the possibility that NKCF may be closely related in some way to the granule cytolysin. There is evidence for a binding step in the interaction of NKCF with target cells, since NKCF can be shown to be absorbed out by susceptible targets (21, 22) and the target cells will then go on to lyse without any further exogenous exposure to granules or NKCF (22).

Clearly, this model is an oversimplification of a complex sequence of events, and Targan, et al. (23) have carefully dissected out several distinct phases which are subsequent to the binding of NK cells to target cells. However, this model does provide a framework for classifying the mechanism of action of the many factors which have been known over the last several years to either increase or inhibit NK activity. It is now possible to arrange most of these factors at least in a sequence as to whether they are affecting binding or one of the later events (22, 23).

In addition to the need to define the NK recognition receptor, a major challenge in the field has been to determine the nature of the target structure which is recognized by the NK cell and which is responsible for conjugate formation. The approach originally used by Roder, et al. (24) to define the mouse NK target structure on YAC-1 cells, by examining the ability of subcellular materials from the target cells to inhibit conjugate formation between NK cells and target cells, appears to be the most direct procedure available. Lewis, et al. (25) have utilized this approach to analyze the target structure on K562 cells which is recognized by human NK cells. They have performed assays of binding between K562 and purified human LGL to determine which subcellular material from K562 might be able to inhibit binding. They have demonstrated that conjugate formation could be inhibited by a glycoprotein, which bound to Con A-Sepharose and could be eluted by alpha-methylmannoside. The soluble inhibiting material had a specificity which was consistent with the specificity of the intact targets and appered to have a fairly high molecular weight. The separation techniques which have been utilized thus far have resulted in about a thousand-fold purification (25). Although they clearly don't yet have a homogeneous molecule, this study represents substantial progress towards understanding of at least one of the target structures involved in recognition by NK cells.

There is as yet less insight into the nature of the receptor on the target cells for binding of the granule cytolysin or of NKCF. However, there are some indications that the target structure for binding of NKCF is glycosolated. Treatment of NK-susceptible targets with tunicamycin was found to block the interaction of the targets with NKCF (22) and this may account for the ability of phosphorylated monosaccharides to inhibit NK activity.

The final step in the pathway of cytotoxicity is the actual lysis of the target cell. The details of the lytic event remain to be elucidated. However, an intriguing possibility, first suggested by Henkart, et al. (26) and then by Podack, et al. (27), is that the cytolysin causes a channel or ring-like hole to be introduced into the target structure. Support for this possibility has come from studies with the cytolysin isolated from LGL granules (20). The cytolysin was found to cause lysis of liposomes and erythrocytes as well as NK-susceptible target cells. Electron microscopy of lysed liposomes showed rings deposited on the surface. These rings were not seen on intact liposomes. These results suggest that the granule cytolysin is a highly lipophilic protein which is able to insert itself into membrane lipid and produce a hole, which may be sufficient to cause the loss of integrity of the cell. Alternatively, the cytolysin might make a channel which then allows a toxin to enter the target cell and induce the loss of viability.

In addition to the mechanism of cytotoxicity, another issue of concern in this volume is the regulation of NK activity. Until recently, the positive regulation of NK activity was thought to be mediated mainly, if not entirely, by interferons of various types. However, with the advent of studies with recombinant interleukin 2 (IL 2), it has become clear that this is an alternative and potent pathway for augmentation of NK activity. An important question is whether the augmentation of NK activity by IL 2 is dependent on the ability to stimulate gamma interferon production, with the actual positive signal for augmentation being interferon rather than IL 2. There has been some suggestive evidence in support of this possibility. The boosting of human (28) and mouse (29) NK activity by IL 2 has been reported to be inhibited by antibodies to gamma interferon. However, other investigators have failed to block IL 2-mediated boosting of human NK activity by monoclonal antibodies to gamma interferon (J. Djeu; G. Trinchieri, personal communication). It now appears that the inhibition observed with some anti-interferon antisera or monoclonal antibody preparations is due to nonspecific inhibition by materials in serum or ascites fluid rather than to the antibodies (T. Sayers and J. Ortaldo, personal communication). Another line of evidence for the distinctness of the mechanisms for augmentation of NK activity by IL 2 and gamma interferon is the observation that IL 2 is considerably more potent in boosting human NK activity than even high concentrations of gamma interferon (J. Ortaldo, A. Maluish, and R. Herberman, manuscript in preparation).

Another interesting aspect of the studies of augmentation of NK activity by IL 2 has been the examination of the receptor on the NK cells for interaction with IL 2. The only defined receptor for IL 2 is recognized by the monoclonal antibody against TAC. Anti-TAC has been shown to inhibit the proliferation of LGL as well as T cells in response to IL 2 (28). However, this monoclonal antibody has had no detectable effect on boosting of human NK activity by IL 2 (28). This raises the question of whether the boosting of NK activity by IL 2 is dependent on a separate, TAC-independent receptor for IL 2. In contrast to TAC, this postulated distinct IL 2 for NK boosting seems to be expressed on cells in G_0 (28).

Although boosting of NK activity by IL 2 *in vitro* seems independent of cell proliferation, we have recently observed that IL 2, including homogeneous recomibnant IL 2, without any additional signal, will stimulate the proliferation of either mouse spleen cells or purified human LGL (J. Talmadge, J. Ortaldo and R. B. Herberman, unpublished observations). Human recombinant IL 2 was found to cause increased thymidine incorporation over the first several days in culture, although the addition of PHA resulted in substantially higher levels of thymidine incorporation. In contrast to the NK boosting effect of IL 2, this stimulation of proliferation by IL 2 was shown to be completely inhibited by anti-TAC antibodies. These results indicate that although it is difficult to detect TAC on the surface of unstimulated LGL, there must be at least a subset of LGL which are spontaneously expressing the receptors for IL 2. This allows those cells to proliferate at least for a while without any additional signal being added.

One might explain the above observations by postulating that some LGL in normal individuals are stimulated *in vivo* to express IL 2 receptors. Endogenous production of IL 2 might in turn lead to expansion of the LGL population. Evidence for such *in vivo* proliferation of NK cells has come from studies of Biron, *et* al. (30) and Santoni, et al. (31). Administration of immunomodulators or interferon to mice not only increased the NK activity but also increased the number of LGL (31). In addition to such an expansion of LGL in the spleen (31) and liver (6), an increase in the number of LGL has been seen in all sites examined, including the peritoneal cavity and bone marrow (A. Santoni and R. B. Herberman, manuscript in preparation). In each of these sites, treatment with various immunomodulators or interferon caused between a three to twenty-fold increase in the number of detectable LGL as well as an increase in NK activity. These expanded numbers have been found to have a lower density than normal LGL, with at least 20 or 25 percent of the stimulated cells in S phase (31). It has been quite surprising that interferon could cause such effects since it is usually an antiproliferative signal for lymphoid cells. A possible explanation of these data is that interferon affects the subset of LGL which are already expressing receptors for IL 2, by stimulating an increased expression of these receptors. Such an increase in IL 2 receptors on some LGL might result in a synergistic interaction with low levels of endogenous IL 2, to result in considerable proliferation of the LGL population.

Another dimension of immunoregulation to consider in regard to NK cells is that these cells may not only be killer cells but may also have the ability to secrete various cytokines. A few years ago, several groups reported that NK cells appeared to have the ability to secrete interferon (e.g. 32, 33). Further studies confirmed that highly purified human LGL could make various types of interferon (34, 35). More recently, it has become apparent that a variety of other cytokines also could be produced by LGL. Highly purified human LGL, repleted of all detectable T cells or monocytes, could be stimulated to secrete substantial amounts of IL 2 (36), IL 1 (37), B cell growth factor (38), colony stimulating factor (36), and lymphotoxin (39). It appears that for each cytokine, the subset of LGL responsible for the production is overlapping but also a bit different. On the one hand, all of the cytokines including IL 1 are produced by T11 positive cells (38), a marker which corresponds to the receptor for sheep erythrocytes. Different signals have been found to elicit a varying array of cytokines, but the main stimuli identified are either mitogens or NK-susceptible target cells. Release of IL 1 by LGL is stimulated by silica or endotoxin, essentially the same signals that are associated with production of IL 1 by monocytes. The subset of LGL which make IL 1 are also DR antigen positive, raising the possibility that some LGL could have yet another function, and act as accessory cells. In fact, a $T11^+$ DR^+ subset of human LGL has been shown to act as accessory cells for mitogens and also for the antigen streptolysin O (G. Scala, manuscript in preparation). Similarly, J. Djeu (personal communication) has observed that human LGL act as accessory cells for specific responses to influenza virus. One might argue that the LGL with NK activity are different from the ones that have these other functions. One approach to answer this question is to determine whether the various phenotypic subsets of LGL have parallel expression of NK activity and some of these other functions. The results have indicated that the populations that have high NK activity also have a strong ability to make the various cytokines (38). As a more convincing demonstration that the same cell might be able to both have NK activity and make one or more cytokines, it has been possible to study clones of LGL, only some of which have NK activity (38), and examine them for their ability to make various cytokines. It was found that the cytotoxic clones as well as the noncytotoxic clones could make quite appreciable levels of various cytokines. Thus, it seems likely that the same cell has multiple functional abilities. These findings suggest that the activity of NK cells may be subject to autoregulation. LGL might be stimulated by various stimuli, such as tumor cells or viruses, which could lead to cytokine production, and that in turn could activate their cytotoxic activity. This could be a very efficient mechanism by which only the cells which are specifically stimulated by a foreign material would be turned on for cytotoxic activity.

To summarize the above remarks and to serve as an introduction to the more detailed manuscripts that will follow, there has been considerable progress recently in the characterization of NK cells and in our understanding of the factors involved in the regulation of NK activity. Most impressive has been the rapid advances in elucidation of the mechanism(s) leading to cytolysis of target cells. However, despite these new insights, there remain many major questions which remain unanswered or controversial, including the following: a) What is the nature of the NK receptor for recognition of target cells and what are the structures on target cells which are recognized by NK cells? b) What are the intracellular events which transduce the cell surface signals and result in activation and/or regulation of cytolytic activity? c) What determines the sensitivity or resistance of various target cells to lysis by the LGL granule cytolysin or NKCF and in particular, how is the NK cell itself protected from lysis by these potent cytotoxic factors? d) If NKCF is related to the mechanism of lysis by NK cells and to the granule cytolysin, why are the kinetics of lysis by NKCF so slow in comparison with the rapid lysis by the other materials? e) Is the mechanism for lysis by NK cells unique or do various types of effector cells use the same or similar pathways of target cell destruction? f) Do the various situations in which effector activity develop upon culture represent models for the development of NK activity or do they indicate that a variety of unrelated cell types can develop functional activities which only superficially resemble each other?

Hopefully, we will soon have answers to at least some of these important issues.

REFERENCES

1. Chang, Z. L., Hoffman, T., Bonvini, E., Stevenson, H.C., and Herberman, R. B. (1983). Scand. J. Immunol. 18, 439.
2. Luini, W., Boraschi, D., Alberti, S., Aleotti, A., and Tagliabue, A. (1981). Immunol. 43, 663.
3. Kumagai, K., Itoh, K., Suzuki, R., Hinuma, S., and Saitoh, F. (1982). J. Immunol. 129, 388.
4. Timonen, T., Ortaldo, J. R., and Herberman, R. B. (1981). J. Exp. Med. 153, 569.
5. Reynolds, C. W., Timonen, T., and Herberman, R. B. (1981). J. Immunol. 127, 282.
6. Wiltrout, R. H. Mathieson, B. J., Talmadge, J. E., Reynolds, C. W., Zhang, S. R., Herberman, R. B., and Ortaldo, J. R. (In press). J. Exp. Med.
7. Ward, J. M., Argilan, F., and Reynolds, C. W. (1983). J. Immunol. 131, 132.
8. Shortman, K., Wilson, A., and Scollay, R. (1984). J. Immunol. 132, 584.
9. Rosenstein, M., Yron, I., Kaufmann, Y., and Rosenberg, S. A. (1984). Cancer Res. 44, 1946.
10. Binz, H., Fenner, M., Frei, D., and Wigzell, H. (1983). J. Exp. Med. 157, 1252.

11. Brooks, C. G. (1985). In "Mechanisms of Cytotoxicity by NK Cells" (R. B. Herberman, ed.), Academic Press, New York.
12. Fisher, D. G., Hubberd, W. J., and Koren, H. S. (1981). Cell. Immunol. 58, 426.
13. Allavena, P. and Ortaldo, J. R. (1984). J. Immunol. 132, 2353.
14. Grossman, Z. and Herberman, R. B. (1982). In "NK Cells and Other Natural Effector Cells" (R. B. Herberman, ed.) p. 229. Academic Press, New York.
15. Hedrick, S. M., Bonyhadi, M., Young, M. A., and Reynolds, C. W. (Submitted for publication).
16. Lavzon, R. J., Kiyohara, T., and Roder, J. C. (1985). In "Mechanisms of Cytotoxicity by NK Cells" (R. B. Herberman, ed.) Academic Press, New York.
17. Hercend, T., Meuer, S., Brennan, A., Edson, M. A., Acuto, O, Reinherz, E., Schlossman, S. F., and Ritz, J. (1983). J. Exp. Med. 158, 1547.
18. Saksela, E., Carpen, O., and Virtanen, I. (1982). In "NK Cells and Other Natural Effector Cells" (R. B. Herberman, ed.), p. 983. Academic Press.
19. Dennert, G., Jupfer, A., and Singer, J. S. (1985). In "Mechanisms of Cytotoxicity by NK Cells" (R. B. Herberman, ed.), Academic Press, New York.
20. Henkart, P., Henkart, M., Millard, P., Blumenthal, R., and Reynolds, C. W. (1985). In "Mechanisms of Cytotoxicity by NK Cells" (R. B. Herberman, ed.), Academic Press, New York.
21. Wright, S. C. and Bonavida, B. (1981). J. Immunol. 126, 1516.
22. Ortaldo, J. R., Blanca, I., and Herberman. (1985). In "Mechanisms of Cytotoxicity by NK Cells" (R. B. Herberman, ed.), Academic Press, New York.
23. Targan, S. and Deem, R. L. (1985). In "Mechanisms of Cytotoxicity by NK Cells" (R. B. Herberman, ed.), Academic Press, New York.
24. Roder, J. C., Rosen, A., Fenyö, E. M., and Troy, F. (1979). Proc. Natl. Acad. Sci. USA 76, 1405.
25. Henkart, P. A., Lewis, J. T., and Ortaldo, J. R., Submitted for publication.
26. Henkart, M. P. and Henkart, P. A. (1982). Adv. Exp. Med. Biol. 146, 227.
27. Podack, E. R. and Dennert, G. (1983). Nature 302, 442.
28. Ortaldo, J. R., Mason, A. T., Gerard, J. P., Henderson, L. E., Farrar, W., Hopkins, III, R. F., Herberman, R. B., and Rabin, H. (1984) J. Immunol. 133, 779.
29. Weigent, D. A., Stanton, G. J., and Johnson, H. M. (1983). Infect. Immun. 41, 992.

30. Biron, C. A. and Welsh, R. M. (1982) J. Immunol. 129, 788.
31. Santoni, A., Piccoli, M., Ortaldo, J. R., and Herberman, R. B. (1985) J. Immunol., in press.
32. Trinchieri, G., Santoli, D., Dee, R. R., and Knowles, B. B. (1978) J. Exp. Med. 147, 1229.
33. Saksela, E., Timonen, T., Virtanen, I., and Cantell, K. (1980). In "Natural Cell-Mediated Immunity Against Tumors" (R. B. Herberman, ed.), p. 645. Academic Press, New York.
34. Djeu, J. Y., Stocks, N., Zoon, K., Stanton, G. J., Timonen, T., and Herberman, R. B. (1982). J. Exp. Med. 156, 1222.
35. Djeu, J. Y., Timonen, T., and Herberman, R. B. (1982) In "NK Cells and Other Natural Effector Cells" (R. B. Herberman, ed.), p. 669. Academic Press, New York.
36. Kasahara, T., Djeu, J. Y., Dougherty, S. F., and Oppenheim, J. J. (1983). J. Immunol. 131, 2379.
37. Scala, G., Allavena, P., Djeu, J. Y., Kasahara, T., Ortaldo, J. R., Herberman, R. B., and Oppenheim, J. J. (1984). Nature 309, 56.
38. Herberman, R. B., Allavena, P. Scala, G., Djeu, J., Kasahara, T., Domzig, W., Procopio, A., Blanca, I., Ortaldo, J., and Oppenheim, J. J. (1984). In "Natural Killer Activity and Its Regulation" (T. Hoshino, H. S. Koren, and A. Uchida, eds.) p. 409. Excerpta Medica Ltd., Tokyo.
39. Evans, C. H. and Ransom. J. H. (1984). In "Thymic Hormones and Lymphokines '83" (A. Goldstein, ed.) p. 357. Plenum Press, New York.

Part I.

Purification and Characterization of NK Cells

PURIFICATION AND CHARACTERIZATION OF NK CELLS

Denis M. Callewaert[1]

Department of Chemistry
Oakland University
Rochester, Michigan 48063

I. INTRODUCTION

The presentations and discussions in this session focused on (A) methods that can be used to enrich for NK cells, (B) surface markers on NK cells with emphasis on the application of monoclonal antibodies, (C) functional characterization of NK cells, and (D) NK cell development and the characteristics of NK like cells in activated lymphocyte cultures.

II. PURIFICATION OF NK CELLS

It is important to recognize that, although essentially all workers in this field agree that the majority of the NK activity is associated with lymphocytes that have a large granular morphology, most will hasten to note that LGL and NK cells should not be considered synonomous. About 10% of peripheral lymphocytes are LGL, whereas various functional assays indicate that about 1 to 5% of the total lymphoid population are active NK cells.

---

[1]Research Career Development Awardee (CA-00744) of the National Cancer Institute.

ISBN 0-12-341370-2

Nevertheless, LGL purified by density gradient centrifugation offer a ready source of lymphocytes that are highly enriched for NK cells.

Another technique that has previously been used to enrich for NK cells involves "panning" lymphocytes on petri dishes that have been pretreated with an agent to which NK cells may bind. Koren has previously reported that target cell monolayers can be used to enrich for NK cells, with the bound cells recovered after first washing off unbound lymphocytes. In this session, Koren and Argov described two additional panning methods for NK enrichment. The first is based on the finding that the majority of non-adherent mononuclear cells in human peripheral blood possess receptors for a lectin from Erythrina Cristagalli (ECA), a glycoprotein that is mitogenic for T but not B cells. However, NK cells do not appear to express the receptor for this lectin. By panning peripheral lymphocytes on petri dishes pretreated with ECA, they are able to achieve a 2-3 fold enrichment in NK and ADCC activity in the non-adherent cells. Additional enrichment could be obtained by performing percoll gradient centrifugation prior to panning on ECA coated plates. Analysis of the ECA- cells revealed approximately a three fold increase in cells bearing HNK-1 (Leu 7), B73.1 and Leu 11b markers. About one-half of the ECA- cells have LGL morphology. This is a fairly rapid, easy and inexpensive method for enriching for NK cells that is based on a novel property and thus may prove quite useful.

Argov reported that, after incubation of human peripheral lymphocytes for 45 minutes with 100 nM phorbol dibutyrate (PDBU) a portion (5 to 30%) acquired the ability to adhere to plastic plates. The adherent cells, recovered by treatment with EDTA, showed substantially enhanced activity in NK and ADCC assays when compared to unseparated control lymphocytes. The cells purified using this procedure resembled LGL, and the percentage of cells bearing NK markers increased. Concern was expressed over the fact that 70 to 90% of the cells recovered using this procedure are LGL, but the increase in cytotoxicity is only about 2 fold. Klein stated that this may, in part, be due to the fact that phorbol esters make lymphocytes very sticky and may impede recycling. Argov agreed, and noted that simply adding PDBU directly into the NK assay can result in 50% inhibition.

Attempts to purify NK cells with defined target cell specificities by affinity chromatography using formaldehyde-fixed NK target cells attached to CNBr treated Sepharose 6MB were described by Callewaert. Peripheral lymphocytes were allowed to interact with the affinity resin for one hour at 4°, the column washed with medium, and bound cells eluted by agitation. Although fractions enriched 4 to 5 fold in NK activity could be obtained, the target cell specificity of the eluted cells was not substantially different from that of the unpurified control lymphocytes.

III. SURFACE MARKERS ON NK CELLS

A variety of antisera and monoclonal antibodies have previously been reported to bind to antigenic determinants on NK cells. However, since it has not been possible thus far to identify a unique antigen that is specific for NK cells, the available reagents vary substantially in their selectivity for NK cells.

A. Human NK Cells

Studies conducted using a monoclonal antibody, NK-9, produced using human LGL as the immunogen were described by Salseka. Briefly, about 30-40% of all peripheral lymphocytes and 40% of bone marrow lymphoid cells react with NK-9. The reactive cells include approximately 90% of the LGL in peripheral blood as well as a low percentage of small lymphoid cells. Pretreatment of lymphocytes with NK-9 does not block NK cytotoxicity. On the other hand, cells that do not express NK-9 are not cytotoxic and cannot be induced to become cytotoxic in NK assays. Thus, this antibody is of interest, but it is far from specific for NK cells. Saksela has demonstrated that NK-9 also reacts with a low percentage of small lymphocytes that are non cytotoxic, but can be induced to become cytotoxic by incubation with an appropriate stimulus such as Raji cells. Additional investigations in which NK-9 was used to study the development of NK activity in bone marrow are described below.

Trinchieri compared the specificity of antibodies reported to be specific for NK cells, concentrating mainly on antibodies directed against low affinity receptors for aggregated immunoglobulins (FcRII). Those directed against the human FcRII receptor (including B73.1, Leu 11, and 3G8) are highly specific for NK cells when tested against mononuclear cells. They recognize an antigen that seems to be composed of two bands (about 50 and 70kD) which appears to be analogous to the FcRII in the mouse. With the exception of B73.1, all of these antibodies react equally with Fc receptors on NK cells and on neutrophilic granulocytes. They appear to react with different antigenic determinants on the same molecule. The B73.1 competes with 3G8 for binding to the cells, but not with any of the other antibodies. On the other hand, 3G8 can cross compete with all the other antibodies. He also noted that the 3G8 has a much higher affinity for the cells. He presented data showing that B73.1 reacts strongly with a small portion of lymphocytes, but weakly if at all with granulocytes, whereas 3G8 reacts with the same population of lymphocytes, but also with about 90% of the granulocytes. It was brought up in the discussion that there is some activation of NK cells and some IFN production induced upon treatment with these antibodies in some (about 50%) of donors tested. This effect may be due to a contaminant in the antibody preparation. Also of interest is the fact that B73.1 reacts with less than 1% of bone marrow cells.

Ades presented strong evidence that the NK activity of Leu 11+ cells is enhanced by treatment with interleukin 2. In contrast, treatment of the Leu 11- cell population obtained using a FACS with IL2 does not induce significant levels of NK activity. He also performed experiments to determine if antibody to the TAC antigen blocked IL2 activation of NK cells, and reported that anti-TAC (at concentrations that inhibit proliferation of blastogenic cells) does not block natural cytotoxicity by Leu 11+ cells or activation of cytotoxicity by IL2.

Evidence suggestive of the HNK-1 (Leu 7) antigen on NK cells being involved in multiple sclerosis was presented by Roder. HNK-1 is specific for a small subset of peripheral lymphocytes, most of which are NK cells. When the central nervous system was investigated using the monoclonal antibody HNK-1, the oligodendrocytes that are responsible for myelination were stained. When proteins isolated from human myelin sheath were subsequently subjected to SDS electrophoresis and immunoblotted, the lane that reacted with anti HNK-1 also reacted with rabbit antiserum to myelin associated glycoprotein (MAG). MAG also has about the same molecular size (110 kD) as the protein recognized by HNK-1 on NK cells. Previous reports indicate that NK activity in at least some MS patients is lower than in normal blood donors. Roder then postulated that at least some MS patients may produce an autoantibody to NK cells that is, in turn, responsible for both the demyelination and the decreased NK activity in these patients. When asked if there is any relationship between NK activity and disease status, Roder responded that he had not done any longtitudinal studies but that work of others suggests little or no correlation. In fact, it is difficult to determine NK activity with the precision needed to investigate a possible correlation with MS demyelinating episodes. In response to other questions, Roder noted that MS patients actually appear to have more HNK-1 positive cells than normal individuals, but these cells appear to be functionally impaired - resulting in lower overall NK activity. Several other questions were raised regarding the presence of an autoantibody in the serum or CSF of MS patients and the possiblity of decreased NK activity in other demlelinating diseases. Unfortunately, answers to these questions must await further experimentation.

Ortaldo reported on results obtained with a new technique that allows for the determination of internal expression of antigens. The technique involves permeabilizing cells with lysolecithin prior to addition of antibody, and analysis of the stained cells using flow cytometry. Using this technique, he found that purified LGL contain some internal antigens that are expressed on the surface of other lymphoid cells but not unstimulated NK cells. These include Leu 1, MO-2, and (in about 50% of the LGL) B2. He was also able to detect MO-2 and B2 in T cells using this technique. However, T3, T4, T8, Leu-6 and B1 were not detectable on the inside or outside of LGL. Thus, many antigens that have been described as lineage specific may, in fact, be synthesized but not transported to the cell surface. These results may also explain why T cells cultured with IL2 quickly become positive for T101, Leu-1 and OKT-3 antigens. Asked whether the reactivity of the treated cells could be artifactual, resulting from degradation or modification, Ortaldo replied that he was very concerned about this possibility as well as possible binding of the antibodies to the Fc receptor. However, he was able to demonstrate by immunoprecipitation that the internal antigen recognized in several cases had the same molecular size as that reported for the antigen isolated from the cell surface of positive cells. When asked about internal expression of Leu-7 by LGL, he noted that Leu-7 is an IgM antibody and thus difficult to investigate using this technique, which is generally restricted to molecules smaller than 200 kD.

B. Mouse NK Cells

Pollack reported on investigations aimed at determining whether the alloantigens recognized by NK 1.1 ((BALB/c x C3H)F_1 antiserum to CE spleen cells) and NK 2.1 (NZB antiserum to BALB/c) are associated with functionally distinct subsets of NK cells. By depletion of NK cells using these antisera in conjunction with a fluorescent activated cell sorter or complement, one can identify mouse strains in which NK cells express one or both of these markers. When spleen cells from C57BL/6 mice, which contain NK cells positive for NK 1.1 and NK 2.1 antigens were tested against a panel of NK target cell lines, she found that both antisera inhibit cytotoxicity of some cell lines (YAC-1, B16 and L5178cl27v). However, the lysis of other cell lines was inhibited by only one of the sera. For example, lysis of FBL-3 is inhibited by NK 2.1 but not NK 1.1, indicating functional heterogeneity among mouse NK cells. In response to several questions, Pollack noted that NK 1.1 reacts with a glycoprotein with a molecular size of about 45 kD and that this antiserum appears to more NK specific than NK 2.1, which also has some effect on cytotoxic T lymphocytes. Attempts to obtain a monoclonal antibody with similar specificity have not been fruitful to date.

Trentin (Houston) compared the ability of different NK 1.1 antisera, produced using bone marrow, spleen or thymus of CE mice as the immunogen, to inhibit lysis of YAC-1 cells by Poly I:C pretreated C57BL/6 mouse spleen cells, and found that the inhibition observed was significantly higher for antisera produced using bone marrow compared to spleen or thymus cells. He had not yet tested whether the results obtained with the different sera can be ascribed to their reacting with different cell types, or whether they differ only in titer. Pollack added that NK 1.1 antiserum has also been produced in her laboratory using CE bone marrow as the immunogen, and that this antiserum binds to the same proportion of C57BL/6 spleen cells (about 5%) as other NK 1.1 sera, but appears to have a higher titer. She also pointed out that, when tested with bone marrow cells, NK 1.1 reacts with granulocytes.

Trinchieri added that there is an antibody, 24G2, that reacts with what appears to be the FcRII in the mouse, an antigen of the same molecular weight and with the same or very similar cell type distribution as is recognized by B73.1 in humans. However, only about 50% of the murine NK cells are positive for this antigen. So, the distribution of this receptor appears to be very different in the human compared to the mouse.

IV. FUNCTIONAL CHARACTERIZATION OF NK CELLS

What is an NK cell? The answer to this question is not as simple as it may seem. Many, but not all, LGL have NK activity. Many, but not all, cells bearing the B73.1 marker have NK activity. Some, but certainly not all, cells that bind to NK sensitive target cells have NK activity. Until the discovery and extensive documentation of a truly NK specific marker, I maintain that the ultimate criteria for an NK cell is its ability to lyse cultured tumor cells (generally, but not exclusively, YAC-1 for the mouse system and K-562 for the human system) in a short term assay without prior stimulation or sensitization. While many will agree with this definition, there is a temptation to equate NK cells with LGL, with Leu 7+ or B73.1+ cells, or with cells that form conjugates with NK target cells. For now, all of these characteristics should be regarded as valuable tools that will aid in furthering our understanding of NK cells and their function(s).

A. Target Binding Cells And The Cell Cycle

While it is known that NK cells are derived from bone marrow, and it has previously been reported that functionally mature NK cells are not in the cell cycle. Pollack reported on experiments designed to determine whether cells capable of binding to NK sensitive target cells are proliferating. The experiments employed C57BL/6 mice, which were injected with ^{3}H-Thymidine. At various time (2 to 48 hours) animals were sacrificed, bone marrow and spleen cells obtained, B cells depleted by treatment with anti IgM + C', and granulocytes and erythroid cells depleted with the aid of a cell sorter. The resultant lymphoid pool (about one-third of the untreated bone martrow cells) was then mixed them with FITC-labeled YAC-1 cells and cytospots examined for target binding cells and for thymidine uptake by autoradiography. Even at the earliest time point examined (2 hours) she found that about 20% of the large (over 10 microns in diameter) target binding lymphocyes had taken up a significant quantity of the isotope. This proportion stayed constant or dropped slightly over 48 hours. Fewer (13%) of the small (8-10 microns in diameter) target binding cells took up appreciable amounts of the isotope in the first two hours, but 28% showed significant uptake within 24 hours. Pollack interpreted this data in terms of a model in which the larger cells are a self-sustaining pool that feeds into a population of smaller target binding cells. Neither population of cells have thus far been examined for lytic activity against NK target cells, nor have they been examined using the NK 1.1 or 2.1 antisera.

B. Subtypes Of NK Cells

Lopez has previously reported the existence of subtypes of NK cells, and is investigating differences between NK cells that lyse K-562 cells in a short term assay (NK[K-562]) and cells that lyse HSV-1-infected fibroblasts in a 12 hour assay (NK[HSV-FS]. He has previously published on the heterogeneity between NK[HSV-FS] and NK[K562] effector cells, including the fact that they are differentiable using monoclonal antibodies, and that there is a tremendous difference in lytic function, with low NK activity to HSV correlating with susceptibility to infection whereas low NK to K562 does not. In recent experiments he investigated the cytotoxicity of several mouse target cells. Briefly, he finds that the HSV-1 infection makes several cell lines better targets for NC and/or NK cytotoxicity, with the effector cell that kills the virally-infected cells determined by the type of cell that is infected. Asked if the increased kill of HSV-1 infected target cells is due to an increase in the number of conjugates formed with the effector cell population or to an increase in the susceptibility of the target cells to lysis, Lopez replied that infected cells do not seem o be more susceptible to cytolysis since they are not lysed more readily by cytotoxic T cells. On the other hand, neither do they appear to be significantly different with regard to antigenic expression.

He also commented that, while some others have suggested that glycosylation patterns correlate with susceptibility, his results are not in complete agreement with this explanation.

C. Kinetic Parameters For NK Cytolysis

Callewaert reported on the results of recent experiments, involving the application of a calcium pulse technique, to measure the rate constants for lytic programming and killer-cell independent lysis (KCIL) steps in the mechanism for natural cytotoxicity by human NK cells. The rate constant for KCIL was found to be significantly larger than that for lytic programming when donor lymphocytes were tested on various targets including K-562 cells. Values for these rate constants, along with V_{max}, K_M^{app} values and the fraction of lymphocytes that form conjugates with target cells were also determined in parallel experiments and used in conjunction with a multistep kinetic model in an attempt to determine the significance and interpretation of kinetic parameters for natural cytotoxicity. Briefly, V_{max} remains a valid quantitative estimate of the overall cytotoxic capacity of a given effector cell population. Experimentally-determined K_M^{app} values tend to vary with the frequency of NK effector cells in the effector cell population tested, but appear to be more directly related to the frequency of target binding cells than functional NK cells. Determination of rate constants for lytic programming and KCIL may prove useful in other studies aimed at further elucidating the mechanism of NK cytolysis.

D. Noncytotoxic Functions Of NK Cells

In addition to their cytotoxic activity, NK cells appear to have additional *in vivo* roles. In this session, Djeu reported on the ability of human NK cells to support the development of cytotoxic T cells. In a later session, Kumagai discussed the ability of murine NK cells to suppress B cell differentiation and augment cytotoxic T cell responses. Djeu reported on investigations on the involvement of NK cells in the generation of infuenza specific CTL in an autologous system. The stimulator cells used were monocytes infected with influenza A Bangkok, which were incubated with buffy coat lymphoid cells for six days, after which the supernate was collected and titered for interferons. Autologous virus-infected PHA blasts were also tested as target cells in a 7 hour chromium-release assay. LGL alone do not lyse these target cells, but unseparated mononuclear cells give about 20% lysis at a 25:1 E:T ratio. Interestingly, the small lymphocytes obtained from fractions 6-7 of a Percoll gradient did not lyse the target cells, but if LGL were added (1 LGL:10 small lymphocytes) cytotoxicity was observed. She therefore performed experiments in a chamber in which the small T cell were separated from the LGL by a Millipore filter. Cells on both sides of the filter developed cytotoxic activity, but on the side containing LGL it was NK activity with no viral or HLA restriction, whereas on the T cell side she found both viral and HLA restricted cytotoxicity. She has thus far been unable to substitute for LGLs by addition of highly purified IL1, IL2 or gamma interferon (alone or in combination) to the small lymphocytes, and is presently trying to isolate a new lymphokine from the supernates of these cultures.

V. DEVELOPMENT OF NK AND NK-LIKE CELLS

There is considerable interest in the ontogeny of NK cells and in the relationship between "classical" NK cells and the NK-like cytotoxic activity that is observed after exposure of lymphocyte preparations to various stimuli including culture with fetal calf serum and mixed lymphocyte cultures.

Saksela reported that 90% of the NK-9 positive cells in normal human bone marrow are not LGL, but after incubation with TCGF for 3-4 weeks most are LGL. He also found that the NK-9 positive, non-LGL cells from bone marrow can be stimulated to exhibit NK-like activity, and that this pool of cells also includes cells that can be stimulated by MLC to become allospecific CTL. These data suggest that NK-9$^+$ small lymphoid cells in bone marrow are a common precursor pool for cytotoxic cells. Saksela also raised the more provocative suggestion of a common precursor cell for NK and CTL. He showed that when NK-9$^+$ bone marrow cells are used as responder cells in MLC, nonspecific lysis is observed prior to appearance of allogeneic cytotoxicity. However, it is important to recognize that the two cytotoxic activities observed need not be ascribed to the same cells. Similar results have previously been reported using unfractionated peripheral lymphocytes as responder cells, but with correlary studies indicating that the different cytotoxic activities were due to different cells.

Experiments designed to investigate the relationship between NK-like cells and LGL were reported by Timonen. He examined different percoll fractions for cytotoxic activity in response to various stimuli. When stimulated with K-562 cells, most of the increase in cytotoxicity occurred in the LGL fraction, a phenomenon which is known to be interferon mediated. However, when Raji cells (which are generally resistant to NK) were used as stimulators, the peak of cytotoxic activation occurred between

6 and 24 hours in the intermediate density fractions. He has previously shown that these medium density cells are not LGL and that they have T cell markers - consistent with the surface phenotype of various NK-like cells that have been described in the literature. He then tried various treatments in an attempt to induce these cells to differentiate into LGL. To date, only one approach has been successful. This involves coculturing autologous macrophages (over 95% pure by latex bead ingestion) with intermediate density cells that have previously been activated for 24 hours and then washed. The number of LGL in these fractions increases during 12 hours of incubation with the macrophages, which is presently ascribed to the differentiation of intermediate density cells into LGL. He also investigated the phenotype of the precursors of the new LGL by cell sorting and found them to be very heterogeneous (10 to 34% were T3 positive, some were T8 positive, and some T4 positive). All of the precursors were NK-9 positive. Based on these data, he presented a model in which a pool of noncytotoxic, $NK\text{-}9^+$ precursor cells, which are not LGL in bone marrow or peripheral blood respond to activation by allogeneic cells, virus infected cells, interferon, IL-2, etc. and become sticky, and are thereby trapped in the reticuloendothelial system where they interact with macrophages and differentiate into LGL. He speculated that they probably undergo changes in specificity during this process and also acquire the Leu 7 marker. In the discussion following his presentation, Timonen noted that the noncytotoxic precursor cells do not bind to target cells, but when activated by culture with Raji cells they become adherent and cytotoxic. These cells need not differentiate into LGL in order for these properties to be expressed. When asked why Raji appears to be such a good stimulator for this response, while K562 is not, and if any other stimulus produces similar results, he responded that the response is probably due to IL2 since Raji is a better inducer of IL2 than is K-562. Addition of IL2 alone results in generation of cytotoxicity, macrophages appear to be required to achieve the morphological changes (i.e. LGL generation). In response to several questions regarding the morphology and characteristics of the activated cells obtained without exposure to macrophages, Timonen stated that they are medium sized and agranular, i.e. definitely not LGL.

Klein pointed out that, for allogeneic MLC performed with unseparated peripheral lymphocytes as the responder cells, the specificity of the NK-like cytotoxicity observed does not appear to be related to the stimulator cells used. She then presented data obtained in experiments in which high vs low density cells were tested as responder cells in MLCs for which both proliferation and cytotoxicity were monitored. She found that the high density (smaller) cells proliferate better, whereas the low density (larger) lymphocytes exhibit the stronger cytotoxic response (K-562 and Daudi were used as targets for NK-like cytotoxicity in addition to allogeneic and autologous target cells). She noted that even control cultures show some allogeneic cytotoxicity, presumably resulting from stimulation by components of the fetal calf serum used in the culture system. Based on these and other results, she argued that the even a nonspecific stimulus may cause the development of specific cytotoxicity by low density cells that probably already express the receptor for the specific antigens of the target presented to them, and therefore refers to these cells as CTL.

Since it is known that the Leu 11 monoclonal antibody reacts with most or all NK cells in fresh blood, Lanier sought to determine if his antibody could be used to discriminate between classical CTL and NK-like cells after performing mixed lymphocyte cultures. The B lymphoblast cell line SB was used as the stimulator in 5 day MLCs, after which the responding cells were stained with Leu 11. About 10 to 15% of the total viable blast cells recovered were Leu 11+, with 50% to 67% of them also expressing the HLA-DR and DC antigens. He found that the cells which coexpressed HLA-DR and Leu 11 lysed K-562 target cells much better than Leu 11+ DR- cells. Both populations had a LGL morphology, although the cells that expressed the DR antigen were more pleomorphic and had a lot of mitotic figures. The DR- cells were more similar to conventional LGL. He also mentioned that treatment of these cells with Leu 11 plus complement abrogates essentially all NK-like activity (using K-562), whereas cytotoxicity of the stimulator cell (SB) is unaffected or enhanced. In reciprocal experiments Leu 1 plus complement was used to eliminate cytotoxicity of SB cells, with concommitant enhancement of K562 cytotoxicity. In unstimulated peripheral blood lymphocyte preparations and in lymphoblast preparations obtained after MLC cultures, Lanier observes no overlap between cells that are Leu 11+ and cells which express Leu 4 or Leu 3. When he used K-562 rather than SB as MLC stimulator cells he found, in agreement with previous reports by Eva Klein, that the majority of the cells obtained after culture are NK cells (i.e. 70% are Leu 11+ and have LGL morphology) and are strongly cytotoxic for K-562, and also lyse CEM, SB and HL-60 target cells. About 50% of the cells generated in response to K-562 express class II histocompatability antigens and very few of them express the Leu 7 antigen. They do not express convenional T cell markers. When asked if the class II MHC antigens could be picked up from the stimulator cells, Lanier responded that he has obtaind similar results using a DC- cell line as the stimulator, so that the class II antigens are either produced by the NK cells generated or picked up from T cells present in the effector cell population. In response to other queries, Lanier suggested that SB stimulation of NK activity is probably mediated by IL2 or interferon produced in the culture as a result of the strong T cell activation by SB. Experiments in which lymphocytes were cultured for 12 hours with IL2 produced NK cells with properties similar to those obtained when SB cells were used as stimulators in the MLC.

VI. CONCLUDING REMARKS

The nature and diversity of presentations in this session illustrate the diverse functions and markers associated with NK cells. They also emphasize the fact that, while several fairly powerful tools have been developed for the isolation and characterization of NK, at present no single assay system, target cell, purification method or antibody appears sufficiently specific to warrant exclusion of other approaches. On the contrary, continued progress in this area will most certainly depend on studies in which several available tools are applied and new, more specific ones, developed.

SURFACE PHENOTYPE AND FUNCTIONS OF HUMAN NATURAL KILLER CELLS[1]

Giorgio Trinchieri, Gino Degliantoni,
Michiko Kobayashi, Lucille London, and Bice Perussia

The Wistar Institute for Anatomy and Biology
Philadelphia, Pennsylvania

The existence of a cell type able to mediate non-specific cytotoxicity spontaneously against various target cell types has been known for more than a decade; however, only recently has it become possible to identify these natural killer (NK) cells as a discrete leukocyte subset, with distinctive morphological, phenotypic and functional properties. One of the first lines of evidence pointing to the assignment of NK cells as a discrete lymphocyte subset was the characterization of their morphology as large granular lymphocytes (LGL). The use of discontinuous Percoll gradients has made it possible to obtain almost pure preparations of LGL (1), which consist largely of cytotoxic NK cells. However, it has become apparent that LGL preparations contain several cell types other than NK, and that LGL morphology is not unique to NK cells. More recent advances in the analysis of the surface phenotype of NK cells have helped to define the NK cell subset and to study functions mediated by this cell type. The lack of antigenic specificity and immunological memory of the NK cells makes these effector cells part of the non-adaptive immune system, together with monocyte-macrophages and polymorphonuclear leukocytes (PMN). The activity of NK cells, like that of other effector cell systems of non-adaptive immunity, can be enhanced by various humoral factors, notably interferon (IFN) and interleukin 2 (IL2). However, terminally differentiated monocyte-macrophages and PMN have only limited

---

[1] Supported in part by U.S. Public Health Service Grants CA05022, CA10815, CA20833, CA32898 and CA37155.

ISBN 0-12-341370-2

or no proliferative ability, and an increased effector cell number is obtained primarily by local recruitment from other sites of the organism or by stimulation of precursor cell growth and differentiation in the bone marrow. By contrast, mature resting NK cells, like T and B cells, proliferate *in vivo* or *in vitro* in response to stimuli such as IFN, virus infection and culture with various cell lines (2,3). The studies on the hematopoietic lineage of NK cells have been hampered by the lack of a precursor cell assay for NK cells, in which the characteristics of committed NK progenitors can be compared with those of the precursor cells of other lineages. No surface antigen unique to NK cells has been identified yet, but rather only a combination of antigens that are shared with other cell types, mostly T-cells and myelomonocytic cells. Often these surface markers are not present on all NK cells, suggesting some heterogeneity within the NK population.

I. MARKERS SHARED BY NK CELLS AND MYELOMONOCYTIC CELLS

The first surface marker identified on human NK cells was the receptor for the Fc fragment of IgG (FcR) (4). FcR are present on a variety of cell types and two major FcR types are identified on myelomonocytic cells. FcR1 is present on human monocytes and macrophages, and it binds with high affinity monomeric IgG. FcR2, present on PMN but not on fresh human monocytes, has lower affinity for Fc and binds aggregated or complexed IgG of almost all subclasses. The FcR on human NK cells does not bind monomeric IgG and is antigenically identical to the FcR2 for aggregated Ig present on human PMN. The first monoclonal antibody shown to react with NK cell FcR was the IgG1 antibody B73.1 (5,6), which inhibits FcR functions on NK cells and on PMN. This antibody reacts strongly with an average of 15% of peripheral blood lymphocytes (PBL) and, at low intensity, with the majority of PMN from about 50% of donors. Two previously described monoclonal antibodies 3G8 and 4F7 (7) with specificity for PMN FcR, also react with NK cells (8). B73.1 and 3G8 precipitate a molecule from NK cells and PMN, respectively (6,7), consisting of two poorly resolved peptides of m.w. 50,000 to 70,000. However, antibodies B73.1 and 3G8 react with different epitopes on the same FcR molecule: they only partially cross-compete for binding to the positive cells, though modulation of the receptor induced with one of the antibodies completely abolishes reactivity of the cells with the other one. Like antibody 3G8, three other antibodies, VEP13 (9), NKP-15 (anti-Leu11a) and GO22 (anti-Leu11b) (10) react strongly with PMN and, at lower intensity, with NK

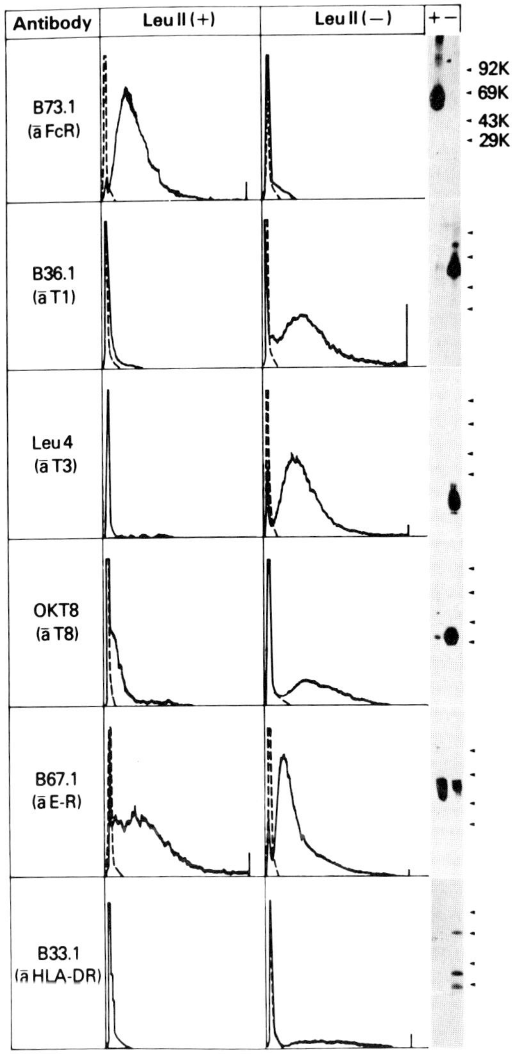

Figure 1. Surface phenotype of NK cells. Lymphocytes were incubated with biotin-labeled anti-Leu11a (detected by XRITC-avidin) and with the indicated antibodies labeled with FITC. The histograms show the distribution (green fluorescence) of the FITC-labeled antibodies (x-axis, fluorescence; y-axis, number of cells) separately on Leu11(+) and (−) cells, electronically gated (11). The SDS gels show the immunoprecipitate obtained with the indicated antibodies from radiolabeled B73.1(+) (left lanes) and B73.1(−) PBL (right lanes), separated by the indirect rosetting method (6).

cells. Although the antigen precipitated by the three antibodies has not been characterized yet, indirect evidence indicated that these antibodies also react with FcR on NK cells, but possibly with epitopes other than those detected by B73.1 and/or 3G8 (11).

B73.1 and antibodies of similar specificity seem to be the best available reagents to obtain almost homogeneous preparations of human NK cells: 1) virtually all spontaneous and antibody-dependent (Ab-CMC) cytotoxic activity is recovered in the B73.1(+) subset (5); 2) spontaneous cytotoxic activity of B73.1(+) cells is enhanced by treatment with IFN (5), TPA (12) or IL2 (3), whereas cytotoxic activity is never induced in similarly treated B73.1(-) cells; 3) more than 60% of B73.1(+) cells and up to 100% of them when pretreated with IFN bind to K562 target cells (5); 4) more than 70% of the B73.1(+) cells forming conjugates with K562 cells lyse the bound target cell in single-cell cytotoxicity assay in agarose (13); 5) the B73.1(+) population is completely nonoverlapping with either B or T cells, as defined by specific surface antigens (5,11); 6) the surface markers detected on and immunoprecipitated from B73.1(+) cells correspond to those present on cells with spontaneous cytotoxic activity (5,11); and 7) the large majority of B73.1(+) cells are LGL (5).

Figure 1 illustrates the use of anti-FcR2 antibodies to analyze surface markers of the NK subset by two-color immunofluorescence (11), or by immunoprecipitation of radiolabeled surface antigens from separated B73.1(+) and (-) peripheral blood lymphocytes (PBL) (6): as discussed below, NK cells have been shown by both methods to express T8 and T11 but not T1, T3 and HLA-DR antigens.

A link between NK cells and cells of the myelomonocytic lineage was suggested on the basis of the reactivity of NK cells with antibodies OKM1, Mac 1 and others of equivalent specificity (14). The antigen recognized by these antibodies has been identified as the receptor for C3bi (CR3), which appears beginning at the promyelocyte stage, and is present at high density on monocytes and myeloid cells and, at a much lower density, on about 80% of NK cells and some T cells.

Antibody N901 (15), derived from mice immunized with cells from a chronic myeloid leukemia patient in blast crisis, detects an antigen that is present at low density on virtually all NK cells and on no other subpopulation of normal peripheral blood cells, but that is expressed at high density by the majority of cells from some patients with acute myelogenous leukemia. The antigen detected by N901 has not been characterized biochemically, but it appears to be different from the FcR and CR3. Reaction of N901 with NK cells does not induce modulation of cytotoxic activities, suggesting that the antigen recognized is not involved in the process of NK cell recognition or killing of the target cells.

With the exception of FcR2, CR3 and N901 antigen, none of a series of other antigens present on myelomonocytic cells at various stages of differentiation, be they myeloid- or monocyte-specific, or shared by the two cell types, are expressed on NK cells. Moreover, these same markers were not induced on NK cells upon treatment with various chemical or naturally occurring substances that induce maturation and expression of differentiation markers on immature myelomonocytic cells.

II. SURFACE MARKERS SHARED BY NK CELLS AND T CELLS

NK cells do not express the two surface proteins of 69-71,000 m.w. (recognized by antibodies OKT1 and anti-Leu 1) or 20-22,000 m.w. (recognized by antibodies OKT3 and anti-Leu4) characteristic of all T lymphocytes. A proportion of NK cells expresses low-affinity receptors for sheep erythrocytes (E-R), and 80-90% of NK cells react with anti-E-R monoclonal antibodies, such as 9.6, OKT11 or B67.1 (6,16). The density of the receptor, as detected by these antibodies, is similar on NK and T cells or higher on NK cells (Fig. 1). Thus, it seems likely that the lower binding level of NK cells with sheep erythrocytes is due to differences in membrane characteristics of NK and T cells and not to a lower number of E-R on NK cells.

Antibodies OKT8, anti-Leu 2a and others recognize an antigen composed of two polypeptides of 32,000 and 30,000 m.w. on a subset of T cells with suppressor-cytotoxic activity, and also react, although with a much lower intensity than on T cells, with 30 to 50% of NK cells (17). Antibody OKT8 precipitates the same antigen from NK and T cells (6).

The 46,000 m.w. protein recognized by antibody OKT10 on thymocytes and activated T cells is also expressed at low density on the majority of NK cells, and is absent from quiescent peripheral blood T cells (18).

A complex reactivity is displayed by the monoclonal antibody HNK-1 (anti-Leu 7), originally described as NK cell-specific (19). Antibody HNK-1 (IgM) precipitates an antigen of apparent m.w. 110,000 from PBL, and reacts with 20 to 70% of peripheral blood NK cells, with variability among donors; HNK-1 also reacts with T and pre-B cell lines and with the myelin-associated glycoprotein. Most, but not all, HNK-1(+) cells have LGL morphology. In lymphocyte preparations from cord blood, HNK-1 fails to react with NK cells and with LGL. HNK-1 reacts with a variable proportion of OKT3/OKT8(+) or OKT4(+) and sometimes HLA-DR(+) lymphocytes, which have low or null spontaneous cytotoxic activity. Our preliminary results show that the proportion of HNK-1(+) cells is increased in some patients (leukemic patients in remission, some kidney transplant patients), to include almost all OKT8 or OKT4(+) cells, whereas NK activity is restricted to the small proportion of B73.1(+) cells present. The HNK-1 antigen, therefore, is preferentially associated with a proportion of LGL and NK cells, but it is also present on other cell types, including those unrelated to hematopoietic lineages. In some cases this antigen appears to be an activation antigen on various lymphocyte subsets.

III. ACTIVATION ANTIGENS ON PROLIFERATING NK CELLS

Human peripheral blood NK cells, positively enriched using antibody B73.1, can be effectively induced to proliferate in vitro in short-term culture. B73.1(+) NK cells, cultured in the presence of IL2 and irradiated lymphoblastoid cells, show significant proliferation starting at day 4. At that time, a large proportion of the cells express HLA-DR and 4F2 antigens, and IL2 and transferrin receptors, detectable on all the proliferating cells at high density (20). The same antigens are absent on fresh NK cells and in the proportion of non-proliferating NK cells present in the induced culures. Monoclonal antibodies against the IL2 receptor (Tac antigens) and the transferrin receptor, as well as antibodies to IL2, effectively prevent NK cell proliferation. Proliferating NK cells still bear FcR, as detected by antibody B73.1. T10 and N901 antigens, expressed at low density on the majority of NK cells, become expressed at high density on the proliferating cells. Expression of T cell markers absent from fresh NK cells, such as T1, T3 or T4, have never been detected on proliferating NK cells. IL2-dependent lymphocyte clones with NK activity have been derived in several laboratories, and some of them express T cell markers. Considerable controversy surrounds the identification of the various clones as derived from fresh NK cells or from other contaminant cell types (e.g. T3(+) T cells) that enjoyed some proliferative advantage. Thus, for now, little information can be derived from these clones for the study of NK phenotype. In one study (21), however, clones have been obtained from PBL preparations highly enriched for B73.1(+) or (-) cells. Nine of 10 clones originated from B73.1(+) cells were able to mediate spontaneous and Ab-CMC and maintained the phenotype of fresh NK cells, plus the activation antigens described above. The clones originated from B73.1(-) cells, on the contrary, lacked spontaneous cytotoxicity and displayed the phenotype of T-cell clones.

IV. PATHOLOGICAL EXPANSION OF THE NK LEUKOCYTE SUBSET

True NK cell leukemias in humans have been difficult to demonstrate due to the lack, at least until recently, of specific markers for these cells. Only one case of acute lymphocytic leukemia (ALL) with NK characteristics has been described in a 9-year-old Japanese girl (22). The leukemic cells of this patient were E-R/FcR/HLA-DR/OKM1(+) large lymphocytes with abundant cytoplasm that lacked the granules characteristic of LGL and other specific markers of T cells. The leukemic cells had strong NK activity against K562 target cells, but no Ab-CMC activity, and two cell lines with NK activity have been established from them.

The cells of about half of all patients with chronic T-cell lymphocytosis have LGL morphology and some of the markers or functions of NK cells (23). The cells from almost all these cases are E-R/OKT11/OKT3(+), but, unlike most T-ALL, are usually OKT1/Leu1/3A1(-). Cells from most cases of LGL lymphocytosis are OKT8/FcR/B73.1/HNK-1(+) and cells of a proportion of cases are OKM1(+) or HLA-DR(+) (23). Cells from most patients mediate Ab-CMC, but only rarely are the cells spontaneously cytotoxic against K562 target cells or suppressive for B cell differentiation or T cell proliferation. The nature of the LGL present in these patients remains unclear. Although these cells have many NK cell characteristics, they also bear markers not expressed by fresh NK cells.

V. FUNCTIONS MEDIATED BY NK CELLS

In addition to their cytotoxic effect, NK cells have recently been shown to exert regulatory function on the hematopoietic and the immune systems, either by direct cellular interaction or by production of different types of lymphokines secreted by NK cells. It is extremely important to use the most stringent criteria for NK cell identification before assigning a specific function to these cells. In particular, caution should be exerted in reaching conclusions derived solely from data obtained with LGL-enriched preparations. Variable, and occasionally very high, proportions of cells bearing markers not expressed on fresh NK cells are present in these fractions, and several types of low-density cells copurify in LGL-enriched fractions, including CFU-C, early myelomonocytic cells and blast "activated" T cells, all cells likely to resemble large lymphocytes morphologically. Several criteria, when possible, should be used to identify the NK leukocyte subset; the use of monoclonal anti-FcR antibody, such as B73.1, appear currently

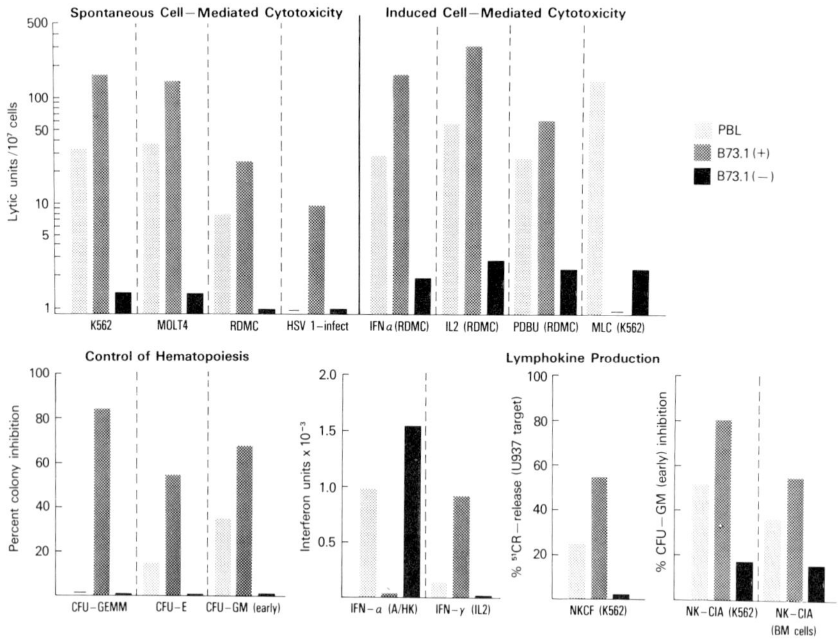

Figure 2. Functions mediated by NK cells: analysis with lymphocyte preparations separated using antibody B73.1. The different experiments show the ability of total PBL, B73.1(+) and (-) lymphocytes, separated by various methods, to mediate several functions attributed to NK cells. Dash indicates not done. Spontaneous cytotoxicity of fresh lymphocytes was tested on an erythromyeloid (K562), a T-cell (Molt 4), and a rhabdomyosarcoma-derived (RDMC) target cell line (5) and HSV1-infected autologous lymphoblasts (23). Cytotoxicity, mediated by IFNα-(5), IL2-(3) and PDBu-(12) activated lymphocytes, was tested on RDMC target cells; cytotoxicity of lymphocytes activated in MLC was tested on K562 target cells (24). Inhibition of hematopoietic colonies in allogeneic normal human bone marrow cells was tested, according to standard methods, after an 18-h incubation of lymphocytes with bone marrow cells at a ratio 2:1 (manuscript in preparation). IFNα was induced infecting PBL with the Hong Kong strain of influenza A virus (A/HK) (26) and IFNγ by recombinant IL2 (3) (18-h incubation in both cases). NK-derived cytotoxic factor (NKCF) was induced by 6-h culture of PBL with K562 cells and tested in an 18-h ^{51}Cr release assay on U937 cells. NK-derived colony inhibiting activity (NK-CIA) was induced by 6-h culture of PBL with K562 cells or with HLA-DR(+) bone marrow cells and tested on 14 day granulocyte-macrophage colony forming units (early CFU-GM).

as the best available tool for NK cell identification. Figure 2 illustrates some of the work in our laboratory and that of others, in which antibody B73.1 has been used to confirm or disprove the assignment of a cellular function to NK cells.

Spontaneous cytotoxicity against a variety of cell lines, both NK-sensitive and relatively insensitive, is mediated by B73.1(+) but not B73.1(-) PBL. The cell lines tested included non-adherent hematopoietic cell lines and substrate-adherent cell lines, both of normal and tumoral origin (5). HSV1-infected autologous lymphoblasts were also lysed by B73.1(+) PBL (24). Treatment of lymphocytes with IFN (5), IL2 (3) and the phorbol ester PDBu (12) also enhances the cytotoxicity of B73.1(+) but not of the B73.1(-) PBL. Removal of B73.1(+) PBL almost completely suppresses tne generation of cells cytotoxic for K562 target cells in mixed lymphocyte cultures (MLC) (25). Suppression of mixed hematopoietic (CFU-GEMM), erythroid (CFU-E) and early granulocyte-macrophage (CFU-GM) colonies is also mediated by B73.1(+) but not B73.1(-) cells. Results from experiments in which PBL were separated using antibody B73.1 do not support a previous conclusion (26), based on Percoll separation experiments, that IFNα is produced by NK cells upon virus infection: B73.1(-) lymphocytes are IFNα producers, whereas B73.1(+) lymphocytes are not (27). On the contrary, IFNγ production by resting lymphocytes in response to IL2 is a characteristic of B73.1(+) cells (3). Production of NK-derived cytotoxic factor (NKCF) during culture of PBL with K562 cells is also restricted to B73.1(+) cells (manuscript in preparation). A colony-inhibiting activity (CIA), suppressing hematopoietic colonies and probably identical to NKCF, is produced by B73.1(+) lymphocytes in response to culture with K562 cells or with normal bone marrow cells.

REFERENCES

1. Timonen, T., Ortaldo, J. R., and Herberman, R. B. (1981). J. Exp. Med. 153, 569.
2. Biron, C. A., and Welsh, R. M. (1982). J. Immunol. 129, 2788.
3. Trinchieri, G., Kobayashi-Matsumoto, M., London, L., and Perussia, B. (1984). J. Exp. Med. in press.
4. Jondal, M., and Pross, H. (1975). Int. J. Cancer 15, 596.
5. Perussia, B., Starr, S., Abraham, S., Fanning, V., and Trinchieri, G. (1983). J. Immunol. 130, 2133.
6. Perussia, B., Acuto, O., Terhorst, C., Faust, J., Lazarus, R., Fanning, V., and Trinchieri, G. (1983). J. Immunol. 130, 2142.

7. Fleit, H. B., Wright, S. D., and Unkeless, J. C. (1982). Proc. Natl. Acad. Sci. USA 79, 3275.
8. Perussia, B., and Trinchieri, G. (1984). J. Immunol. 132, 1410.
9. Rumpold, H., Kraft, D., Obexer, G., Bock, G. and Gebhart, W. (1982) J. Immunol. 129, 1458.
10. Lanier, L. L., Phillips, J. H., Warner, N. L., and Babcock, G. F. (1984). J. Leuk. Biol. 35, 11.
11. Perussia, B., Trinchieri, G., Jackson, A., Warner, N. C., Faust, J., Rumpold, H., Kraft, D., and Lanier, L. L. (1984). J. Immunol. 133, 180.
12. Trinchieri, G., O'Brien, T., Shade, M., and Perussia, B. (1984). J. Immunol. in press.
13. Freundlich, B., Trinchieri, G., Perussia, B., and Zurier, R. B. (1984). J. Immunol. 132, 1255.
14. Kay, H. D., and Horwitz, D. A. (1980). J. Clin. Invest. 66, 847.
15. Griffin, J. D., Hercend, T., Beveridge, R., and Schlossman, S. F. (1983). J. Immunol. 130, 2947.
16. Fast, L. D., Hansen, J. A., and Newman, W. (1981). J. Immunol. 127, 448.
17. Perussia, B., Fanning, V., and Trinchieri, G. (1983). J. Immunol. 131, 223.
18. Ortaldo, J. R., Sharrow, S. O., Timonen, T., and Herberman, R. B. (1981). J. Immunol. 127, 2401.
19. Abo, T., Cooper, M. D., and Balch, C. M. (1982). J. Immunol. 129,1752.
20. London, L., Perussia, B., and Trinchieri, G. (1984). J. Immunol. in press
21. Bolhuis, R. L. H., Van der Griend, R. J., and Routeltap, P. M. (1984). Nat. Immun. Cell Growth Regul. 3, 61.
22. Komigama, A., Kawai, H., Miyagawa, Y., and Akabane, T. (1982). Blood 60, 1429.
23. Van der Griend, R. S., van der Reijden, H. J., Bolhuis, R. L. H., Melief, C. J. M., Van dem Borne, A. E. G. K. R., and Roos, D. (1983). Blood 62, 669.
24. Yasosawa, M., and Zarling, J. M. (1983). J. Immunol. 131, 2011.
25. Strassman, G., Bach, F. H., Zarling, J. M. (1983). J. Immunol. 130, 1556.
26. Djeu, J. Y., Stocks, N., Zoou, K., Stanton, G. J., Timonen, T., and Herberman, R. B. (1982). J. Exp. Med. 156, 1222.
27. Perussia, B., Fanning, V., and Trinchieri, G. (1983). In "Natural Killer Activity and its Regulation" (T. Hoshino, ed.), p. 107. Excerpta Medica, Tokyo.

CHARACTERISTICS OF MOUSE NK CELLS AND THEIR PRECURSORS[1]

Sylvia B. Pollack

Department of Biological Structure SM-20
University of Washington
Seattle, Washington 98195

I. INTRODUCTION

Since the first reports of murine naturally cytotoxic lymphocytes in 1975 by Herberman *et al*. (1), Kiessling *et al*. (2) and Sendo *et al*. (3) there has been controversy regarding the characteristics of these cells. Recent studies have made it clear that there is no one "prototype" natural killer (NK) cell but that NK cells are a heterogeneous group of cells. Their heterogeneity extends from their morphologic and phenotypic characteristics to their functional roles (reviewed in 4). In this paper some data on the physical characteristics and surface markers of mature murine splenic NK cells will be considered as well as recent evidence on the production of NK cells from their precursors in the bone marrow.

II. PHYSICAL CHARACTERISTICS OF NK CELLS

Initial reports on NK cells indicated that they are small to medium-size cells with a velocity sedimentation rate of approximately 4.5 mm/hr at unit gravity (5,6). Morphologically, they appear to be small to medium-size

[1]Supported by NIH Grant CA32553 and DOE Contract EV 10270.

ISBN 0-12-341370-2

lymphocytes (7-9).

We have confirmed that endogenous NK activity of unstimulated mice is due to small to medium-size cells in the following experiments. Spleen cells from 6-8-week-old C57BL/6J (B6) mice were passed through nylon wool to deplete B cells and enrich NK cells and then labeled with anti-NK-1.1 alloantiserum and FITC-goat anti-mouse Ig as described previously (10). The labeled cells were then sorted into positive and negative populations with the fluorescence activated cell sorter (FACS). After sorting, the two subpopulations were (1) reanalyzed for size and fluorescence and (2) tested for NK activity on ^{51}Cr-labeled YAC-1 target cells. As shown in Figure 1, the NK-1.1$^+$ cells were the small to medium size splenic cells whereas the NK-1.1$^-$ cells were larger cells. As shown in Table II, NK activity was highly enriched in the NK-1.1$^+$ fraction, while the NK-1.1$^-$ fraction had negligible activity.

How can such data be reconciled with the findings (11-13) that murine NK activity is associated with large lymphoid cells analogous to the large granular lymphocytes (LGL) described in humans (14) and rats (15)? Part of the explanation may lie in the differences between endogenous and activated NK cells. For example, Tai and Warner (16) compared the velocity sedimentation rates of NK effector cells from normal mice with those from mice stimulated with poly I:C or with lymphocytic choriomeningitis virus (LCMV). In normal spleen cells, NK activity was mainly in small to medium size cells. However, NK activity in spleen cells from stimulated mice was found to shift to larger cells. Tai and Warner suggested that the shift to larger cells may reflect a blastogenic response of the NK cells or the emergence of a new NK cell population. Recent studies by Biron *et al.* (17) have shown that LCMV does induce a blastogenic response in splenic NK cells. It seems likely that any stimulus, from experimenter-introduced agents such as poly I:C or LCMV to environmental agents such as pathogenic virus or bacteria in a mouse colony, may induce the appearance of larger NK cells.

It seems likely that in the "normal" steady-state the spleen contains a heterogeneous assortment of NK cells including newly arrived small NK cells from the bone marrow (see below), as well as resident NK and recirculating NK cells, both of which may have encountered stimuli that lead to activation and/or blastogenesis. Thus, both small and large (LGL and large agranular lymphocytes) appear to mediate NK activity in the mouse.

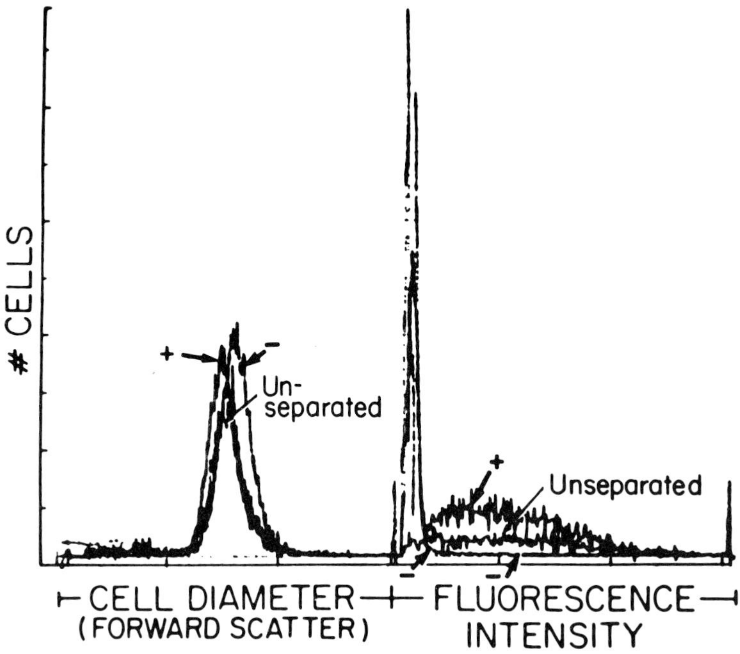

Figure 1. Nylon column-passed B6 SC labeled with anti-NK-1.1 and FITC-anti-mouse Ig were sorted into NK-1.1$^+$ and NK-1.1$^-$ populations. The + cells were shifted to the left (smaller) relative to the unseparated cells, whereas the - cells were shifted to the right (larger) as shown in the left panel. The fluorescence profiles of the 3 populations are shown in the right panel.

TABLE I

LYTIC ACTIVITY OF FACS-SORTED NK-1.1$^+$ AND NK-1.1$^-$ CELLS[a]

Population	$LU_{25}/10^7$
Unsorted	23.0
NK-1.1$^+$	64.1
NK-1.1$^-$	< 1.0

[a]Cells from the experiment depicted in Figure 1 were tested for their ability to lyse ^{51}Cr-YAC-1. Lytic units (LU) based on 25% lysis of 10^4 YAC-1 were calculated by regression analysis of the data.

III. PHENOTYPIC CHARACTERISTICS OF NK CELLS

The association of T cell antigens with murine NK cells has been observed repeatedly. Some NK cells express Thy-1 and most express Qa-4 and Qa-5. Few, if any, endogenous NK cells express Ly-1 or Lyt-2. In addition, a number of allo-antisera have been described that appear to have specificity for NK cells (e.g., NK-1.1, NK-2.1). No one phenotype has been described that includes all murine NK cells: they are phenotypically heterogeneous. Studies that support this conclusion have been reviewed in References 4, 18 and 19, among others. The major findings are summarized in Table II.

We have analyzed whether the two alloantigens, NK-1.1 and NK-2.1 are associated with functional subsets of NK cells (20). We had previously shown that NK-1.1 antiserum ((BALB/c x C3H)F_1 anti-CE) and NK-2.1 antiserum (NZB anti-BALB/c) detect genetically distinct alloantigens (21). Both are expressed in B6 mice. For the studies described here, we took nylon wool nonadherent B6 spleen cells (SC) and treated them with complement (C) and NK-1.1 or NK-2.1 anti-sera to deplete NK-1 or NK-2 antigen-bearing cells. The treated cells were then tested for NK activity against a panel of tumor targets in 6 and 19 hr ^{51}Cr release assays. NK activity to the lymphoma lines YAC-1 and L5178c127v, to the B16 melanoma and to the human polyploid liver line Chang was reduced to a similar degree by either antisera (Table III). However, NK activity to FBL-3 and to K562 was sensi-tive to NK-2.1 antiserum and C but not to NK-1.1 antiserum and C. Neither serum significantly affected natural killing of E♂G2 or RBL-5. We then attempted to separate NK-1.1$^+$ from NK-2.1$^+$ effector cells by antibody and C depletion or by sorting on the FACS. In neither case were we able to obtain a population expressing only one of the alloantigens. Our results suggest that most murine splenic NK cells express both NK-1.1 and NK-2.1 alloantigens but that the density of expression of these antigens may vary among the functionally different subpopulations.

IV. PRECURSORS OF NK CELLS

It is well established that NK cells are bone marrow derived and bone marrow dependent for their proliferation and/or differentiation (22,23). However, little is known about the morphology, phenotype, or cellular kinetics of NK precursor cells. We have begun studies to test whether the

TABLE II

Phenotype of Murine NK Cells[a]

Antigen	Expression
Thy-1	+/- (50-70% +)
Ly-1	-
Lyt-2	-
Ly-5 (T200)	+
Qa-1	-
Qa-2,3	-
Qa-4	+
Qa-5	+
NK-1.1	+
NK-2.1	+
asialo GM_1	+
sIg	-
ThB	-
FcR	+

[a]Based on data reviewed in 4, 18, 19

TABLE III

EFFECT OF DEPLETION WITH ANTI-NK-1.1 OR ANTI-NK-2.1 AND C ON NK ACTIVITY TO DIFFERENT TARGETS

Effector Cells Depleted by C and	Targets
NK-1.1 or NK-2.1	YAC-1, Chang, B16 L5178cl27v
NK-2.1	FBL-3, K562
Neither	E♂G2, RBL-5

production and differentiation of NK cells from their precursors follows the pattern of other cell types produced in the marrow such as B cells or granulocytes. The general pattern includes development from a pluripotent stem cell to committed progenitors that, in turn, give rise to a "maturational compartment" of cell proliferation and differentiation that leads to the production of postmitotic mature cells that are exported from the marrow.

Several groups have presented evidence for an NK cell progenitor in the marrow. Both Claesson and Olsen (24) and Koo *et al*. (25) have grown colonies of NK cells from bone marrow-derived precursor cells *in vitro*. Evidence regarding the production rate of NK cells *in vivo* was obtained by Miller (26) using the cell cycle-specific drug hydroxyurea. The data, based on NK cytolytic activity, indicate that NK cells are produced rapidly in the marrow with a T1/2 of 7.9 hr. Maturation to a functionally active NK cell may occur during a postmitotic stage.

We are using a classic cell kinetic approach to characterize NK precursors and their maturational stages in the marrow. Since NK cells acquire the ability to bind to targets prior to acquisition of lytic activity, we have used target binding as a marker for presumptive NK cells. Six-week-old B6 mice were injected with 1 μCi ^{3}H-TdR/gm body weight. Femoral marrow was obtained 1, 12, 24 or 48 hr after a single pulse of ^{3}H-TdR. In order to enrich for NK cells, surface μ^+ B cells were eliminated by treatment with anti-Ig and C. Erythroid and granulocytic cells were then eliminated by flow cytometry on the basis of their forward and right angle scatter patterns. The remaining cells (15-20% of the initial bone marrow cells) had lymphoid morphology. The characteristics of these BM "lymphoid" cells are shown in Table IV.

^{3}H-TdR labeled BM "lymphoid cells" were mixed with FITC-labeled YAC-1 cells in a standard target binding cell (TBC) assay. Cytospots were prepared, radioautographed and then examined with phase contrast fluorescence microscopy. Two size classes of TBC were apparent: small-medium TBC, < 9 mm in diameter and large TBC, > 9 mm. The labeling indices of the total "lymphoid" population and of the small and large TBC are given in Table V. The increase in labeling index of the small TBC during the first 24 hours after a single pulse and the subsequent decrease indicates that small TBC are newly produced cells with a rapid turnover. In contrast, the large TBC appear to be a self-maintaining precursor pool that gives rise to the small TBC.

Only a small percentage (< 5%) of splenic TBC were labeled at any of the times tested, suggesting that most NK

TABLE IV

Characterization of B-Depleted BM "Lymphoid" Cells

$NK\text{-}1.1^+$	= 75%
Target binding cells (TBC)	= 23%
$NK\text{-}1.1^+$, TBC^+	= 19%
% of $NK\text{-}1.1^+$ that are TBC^+	= 25%
% of TBC^+ that are $NK\text{-}1.1^+$	= 81%

TABLE V

Labeling of BM "Lymphoid" Cells and Target Binding Cells (TBC)

	2 hr	12 hr	24 hr	48 hr
Total "lymphoid" population[b]	15.8	32.1	37.5	22.2
Small TBC[c]	8.0	16.3	27.1	16.0
Large TBC[d]	24.0	18.4	20.7	13.7

[a]Bone marrow was obtained at the times indicated after injection of B6 mice with a pulse of ^{3}H-TdR (1 μCi/gm body weight).

[b]Total ^{3}H-TdR labeled/total cells in lymphoid fraction.

[c]Number ^{3}H-TdR labeled small TBC/total number TBC.

[d]Number ^{3}H-TdR labeled large TBC/total number TBC.

cells in the spleen are immigrant from the bone marrow. Some, however, may be the progeny of cells replicating in the spleen (cf. discussion above and Refs. 16 and 17).

V. CONCLUSIONS

Murine NK cells, like their human counterparts, are heterogeneous and vary in their size and morphology, as well as their phenotype. Our data and that of Tai and Warner (16) show that endogenous NK activity of unstimulated mice is mainly in small-medium size lymphocytes. Upon stimulation there is a shift to larger cells. Both the state of stimulation of the cells and the separation conditions used to obtain NK cells may affect their apparent size. Whether phenotypic heterogeneity relates to functional heterogeneity remains to be more fully explored. Our studies with NK-1.1 and NK-2.1 antisera have shown that functional subpopulations can be delineated on the basis of reactivity with one, both or neither antisera. It appeared that NK cells differ in their relative expression of the two alloantigens.

Markers such as NK-1.1, NK-2.1 and the ability to bind NK-sensitive targets are proving to be useful in studies of NK precursors in the marrow. Our studies suggest that NK-1.1 is expressed on bone marrow NK precursor cells prior to the acquisition of the ability to bind targets. Upon further maturation, NK cells acquire lytic activity. The very low lytic activity of BM TBC suggests that this latter step may occur in the periphery, after export of NK cells from the marrow.

REFERENCES

1. Herberman, R. B., Nunn, M. E., and Lavrin, D. H. (1975). Int. J. Cancer 16, 216.
2. Kiessling, R., Klein, E., and Wigzell, H. (1975). Eur. J. Immunol. 5, 112.
3. Sendo, F., Aoki, T., Boyse, E. A., and Buafo, C. K. (1975). J. Natl. Cancer Inst. 55, 603.
4. Pollack, S. B. (1984). In "Immunobiology of Natural Killer Cells" (E. Lotzova and R. B. Herberman, eds.) CRC Press, Boca Raton, FL. (In Press)

5. Herberman, R. B., Bartram, S., Haskill, J. S., Nunn, M., Holden, H. T., and West, W. H. (1977). J. Immunol. 119, 322.
6. Paige, C. J., Figarella, E. F., Cuttito, M. J., Cahan, A., and Stutman, O. (1978). J. Immunol. 121, 1827.
7. Kiessling, R., Klein, E. and Wigzell, H. (1975). Eur. J. Immunol. 5, 117.
8. Roder, J. C., Kiessling, R., Biberfeld, P., and Andersson, B. (1978). J. Immunol. 121, 2509.
9. Herberman, R. B., and Holden, H. T. (1978). Adv. Cancer Res. 27, 305.
10. Tam, M. R., Emmons, S. L., and Pollack, S. B. (1980). J. Immunol. 124, 650.
11. Lotzová, E. (1980). In "Natural Cell-Mediated Immunity Against Tumors" (R. B. Herberman, ed.) p. 131. Academic Press, New York.
12. Luini, W., Boraschi, D., Alberti, S., Aleotti, A., and Tagliabue, A. (1981). Immunology 43, 663.
13. Kumagai, K., Itoh, K., Suzuki, R., Hinuma, S., and Saitoh, F. (1982). J. Immunol. 129, 388.
14. Timonen, T., Ortaldo, J., and Heberman, R. (1981). J. Exp. Med. 153, 569.
15. Reynolds, C. W., Timonen, T., and Herberman, R. B. (1981). J. Immunol. 127, 282.
16. Tai, A. and Warner, N. L. (1980). In "Natural Cell-Mediated Immunity Against Tumors" (R. B. Herberman, ed.) p. 241, Academic Press, New York.
17. Biron, C. A., Turgiss, L. R., and Welsh, R. M. (1983). J. Immunol. 131, 1539.
18. Stutman, O., Lattime, E., and Figarella, E. (1981). Fed. Proc. 40, 2699.
19. Herberman, R. B., and Ortaldo, J. R. (1981). Science 214, 24.
20. Emmons, S. L., and POllack, S. B. (1984). Submitted
21. Pollack, S. B., and Emmons, S. L. (1982). J. Immunol. 129, 2277.
22. Haller, O., Kiessling, R., Orn, A., and Wigzell, H. (1977). J. Exp. Med. 145, 1411.
23. Kumar, V., Ben-Ezra, J., Bennett, M., and Sonnenfeld, G. (1979). J. Immunol. 123, 1832.
24. Claesson, M. H., and Olsson, L. (1980). Nature 283, 578.
25. Koo, G. C., Peppard, J. R., and Mark, W. H. (1984). J. Immunol. 132, 2300.
26. Miller, S. (1982). J. Immunol. 129, 2282.

ENRICHMENT METHODOLOGY FOR MURINE NK ACTIVITY

T. Juhani Linna
Mahadev R. Patel

Central Research and Development Department
Glenolden Laboratory
E. I. du Pont de Nemours and Company
Glenolden, Pennsylvania

I. INTRODUCTION

Many laboratories are using discontinuous poly-(vinylpyrrolidone) silica (PVP-silica, Percoll®) gradients, originally introduced by Timonen and colleagues, for enrichment of human and rat natural killer (NK) cells (1,2,3). This method, as originally described, does not result in significant enrichment of murine NK activity in many laboratories, including ours. We will report here a modification of the discontinuous Percoll® gradient method, which also allows significant enrichment of murine splenic NK activity. We will also document a simple, but relatively efficient way of enriching for murine NK activity. In addition, we have done a morphological assessment of cell populations present in the NK-enriched and NK-depleted fractions.

II. MATERIALS AND METHODS

Mice Several mouse strains have been used successfully for these enrichment studies, including strains with relatively low NK activity in unenriched spleen cell preparations. Most studies were done with C57BL/6J mice.

Spleen Cells A single cell suspension was made, washed twice and resuspended at $0.5\text{-}1.5\times10^8$ cells/ml in assay medium (RPMI 1640 medium containing 5% heat-inactivated fetal bovine serum (FBS)). These cells were used for

ISBN 0-12-341370-2

enrichment experiments. An aliquot was used for determination of NK activitiy before enrichment and for Giemsa staining after cytocentrifugation.

PVP-Silica (Percoll®) gradients Solutions of Percoll® (Pharmacia Fine Chemicals, Uppsala, Sweden) with osmolarity ranging from 310-340 mOsm/kg, similar to that in mouse blood and tissues (4,5) were made by mixing calcium- and magnesium-free 10x Dulbecco's phosphate buffered saline (DPBS) with Percoll® and distilled water. The density for each solution was measured by pycnometry, and osmolarity by freezing point depression using a μ-osmette (Precision Systems, Inc., Sudbury, MA). Such a typical four step gradient consisted of 44.2%, 50.8%, 55.0% and 66.7% Percoll®, with densities of 1.062, 1.070, 1.077 and 1.092 g/ml, in the respective steps. With this method of gradient preparation, higher density solutions will have a slightly higher osmolarity due to the higher Percoll® content. Therefore, other gradients were prepared with the same ingredients or with RPMI 1640 plus 10% FBS, keeping the osmolarity constant at 320 mOsm/kg. The gradients are described in greater detail elsewhere (6). One to 2 x 10^8 spleen cells were carefully placed on top of the gradient in a 15 ml conical centrifuge tube, and the gradient centrifuged at 550x g for 30 min. at room temperature. Four distinct bands and a pellet containing mostly red cells was obtained. The cells from each band were carefully retrieved with a Pasteur pipet, washed and resuspended in assay medium. The viability in bands 2-4 was usually over 95%, and 70-90% in band 1 (top). All bands and the unseparated population were tested simultaneously in a ^{51}Cr release assay against YAC-1 and other target cells. Cytocentrifuge preparations were also made, air dried and fixed in 95% methanol for Giemsa staining.

Ficoll-Paque® Separation Four ml of Ficoll-Paque® was layered under one to 2x10^8 spleen cells (Pharmacia Fine Chemicals, Uppsala, Sweden), density 1.077 g/ml in a 15 ml conical centrifuge tube and centrifuged at 400x g for 30 min at room temperature. An interphase layer and a pellet were obtained. These cell populations were washed, resuspended in assay medium, and used as effectors in the ^{51}Cr release assay or for Giemsa staining.

^{51}Cr Release Assay The lymphoma YAC-1 target cells were used in the ^{51}Cr release assay. Unseparated spleen cells, or spleen cells from the above mentioned populations were used as effectors. Three to 10x10^6 YAC-1 target cells in a Tris-phosphate buffer were incubated with 100 μCi of ^{51}Cr as sodium chromate (1 mCi/ml, Du Pont - NEN, Boston, MA) at 37°C for 45-60 min. After incubation, the cells were washed repeatedly. Ten thousand labeled targets were added to varying numbers of effector cells to give at least five different effector:target ratios. Duplicate samples were centrifuged at 550x g for 5 min and a portion of the supernatant fluid harvested. Radioactivity in each tube was counted in a gamma counter. Total counts in the target cells, maximum releasable counts, and spontaneous release were also determined. Spontaneous release was usually about 10% of maximum release, and in no case reached 25%. Cytolytic activity for each sample was determined as

$$100\times \frac{\text{test release} - \text{spontaneous release}}{\text{maximum release} - \text{spontaneous release}}$$

Cytolytic activities are reported as % ^{51}Cr release according to the above formula, or as lytic units (LU_{30}) per 10^7 effector cells, one lytic unit being the amount of effector cells required for 30% lysis of 10^4 target cells.

Staining technique and evaluation of slides The fixed cytocentrifuged preparations were stained in Giemsa stain (Fisher Scientific Co., King of Prussia, PA) and gently washed in buffer for about 2 min. The preparations were air dried, mounted with Permount® and covered with a coverslip for light microscopic evaluation. One thousand mononuclear cells were counted in each preparation, and their morphology including size and presence of azurophilic granules evaluated and recorded. Large granular lymphocytes (LGL) were defined with the usual critiera of a relatively high cytoplasmic/nuclear ratio, an often kidney-shaped nucleus, and the presence of azurophilic granules (7). Large agranular lymphocytes (LAL) had similar morphology but lacked granules. Other lymphocytes not fitting into the above categories were recorded separately.

III. RESULTS

A. Enrichment of NK cell activity in the four step high osmolarity density gradient

There was a high level of NK enrichment in high osmolarity gradients (average 10-fold; range 4-25-fold) when compared to the simultaneously assayed non-separated populations. The enrichment occurred in band 3 of the gradient, above the 1.077 g/ml Percoll® density layer (Fig. 1). Thus, it appeared that mouse splenic NK cells have higher density than human or rat PBL derived NK cells where the highest activity is found at 1.062 g/ml density.

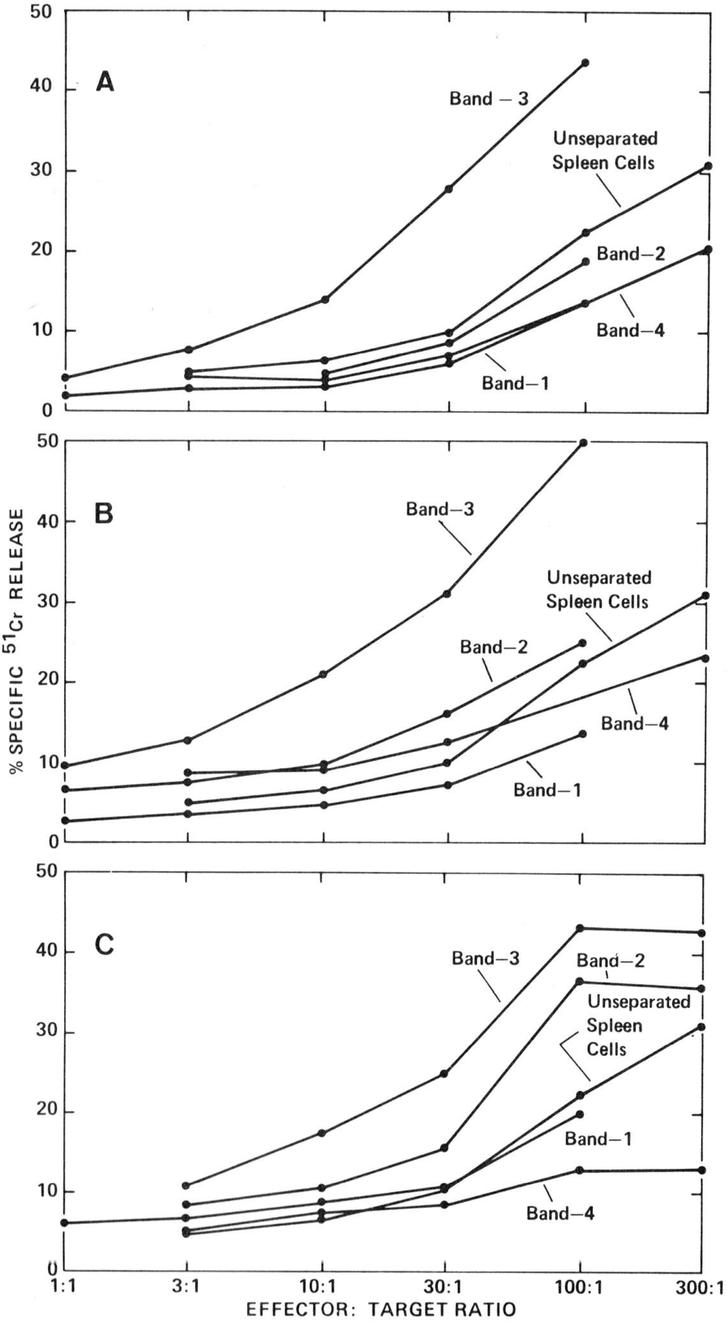

A
Band – 3
Unseparated Spleen Cells
Band–2
Band–4
Band–1
B
Band–3
Unseparated Spleen Cells
Band–2
Band–4
Band–1
C
Band–3
Band–2
Unseparated Spleen Cells
Band–1
Band–4
% SPECIFIC ^{51}Cr RELEASE
EFFECTOR: TARGET RATIO
1:1
3:1
10:1
30:1
100:1
300:1
0
10
20
30
40
50

Fig. 1 A, B, C.

Band 3 contained on the average 9-10% of the recovered cells. Most cells (75% or more) were found in band 4, but this band was consistently depleted for NK activity (Fig. 1). The results were similar in gradients with varying osmolarity (310-340 mOsm/kg) and with uniform osmolarity 320 mOsm/kg). (Fig. 1A, B) The presence of FBS in the medium did not impair the enrichment (Fig. 1C).

B. Ficoll-Paque® Separation

As most NK activity was localized above the 1.077 g/ml density layer in the gradients, we decided to see whether an enrichment of NK cells could be accomplished simply by centrifugation of spleen cells in Ficoll-Paque®, which is normally used for human peripheral blood mononuclear cell separation (8). This simple procedure resulted in a significant enrichment of NK activity, which was on the average somewhat lower than with the high osmolarity Percoll® gradients (Table I). This fraction also contained a greater percentage of the recovered cells than band 3 after high osmolarity Percoll® gradient separation (14% vs 9%), arguing for some more dilution with non-NK cells in the Ficoll-Paque® separated population. The pellet after Ficoll-Paque separation contained most of the lymphoid cells, but was depleted of NK activity (Fig. 2).

C. Morphology

In the comparison between the NK-enriched band 3 cells, the NK-depleted band 4 cells and the unseparated spleen cell population, an enrichment for agranular lymphocytes without obvious identifying features was noted. As shown in Table II, enrichment occurred in the larger lymphocyte populations, i.e., 10 μm and above, and was particularly striking for cells with a diameter of 12 μm or greater. No particular enrichment of large granular lymphocytes (LGL) or of large agranular lymphocytes, sharing LGL morphology but lacking granules, could be noted.

Fig. 1. NK activity enrichment in different conditions at higher osmolarity: The efficiency of NK enrichment in modified gradients was comparable in four-step gradients: A, with varying osmolarity (310-340 mOsm/kg); B, uniform osmolarity of 320 mOsm/kg throughout, using DPBS; and C, in the presence of RPMI 1640 medium and FBS at 320 mOsm/kg throughout.

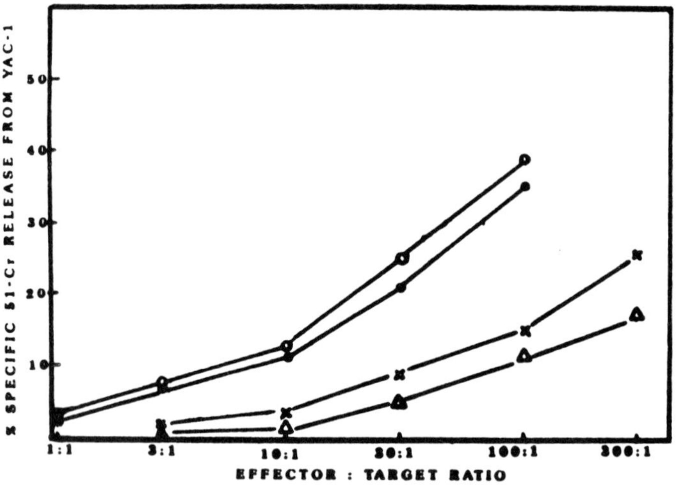

Figure 2. Typical experiment.
o——o Band 3 (density ≤ 1.077 g/ml), four-step Percoll® gradient.
●——● Interphase (density ≤ 1.077 g/ml), Ficoll-Paque® centrifugation
x——x Unseparated population
Δ——Δ Pellet, Ficoll-Paque® centrifugation.
Eight-fold enrichment of NK activity in Ficoll-Paque® band, ten-fold in band 3 after Percoll® separation.

IV. DISCUSSION

There are some reports of successful enrichment of NK-activity of mouse splenic and other lymphocytes using PVP-silica gradients which can enrich for human or rat NK activity and large granular lymphocytes (9,10,11). However, this method enriches poorly (typically less than four-fold) for mouse NK cells in our hands. Therefore, we have attempted to find ways to obtain a more significant enrichment of murine NK cell activity. We found that this could be accomplished simply by raising the density range as well as the osmolarity of the gradient from 285 mOsm/kg to 310-340 mOsm/kg, to correspond better to the osmolarity in mouse blood and tissues (4,5). The appropriate changes in the densities in the step gradient, allowed enrichment

TABLE I. Comparison of enrichment of NK cell activity: Four-step PVP-silica gradient and one step centrifugation

Method	Number of Experiments	% of Recovered cells [a,b]	LU_{30} [b,c] Fold Enrichment band/unseparated
Four-step Percoll® gradient (310-340 mOsm/kg)	10	8.7±3.2 (band 3)	10.4±5.8
One step Ficoll Paque® (285 mOsm/kg)	14	14.8±9.5 (interphase)	8.2±3.1

[a]The separated cells have density ≤ 1.077 g/ml.
[b]Mean ± standard deviation of the mean.
[c]LU_{30} defined in Materials and Methods.

corresponding to or better than what is commonly accomplished for enrichment of human or rat NK activity with the 285 mOsm/kg step gradient. Using this method, most of the NK active cells were found at a higher density (≤ 1.077 g/ml in the step gradient than what is the case for human or rat NK cells (≤ 1.062 g/ml). As the density, where mouse NK activity is found, is similar to that of in Ficoll-Paque® in common use for human lymphocyte separation, we attempted this method as a quick and simple way of murine NK enrichment. Indeed, we found that it works, although the enrichment was on the average somewhat lower, probably due to dilution with non-NK cells in the Ficoll® band. Ficoll-Paque® has an osmolarity of 285 mOsm/kg. Thus, a lower osmolarity *per se* is no obstacle to murine NK enrichment and activity. It may well be associated with some compensatory lymphocyte "swelling" in the relatively hypotonic environment. However, as there are no other density steps than 1.077, there is no significant redistribution of the perhaps slightly less dense NK cells. There may be some admixture in the NK-enriched layer of non-NK cells which would fall below the 1.077 g/ml layer under iso-osmolar conditions, as there are more cells and somewhat less enrichment in terms of NK activity than in band 3 of the Percoll® step gradient. However, the extreme simplicity of the Ficoll-Paque® separation should make it a useful tool in murine NK work.

Table II. Morphological Determinations. Band 3 from PVP-silica gradients (310-340 mOsm/kg) is enriched for agranular lymphocytes with diameter $\geq 10\ \mu m$, particularly those with diameter $\geq 12\ \mu m$. [Mean Activity enrichment = 10-fold (range 5-22), n = 7 experiments]

Cell Morphology	Percentage (mean ± S.D.) of cells with diameter in micrometers				
	7-9	10-11	12-13	>13	Total
Unseparated cells					
LGL	0	1.7±0.4	1.7±0.9	0.1±0.2	3.5±1.0
LAL	0	2.3±0.4	2.4±1.0	0.3±0.3	5.0±1.1
Other Granular Lymphocytes	7.8±4.1	2.9±1.6	0.1±0.1	0	10.8±5.0
Other Agranular Lymphocytes	65.0±7.2	13.1±1.7	2.1±0.5	0.1±0.1	79.0±5.4
Total	72.6±4.2	20.1±3.2	6.6±2.0	0.5±0.3	
NK Enriched (band 3)					
LGL	0	1.3±0.5	1.6±0.5	0	3.0±0.7
LAL	0	3.1±0.7	3.7±1.2	1.3±0.8	8.2±1.2
Other Granular Lymphocytes	2.2±1.2	3.4±1.8	1.4±0.8	0	7.1±3.6
Other Agrandular Lymphocytes	24.9±6.0	42.1±6.3	13.0±2.8	1.4±0.6	81.5±3.8
Total	27.1±5.6	50.0±6.0	19.7±4.6	2.8±1.2	
NK depleted (band 4)					
LGL	0	1.0±0.7	0.6±0.4	0	1.7±1.0
LAL	0	1.2±0.7	0.7±0.4	0	2.0±1.1
Other Granular Lymphocytes	6.3±2.8	2.6±2.0	0.1±0.2	0	9.1±4.7
Other Agranular Lymphocytes	76.6±14.1	10.1±7.7	0.4±0.3	0.1±0.1	87.3±6.5
Total	83.0±11.5	15.1±10.3	1.9±1.0	0.1±0.2	

Murine NK cell morphology has remained an unresolved issue. While there is little doubt that at least a significant proportion of human and rat NK activity is associated with LGL morphology, many laboratories including ours have not noted the same association in the mouse. However, there are some reports associating LGL with murine NK type activity in the blood and spleen (9,10,11) and in intestinal lamina propria and epithelium (12). In the present study, we found that enrichment for NK activity was associated with enrichment of a population of lymphocytes lacking granules and other LGL characteristics, and having a diameter of 10 μm or greater. The greatest enrichment was seen of such relatively non-descript cells in the Giemsa stain with a diameter of 12 μm or greater. Similar findings were made in the Ficoll-Paque® enriched cells. The Giemsa stain does not provide further morphological markers for these cytocentrifuged cells.

V. SUMMARY

A discontinuous PVP-silica (Percoll®) gradient suitable for enrichment of murine NK activity, is described. NK activity was found to be most enriched at density $\leq$ 1.077 g/ml. Therefore, centrifugation of murine spleen cells in Ficoll-Paque®, density 1.077 g/ml, was also tried and found to be a very simple and relatively efficient way of enhancing for murine NK cell activity. Enrichment of activity was associated with enrichment of relatively large lymphocytes, which did not have LGL or LAL characteristics in Giemsa stained preparations.

REFERENCES

1. Timonen, T., and Saksela, E. (1980). J. Immunol. Meth. 36, 285.
2. Timonen, T., Ortaldo, J.R., and Herberman, R.B. (1982). In "NK Cells and Other Natural Effector Cells" (R.B. Herberman, ed.), p. 9. Academic Press, New York.
3. Timonen, T., Reynolds, C.W., Ortaldo, J.R., and Herberman, R.B. (1982). J. Immunol. Meth. 51, 269.

4. Dettman, G.L., and Wilbur, S.M. (1980). In "Selected Methods in Cellular Immunology" (B.B. Mishell and S.M. Shiigi, ed.), p. 197. W.H. Freeman and Company. San Francisco.
5. Silverstein, E., Sokoloff, L., Mickelson, O., and Jay, G.E. (1961). Amer. J. Pathol. 38, 143.
6. Patel, M.R., and Linna, T.J. (1984). Immunology, In Press.
7. Timonen, T., Saksela, E., Ranki, a., and Hayry, P. (1979). Cell. Immunol. 48, 133.
8. Boyum, A. (1968). Scand. J. Clin. Lab. Invest. 21, Suppl. 97, 1.
9. Luini, W., Boraschi, D., Alberti, S., Aleotti, A., and Taglibue, A. (1981). Immunology 43, 663.
10. Kumagai, K., Itoh, K., Suzuki, R., Hinuma, S., and Saitoh, F. (1982). J. Immunol. 129, 388.
11. Itoh, K., Suzuki, R., Umezu, Y., Hanaumi, K., and Kumagai, K. (1982). J. Immunol. 129, 395.
12. Tagliabue, A., Befus, A.D., Clark, D.A., and Bienenstock, J. (1982). J. Exp. Med. 155, 1785.

Part II.

Characterization of NK Target Structures and Recognition Sites on NK Cells

MOLECULAR CHARACTERIZATION OF NK TARGET ANTIGENS AND TARGET RECOGNITION SITES ON NK CELLS

Joseph Kaplan

Department of Pediatrics
Wayne State University School of Medicine
Detroit, Michigan 48201

I. INTRODUCTION

The nature and diversity of target structures recognized by NK cells has been a matter of considerable controversy. Some investigators favor the possibility that NK cells have a very limited range of specificity, and recognize only a small number of ubiquitous cell surface moieties. Others favor the view that NK cells, like T cells, comprise a polyclonal population, each clone of which recognizes one of multiple unique antigens. Another more recently developed view is that individual NK cells can have both "T cell-like" monospecificity for unique antigens, and "NK-like" polyspecificity for ubiquitous target structures.

II. STEM CELL TARGET ANTIGENS

There is general agreement that NK target structures are present on normal cells as well as on tumor cells. This, and the fact that immature rapidly dividing cells such as fetal cells and immature hemopoietic cells are particularly susceptible to NK lysis has led to the suggestion that NK cells

ISBN 0-12-341370-2

play a physiological regulatory role in the control of stem cell populations. In support of this, Kiessling reported that human and murine NK cells inhibit the in-vitro proliferation of autologous and allogeneic bone marrow stem cells. As has been shown for in-vivo bone marrow resistance, allogeneic and semiallogeneic bone marrow cells were found to be more sensitive to in-vitro inhibition by splenic NK cells than were autologous or syngeneic bone marrow cells.

One of the first clues that NK cells might play a role in controlling hemopoietic stem cells was the observation that the effectors of in-vivo F_1 hybrid resistance to bone marrow grafts shared many properties with NK cells, e.g., thymic independence, marrow dependence, radioresistance, and age of appearance (Eur. J. Immunol. 7:655, 1977). Direct evidence that NK cells are the effectors of hybrid resistance was presented at this meeting by Nakamura. He reported that asialo $GM1^+$ low density spleen cells from F_1 hybrid mice heterozygous at $H\text{-}2D^b$ preferentially inhibit in-vitro and in-vivo proliferation of $H\text{-}2D^b$ homozygous parental bone marrow cells as compared to syngeneic bone marrow cells. Just as observed for in-vivo hybrid resistance, the target antigens for in-vitro NK cell-mediated hybrid resistance appear to be distinct antigens which map to the MHC region at or near $H\text{-}2D^b$. It would appear that if these antigens are located on $H\text{-}2D^b$ molecules, they are either unique $H\text{-}2D^b$ epitopes, or molecules distinct from $H\text{-}2D^b$ molecules, since they are clearly distinct from the determinants detected by $H2\text{-}D^b$-specific alloantibodies or by allospecific cytotoxic T cells. Consistent with the findings reported by Kiessling, Nakamura observed that NK cells from F_1 hybrids also inhibit growth of syngeneic bone marrow cells. However this requires higher ratios of NK cells to bone marrrow targets than those needed for inhibition of $H\text{-}2^b$ homozygous parental strain bone marrow cells.

The provocative implication of the findings presented by Nakamura is that the NK cells which mediate hybrid resistance are exquisitely antigen specific. Dennert reported data which suggest that this specificity may not be due to antigen-specific receptors intrinsic to NK cells, but rather to natural antibodies which, via Fc receptors, allow NK cells to inhibit bone marrow target cells by the mechanism of antibody-dependent cell-mediated cytotoxicity (ADCC). Dennert presented evidence that the H-2 specificity of bone marrow allograft resistance mediated by adoptively transferred NK clones was due to H-2 specific natural antibody present in "responder" bone marrow recipients: serum from such "responders" adoptively transferred bone marrow allograft resistance to "non-responders".

In commenting on Dennert's findings, Nakamara pointed out that although ADCC may well account for the ability of NK clones to adoptively transfer resistance to allogeneic bone marrow cells, if it accounts for F_1 hybrid NK cell mediated in-vitro inhibition of parental bone marrow proliferation, one has to assume that inhibitory F_1 hybrid NK cells are armed in-vivo by natural antibodies to Hh antigens. There are several reasons for thinking this unlikely. First, it is known that by comparison to Fc receptors on human NK cells, the affinity for Ig molecules of murine NK cell Fc receptors is quite low. Thus, even if natural anti-Hh antibodies bind in-vivo to NK cells by their Fc receptors, it seems unlikely that these antibodies would remain attached to NK cells throughout the extensive processing they are subjected to prior to the in-vitro inhibition assays. Secondly, the natural antibody hypothesis predicts that the efficiency of cold target inhibition of hybrid resistance by irradiated tumor cells injected in-vivo should be related to the total number of irradiated cells injected rather than to the ratio of irradiated to non-irradiated cells injected. Nakamura has not found this to be the case. For example, he has observed that a quantity of irradiated cells which totally abrogates in-vivo resistance to 2×10^6 bone marrow cells has no effect on the resistance to 5×10^5 bone marrow cells. Thirdly, the natural antibody hypothesis predicts that procedures which reduce in-vivo antibody levels should diminish hybrid resistance. In fact, it has been reported that F_1 hybrid mice rendered antibody deficient by chronic injection of anti-mu chain antibody, if anything, show stronger hybrid resistance than control animals (J. Reticulo. Soc. 30:283-289, 1981). If natural antibody to Hh antigens do, in fact, play a role in hybrid resistance, it should be possible to demonstrate that inhibition of H-2D^b homozygous bone marrow cells by syngeneic NK cells is increased following preincubation with serum from "resistant" H-2D^b heterozygotes, or by showing that such serum can adoptively transfer hybrid resistance to other non-resistant H-2D^b heterozygotes. At this time, no such studies have been reported.

III. NK RECOGNITION OF ALLOANTIGENS

The possibility raised by hybrid resistance phenomena that some NK cells specifically recognize MHC-encoded antigens was a major factor which prompted Kaplan et al (in "Natural Cell Mediated Immunity Against Tumors", Academic Press, 1980, p. 893)to postulate that NK cells are germ line V-gene-encoded prothymocytes specific for self and non-self histocompatibility antigens. One of the predictions of this hypothesis is that at least some NK cells are precursors of allospecific cytotoxic T lymphocytes. At this meeting Kaplan reported findings consistent with that prediction, i.e., that a subpopulation of murine allospecific cytotoxic T cell precursors appears to have the asialo $GM1^+$ Thy 1^- phenotype of NK cells. MLC responder cells depleted of Thy 1^+ cells were able to generate allo-CTL if Con A-conditioned medium was added to the MLC cultures, but were not able to do so if also depleted of asialo $GM1^+$ cells. This latter effect did not appear to be due to removal of an asialo $GM1^+$ accessory cell since the effect of removing asialo $GM1^+$ cells was not reversed by addition to the MLC cultures of responder type mitomycin-C-treated spleen cells.

The hypothesis put forth by Kaplan asserts that NK cells have germ-line V-gene-encoded T cell receptors. Preliminary evidence consistent with that notion was presented by Roder. Results of recent experiments he conducted in collaboration with Mak (Toronto) demonstrate the existence of T cell receptor messenger RNA in 2 NK clones, and, more importantly, in purified poly I:C-induced murine NK cells. If confirmed, these findings would represent the strongest evidence currently available that NK cells are in the T cell linage and are antigen specific.

If in some instances NK cell recognition of target cell MHC antigens results in activation of the NK cell lytic machinery, in other instances such recognition may result in inhibition of NK cell activation. Thus, based on evidence his group has obtained (a) of an inverse correlation between target cell expression of H-2 antigens and target cell susceptibility to in-vivo or in-vitro NK cell killing or growth inhibition, and (b) a lack of correlation between target cell H-2 expression and ability to act as cold target inhibitors of NK cell lysis, Karre proposed that self-MHC molecules on target cells act as signals which inhibit a post-binding stage of NK-mediated lysis. According to this hypothesis, F_1 hybrid resistance occurs because NK-susceptible bone marrow cells of parental donor type lack some of the "self" MHC antigens of the recipient, i.e., those conferred by the other parental type. The resistance of some $H\text{-}2D^b$ heterozygotes (e.g., $H\text{-}2D^{b/x}$) but not other $H\text{-}2D^b$ heterozygotes (e.g., $H\text{-}2D^{b/y}$) to $H\text{-}2D^{b/b}$ homozygous bone marrow cells is then explained by the existence of cross-reactivity between $H\text{-}2D^b$ and $H\text{-}2D^y$ but not $H\text{-}2D^x$. However, it is not clear how this hypothesis explains the frequent occurrance of $H\text{-}2D^{b/x}$ heterozygotes which reject $H\text{-}2D^{b/b}$ homozygous bone marrow cells but accept similar numbers of $H\text{-}2D^{x/x}$ homozygous bone marrow cells. If acceptance of $H\text{-}2D^{x/x}$ bone marrow cells by $H\text{-}2D^{b/x}$ animals is ascribed to cross-reactivity between $H\text{-}2D^b$ and $H\text{-}2D^x$, then it becomes difficult to explain why such F_1 animals reject $H\text{-}2D^{b/b}$ homozygous cells, since such cells should be viewed by the F_1 as expressing "self" $H\text{-}2D^x$ as well as "self" $H\text{-}2D^b$.

IV. NORMAL LYMPHOCYTES AS NK TARGETS

Normal as well as neoplastic lymphoid cells are among the most susceptible of NK targets. This suggests that in addition to a physiological regulatory effect on hemopoietic stem cells, NK cells may also play a role in controlling the proliferation and differentiation of lymphoid populations. Several lines of evidence supporting this possibility were presented at this session.

A. B Lymphocytes

Pollack reported that the primary in-vivo antibody response to sheep red blood cells in mice is enhanced following ablation of in-vivo NK activity using NK 1.1 alloantiserum, and is suppressed following in-vivo activation of NK activity with Poly I:C. Similarly Kumagai reported that in-vivo ablation of murine NK activity using anti-asialo GM_1 results in increased numbers of splenic B cells, increased responses to B but not T-cell mitogens, and increased PFC responses to pokeweed mitogen.

The down-regulatory effects of NK cells on B cell responses are probably due, in large part, to NK-mediated lysis of activated B lymphoblasts. Thus Pollack presented evidence that splenic NK cells bind and lyse LPS-induced B blasts. In addition, Targan reported that monoclonal antibodies which block the post-binding lytic triggering signal block NK-mediated suppression of B lymphoblast antibody production. Moreover, NK-derived soluble lytic factors (NKCF) can, by themselves, inhibit in-vivo antibody production. Targan also presented evidence that the transferrin receptor expressed by activated B cells may be involved in NK-mediated inhibition of B cell function. He reported that antibodies to transferrin receptor by themselves partially inhibit in-vitro antibody secretion by activated B cells, and that this is reversed by addition of free Fe ions. Moreover, under these conditions the inhibition of antibody secretion mediated by inferferon-activated NK cells is no longer observed. Kay presented evidence that transferrin itself may be involved in this process. He showed that transferrin inhibits NK lysis by inhibiting production of NKCF, but does not inhibit the lytic effect of NKCF on target cells. These findings raise the possibility that transferrin receptors are important not as target structures for NK recognition, but rather as cell surface molecules which play a role in post-binding events regulating the NK lytic machinery.

Pollack pointed out that if NK cells down-regulate antibody production they may play a role in controlling auto-antibody production. In this light it is of interest that certain NK-deficient strains of mice, e.g. "Motheaten" and SJL, have a high incidence of autoantibody formation, and that NK deficiency is common in patients with such autoimmune diseases as systemic lupus erythematosis, Crohn's disease, and rheumatoid arthritis.

B. T Lymphocytes

The possibility that NK cells may also have a regulatory effect on T cell proliferation and differentiation was raised by Kumagai. He reported that in-vivo or in-vitro ablation of murine NK activity following treatment with anti-asialo GM1 markedly impairs the ability of spleen cells to generate allospecific cytotoxic T cells. This can be corrected in-vitro by addition of either interferon or IL-2 suggesting that generation of allo-CTL requires NK cell release of these lymphokines. These findings contrast with Kaplan's observation that anti-asialo GM1 depletion of spleen NK activity has little if any inhibitory effect on generation of allo-CTL. Moreover, others reported that spleen cells depleted of NK activity by complement lysis with anti-QA5 or anti-NK 1.1 show a normal ability to generate allo-CTL. The reason for these discrepancies is unclear. One possibility is that NK cells which produce IL-2 and interferon represent a minor QA5 NK 1.1^- subpopulation which expresses low amounts of asialo GM1, and that the anti-asialo GM1 treatment used by Kumagai more efficiently depleted these cells than the treatment we used.

V. VIRUS-INFECTED CELLS AS NK TARGETS

It has been well established that although certain normal cells are susceptible to NK lysis, many normal cells are resistant to lysis unless they are infected with a virus. The mechanism underlying this enhanced susceptibility of virus-infected cells was discussed by Lopez. He reported that NK cells form 3 times more conjugates with Herpes simplex virus infected fibroblasts than with uninfected fibroblasts. An increase in the frequency of lytic conjugates was also observed on occasion. These findings suggest that viral infection either increases the density of NK target structures thereby increasing the efficiency in target cell binding by NK cells, or induces additional NK target structures not present on uninfected cells. Virus infection may also increase lytic efficiency. This does not appear to be due to enhanced interferon production since anti-interferon antibodies do not block the increased NK lysis of virus-infected cells.

Experimental support for the possibility that enhanced NK susceptibility of virus-infected cells is due to NK recognition of viral-induced antigens was reported by Trentin. He described experiments of others which demonstrated a correlation between NK susceptibility and expression of a nuclear "T" antigen in adenovirus type 2-infected cells. Apparently, some "T" antigen is also expressed on the cell surface, as would have to be case if such a structure were to act as an NK target.

VI. TUMOR CELLS AS NK-TARGETS

In addition to virus infected cells, many tumor cells are particularly sensitive to NK lysis. A general correlation has been observed between tumor cell NK sensitivity and in-vivo tumorogenicity, and a similar correlation exists between NK levels and in-vivo resistance to adoptively transferred tumor cells. However, the role of NK cells in resistance to spontaneous tumors remains uncertain. To examine this, Roder conducted a long term study in NK-deficient beige mice. He reported that although the overall incidence of tumors in the two groups was the same, the rate of tumor development was greater in homozygous bg/bg than in heterozygous bg/+ littermates. These findings are consistent with the hypothesis that NK cells play a role in surveillance of spontaneous tumors in vivo.

VII. NK CELL SURFACE MOLECULES IMPORTANT IN CYTOLYTIC PROCESS

NK lysis of target cells is a multistep process, and cell surface molecules other than target cell antigens and NK receptors for these antigens may greatly influence the outcome of the initial binding of NK cells to their targets. For example, Targan reported that pretreatment of NK target cells with phospholipase A-2 enhanced NK-mediated lysis, and that this is associated with increased effector-target conjugate formation, increased lytic conjugates, and enhanced production of NKCF. He suggested that by stabilizing NK-target binding, enzyme treatment may enhance a binding-dependent-processing of NKCF; such processing may be necessary for NKCF to become a lytic moiety. In further support of the notion that binding structures play a role in post-binding lytic events, he presented evidence that a glycopeptide isolated from plasma membranes of K562 cells is capable of dissociating NK-K562 conjugates, and blocks killing in an NKCF assay.

Evidence for a new post-binding NK triggering receptor was presented by Roder. He reported that alpha-mannosidase pretreatment of effector cells inhibited NK-mediated cytolysis by affecting a pre-Ca^{++}-dependent (i.e. early) post-binding event in the activation of lysis.

An interesting approach to identifying and characterizing target binding structures on NK cells was reported by Saksela. He found that effector-target conjugates were resistant to detergent treatment which removes most membrane proteins not involved in conjugate formation. After internal labelling of effector cell membrane proteins with radioactive methionine, those proteins which presumably play a role in target binding were characterized. A 95K protein was obtained which inhibited effector-target binding. Whether such a protein is similar to the 90K heterodimer T cell receptor awaits further study.

In addition to antigen-specific receptors, non-specific receptors such as complement receptors and E-receptors may also play an important role in NK binding and lysis. Klein presented evidence that C3 molecules can bridge C3-receptors present on both NK cells and certain target cells (e.g. Raji) thereby enhancing binding. This strengthened binding may, in turn, increase the frequency of lytic interactions. A similar role for E-receptors in strengthening the bond between NK cells and their targets was proposed by Schlessinger. He reported that monoclonal antibodies to the E receptor blocked human NK lysis of K562 in the absence of complement. Similar effects were seen with sugars known to inhibit E-rosette formation.

VIII. CONCLUSIONS

Clearly, it remains to be determined to what extent specific vs. non-specific interactions between NK receptors and target cell structures contribute to NK lysis. Undoubtedly, much of the work presented here will serve as the basis for future studies addressing these questions.

ANALYSIS OF NK TARGET STRUCTURES. K562 MEMBRANE COMPONENTS ASSOCIATED WITH SUSCEPTIBILITY TO NK ATTACK

Dorothy Hudig

Department of Microbiology
School of Medicine
University of Nevada, Reno
Reno, Nevada

Lory Minning

UC San Diego Cancer Center
University of California, San Diego
San Diego, California

I. INTRODUCTION

How natural killer (NK) lymphocytes recognize the cells which they kill (i.e., their target cells) is still unknown despite considerable experimental effort. The results of these experiments can support several different interpretations. For example, many different NK target cell lines can compete the natural killer activity of freshly isolated NK lymphocytes to a given NK target. These experiments provide the basis for proposals that either 1) there is an NK receptor common to all NK lymphocytes and also an NK target structure shared by all potential NK target cells, or alternatively, 2) there are several receptor-like molecules that are co-expressed on many cells with NK activity and that there are several NK target structures, or even, 3) that there are many receptor-like molecules which segregate uniquely to individual NK cells but which recognize different

ISBN 0-12-341370-2

regions of a ubiquitous NK target structure. Thus, one tumor line might be expected to compete the NK directed towards another tumor cell line sharing a common epitope, or if a killer cell had receptors for two target structures, then a tumor line with one NK target structure could sterically compete the killing of the another line bearing an unrelated target structure. This dilemma remains to date, although experiments with target specificity of cloned human NK cells (1) suggest that there is a heterogeneity of recognition and presumably of clonotypic NK receptors among the clonable lymphocytes capable of mediating NK. The nature of the target structure for human NK also remains unclear, although experiments by Vodinelich et al. (2) indicate that one target structure may be a human transferrin receptor.

Our laboratory has utilized simple approaches to the NK target structure of K562 cells to address the following issues: 1) whether mycoplasmal components are essential for eliciting or accepting human NK activity; 2) whether the K562 target structures are protein or carbohydrate in nature and 3) if there a normal source of the K562 target structure which could be exploited in future experiments. Our data indicate 1) that mycoplasma are not essential for either eliciting or accepting the NK activity of freshly isolated human peripheral blood lymphocytes, 2) that the K562 target structure is expressed on a cell surface protein and is not a readily digestible carbohydrate molecule, and 3) that NK to K562 could be competed with rabbit reticuloycte preparations. This latter observation is consistent with the presence of a shared NK target structure on these normal, immature red blood cells and on K562 cells.

II. MATERIALS AND METHODS

A. Tests for Mycoplasma

The radioactive uridine/uracil uptake assay (3) was used as a preliminary screen for the presence of mycoplasma. A more sensitive assay utilizing a radioactive cDNA probe for mycoplasmal ribosomal RNA, developed and performed by David E. Kohne, Ph.D.,

Gen-Probe, 9620 Chesapeake Dr., San Diego, CA. 92123, was subsequently employed.

B. NK Assays

We used standard ^{51}Cr release assays with freshly prepared human peripheral blood mononuclear cells as effectors (4), and K562 cells obtained directly from the Drs. Lozzio (5) in all assays unless otherwise indicated. These assays were always performed at several effector lymphocyte to K562 target (E:T) ratios. Details may be found in reference 6.

C. Enzymatic Stripping of K562 Cell Surface Structures

We used proteinases to digest proteins and glycosidases to digest carbohydrate residues from K562 cells. In both cases, the K562 cells were pretreated for 1 hr at 37° C with 1×10^{-5} M emetine (Sigma Chemical Co., St. Louis, MO) to prevent *de novo* synthesis and reexpression of cell surface proteins which could also reintroduce carbohydrate structures bound to proteins. Chymotrypsin, trysin, papain, and *V. cholera* neuraminidase were obtained from Sigma Chemical Co. Pronase was from Cal-Biochem, La Jolla, CA. Endo-beta-galactosidase, α-mannosidase, β-galactosidase and β-hexosaminidase were the generous gifts of Drs. M. and M.N. Fukuda of the La Jolla Cancer Research Foundation, La Jolla, CA.

Special buffers were needed for some of these enzymes. A buffer containing 5 mM cysteine, 10 mM EDTA, 10 mM HEPES, and 0.05 mM 2-mercaptoethanol in Dulbecco's MEM pH 6.2 was used for papain digestions. A pH 6.0 10 mM cacodylate buffer containing 10 mM $CaCl_2$, 5 mM glucose, 0.5 mM $MgCl_2$, and 0.27 M NaCl was determined to be the best buffer representing a compromise of the lowest pH tolerable to the K562 cells and closest to the acidic pH optima of the glycosidases.

D. Competition with Erythroid Cells

Adult normal rabbit, sheep and human blood, human umbilical cord blood, and blood from phenylhydrazine-treated rabbits was collected into citrated Alsever's solution with 0.1% NaN_3 as a metabolic poison to block *in vitro* cell maturation.

Blood was stored at 4° C. The cells were washed three times with citrated saline and the buffy coat of leukocytes was removed. The red cells were then resuspended in NK assay medium. These cells were included in the NK assays in twofold serial dilutions with the highest final concentration being 5% rbc (vol./vol.). Rabbits were induced to form reticulocytes by five daily injections of 6.25 mg/kg phenylhydrazine. Blood was collected five days after the last phenylhydrazine injection.

III. RESULTS

A. NK Activity To K562 Cells Is Not Dependent upon the Presence of Mycoplasma

Although K562 cells from several sources contained mycoplasma which could be detected both by the uridine/uracil incorporation technique and the cDNA probe, the K562 cells obtained directly from the Lozzios appeared by the cDNA assay to be free of contamination. Since this assay could detect as few as one mycoplasma per 10^4 cells in a cell sample of 10^7 cells, it seems unlikely that mycoplasmal antigens contribute to the NK target structure, particularly as nearly all the K562 cells could be lysed. Furthermore, it seems unlikely that mycoplasma were present at all since culture in antibiotic-free medium should have expanded any contamination into a detectable range. Thus, it is unlikely that mycoplasma are even necessary to activate natural killer lymphocytes. However, it should be noted that mycoplasmal contamination may still influence NK activity towards K562. We noticed one heavily infected K562 cell line that was killed more rapidly than any of the other K562 lines.

B. K562 Cell Surface Proteins Are Needed for NK-Mediated Lysis. However, K562 Carbohydrate Structures are Not Readily Implicated in the NK Process.

When K562 cells were pretreated with 1 mg/ml of pronase, papain, chymotrypsin or trypsin at 37° C for 1 hr., they were not able to be lysed by NK cells (fig.

1). These proteinase-stripped K562 cells also did not function as cold target competitors in NK assays, even at ratios of 20:1 proteinase-treated K562 cells: ^{51}Cr K562 cells. Thus it appears that a K562 cell surface protein is necessary for binding of K562 cells to NK effectors. Concentrations of proteinases as low as 0.1 ug/ml of trypsin affected the ability of K562 cells to serve as targets (data not illustrated).

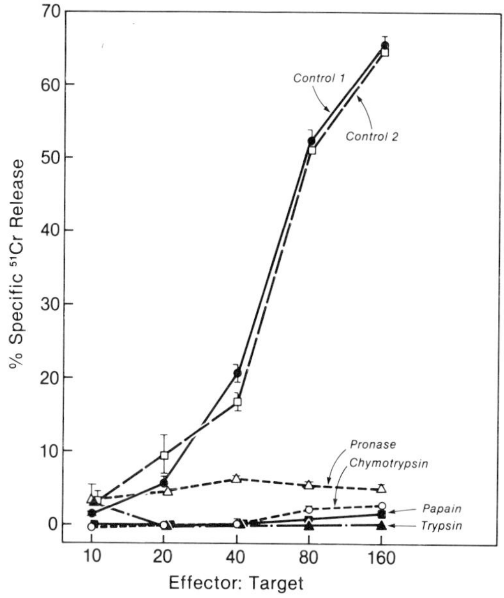

Figure 1. ^{51}Cr labeled K562 cells were unable to serve as NK targets after digestion with either pronase, trypsin, chymotrypsin or papain.

Treatment of K562 cells with glycosidases had no effect beyond the potentiating effect of neuraminidase treatment. When K562 cells were digested with 5 µ/ml neuraminidase to expose sugars for further digestion in the combined presence of the exoglycosidases α-mannosidase, β-galactosidase, and β-hexosaminidase, there was no effect relative to controls treated with neuraminidase alone. Also, there was only an enhancing effect when K562 cells were pretreated with the endoglycosidase endo-beta-galactosidase. This enhancing effect was similar in magnitude to the effects of neuraminidase. These results are illustrated in Table I.

Table I. Effects of Glycosidase Treatments on the Ability of K562 Cells To Be Lysed

Exp't No.	Treatment of K562 Cells	NK Activity (Lytic units/10^7 PBLs)
373*	control	67
	neuraminidase + mixed glycosidases	119
	endo-β-galactosidase	77
378*	control	16
	neuraminidase alone	22

*Representative experiments

C. Competition of NK with Normal Erythroid Cells

These experiments were initiated to test the idea that the fetal i unbranched carbohydrate moiety might be the NK target structure, since K562 cells are a rich source of the i carbohydrate (7). We found that human newborn cord red blood cells inhibited NK much more than either adult human rbc's (autologous or allogeneic to the donor of the NK cells) or xenogeneic sheep rbc's (fig. 2).

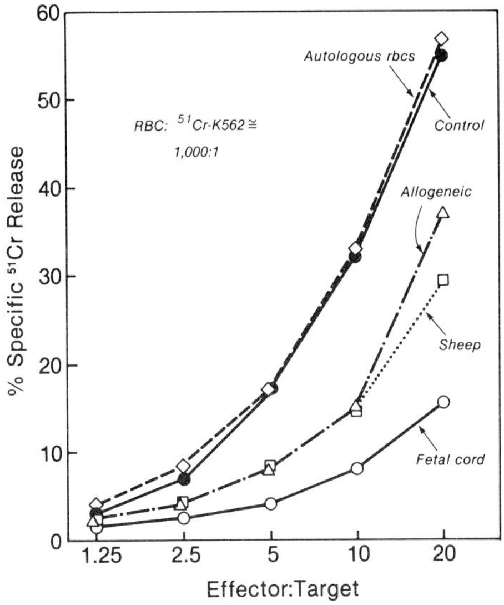

Figure 2. Fetal cord red blood cells competed NK better than either sheep rbc's or adult human rbc's allogeneic or autologous to the donor of the NK lymphocyctes. Rbc's were 5% vol/vol in the experiment illustrated. Competition by 10:1 unlabeled K562: ^{51}Cr K562 cells (not illustrated) was similar to the fetal cord rbc competion above. Thus, approximately 100 total cord rbc's (with approx. 3% reticulocytes) competed as well as 1 K562 cell.

In subsequent experiments, we found that rabbit rbc's also competed NK. We also found that preparations of rbc's from rabbits after phenylhydrazine treatment were enriched for reticulocytes and competed NK more effectively than did normal rbc's from the same rabbit (fig. 3A). Both of these blood samples were stored in citrated Alsever's solution with azide prior to use. To find out whether rabbit reticulocytes bound to human lymphocytes, we performed coded rbc-binding assays under conditions like those used for sheep rbc rosetting of human peripheral blood T cell. Under these conditions, no true rosettes were observed. However, more lymphocytes were associated with rabbit rbcs from the reticulocyte preparation than with the rabbit rbc's from the rabbit prior to treatment (fig. 3B). We were unable to deplete NK activity in attempts to remove reticulocyte-rosetting cells by centrifugation over Ficoll-hypaque layers.

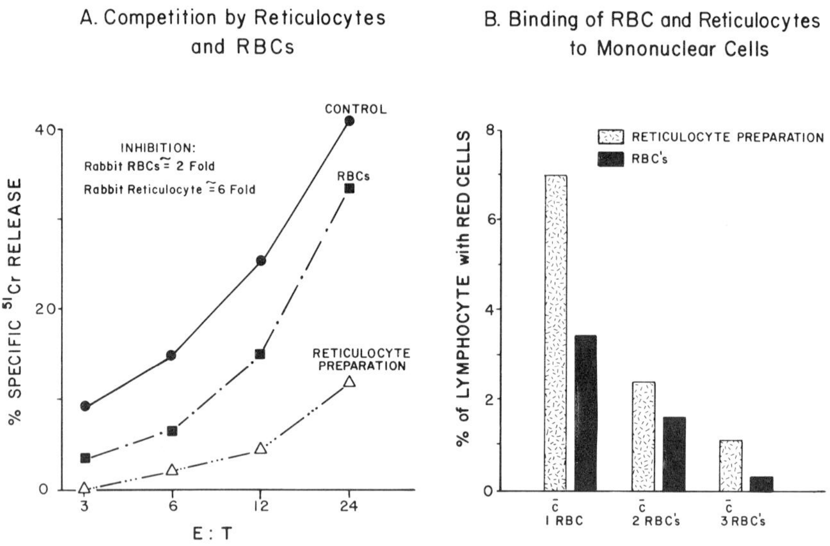

Figure 3. Rabbit rbc and rabbit reticulocyte inhibition of human NK and binding of rabbit reticulocyte-enriched rbc's to lymphocytes. A. Five % vol/vol rabbit rbc's or rbc's containing 37% reticulocytes were included in the NK assays. B. Rosetting with rabbit reticulocytes was performed at a 50:1 rbc to lymphocyte ratio, and coded samples were evaluated.

In subsequent experiments, rbc's from a patient with sickle cell anemia were used to compete NK of another donor. This rbc preparation containing 16% reticulocytes did not compete NK better than normal adult allogeneic or autologous rbc's. Thus there is no simple, direct correlation of the degree of inhibition of NK with the percentages of reticulocytes found in the human cord blood, in the rbc's from a patient with sickle cell anemia, or in the varying preparations of rabbit reticulocytes used in these competition experiments.

IV. DISCUSSION

Our data indicate that mycoplasma are not necessary for the induction of human NK or the ability of K562 cells to be lysed by NK cells. In addition, we oberved that a cell surface protein was necessary for K562 cells to be lysed and to act as cold target competitors. Although this cell surface protein is very likely to be a glycoprotein, we found no evidence for a requirement for carbohydrate structure for killing. Furthermore, our experiments indicate that normally occurring cells may share the NK target structure of K562 cells.

The mycoplasmal experiments are significant because they undermine the niggling possibility that NK cells recognize only exogenous antigens. The experiments with the proteinases indicate that the human NK target structure is unlikely to be a glycolipid, or a carbohydrate moiety which is also expressed on glycolipids. Similar experiments were reported several years ago for mouse NK to YAC1 targets (8).

The experiments with reticulocytes as competitors are not easy to interpret. On the one hand, if reticulocytes bind effectively to the lymphocyte receptors for erythrocytes, then all inhibition might be attributable to trivial hindrance of access of NK lymphocytes to the radiolabeled targets. However, if the reticulocytes are enriched for the NK target structure (as they are enriched for transferrin receptors), they might represent a naturally occuring form of the NK target structure. Since the reticulocytes of the patient with sickle cell anemia were not effective competitors, the issue remains confusing.

Additional information that would aid these studies would be quantitation of the number of transferrin receptors on the various reticulocytes. Also, one would like to know if a second variable is essential for NK-mediated lysis. This could be either a signal from the target to induce killing or binding of higher avidity, or inability of the target cells to repair after attack (9). Newton et al. (10) found that transfected mouse L cells expressing the human transferrin receptor can compete NK but are not themselves lysed. Perhaps these mouse L cells lack a second signal or a vulnerability to NK attack. One of the most frustrating aspects of the pursuit of the NK target structure is the disquieting thought that it will all seem very simple after it is delineated.

V. ACKNOWLEDGEMENTS

We would like to thank the Drs. M. and M.N. Fukuda for the generous gifts of glycosidases which made those experiments possible, Drs. Ann Rearden, Doug Redelman and Ajit Varki for helpful discussions, and Ms. Phyllis Stookey for illustrations. This work was supported by NIH CA38942.

REFERENCES

1. Hercend, T., E.L. Reinherz, S. Meuer, S.F. Schlossman, and J. Ritz. (1983) Nature 301: 158.

2. Vodinelich, L., R. Sutherland, C. Schneider, R. Newman, and M. Greaves. (1983) Proc. Natl. Acad. Sci. USA 80:835.

3. Schneider, E.L., E.J. Stanbridge, and C.J. Epstein. (1976) Exp. Cell Res. 84:311.

4. Boyum, A. (1968) Scand. J. Clin. Lab. Invest. 21 (Suppl. 97):77.

5. Lozzio, C.B., and B.B. Lozzio. (1975) Blood 45:321.

6. Hudig, D., D. Redelman, and L. Minning. (1984) J. Immunology 133:408.

7. Turco, S.J., J.S. Rush, and R.A. Laine. (1980) J. Biol. Chem. 255:3266.

8. Durnik, J.M., B.N. Beck, and C.S. Henney. (1980) P 805. In, Natural Cell-Mediated Immunity Against Tumors, edited by R.B. Herberman. Academic Press, N.Y.

9. Hudig, D., M.D. Djobadze, D. Redelman, and J. Mendelsohn. (1981) Cancer Res. 41:2803.

10. Newman, R.A., J.F. Warner, and G. Dennert. (1984) J. Immunol. 133:1841.

ROLE OF TARGET HISTOCOMPATIBILITY ANTIGENS IN REGULATION OF NATURAL KILLER ACTIVITY: A REEVALUATION AND A HYPOTHESIS[1]

Klas Kärre

Department of Tumor Biology
Karolinska Institute
Stockholm, Sweden

I. INTRODUCTION

Considerable progress has been made in the analysis of the natural killer cell system since the first systematic studies of this phenomenon around 1975 (1,2). This is particularly clear for the charachterization of the effector cell, its regulation by various soluble factors and lymphokines, and the lytic mechanism, as described elsewhere in this volume. However, one major question that remains unresolved is the basis of the target selectivity of NK cells, and thereby their immunobiological function. NK cells seem to preferentially kill certain malignant, undifferentiated, infected and nonautologous cells (3), but we do not know any single molecular change that converts a cell from the NK sensitive to the NK resistant phenotype, or *vice versa*. Which are the genes and products that help NK cells to discriminate the mature from the immature, the normal from the transformed and infected, or indeed "self" from "non self"? Is there a pattern behind these discriminative capacities of NK cells that makes them unique, deserving recognition as a separate functional component in the immune system?

---

[1]Based on a doctoral thesis at The Karolinska Institute, supported by grants from the Karolinska Institute, The Swedish Cancer Society, The National Cancer Institute (US DHEW) and The Engwall Foundation.

ISBN 0-12-341370-2

One common answer to these questions is that NK cells function in principle like T cells (or even, represent subpopulations of T cells), by recognizing the presence of abnormal target antigens, except for the difference that NK cells are not guided by MHC gene products in their effector function. This paper presents a hypothesis based on the opposite assumptions, i e that NK cells represent a defence system triggered by absence of the normal - self MHC - rather than presence of the abnormal, and that they therefore in a sense are guided by histocompatibility genes in their target selectivity. This model is based on an analysis of the literature on natural killing in autologous and nonautologous effector/target combinations, considering *in vivo* as well as *in vitro* studies, and with special emphasis on the role of the MHC. Some immunobiological, evolutionary and mechanistic aspects of the model are discussed, as well as some testable predictions. It should be emphasized that in this presentation, some simplifications and generalizations must be made for the sake of space and style. The model should not be viewed as a claim for final solutions, but rather as an example of one novel approach to an important problem in the field. Hopefully, it may serve as a basis for further discussions, experiments, refutations and novel developments.

A. NK target selectivity and histocompatibility antigens; some previous studies

The models proposed to account for the target selectivity of NK cells usually focus on the primary binding event, and are of two general types. Some authors suggest that NK recognition depends on interactions between specific receptors and acceptors in the membranes of effector and target cells (e g 4,5). Other investigators stress the alternative that it may not be possible to define the basis of NK recognition in terms of specific, interacting macromolecules (6). According to the latter type of model, the initial effector/target interaction could depend on more diffuse biophysical properties of cell mambranes, such as the distribution of charged groups or the three dimensional configuration over large areas.

The NK lytic target selectivity pattern for murine *in vitro* grown lymphomas correlates well with the binding pattern observed in conjugate assays with the same targets and nylon wool enriched effector cells. However, the same preferential binding pattern is seen with thymocytes and adherent peritoneal cells devoid of NK activity (7). There are also several human and murine systems where the lymphocyte binding to sensitive and insensitive targets is similar (8,9). Taken together, these observations suggest that the initial target binding event may be overemphasized as a specific determinator of NK target selectivity - the binding pattern is not unique for NK cells, and the final control may often be exerted at the post-binding level.

As to the role of histocompatibility antigens, Callewert and Kaplan interpreted their studies on human spontaneous cytotoxicity against autologous cell lines to suggest that NK-cells represent prothymocytes specific for self and non self histocompatibility antigens (10). Several reports have shown that human and murine NK cells lyse both syngeneic and allogeneic targets (11-23). In some systems NK activity is consistently more efficient against the allogeneic than against the syngeneic cells, summarized under A. in Table I. The preferential killing of allogeneic cells has been particularly emphasized by E. Klein and F. Vanky *et al*, who studied lysis of freshly explanted human tumors, and suggested that a part of the operationally defined natural killing is mediated by polyclonally activated T cells with specificity for foreign MHC antigens (12,13).

TABLE I. NK sensitivity of autologous *versus* non-autologous targets - role of histocompatibility antigene genotype?

Nature of target cell, experimental system	Reference
A. Systems where killing is seen predominantly or exclusively with non-autologous targets:	
Freshly explanted human tumor cells	12,13
Human fibroblasts	14
Human PBL (all bloud group A)	15
Rabbit lymphocytes	16,17
Murino peritoneal exsudate cells	18
Murine L1210 leukemia cells (*in vivo*)	19
B. Systems where efficient killing is seen in autologous as well as in non-autologous effector/target combinations:	
Murine immature thymocytes	18
Murine teratocarcinoma cells	22,23
Murine YAC-1 lymphoma cells (tested with H-2 congenic strains on A/Sn background)	21
Human bone marrow cells	20

Kiessling and Welsh reported that NK cells from LCMV infected mice killed peritoneal exsudate cells from allogeneic mice, but not those from the syngeneic strain; there was no such pattern when thymocytes from suckling mice were used as targets in the same study (18). Similarly, Hansson *et al* observed equal killing of autologous and allogeneic bone marrow cells from human donors, tested with PBL effectors (20).

A third line of observations by Gidlund *et al* argues against the "NK recognition of self MHC" concept as well as against the "NK recognition of allogeneic MHC" concept: immature cells like teratocarcinoma cells, which do not express any cell surface H-2 at all, are highly susceptible NK targets (22,23). When such cell lines are induced to express more differentiated features, including H-2, their susceptibility to NK-lysis decreases (23). This indicates that NK lysis does not require target cell expression of MHC gene products, which are so vital for recognition by CTL (24). Indeed, when considering the evidence from several studies on murine targets (TABLE II), an inverse correlation between NK susceptibility and sensitivity to allospecific CTL (or expression of serologically detectable H-2) appears.

TABLE II. Divergent target preferences of NK cells and allospecific CTL - role of histocompatibility antigen phenotype?

Target cells	NK sensitivity	AlloCTL sensitivity (serological H-2)	Ref.
YAC lymphoma cells:			
- ascites line	low	high	
- *in vitro* line	high	low	25
- *in vitro* line, interferon-beta pretreatment	low	high	26 27
Thymocytes:			
- cortical, immature subset	high	low	
- medullary, mature subset	low	high	28
Teratocarinoma cells:			
- undifferentiated	high	negative	22,23
- differentiated	low	positive	

A reductionistic interpretation of the reviewed observations would therefore suggest that *reduced* expression (or complete absence) of the histocompatibility antigens expressed by the host/effector cell is a common denominator for targets that are efficiently killed by NK cells. All cells with an immature phenotype, such as the targets under "B" in Table 1 and the sensitive targets in Table 2, would be killed regardless of their genotype in relation to the effector cells. The MHC genotype of the target cells would only be important for high expressor phenotypes, where absence of effector cell histocompatibility antigens would be restricted to nonautologous cells, thus explaining the preferential killing of these. The studies demonstrating this (Table 1, A) did not include direct tests on the levels of histocompatibility antigens expressed by the targets. However, it is expected from the literature that these cell types should express at least MHC class I gene products readily (29). In any case, no attempts were made to exclude the possibility that preferential killing of nonautologous cells was due to "absence of self" rather than "presence of non self" on the targets. This problem can be approached by studies of MHC heterozygous (A x B) effectors and parental, MHC homozygous (B x B) targets. Relevant information can be derived from several studies, *e g* by Harmon *et al* (30), using crosses between H-2 congenic strains with the EL-4 lymphoma, and by Klein *et al*, using backcross mice segregating for H-2 with the RBL-5 lymphoma (31). These two $H\text{-}2^{b/b}$ tumors were both killed by $H\text{-}2^{d/b}$ spleen cells, but not by MHC identical $H\text{-}2^{b/b}$ effector cells. This pattern correlated with the genetic control of *in vivo* resistance against small grafts of the lymphomas (30,31). Conversely, Carlson *et al* observed that $H\text{-}2^{b/b}$ as well as $H\text{-}2^{d/b}$ mice rejected L1210 leukemia cells ($H\text{-}2^{d/d}$) by rapid natural killing (18 h) *in vivo*, while the leukemia cells survived in MHC identical mice (19).[3]

These observations on natural resistance against H-2 semi-syngeneic cells adds further support to the concept of efficient natural killing in situations where targets express no or reduced levels of the MHC genes expressed by the recipient or effector cell they are confronted with. The general conclusion from this review and reevaluation is therefore that the preferential lysis of allogeneic, semisyngeneic and immature syngeneic targets may actually be controlled by target MHC genes; however it is not necessarily based on what these targets commonly express, but rather on what they all fail to express. This is the basis of the hypothesis presented below.

---

[3]See also contribution by Nakamura *et al*, this volume.

II. IMMUNE SURVEILLANCE TRIGGERED BY REDUCED SELF EXPRESSION; A MHC RESTRICTED VIEW OF NATURAL KILLER CELL BEHAVIOR

A. Immunobiological and evolutionary aspects

Multicellular organisms need defense systems against invasive destruction of their tissues by foreign or inappropriate elements. In theory, such defense systems can operate through two distinct strategies:

1) By elimination of elements expressing evidence for foreigness or "altered self" (positive identification).

2) By elimination of elements failing to express sufficient evidence for "self" (negative identification).

The latter principle applies in certain invertebrate species, where it operates in the rejection of allogeneic cells as well as in the prevention of self fertilization (32). The mammals have evolved defence systems based on the first strategy. The clonally distributed receptors of the T- and B-lymphocytes are generated partly by a random process, and the receptor repertoire is then somatically selected to positively identify foreigness in the context of self MHC. Here it is argued that mammals also keep a defence system of the second type, since this can cover up in certain situations where the first system may fail, and that natural killer cells are the principal effector in this second defence. It is thus suggested that *T-cells and NK cells represent two functionally distinct, but complementary systems for cell mediated immune surveillance; while the T-cells are called into action by the presence of of "altered" or "non self" MHC products, NK cells are specifically trigged by the absence or reduced expression of "self" MHC products.*

Why do we need a defense system triggered by the reduced expression of self MHC genes? Embryonic cells and immature hematopoetic cells lack or express only very low levels of major histocompatibility antigens (29). Such cells also have the capacity to proliferate rapidly; if they were seeded and spread outside of their normal compartments, they might escape their physiological control, form ectopic tissues or even become neoplastic. Furthermore, consider an intracellular invader (parasite, bacterium, virus or even a misplaced cellular gene fragment - causing lytic, latent or transforming infection)- which had the capacity to turn off the MHC expression in host cells, or which selectively infected cells that lack histocompatibility antigens. Such infected cells would escape the T cells, since the latter can only recognize foreigness in the context or self MHC (24). The threat represented by this possible escape route may thus have exerted a selective pressure to conserve the invertebrate defence strategy throughout mammalian evolution, modified to specifically detect and eliminate cells failing to express MHC gene products adequately. This way of discriminating the normal from the abnormal might also have proved useful as a first line defence with a certain selectivity; there are examples of lytic as well as of transforming viruses that cause a profound and selective reduction of histocompatibility antigens in infected cells (33,34).

One interesting feature of the model is that the sparing of normal, syngeneic cells in infected tissues would be garanteed by an increase in their MHC gene expression as a consequence of regulation by interferons. The latter are known to stimulate an increase in cell surface expression of H-2 and HLA antigens (35) as well as to reduce the NK susceptibility of many different cell lines (27). It is suggested that the latter phenomenon may be at least partly dependent on the former, and that foreign, variant, undifferentiated and perhaps also infected cells would be among those failing to present adequate "self" signals in response to interferons.

B. Mechanistic aspects

An immunocyte evolved to be triggered when confronted with cells that *fail* to express certain "self" marker molecules cannot be dependent on the same molecules for the initial target binding event. A mechanism for this initial contact is a crucial requirement for any model designed to account for the regulation of NK lysis. The present hypothesis allows that this first binding step is less selective, involving a variety of membrane properties that act to promote cell to cell contacts in general, without requirements for immunological specificity. Interactions between single specific molecules are not excluded, nor are binding forces dependent on more diffuse membrane properties (hydrophobicity, 36, charge distribution dependent on sialylation of glycoconjugates, 37). None of these mechanisms are necessarily unique to the NK system; they might very well be shared with other leucocytes (7), thus explaining the "NK like selectivity" aquired by specific CTL under certain culture conditions (38). The binding mechanisms would allow NK cells to form short lived "conjugates" with most cells in the body (8), even if certain cultured hematopoetic lines bind extremely well (7), making them useful for short term *in vitro* assays.

It is not excluded that the initial target binding results in the first step towards triggering of the lytic mechanism; alternatively, this could take place through separate (MHC <u>in</u>-dependent) target cell molecules. The crucial point of the model presented here is that *the ultimate control determining delivery of the lytic hit is an inhibitory signal, associated with target expression of self MHC class I gene products*. The latter may interact directly with the effector cell; alternatively the MHC genes may be involved by indirectly influencing the expression or recognition of other molecules on the target cell surface.

The reason for specifying the class I molecules of the MHC as responsible for inhibition of NK triggering is that the genes in control of rapid rejection of nonautologous hematopoetic grafts map to class I loci (19,30,31). Moreover, the inverse relation between NK susceptibility and MHC gene expression discussed above applies to class I products, since Ia and DR antigens are not necessarily expressed on mature cells even after treatment with interferon-beta (35).

Several previously suggested models may be cited for the reader who is interested in a possible molecular mechanism for the proposed role of MHC class I genes. To explain the phenomena of hybrid resistance and rapid rejection of nonautologous hematopoetic grafts, Snell proposed that class I molecules of radioresistant host effector cells would trigger an "alarm reaction" when they "mismatched" rather than "matched" with class I products of the graft (39). There is no evidence that MHC gene products can act as self complementary receptors, although it has been reported that "auto-rosetting" between lymphoid cells and erythrocytes in the mouse depends on a protein-carbohydrate interaction controlled by the H-2 D/L region (40,41). MHC gene regulation of glycosyltransferases specific for self MHC sugar sequences has also been proposed in a model for T-cell restriction (42). At this stage, there is no point in discussing these and other possible mechanisms extensively, since this would merely lead to speculations superimposed on an already speculative hypothesis. It may be more pertinent to discuss some specific predictions derived from the general model; experimental tests of these may either give a firmer basis for molecular analysis, or indeed disprove the concept altogether.

C. Some testable predictions

-Any target cell that is killed *in vitro* or rejected *in vivo* by NK cells should become resistant after induction of high expression (at least equal to that of normal mature thymocytes) of MHC class I genes expressed by the effector/host. This could be achieved by DNA mediated gene transfer, membrane glycoprotein transplantation, and for autologous effector/target combinations, also by selection for $H\text{-}2^{++}$ or HLA^{++} phenotypes as well as by interferon modulation of target cells.

-Conversely, such $H\text{-}2^{++}$ NK resistant cells should become susceptible to NK cells after loss of H-2, whether this occurred spontaneously or after specific selections; the latter must then be performed *in vitro*, in the absence of selection pressure from NK cells and other immune mechanisms which could otherwise lead to fixation of additional changes in the selected cells. It is also important that the predicted alterations in susceptibility to lysis following changes in H-2 expression do not necessarily apply to "NK lines" that have been derived from CTL which have lost their antigenic specificity during *in vitro* culture (38).

-In tumor immunology, the model predicts a selection for increased H-2 expression during tumor progression and metastasis, except for those allelic products that are used as restriction elements by CTL against highly antigenic tumors (43). A correlation between metastatic potential and an increased expression of certain or all MHC class I products has already been observed in some experimental systems (44,45). It is further predicted that manipulations resulting in increased MHC gene expression should also lead to an enhanced malignant potential, due to escape from NK mediated elimination. Conversely, selection for loss of H-2 should result in reduced malignancy. Highly immunogenic tumors, for which T-cells represent the dominant immunoselective pressure *in vivo* would however be exceptions, at least when tested i T-cell immunocompetent animals. To avoid problems with interpretations, control experiments should be performed in NK depleted as well as in T-cell depleted hosts.

-In transplantation immunology, natural killing of MHC disparate bone marrow and lymphoid grafts - including the F_1 hybrid resistance phenomenon (46) as well as early elimination of allogeneic lymphocytes (47) - may be interpreted to reflect recognition of incomplete self MHC: allogeneic grafts lack MHC genes of the recipient completely, and parental grafts fail to match one haplotype, *i e* they may lack more than 50% of the total self MHC molecules expressed in the F_1 environment (29). This predicts that the critical level of reduction in H-2 expression required to make a cell eliminated by NK cells would be around 50% of the total amount expressed in normal cells. Furthermore, it should be possible to to rescue nonautologous hematopoetic grafts from rapid (T-cell independent) rejection by introduction of the relevant recipient MHC genes in the graft cells prior to transplantation.

One cautionary note: the model does *not* predict a general correlation between MHC gene expression and resistance to NK-lysis in comparisons of unrelated cells or lines; it is not even predicted that all complete MHC loss mutant lines should be more sensitive than any of the cells in a panel of unrelated histocompatibility antigen positive lines. The reason for this is that selection pressures during tumor progression as well as during phylogeny of a species may have fixed additional changes in specific cells, permitting them to escape NK cells in spite of reduced MHC expression. These alterations could interfere with initial effector/target binding or with the general sensitivity to the lytic mechanism, and they might occur in tumor cells (34,43) as well as in certain normal cells , *e.g.* trophoblast cells or erythrocytes. It is therefore stressed that the MHC related changes in NK sensitivity are expected mainly in studies of variants, clones or subpopulations derived from one cell or one cell lineage. In our own studies along these lines, we have so far observed a reduction in tumorigenicity and metastatic potential accompanying the loss of H-2 in a RBL-5 lymphoma variant (Ljunggren *et al*, to be published) and in a B16 melanoma subline (Taniguchi *et al*, to be published). In parallel, the cells switched from the NK resistant to the NK susceptible phenotype.

As noted above, the model also suggests that at least a part of the protection from NK lysis seen after interferon pretreatment of targets (27) may be due to increased expression of MHC class I gene products. This does not exclude that other, MHC independent effects of interferons also contribute. So far, we have observed that a YAC-1 lymphoma subline selected for loss of H-2 expression became completely refractory to interferon mediated protection from NK lysis, while the wild type cells switched from $H\text{-}2^{low}$ to $H\text{-}2^{high}$ phenotype and became NK resistant in response to interferons (beta or gamma, Piontek *et al*, to be published). Although these and the other cited observations are suggestive, they need to be confirmed with independently derived cell lines before any conclusions can be drawn. Specific interference with H-2 gene expression at the molecular level will be necessary for a more critical examination of the model.

The purpose of the present paper was to stimulate similar and novel approaches to other sublines differing in MHC gene expression, in experimental systems in the hands of other investigators. Admittedly, the presented model is based on a narrow, MHC restricted view of natural killer cell behavior. However, we still lack knowledge about single genes and products that can control, at the target cell level, whether death or escape will follow after a confrontation with a natural killer cell. Even if the major histocompatibility complex may not provide the final answer, it is not such a bad place to start the search, in view of its tremendous importance for self/non self discrimination in adaptive immunity, and the fact that the genes have been very well characterized and are available for experimental manipulations (48). The MHC may provide new tools, leads and surprises for investigators of the natural killer system.

ACKNOWLEDGEMENTS

Drs Gunnar O. Klein, Gerald E. Piontek, Kazuto Taniguchi, Hans-Gustaf Ljunggren and other colleagues at The Karolinska Institute are gratefully acknowledged for stimulating criticism and discussions. Special thanks to Drs Rolf Kiessling, George Klein and Eva Klein for support, and to Dr Hans Wigzell for constructive criticism.

REFERENCES

1. Kiessling, R., Klein, E., and Wigzell, H. (1975). Eur. J. Immunol. 5, 112.
2. Herberman, R.B., Nunn, M.E., Holden, H.T. and Lavrin, D.H. (1975). Int. J. Cancer 16, 230.
3. "NK cells and Other Natural Effector Cells" (R.B. Herberman, ed). Academic Press, New York (1982).
4. Roder, J.C., Rosén, A., Fenyö, E.M., and Troy, F.A. (1979) Proc. Natl. Acad. Sci. USA 76, 1979
5. Ortaldo, J.R., Philips, W.H., and Herberman, R.B. (1981). J. Immunol. 125, 2322.
6. Kiessling, R., and Wigzell, H. in "Natural Resistance to Tumors and Viruses" (O. Haller, ed.), p.107 Springer Verlag (1981).
7. Piontek, G.E., Grönberg, A., Ährlund-Richter, L., Kiessling , R., and Hengartner, H. (1982). Int. J. Can. 30, 225.
8. Collins, J.L., Patek, P.Q., and Cohn, M.J. (1981). J. Exp. Med. 133, 84.
9. Ullberg, M., and Jondal, M. (1982). J. Clin. Lab. Imm. 9, 177.
10. Kaplan, J., and Callewaert, D.M. (1980) in Natural Cell Mediated Immunity (R.B. Herberman, ed.) p. 893. A.P. NY.
11. Becker, S., Fenyö, E.M., and Klein, E. (1976) Eur. J. Immunol. 6, 882.
12. Vanky, F., Argov, S., Einhorn, S., and Klein, E. (1980). J. Exp. Med. 151, 1151.
13. Vanky, F., Argov, S., and Klein, E. (1981). Int. J. Cancer 27, 273.
14. Timonen, T., and Saksela, E. (1977). Cell. Immunol. 33, 340.
15. Rooney, C.M., and Munro, A.J. (1984). Immunology 193, 51.
16. Behelak, Y., and Richter, M. (1972). Cell. Immunol. 3, 542.
17. Behelak, Y., and Richter, M. (1973). Transplant. 18, 123.
18. Kiessling, R., and Welsh, R. (1980). Int. J. Cancer 25, 611.
19. Carlson, G.A, Melnychuk, D., Meeker, M.J. (1980) Int. J. Cancer 25, 111.
20. Hansson, M., Kiessling, R., and Andersson, B. (1981). Eur. J. Immunol. 11, 8.
21. Becker, S., Fenyö, E.M., and Klein, E. (1976). Eur. J. Immunol. 6, 882.
22. Stern, P., Gidlund, M., Örn, A., and Wigzell, H. (1980). 285, 341.
23. Gidlund, M., Örn, *et al.* (1981). Nature 292, 848.

24. Berke, G., McVey, E., Hu, V., and Clark, W. (1981). J. immunol. 127, 782.
25. Becker, S., Kiessling, R., Lee, N., and Klein, G. (1978). J. Natl. Cancer Inst. 61, 1495.
26. Kärre, K., and Seeley, J. (1979). J. Immunol. 123, 1511.
27. Welsh, R.M., Kärre, K., Hansson, M., Kunkel, L.A. and Kiessling, R. (1981). J. Immunol. 126, 219.
28. Hansson, M., Kärre, K., Kiessling, R., Roder, J.C., Andersson, B., and Häyry, P. (1979). J. Immunol. 123, 765
29. Klein, J. (1975). "Biology of the Mouse Histocompatibility-2 Complex" (Springer Verlag).
30. Harmon, R.C., Clark, E.A., O´Toole, C., and Wicker, L.S. (1977). Immunogenetics 4, 601.
31. Klein, G.O., Klein, G., Kiessling, R., and Kärre, K. (1978). Immunogenetics 6, 561.
32. Burnet, M.F. (1971). Nature 232, 230.
33. Hecht, T., and Summers, D. (1972). J. Virol. 10, 578.
34. Schrier, P.I., Bernards, R., Vaessen, J., Houweling, A., and van der Eb, A.J. (1983). Nature 305, 771.
35. Vignaux, F., and Gresser, I. (1977). J. Immunol. 118, 721.
36. Becker, S., Stendahl, O., and Magnusson, K.E. (1979). Immunol. Comm. 8, 73.
37. Yogeeswaran, G., Grönberg, A., Hansson, M., Dalianis, T., Kiessling, R., Welsh, R.M. (1981). Int. J. Cancer 28, 517.
38. Brooks, C.G. (1983). Nature 305, 8.
39. Snell, G. (1976). Transplant. Proc. 8, 147.
40. Sia, D.Y., and Parish, C.R. (1981). Immunogenetics 12, 857.
41. Parish, C.R., Sia, D.Y., and Rylatt, D.B. (1981). Immunol. Today 2, 172.
42. Parish, C.R., O´Neill, H.C., and Higgins, T.J. (1981). Immunol. Today 2, 98.
43. Haywood, G.R., and McKhann, C.F. (1971). J. Exp. Med. 133, 1171.
44. Katzav, S., De Baetselier, P., Tartakovsky, B., Feldman, M. and Segal, S. (1983). J. Natl. Cancer Inst. 71, 317.
45. Hui, K., Grosveld, F., and Festenstein, H. (1984). Nature, 311, 750.
46. Cudkowicz, G. and Bennett, M. (1971). J. Exp. Med 134, 1513.
47. Rolstad, B., and Ford, W. L. (1983). Immunol. Rev. 73, 87.
48. Steinmetz, M., Frelinger, J.,G., Fisher, D., Hunkapillar, T., Pereira, D., Weissman, S.M., Uehara, H., Nathenson, S., and Hood, L. (1981). Cell 24, 125.

ROLE OF NK CELLS IN HYBRID RESISTANCE TO BONE MARROW GRAFTS[1]

Ichiro Nakamura
Claudio Bordignon
John P. Daley

Department of Pathology
State University of New York at Buffalo
Buffalo, New York 14214

During the 1960s and early 1970s, Cudkowicz and his co-workers published a series of papers (1,2,3) on the immunobiology of bone marrow (BM) allograft rejection. They showed that acceptance or rejection of blood forming cells is not only controlled by the classical histocompatibility (H) genes, but also by a class of noncodominant genes, designated Hh for hemopoietic-histocompatibility, most of which are linked to the MHC. The Hh incompatible transplants are the targets of host-versus-graft reactions, which are paradoxically radioresistant and immunogenetically specific. Characteristics of this phenomenon suggested the existence of an effector cell which is radioresistant, naturally occurring, thymus independent, bone marrow dependent, and delayed in maturation (4). However, the nature of the effectors remained undefined until 1977, when Kiessling et al. (5) presented strong evidence relating the effector of *in vivo* natural resistance against marrow grafts to the effectors of natural killer (NK) activity against lymphoma targets *in vitro*. More recently, additional evidence has

---

[1]Supported by U.S. Public Health Service grants AM-13969 and CA-12844. C.B. was supported by the Associazione Italiana per la Ricerca sul Cancro and is currently funded in part by the Ralph Hochstetter Medical Research Fund in honor of Dr. Henry C. and Bertha H. Buswell.

ISBN 0-12-341370-2

strengthened this close relationship (6,7); however, the growing list of properties shared between natural resistance to BM allografts and NK cell activity does not include the single most pertinent property of *in vivo* resistance, i.e., its immunogenetic specificity. This specificity could result from direct target recognition by specific effector cells, or from regulatory mechanisms not related to effector specificity. The latter possibility is indirectly supported by recent observations which include the findings that bone marrow cells (BMC) from normal mice or healthy humans are susceptible to cytolysis *in vitro* by allogeneic or syngeneic NK cells (8), and that resistance to BM allografts can be specifically reconstituted by an NK-like cell line (9). To resolve this dual possibility, we developed an *in vivo* competitive inhibition assay to determine whether the immunogenetic specificity of hybrid resistance (HR) is the consequence of direct target recognition by restricted, determinant specific, effector cells. These putative effectors were then isolated and further characterized by means of an *in vitro* assay for F_1 HR against parental hematopoietic grafts.

I. H-2/Hh-RESTRICTION OF EFFECTOR CELLS

As shown in Fig. 1, the resistance of irradiated $B6D2F_1$ ($H\text{-}2^{b/d}$) mice against B6($H\text{-}2^b$) parental hemopoietic grafts was inhibited by irradiated tumor cells homozygous for $H\text{-}2^b$, but not by cells bearing other H-2 types or heterozygous for the $H\text{-}2^b$ (11). It is of particular interest that YAC-1 cells ($H\text{-}2^a$), an optimal target for NK activity *in vitro*, did not express any capacity for competition. Moreover, pretreatment of recipient mice with poly-I:C, a potent interferon inducer capable of augmenting both NK activity and HR, did not modify the genetic specificity of the competition. Splenic localization frequencies of tumor cells utilized in the experiment of Fig. 1 were comparable (11). Studies were also carried out using normal cells as *in vivo* inhibitors. Since we have demonstrated in a separate series of experiments that target determinants for HR are expressed on differentiating hemopoietic precursors (see below), regenerating splenic hemopoietic cells were employed as the source of competitors. As the results in Fig. 2 suggest, the specificity of competition is controlled by the Hh-1 locus linked to the H-2D region. In fact, hemopoietic regenerating cells

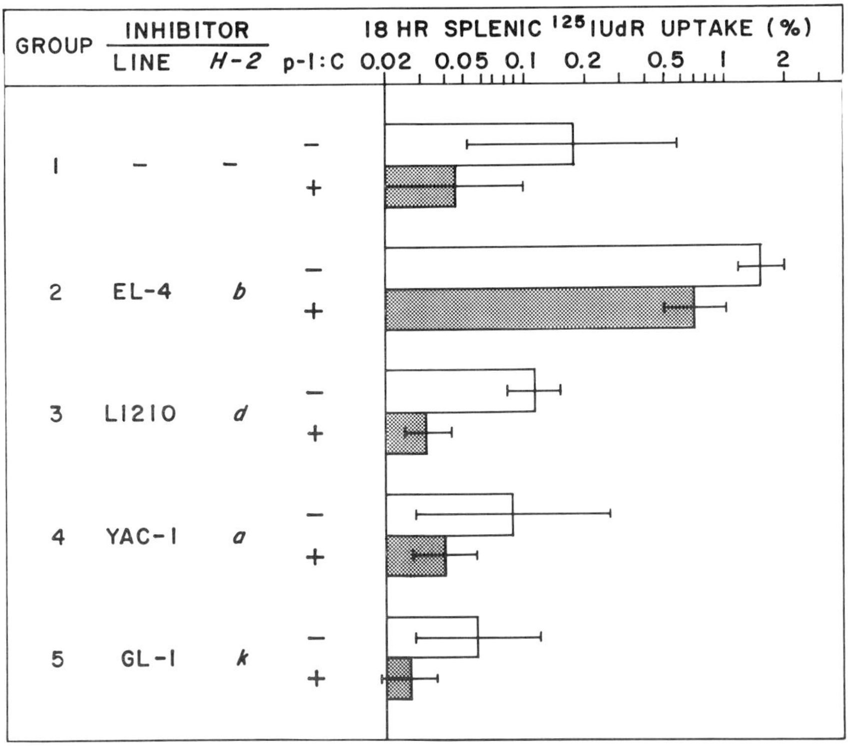

Fig. 1. H-2 restricted inhibition of in vivo HR by lymphoma cells. The prospective recipient mice were lethally irradiated (900 rads) and injected i.v. with 1 x 10^8 irradiated (10,000 rads) competitor cells 3 hr before transplantation of 5 x 10^6 B6 BMC. The splenic ^{125}IUdR incorporation was assessed 5 days later by the standard assay as described before (11). Data are expressed as the geometric mean ± 95% confidence limit. It should be noted that inhibition of resistance results in grafted cells escaping from rejection and hence increased proliferation and ^{125}IUdR uptake.

obtained from mice with the H-2^s haplotype were equally effective competitors as were those of parental B6 origin. This confirms the previous observation obtained from classical genetic experiments that the Hh-1^s determinants are indistinguishable from the Hh-1^b (3). It is important to note that when competitive inhibition experiments were

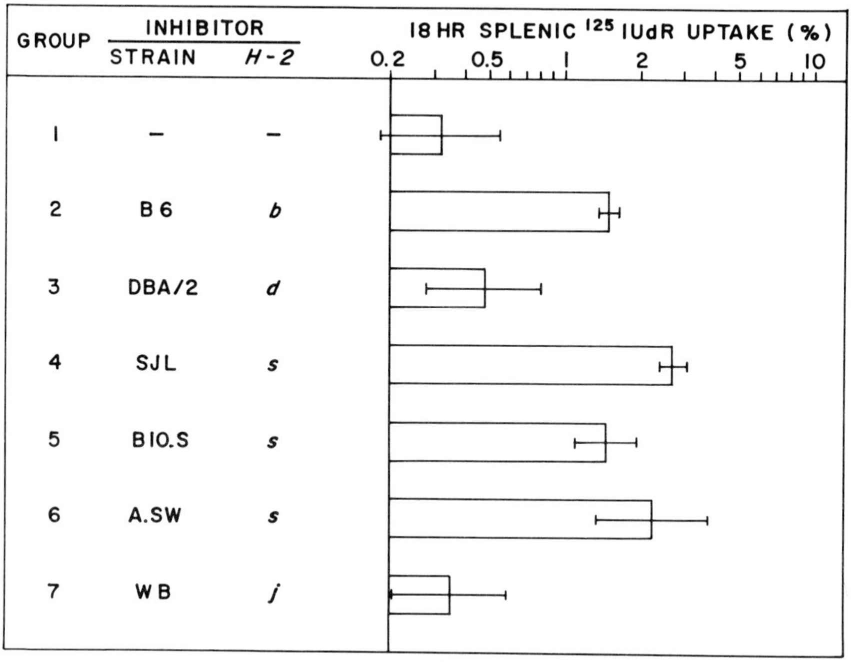

Fig. 2. H-2/Hh-1 restricted inhibition of in vivo HR by regenerating hemopoietic spleen cells. The competitor cells were splenocytes from mice of different H-2/Hh-1 type which had been lethally irradiated (750 rads) and reconstituted with 7 x 10^6 syngeneic BMC 15 to 18 days earlier. The regenerating hemopoietic cells were irradiated (2000 rads) and used without additional treatments as competitors for in vivo resistance. The prospective recipient mice were lethally irradiated (900 rads) and injected i.v. with two doses of 1 x 10^8 irradiated competitor cells 24 hr and 3 hr before transplantation of 5 x 10^6 B6 BMC. The splenic ^{125}IUdR incorporation was assessed 5 days later by the standard assay as described before (11). Data are expressed as the geometric mean ± 95% confidence limit.

carried out with F_1-anti-parental H-2^b cytotoxic T lymphocytes (CTL), cells of the H-2^s haplotype did not compete with H-2^b targets (12). These data demonstrate that the effectors of HR against parental H-2^b BM targets differ in their specificity from F_1 anti-parent CTL. This is further demonstrated by the absence of *in vivo* competitive inhibition by cells obtained from WB mice (Fig. 2), since WB cells are indistinguishable from those of other $H\text{-}2D^b$ strains with regard to CTL-defined specificities (12).

II. SPECIFIC ELIMINATION OF COMMITTED PRECURSORS *IN VIVO*

Since the above results suggest that HR is mediated by a unique population of radioresistant effectors whose recognition specificities are restricted by the putative Hh-1 genes, an attempt was made to develop an *in vitro* model for HR to better characterize the putative effector and effector-target interactions. Such an *in vitro* model should reflect the cytolytic or cytostatic interactions between noninduced NK-like effectors and proliferating normal marrow targets. For this purpose, it was necessary to determine whether the target susceptibility to *in vivo* resistance persists after commitment to a lineage. B6 BMC were injected i.v. into lethally irradiated $B6D2F_1$ and B6 recipient mice, spleens were removed at different times post-transplantation, and the spleen cells assayed for *in vitro* colony forming capacity. Since recipient mice were lethally irradiated, no colonies of recipient origin developed *in vitro*. In spleens removed as late as 6 hr after transplantation, no significant inhibition of parental B6 BFU-E and CFU-GM was observed in $B6D2F_1$ mice, when compared with the number of colonies obtained from the spleens of syngeneic recipients. However, after contact with the F_1 enrivonment for 12 hr or longer, a strong inhibition of parental erythroid and myeloid colony forming capacity occurred (Table I). This inhibition was not observed in syngeneic combinations. Moreover, the reactivity was as NK-dependent as the classical *in vivo* HR against parental hemopoietic stem cells. As shown in Table I, *in vivo* pretreatment of recipient F_1 mice with anti-asialo GM-1 antibody or poly-I:C resulted in abrogation or enhancement, respectively, of the F_1 anti-parental responses. In a separate series of experiments, the treatment of syngeneic B6 recipient mice with anti-asialo GM-1 did not modify the recovery of CFU-GM, but slightly reduced the number of BFU-E.

TABLE I. Elimination of parental (B6) BFU-E and CFU-GM from the spleens of lethally irradiated $B6D2F_1$ mice: Effect of *in vivo* pretreatment of the recipients with anti-asialo-GM-1 antibody or poly-I:C [a]

Recipient		Units/Spleen		Mean % ^{125}IUdR Uptake
Strain	Treatment	BFU-E	CFU-GM	(95% confidence limits)
B6	-	380	930	2.242 (1.492-3.369)
$B6D2F_1$	-	130	480	0.227 (0.216-0.244)
$B6D2F_1$	NRS	140	460	0.340 (0.215-0.528)
$B6D2F_1$	α-AS GM1	400	1030	2.375 (1.601-3.244)*
$B6D2F_1$	poly-I:C	20	110	0.029 (0.011-0.055)

[a]Recipient mice were exposed to 900 rads of γ-rays from a ^{137}Cs source 2 hr before i.v. transplantation of 2 x 10^6 B6 bone marrow cells, recipient spleens were removed 18 hr after marrow transplantation, and the spleen cells were cultured in a colony forming assay as previously described (15). Twenty-four hr before bone marrow transplantation, some of the prospective recipient mice were injected with 20 μl i.v. of normal rabbit serum (NRS) or rabbit anti-asialo GM-1 antiserum (α-AS GM-1) or with 10 ug of polyinosinic:polycytidylic acid (poly-I:C). Proliferation of grafted hemopoietic cells in the spleens of irradiated recipients was assessed 5 days after grafting by measuring the incorporation of ^{125}I-labelled 5-iodo-2'-deoxyuridine (^{125}IUdR) as described previously (11).

*Significant ($p<0.05$, Student's *t*-test) growth over that in untreated or NRS-treated mice.

III. INHIBITION OF PARENTAL BFU-E AND CFU-GM *IN VITRO*

Based on the above results, an *in vitro* model for HR was developed as follows. Nylon wool non-adherent spleen cells from $B6D2F_1$ mice were fractionated on discontinuous Percoll density gradients, as described previously (13,14). Cells from each fraction were irradiated, tested for NK activity against the susceptible YAC-1 target cells, and co-cultivated with parental (B6) or syngeneic ($B6D2F_1$) bone marrow cells in a colony-forming assay. The potential effector cells and bone marrow target cells were either preincubated in liquid medium for 6-8 hr to allow effector-to-target cell contact

Table II. Effects of in vitro interactions between F_1 hybrid spleen cells and syngeneic ($B6D2F_1$) or parental (B6) bone marrow derived hemopoietic progenitors [a]

Effector	NK Activity	BFU-E/Plate±s.d.		CFU-GM/Plate±s.d.	
		$B6D2F_1$	B6	$B6D2F_1$	B6
None (medium)	-	29 ± 3	25 ± 4	112 ± 5	103 ± 5
Unfractionated	5.2	22 ± 1	20 ± 1	120 ± 3	80 ± 7
Fraction 2	55.6	39 ± 2	9 ± 4	101 ± 6	31 ± 3
Fraction 4	2.0	25 ± 2	20 ± 3	119 ± 7	110 ± 9

[a]Spleen cells from normal $B6D2F_1$ mice, either unfractionated or fractionated by Percoll density gradients (13,14) following a passage through a nylon wool column, were exposed to 900 rads of γ-rays. Aliquotes of the cell suspensions were tested for NK activity and cultured with syngeneic or parental B6 BMC following an 8 hr preincubation in liquid medium. The culture condition was as previously described (15). Splenic effector to BM target cell ratio was 4:1. The number of colonies/plate represents the mean ± standard deviation (s.d.) of 2 plates containing initially 2 x 10^5 bone marrow cells/plate. No colonies developed when irradiated effector cells were cultured alone. NK activity is expressed as percent specific lysis of YAC-1 targets in a 4 hr ^{51}Cr-release assay at E:T ratio of 12.5:1.

or mixed immediately before being embedded in the semisolid methyl cellulose-containing culture medium. As the results of a typical experiment show (Table II), the NK-enriched fraction 2, but not T cell-enriched fraction 4, inhibited the development of both BFU-E and CFU-GM of B6 parental origin. Under these experimental conditions, $B6D2F_1$ colonies were not inhibited by syngeneic effector cells, although inhibition of syngeneic CFU-GM was observed at higher E:T ratios. When the coculture experiments were performed without effector-target cell preincubation, or following a preincubation shorter than 6 hr, no significant inhibitory activity was observed, suggesting that *in vitro*, as *in vivo*, several hours of effector-target cell contact is a prerequisite for the expression of inhibitory activity. The similarity of the effectors for *in vivo* and *in vitro* reactions was confirmed by pretreatment of Percoll fraction 2 effectors with anti-asialo GM-1 antiserum plus complement; this treatment of *in vitro* effectors led to a complete abrogation of the colony suppression.

IV. CONCLUDING REMARKS

The present data have several interesting implications with regard to the nature of the effector cells and the target structures recognized in natural resistance to hemopoietic grafts. These results have confirmed the presence, in responder F_1 mice, of radioresistant effectors that specifically recognize and eliminate $H\text{-}2^b/Hh\text{-}1^b$ homozygous BMC. The F_1-anti-parental activity could be blocked by normal hemopoietic cells or tumor cells bearing the $H\text{-}2^b/Hh\text{-}1^b$ or a related Hh-1 allele, i.e., $Hh\text{-}1^s$, but not by cells with different H-2/Hh-1 haplotypes, including the NK susceptible YAC-1 cells. These results also indicate that the effector specificity agrees with the known strain distribution of H-2D/Hh-1 alleles and differs from class I H-2D specificities defined by serological reagents and allogeneic or F_1 anti-parental $H\text{-}2D^b$ CTL. Recently alloreactive CTL clones have been shown to have a dual reactivity, one restricted to MHC and the other directed to cell surface structures of NK sensitive lymphomas (16). The coexistance of the two independent recognition structures on the same cells of these CTL clones has been indicated, since the binding of CTL to target cells through one type of receptor was found to block the binding capacity of the other receptor (17). Such cross-inhibition has not been observed in the effector cells mediating HR *in vivo*.

In the *in vitro* model, the effector cells that specifically inhibit the proliferation of erythropoietic and myelopoietic precursors were found to co-enrich with NK cells and also share a surface marker. A possible explanation for specific reactivities mediated by NK-like effectors has been recently suggested (18), i.e., antibody-dependent cellular cytotoxicity (ADCC) in which the specificity is dictated by naturally occurring antibodies against surface determinants of hemopoietic precursors. This mechanism has also been suggested to account for allogeneic lymphocytes cytotoxicity *in vivo*, a phenomenon sharing several features with natural resistance to bone marrow grafts (19). Since a wide range of evidence indicates that both NK and K cell activities are functions of the same cell type (20), the arguments implicating NK cells as effectors also support the possible involvement of an ADCC mechanism. However, it should be remembered that nude mice (21) and B-cell depleted recipients (22) display a stronger resistance than do their normal counterparts. Moreover, our attempts to transfer specific reactivity from responsive to unresponsive recipients by means of serum have been unsuccessful in both the *in vivo* and *in vitro* models of HR.

The role of NK-like effector cells in self-regulation of hematopoietic homeostasis has been previously hypothesized (23). Of particular interest in this regard is the observation that syngeneic interactions *in vitro* between these effectors and hemopoietic precursors led to contact-dependent suppression of myeloid precursors and contact-independent enhancement of erythroid bursts. Related findings include positive (24) and negative (25) regulation of the erythroid compartment and inhibition of myeloid precursors (8) by human NK cells, negative regulation of both compartments by human NK cytotoxic factor (26), and the inhibitory effect of syngeneic, asialo GM-1-positive, rat large granular lymphocytes on hemopoietic repopulation of lethally irradiated recipients (27).

ACKNOWLEDGMENTS

We thank Cesare Peschle and his colleagues at the Istituto Superiore di Sanita, Rome, for technical guidance and discussion; Paul Evans and Laura Langer for technical assistance; and Julieann Kostyo for preparation of the manuscript.

REFERENCES

1. Cudkowicz, G. (1968). *In* "The Proliferation and Spread of Neoplastic Cells", p. 661. Williams & Wilkins, Baltimore.
2. Cudkowicz, G., and Bennett, M. (1971). J. Exp. Med. *134*, 83.
3. Cudkowicz, G., and Bennett, M. (1971). J. Exp. Med. *134*, 1513.
4. Cudkowicz, G., and Nakamura, I. (1983). Transpl. Proc. *15*, 2058.
5. Kiessling, R., *et al*. (1977). Eur. J. Immunol. *7*, 655.
6. Lotzova, E., *et al*. (1983). Transplantation *35*, 490.
7. Okumura, K., *et al*. (1982). *In* "NK Cells and Other Natural Effector cells", p. 1527. Academic Press, New York.

8. Hansson, M., et al. (1982). J. Immunol. 129, 126.
9. Nunn, M.E., et al. (1977). Int. J. Cancer 20, 381.
10. Warner, J. F., and Dennert, G. (1982). Nature 300, 31.
11. Daley, J. P., and Nakamura, I. (1984). J. Exp. Med. 159, 1132.
12. Nakano, K., et al. (1981). J. Immunol. 127, 173.
13. Iscove, N. N., et al. (1982). J. Cell. Physiol. Suppl. 1, 65.
14. Luini, W., et al. (1981). Immunol. 43, 663.
15. Kumagai, K., et al. (1982). J. Immunol. 129, 388.
16. Brooks, C. G. (1983). Nature (Lond.) 305, 155.
17. Binz, H., et al. (1983). J. Exp. Med. 157:1252.
18. Dennert, G., and Warner, J. (1984). The Second International Workshop on Natural Killer Cells (Abstract). May 29-June 1, Rochester, Michigan.
19. Rolstad, B., and Ford, W. L. (1983). Immunol. Rev. 73, 87.
20. Bloom, B. R. (1982). Nature 300, 214.
21. Campanile, F., et al. (1977). Cancer Res. 37, 394.
22. Brodt, P., et al. (1981). J. RES 30, 283.
23. Cudkowicz, G., and Hochman, P. S. (1979). Immunol. Rev. 44, 13.
24. Pistoia, V., et al. (1984). Blood, In press.
25. Mangan, K. F., et al. Blood 63, 260.
26. Trinchieri, G., et al. (1984). In This Volume.
27. Barlozzari, T., et al. (1984). Immunobiol. 167, Abstract 42.

EVIDENCE THAT NATURAL KILLER CELLS ARE PRE-THYMIC T CELLS WITH RECEPTORS FOR SELF AND NON-SELF H ANTIGENS

Joseph Kaplan

Department of Pediatrics
Wayne State University School of Medicine
Detroit, Michigan 48201

I. INTRODUCTION

The cell lineage to which NK cells belong and the nature of their target cell specificity are issues which remain unresolved. Based on evidence that the NK cell population belongs to the T cell lineage, is polyclonal, and contains subsets which recognize hemopoietic histocompatibility (Hh) antigens, we (Kaplan et al, 1980) previously put forth the hypothesis that NK cells are prothymocytes identical to those which, as Jerne (1971) had postulated, express germ-line V-gene-encoded receptors for either self or non-self histocompatibility (H) antigens. Here, we review some of the currently available evidence relevant to the hypothesis, including the results of recent experiments we have conducted which give support to one of its predictions-that the NK population contains precursors of allospecific cytotoxic T cells.

II. EVIDENCE THAT NK CELLS ARE PRE-THYMIC T CELLS

A. T Cell Markers

Most of the cell markers associated with the T cell lineage have been detected on NK cells. The Thy 1 antigen is expressed, in low density on many resting murine NK cells (Herberman et al, 1978), and is expressed in high

ISBN 0-12-341370-2

density on activated or proliferating NK cell such as those found in-vivo during certain viral infections (Mattes et al, 1979) and following recovery from cytotoxic drugs (Hurme et al, 1983), and which occur in-vitro during the course of mixed lymphocyte culture (Karre et al, 1979). Another murine T cell-associated alloantigen, Lyt 1, has been detected on about 25% of murine NK cells (Koo et al, 1980).

T cell markers detected on human NK cells include sheep erythrocyte receptors (West et al, 1976), xenoantisera-defined human T lymphocyte antigens (HTLA) (Kaplan et al, 1976), and monoclonal antibody-defined T cell antigens T1, T3, T4, T8, T10, and 3A1 (Ortaldo, et al, 1981; Abo et al 1982). NK cells also express the monoclonal-antibody defined antigens MAC-1 (Ault et al, 1981) and M1 (Kay et al, 1980) which are found on macrophages but not on most thymocytes or peripheral T cells. This has led to the suggestion that NK cells belong to the myeloid-lineage or to a lineage separate from those previously defined. However, it is now known that Mac 1 and M1 represent complement receptors (CR3) which are widely distributed on different cell types (Beller et al, 1982). By contrast many of the T cell-associated antigens detected on NK cells, e.g. T3, T8 and HTLA, occur only on T cells, and are not present on B cells or myeloid cells. That NK cells belong to T cell lineage and not the myeloid lineage is also supported by the observations of Fast et al (1983) that T200 molecules expressed by NK cells are identical to those of T cells, but distinct from those of monocytes.

B. IL-2 Responsiveness

Like T cells, NK cells proliferate upon exposure to IL-2 (Suzuki et al, 1983). Although this in itself is not proof of T cell lineage, it fits well with that concept, and receives further support from the fact that when grown in IL-2, NK cells show increased expression of T cell markers. For example, human large granular lymphocytes grown in IL-2 lose M1 but increase their expression of T3 (Timonen et al, 1982); in the mouse, Thy 1^- NK cells induced to proliferate by culture with IL-2 become strongly Thy 1^+ (Brooks et al, 1982).

C. T Cell Receptors

According to our model, NK cells, like mature post-thymic T cells, possess V-gene-encoded antigen receptors. It follows that they should possess cell-surface structure similar to the recently identified T cell receptors for antigen. Clonally restricted 90kD heterodineric cell surface structures analogous to those detected on cloned CTL lines have been detected on cloned NK-like cell lines, and, as with CTL clones, these structures appear to be involved in determining target specificity (Hercend et al, 1983). However, since the relationship of NK-like cell lines to in-vivo NK cells is still controversial, it can be argued that these clones represent T cells which, during in-vitro culture, have acquired NK-like cytotoxicity. More compelling support for the possibility that NK cells have T cell receptors comes from the preliminary report by Roder of studies he performed in collaboration with Dr. Tak Mak which demonstrate the existence in purified murine NK cells of T cell receptor messenger RNA. If confirmed, these findings would represent the strongest evidence that NK cells belong to the T cell lineage, and would further imply that NK cells, like, T cells, are antigen specific.

D. Evidence that NK Cells are Prothymocytes

The evidence above strongly suggests that NK cells belong to the T cell lineage. However, NK cells are clearly not thymus-dependent since high NK levels are found in athymic nude mice. They could, however, represent prothymocytes, i.e., committed pre-thymic T cell precursors. Support for this comes from the observation that NK cells express at least two antigens which are present on prothymocytes but are not found on most thymocytes or mature post-thymic T cells: Ly 11 (Meruelo et al, 1980), and asialo Gm1 (Habu et al, 1982). By contrast, Kumar et al (1982) found that prothymocytes do not express the NK-specific alloantigen NK 1.2. This may mean that a subpopulationn of Ly 11^{+} asialo $Gm1^{+}$ NK 1.2^{-} NK cells function as prothymocytes, and that NK 1.2^{+} NK cells are end-stage cytolytic cells.

If NK cells are prothymocytes, one would expect to find them in the thymus itself. But under normal conditions thymocytes lack NK activity. This does not exclude the possibility that NK cells are present in the thymus, since any attempt to measure thymic NK activity against labeled targets may be blocked by autocthonous cold target inhibition. This possibility is supported by the fact that thymocytes are themselves susceptible to NK lysis (Hansson et al, 1979). In fact, NK activity has been detected in the thymus of rats with graft vs. host disease (Roy et al, 1982). In addition, human thymocytes develop NK activity when cultured in vitro with IL-2 (Toribio et al, 1983).

III. EVIDENCE THAT NK CELLS ARE POLYCLONAL

A. Cold Target Inhibition Analysis

Our view that NK cells, like T cells, comprise a polyconal population, each clone of which recognizes one of multiple unique antigens, was prompted originally by the results of cold target-inhibition studies. These studies showed that inhibition of NK lysis of a given labeled target cell could be accomplished by many different types of unlabeled cells, but that maximal cold-target inhibition almost always occurred with homologous cells (Calleweart et al, 1979). Since then, studies of clones NK cell lines have provided important additional evidence for NK polyclonality.

B. Cloned NK Lines

1. Distinct specificity of different NK clones

The first cloned NK cell lines described (Dennert, 1980) were all found to have the same pattern of specificity when tested against a panel of target cells. This implied that NK cells are not polyclonal, but instead recognize common structures on NK-sensitive targets, or carry multiple receptors each with a different specificity. However, other investigators (Ting et al, 1982; Kedar et al, 1982; Hercend et al, 1983) subsequently found distinct patterns of specificity amongst different NK clones implying that NK cells are, in fact, polyclonal. Conceivably, the cloned NK cells examined by Dennert (1980) had been derived from a parent NK cell line which, at the time of subcloning, was itself a monoclonal population that had selectively grown out from a polyclonal starting population.

2. Clonotypic receptors on NK clones.

If NK cells are clonal they should possess clonotypic antigen recognition structures. The studies of Hercend et al (1983) provide support for this possibility. They developed a monoclonal antibody to a cloned human NK cell line which reacted with a T cell receptor-like 90kD heterodimeric structure, and was able to block the lytic activity of this clone but not that of another NK clone having a different pattern of specificity.

C. Cloned Cytotoxic T Cell Lines with "Dual" Specificity

Antigen-specific CTL clones have been described which, in addition to killing the expected antigen-positive target cells, also kill NK-susceptible targets not thought to express the relevant antigen (Binz et al, 1983; Acha-Orbea et al, 1984; Brooks et al, 1983; Moretta et al, 1984). These findings are open to a number of different models of interpretation concerning the relationship and specificity of T cell and NK cell target recognition structures.

One possibility (Model A) is that "dual-specific" CTL clones have two types of receptors-clonotypic receptors for specific antigen, and non-specific receptors with affinity for structures present in higher density on NK-susceptible targets than on other cells. Non-specific receptors could include such cell surface moieties as E-receptors and complement receptors which may enhance NK-target cell binding and lysis. According to this model, lysis of NK targets would occur only if some threshold level of non-specific receptor-ligand interaction is achieved.

A second possibility (Model B) is that "dual-specific" CTL clones possess only one type of receptor: clonotypic receptors which recognize "private" antigens on "antigen-specific" targets that are cross-reactive with "public" antigens expressed in higher density on NK-susceptible targets than on other cells.

A third possibility (model C) is that "dual-specific" CTL clones have two types of receptors a) clonotypic, and b) nonspecific-and that the clonotypic receptors are cross-reactive with public "NK" antigens. According to this model, lysis of "antigen specific" targets requires only clonotypic receptor-recognition, whereas lysis of NK-targets requires that the cytolytic effectors recognize targets via both clonotypic and non-specific receptors. The assumption is that clonotypic receptor binding to cross-reactive antigens is usually of low avidity, and that lytic interaction requires the enhanced avidity provided by non-specific receptor-ligand interaction.

A fourth possibility (model D) is that "dual specific" CTL clones possess two types of clonotypic receptors, one for private antigens, and the other for public "NK" antigens. Such cells would be exceptions to the rule of allelic exclusion, and would thereby resemble a hypothetical population of pre-thymic germ-line V-gene-encoded cells that Jerne had considered in his theory on somatic generation of diversity (Jerne 1971)-a population which expresses two distinct V-gene-encoded receptors, one specific for an alloantigen, and the other specific for a self H antigen.

Evidence consistant with models A and D and against models B and C is 1) The observation of Binz et al (1983) that anti-idotype antibodies presumably directed against antigen-specific receptors of a "dual-specific" rat CTL clone partially inhibited "antigen-specific" killing but did not inhibit "NK-like" killing, and 2) The observation of Moretta et al (1983) that a monoclonal anti-CTL antibody inhibited allospecific killing but not "NK-like" killing mediated by a "dual-specific" human CTL clone. On the other hand, models B, C, and D fit with the evidence that clonotypic surface structures resembling T cell receptors play a role in recognition and lysis of NK targets by cloned NK cell lines (Hercend et al, 1983). By contrast, the fact that some "dual-specific" CTL clones have arisen in culture from "antigen-specific" CTL clones (Brooks et al, 1982; Acha-Orbea et al, 1983) is difficult to reconcile with models B and D, but can be readily accounted for in both models A and C by assuming that culture of "antigen-specific" cells in conditioned-medium in the absence-of "antigen" induces expression of non-specific receptors. Elucidation of which, if any, of these models is correct remains to be accomplished. Resolution of this issue should provide important insights into the relationship between NK cell and T cell specificity.

IV. EVIDENCE THAT NK CELLS RECOGNIZE SELF AND NON-SELF H ANTIGENS.

That NK cells kill autologous as well as allogeneic and xenogeneic cells indicates that NK target antigens must fall into both the "self" and "non-self" category. As discussed below, some of these NK target antigens appear to be encoded by the major histocompatibility complex (MHC).

A. NK Recognition of Self-MHC-Antigens

1. Hybrid Resistance and F_1 anti-parent cytotoxicity

The resistance of F_1 hybrids to engraftment by parental strain bone marrow cells appears to be mediated by NK cells (Cudkowicz, 1978), and is directed against distinct antigens termed hemopoietic histocompatibility (Hh) antigens many of which are closely linked if not identical to class I MHC antigens. On the assumption that F_1 hybrids recognize

Hh antigens as "non-self" antigens, Cudkowicz (1978) originally suggested that the antigens are encoded by recessive MHC-linked genes. More recently, Drizlikh et al (1984) have proposed that some Hh antigens are associative complexes of class I and class II MHC-encoded antigens which some F_1 hybrids lack because of preferrential association of Ia alpha chains with allogeneic beta chains, or of class I antigens with allogeneic or hybrid class II antigens. It can also be argued, however that hybrid resistance represents auto-recognition of self MHC determinants expressed in higher density on homozygous parental cells than on heterozygous F_1 hybrid cells. Consistent with this possibility is the fact that the effectors of F_1 anti-parent cytotoxicity, an in-vitro analogue of in-vivo hybrid resistance, recognize but do not lyse autologous targets (Nakano et al, 1981).

2. Autocytotoxic CTL

The generation of autocytotoxic CTL has been detected in several different in-vitro systems (Altman et al, 1980; Ishikawa et al, 1980; Ting et al, 1983; Ting et al, 1984), and in some of these system the CTL appear to recognize self H-2 antigens (Altman et al, 1980; Ishikawa et al, 1980). In one system the CTL-precursors have been shown to be Thy 1 asialo $Gm1^+$ NK cells (Ting et al, 1983). Whether they are self-H-2 specific, and whether the self-H-2-specific CTL identified in the other systems represent NK cells remains to be determined.

B. NK Cell Recognition of Allo-MHC-Antigens

1. Allogeneic resistance to bone marrow grafts, and allogeneic lymphocyte cytotoxicity

Like the mediators of F_1 hybrid resistance to parental bone marrow grafts, the cells which mediate resistance to allogeneic bone marrow grafts and those responsible for rapid in-vivo elimination of allogeneic lymphocytes have properties which correspond to those of NK cells: late maturation, radioresistance, thymus-independence, bone marrow-dependence, and expression of asialo GM1 (Heslop et al, 1983; Cudkovicz et al, 1971). Many of the target antigens recognized by these NK-like mediators appear to be MHC-encoded alloantigens. This antigeneic specificity cannot be ascribed to "natural" alloantibodies since allogeneic lymphocyte cytotoxicity and bone marrow resistance are, if anything, stronger in anti-mu chain-treated animals which lack immunoglobulin, and cannot be transferred by serum.

2. A subpopulation of NK-like allospecific CTL precursors

One of the predictions of our hypothesis is that some NK cells, i.e., those with receptors for non-self MHC antigens, act as precursors of allospecific cytotoxic T cells. Evidence consistent with this prediction has been previously reported by Beck et al (1982). They found that generation of allospecific CTL was severely compromised by depleting MLC responder populations of cells expressing the NK cells markers asialo Gm1. In the experiments described below, conducted by Ken Wasserman, we sought to confirm and extend these observations.

Responder spleen cell population were subjected to complement lysis with either anti-asialo Gm1, or anti-Thy 1.2. One aliquot was tested for NK activity against YAC, and another was cultured with mitomycin-C-inactivated allogeneic stimulator cells; 5 days later they were tested for cytolytic activity against ^{51}Cr-labeled Con A blasts of the stimulator strain. As expected, responder cells treated with anti-asialo Gm1 had little if any NK activity against YAC target cells. In contrast to the finding of Beck et al (1982), these same cells showed only a partial and inconsistant reduction in their capacity to generate allospecific CTL compared to cells treated with normal rabbit serum. By contrast, cells treated with anti-Thy 1.2 + C were completely unable to generate allospecific CTL. This implied that most of the allospecific CTL presursors (allo-CTL-P) present in the responder populations we examined were asialo $Gm1^-$. However, it remained possible that a minor asialo $Gm1^+$ subpopulation of allo-CTL-P was present, the removal of which only slightly reduced the number of CTL effectors generated. We reasoned that if these putative asialo $Gm1^+$ allo CTL-P were NK cells, they would likely be resistant to lysis by anti-Thy 1, and that asialo $Gm1^+$ CTL-P might be enriched by anti-Thy + C treatment of responder spleen cell populations. The existence of such anti-Thy 1 resistant asialo $Gm1^+$ CTL precursors might then be demonstrable if they could be successfully induced to generate allospecific CTL. In the absence of Thy 1^+ cells this would likely require an exogenous source of T cell-derived lymphokines, since lymphokines produced by Thy 1^+ cells such as IL-2 and other growth factors are known to be essential for the generation of allospecific CTL (Wagner et al, 1980). Thus, the fact that spleen cells treated with Thy 1 + C failed to generate allospecific CTL could have been due to the combined effect of a) depletion of Thy 1^+ CTL precursors, and b) depletion of Thy 1^+ producers of lymphokines necessary for the generation of allospecific CTL by both Thy 1^+ and the putative Thy 1^- CTL precursors.

To determine whether an exogenous source of lymphokines would promote the generation of allospecific CTL by the putative Thy 1⁻ asialo Gm1⁺ allo-CTL-P population, responder spleen cells treated with anti-Thy 1.2+C or anti-Thy 1/anti-asialo Gm1 + C were cultured with mitomycin-C treated allogenic spleen cells in the presence or absence of 20% V/V Con A-conditioned medium (Con-A-CM) derived from Con A-stimulated rat spleen cells. As expected, responder cells treated with anti-Thy 1+C failed to generate allospecific CTL in the absence of Con A-CM. However, when Con A-CM was included in the MLC reactions, these same Thy 1^- cells did generate allospecific CTL which 1) killed stimulator-strain Con A blasts, 2) did not kill Con A-blasts derived from a third party unrelated strain and 3) were highly sensitive to lysis by treatment with anti-Thy 1 unlike their Thy 1^- percursors.

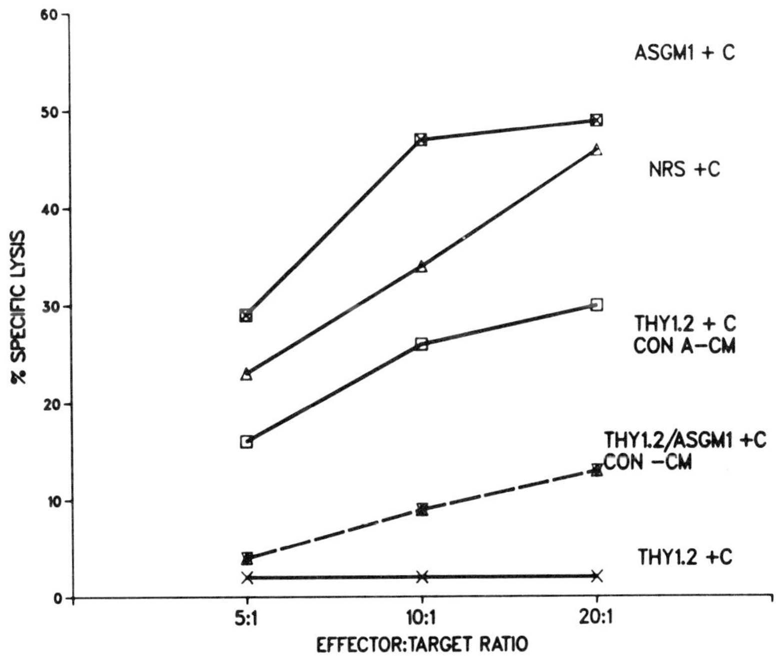

Fig. 1. Generation of alloreactive CTL. B10.D2 responder cells were treated with either anti-Thy 1 alone or with anti-Thy and anti-asialo Gm_1 + C. One aliquot was tested for NK activity against YAC targets, another was cultured with (CBA x SJL)F_1 stimulators for 5 days with or without Con A-CM, and tested for allospecific killing of Con A blasts.

Most importantly, as shown by the results of a representative experiment in Figure 1, the Thy 1^- allo CTL-P detected in this Con A-Cm-dependent system were found to express the NK cell marker asialo Gml: responder spleen cells treated with both anti-Thy 1 and anti-asialo Gml prior to addition of complement showed a markedly reduced ability to generate CTL in MLC supplemented with Con A-CM compared to cells treated with anti-Thy 1 + C alone.

We conclude that a subpopulation of allo-CTL-P has the NK phenotype Thy 1 asialo Gml$^+$, and that this subpopulation is distinct from the predominant Thy 1^+ asialo Gml population of allo-CTL-P. An alternative interpretation, that treatment with anti-asialo Gml non-specifically renders Thy 1^+ CTL precursors more sensitive to lysis by anti-Thy 1+C, seems unlikely based on our finding that cells treated first with anti-Thy 1+C, washed, and then treated with anti-asialo Gml + C showed the same reduction in ability to generate allo-CTL as did cells treated with anti-asialo Gml either before or at the same time as anti-Thy 1. Another possibility, that asialo Gml-bearing cells are not the actual anti-Thy 1-resistant CTL precursors detected in these assays, but act instead as accessory cells, is made unlikely by the observation that the impaired ability of anti-Thy 1.2/anti-asialo Gml-treated responder cells to generate allo-CTL in mixed lymphocyte cultures suppplemented with Con A-CM was not corrected by addition to the MLC of equal numbers of mitomycin C-inactivated non-antibody-treated autologous spleen cells. The latter would likely have provided an adequate source of syngeneic asialo Gml$^+$ accessory cells.

V. CONCLUSIONS

The conceptual basis for our hypothesis that NK cells are germ-line V-gene-encoded prothmocytes specific for self and non-self H antigens is the theory put forth by Jerne (1971) on the somatic generation of diversity. According to this theory, pre-thymic lymphocytes clonally express germ line V-gene-encoded receptors specific for either self or non-self H antigens. Such genes are postulated to have evolved from cell surface structures important in cell-cell interactions regulating morphogenesis. The histocompatibility antigens recognized by these V-gene-encoded receptors would therefore include both MHC antigens and non-MHC H antigens which might play a role in morphogenesis. Based on this concept, our hypothesis predicts that some subsets of NK cells have receptors for self or non-self MHC antigens, whereas others have receptors for self or non-self non-MHC H antigens. Such a receptor repertoire would fit well with the role NK cells appear to play in controlling the proliferation of normal stem cell populations.

We have reviewed much of the currently available evidence that NK cells are a polyclonal population of pre-thymic T cells, and that subsets of NK cells exist which are specific for either self or non-self MHC antigens. We assume that other NK subsets exist which recognize non-MHC H antigens, and that this explains why some NK-susceptible targets, e.g. K562, lack expression of MHC antigens. The experimental model systems which have provided evidence for MHC-specific NK subsets i.e, in-vitro generation of antigen-specific CTL by Thy 1^- precursors, and in-vivo resistance to adoptively transferred bone marrow stem cells or foreign lymphocytes, could provide the tools for identifying NK subsets specific for other antigens.

REFERENCES

Abo, T. Cooper, M.D., Balch, C.M. (1982) J. Immunol. 129, 1752.

Acha-Orbea, H., Groscurth, P., Lang, R., Stitz, L. and Hengartner, H. (1983) J. Immunol. 130, 2952.

Altman, A., and Katz, D.H. (1980) J. Immunol. 125, 1536.

Ault, K., and Springer, T. (1981) J. Immunol. 126, 359.

Beck, B.N., Gillis S., and Henney, C.S. (1982) Transplant., 33:118.

Beller, D.I., Springer, T.A., and Schreiber, R.D. (1982) J. Exp. Med. 156,1000.

Bennett, M., Melvold, W.R., and Kohn, H.I. (1980) Immunogenetics 10,273.

Binz, H., Fenner, M., and Wigzell, H. (1983). J. Exp. Med. 157,1252.

Brooks, C.G., Urdal, D.L., and Henney, C.S. (1983). Immunol. Rev. 72,43.

Callewaert, D.M., Kaplan J., Johnson D.F., and Peterson, Jr., W.D. , (1979). Cell Immunol. 42:103.

Cudkowicz, G., and Bennett, M. (1971) J. Exp. Med. 134, 83.

Cudkowicz, G., Landy, M. and Shearer, G.M., editors. (1978). "Natural Resistance Against Foreign Cells, Tumors, and Microbes." Academic Press, Inc. New York.

Drizlich, G., Schmidt-Sole, J., and Yankelevich, B. (1984) J. Exp. Med. 159,1070

Fast, L.D., Beatty, P., Hansen, J.A., and Newman, W. (1983). J. Immunol. 131, 2404.

Habu, S., and Okumura, K. (1982) in "NK Cells And Other Natural Effector Cells" (R.B. Herberman, ed.) p. 209. Academic Press, New York.
Hansson, M., Karre, K. Kiessling, R., Roder, J., Andersson, B., and Hayry, P. (1979) J. Immunol. 123, 765.
Herberman, R.B., Nunn, M.E., and Holden, H.T. (1978) J. Immunol. 121, 304.
Hercend, T., Reinherz, E.L., Mauer, S.C., Schlossman, S.F., and Ritz, J. (1983a) Nature, 301, 158.
Hercend, T, Mauer, S., Brennan, A., Edson, M.A., Acuto, O., Reinherz, E.C., Schlossman, S., and Ritz, F. (1983b) J. Exp. Med. 958, 1547.
Heslop, B.F., and McNeilage, L.J. (1983). Immunol. Rev. 73,35.
Hurme, M., and Sihvola, M. (1983) J. Immunol. 131, 658.
Ishikawa, J., and Dutton, R.W. (1980). J. Immunol. 125, 656
Jerne, N.K. (1971) Eur. J. Immunol., 1:1.
Kaplan, J. and Callewaert, D.M. (1978) J. Nat. Can. Inst. 60, 901.
Kaplan, J. and Callewaert, D.M. (1980) in "Natural Cell-Mediated Immunity Against Tumors" (R.B. Herberman, ed.) p. 893. Academic Press, New York.
Karre, K., and Seeley, J.K. (1979) J. Immunol. 123, 1511.
Kay, H.D. and Horwitz, D.A. (1980). J. Clin. Inv. 66:847.
Kedar, E.B., Ikejiri B., Sredni, B., Bonavida, B., and Herberman, R.B. (1982) Cell. Immunol. 69, 305.
Koo, G.C., Jacobsen, J.B., Hammerling, J., and Hammerling, G. (1980) J. Immunol. 125, 1003
Kumar, V., Mellen, P.F., and Bennett, M. (1982) in "NK cells And Other Natural Effector Cells" (R.B. Herberman, Ed.) P. 329. Academic Press, New York.
Mattes, M.J., Sharrow, S.O., Herberman, R.B., and Holden, H.T. (1979)) J. Immunol. 123, 2851.
Meruelo, D., Paolino, A., Flieger, N., Dworkin, J., Offer, M., Hirayama, N, and Ovary, Z (1980). J. Immunol. 125:2719.
Moretta, A. Pantaleo, G., Mingari, M.C., Melioli, G., Moretta, L., and Cerottini, J.C. (1983) Eur. J. Immunol. 14, 121.
Morgan, G.M., and McKenzie, I.F.C. (1981) Eur. J. Immunol. 7,62.
Nakano, K., Nakamura, I., and Cudkowicz, G. (1981) Nature 289,559.
Ortaldo, J.R., Sharrow, S.O., Timonen, T., and Herberman, R.B. (1981) J. Immunol. 127, 2401.
Roy, C., Ghayur, T., Kongshaven, P.A.L., and Lapp, W.J. (1982) Transplant. 34,144

Suzuki, R., Handa, K., Itoh, K., and Kumagai, K. (1983) J. Immunol. 130, 98.

Timonen, T., Ortaldo, J., Herberman, R.R. (1982) in "NK Cells And Other Natural effector cells" (R.B. Herberman, ed.) p. 821. Academic Press, New York.

Ting, C-C., Yang, S.S., and Hargrove, M.E. (1982) Cell Immunol. 73,275.

Ting, C-C., and Hargrove, M.E. (1983) J. Immunol. 131,1734.

Ting, C-C., and Hargrove, M.E. (1984) J. Immunol. 133, 660.

Toribio, M.L., DeLandazuri, M.D., and Lopez-Botet, M. (983) Eur. J. Immunol. 13,964.

Wagner, J., Rollinghoff, M., Pfizenmaier, K., Hardt, C. and Jonscher, G. (1980) J. Immunol., 124:1058.

RECEPTORS FOR "SELF" RECOGNITION SHARED BY HUMAN T LYMPHOCYTES AND NK CELLS[1]

M. Schlesinger

Hubert H. Humphrey Center for
Experimental Medicine and Cancer Research
The Hebrew University
Hadassh Medical School
Jerusalem, Israel

E.J. Margalioth

Department of Obstetrics and Gynecology
Hadassah University Hospital
Jerusalem, Israel

J.G. Bekesi
Department of Neoplastic Diseases
Mount Sinai Medical School
New York, New York

I. INTRODUCTION

The nature of the cell structures that enable NK cells to recognize target cells and to bind to them is still unclear. The aim of the present study was to determine to what extent "E-receptors" on the surface of human NK cells enable their interaction with target cells. E-receptors were initially detected on human T-cells, where they mediate

---

[1]Supported by the T.J. Martell Foundation, the Goldhirsch Foundation, the Zelman Cowen University Fund, and the Ministry of Science of Niedersachsen.

ISBN 0-12-341370-2

the binding of sheep red blood cells (SRBC)(1, 2). Monoclonal antibodies have been developed, capable of blocking the interaction of human T-cells with SRBC (3-5). These antibodies have recently been shown to inhibit the binding of unstimulated (6) and of mitogen-activated human T-lymphocytes to autologous erythrocytes (7). Monoclonal antibodies against the E-receptor also inhibit the attachment of human thymus cells and of activated peripheral T-cells to nucleated human target cells (5). E-receptors were demonstrated not only on human T-lymphocytes but also on human NK-cells (8). A distinct subset of NK cells, lacking the OKM 1 antigen, was shown to bind SRBC (9). Monoclonal antibodies against the E-receptor were found to react in immunofluorescent tests with a portion of the human NK cell population (10). Exposure of peripheral blood lymphocytes (PBL) to OKT-11 monoclonal antibody, specific for the E-receptor, in the presence of complement diminished their NK activity (11, 12). Indeed, antibodies specific for the E-receptor seemed to inhibit NK activity even in the absence of complement (13, 14). In the present study PBL of donors from various ages were exposed to antibodies specific for the E-receptor, and their NK activity was assessed under various experimental conditions.

II. MATERIALS AND METHODS

A. Peripheral Blood Lymphocytes (PBL) and Cord Blood Lymphocytes

PBL were obtained from normal adult blood donors of various ages. Venous blood was collected with preservative-free heparin. Mononuclear cells were isolated by Ficoll-Hypaque barrier centrifugation. The cells in the interface were collected, washed twice in normal saline, and resuspended in RPMI-1640 medium supplemented with 10% heat inactivated fetal calf serum (FCS). Cord blood lymphocytes were isolated from the blood of the umbilical vein by a similar procedure. Because of contamination with red blood cells, the cell suspensions obtained after Ficoll-Hypaque barrier centrifugation usually had to be run through a second Ficoll-Hypaque centrifugation.

B. Antibodies

The following monoclonal anti-T reagents were used: Leu 4 (Pan T), Leu 3a (Helper T), Leu 2a (Suppressor/Cytotoxic T), were all purchased from Becton Dickinson (Mountain View, California). In addition, T8, (suppressor/cytotoxic T) and T11 (anti-E receptor) were purchased from

Coulter Immunology (Hialeah, Florida). OKT 9 (anti-transferrin receptor) and OKT 11 (anti-E receptor) were purchased from Ortho Pharmaceutical Corporation, Raritan, N.J. NK distinctive antibodies were obtained from Becton Dickinson. Rabbit antisera to the E-receptor were prepared by a modification of the method of Mendes *et al.* (15). Human thymus cell suspensions, containing 200×10^6 cells/ml were incubated for one hour at 45°C. The suspensions were centrifuged and the supernates were collected. Washed, packed SRBC were suspended in equal volume of supernatant fluid, incubated for 3 hours at 4°C, and were then washed twice in normal saline. One ml alliquots of packed, E-receptor-coated SRBC, were injected i.p. into rabbits, at 2-week-intervals. The antisera used in the present study were obtained by bleeding from the orbital vein one week after the 5th or 6th injection.

C. Interferon

α-interferon was a generous gift from Hoffman La Roche, N.J. It was stored at - 70°C as a stock solution containing 10^5 units/ml. Shortly before use the stock solution was diluted 1:25 in RPMI-1640 medium supplemented with 10% FCS and 50 µl of the diluted interferon solution was added to each well for the stimulation of NK activity.

D. NK Assay

K562 cells suspended at a concentration of 10^6 cells/ml were labeled for 1 hr at 37°C with 150-200 µCi ^{51}Cr (sodium chromate). The labeled cells were washed and resuspended in RPMI-1640 medium, supplemented with 10% heat inactivated FCS. 1.25×10^5 effector cells suspended in 50 µl RPMI-1640 supplemented with 10% FCS were introduced in triplicates into each microplate well. To each well was then added 25 µl of target cell suspension, containing 5×10^3 ^{51}Cr-labeled K562 cells. In each assay, wells containing target cells alone, were kept as controls to determine the spontaneous ^{51}Cr release. Microplates were incubated for 4 hr at 37°C in incubators containing an atmosphere of 5% CO_2. At the end of the incubation the supernatant fluids of each well was removed and the radioactivity of the supernatant fluid determined in a gamma counter (Packard, Auto-Gamma Spectrometer). The total amount of radioactivity added to each well was determined by measuring the radioactivity of 25 µl of the target cell suspension.

E. Inhibition of NK Activity with Antibodies

The inhibitory effect of various antibodies was determined by exposing 50 μl of the PBL suspensions to 5 μl of various antibodies, in individual microplate wells. The mixture was incubated at 37°C for 30 minutes and ^{51}Cr labeled K562 cells were added. The results of the NK assay obtained with antibody-treated lymphocytes were compared with those obtained with untreated lymphocytes. In some experiments the effect of antibodies on interferon-augmented NK activity was determined. In these tests, 50 μl of PBL suspension were first exposed to 5 μl monoclonal antibody for 30 minutes at 37°C, and then 50 μl α interferon was added. In control wells 50 μl RPMI-1640 medium, supplemented with 10% FCS was added. 15 minutes later ^{51}Cr labeled K562 target cells were added. The inhibitory effect of various sugars was determined as described previously (16).

The results of NK assays carried out in the presence of antibodies are expressed as percentage of the specific chromium release obtained in control assays. Alternatively the percentage of inhibition by exposure to antibody was calculated according to the formula:

$$100 - \left(\frac{\%\ \text{specific release (antibody)}}{\%\ \text{specific release (control)}} \times 100\right)$$

III. RESULTS

A. The Effect of Various Antibodies on NK Activity

PBL were exposed to various monoclonal antibodies, in the absence of complement, and tested for their NK activity. Most antibodies tested had no effect on NK activity. These included the Leu 2a, Leu 3, Leu 4, Leu 7, Leu 11, T8 and OKT 9 antibodies. Only exposure to monoclonal antibodies to the E-receptor (T11 and OKT 11) displayed a consistent inhibitory effect. The polyclonal E-receptor antiserum, raised in rabbits, also exerted an inhibitory effect on NK activity.

B. The Effect of Age of the Donors on Susceptibility to Inhibition

PBL from adult donors of various ages were exposed to T11 antibody and the inhibition of NK activity was determined (Table I). No significant difference in the susceptibility to inhibition by T11 was noted among various age groups, although the NK activity of the age group 30-39 years was less susceptible to inhibition by T11 antibody than that of other age groups. Rabbit antiserum to the E-receptor inhibited the NK activity of adult PBL to the same extent as

TABLE I. The Effect of T11 Antibody on the NK Activity of PBL from Individuals of various Ages

Age group (years)	Number tested	Inhibition of NK activity (%)
20 - 29	16	43.07 ± 3.14[a]
30 - 39	16	38.66 ± 3.96
40 - 49	10	48.90 ± 3.42
50 - 59	5	44.70 ± 7.66

[a]Mean ± S.E.

T11 antibody, while OKT 11 antibody had a weaker inhibitory effect (Table II). The results obtained with antibody-treated cord blood lymphocytes were more variable than those obtained with adult PBL. It seemed, however, that OKT 11 inhibited the NK activity of cord blood lymphocytes to the same extent as that of adult PBL. Rabbit antiserum to the E-receptor tended, however, to have a weaker effect on cord blood lymphocytes as compared with adult PBL.

C. The Effect of T11 Antibody on Interferon-Treated NK-Cells

In the experiments shown in Table III, exposure of PBL to α-interferon only slightly augmented their NK activity. When PBL were coated with T11 antibody, the NK activity of interferon-stimulated cells was considerably stronger than that of cells exposed to antibody alone. Thus, α interferon seemed to partially counteract the NK-inhibitory effect of T11 antibody.

TABLE II. The Effect of Antibodies to the E-Receptor on the NK Activity of Adult and Cord Blood Lymphocytes

Antibody	Source of NK cells	Number tested	Inhibition of NK activity (%)
OKT-11	Adult PBL[a]	8	34.63 ± 2.58[b]
"	Cord blood	8	30.54 ± 10.58
Rabbit anti-E receptor	Adult PBL[a]	9	41.79 ± 4.57
"	Cord blood	6	13.68 ± 18.80

[a]Donors ranged in age from 26 to 37 years

[b]Mean ± S.E.

TABLE III. The Effect of T11 Antibody on the NK-Activity of Unstimulated and Interferon-treated PBL

Treatment of PBL	Specific ^{51}Cr release	Inhibition of NK activity (%)
None	21.86 ± 6.98[a]	
T11	12.08 ± 5.66	45.57 ± 6.43
Interferon	24.91 ± 6.19	
Interferon + T11	20.17 ± 3.64	22.54 ± 8.75

[a]Mean (± S.E.) from 7 separate assays

D. The Effect of Simple Sugars and Antibody on NK-Cells

The exposure of PBL to various sugars was found to inhibit their NK activity (*cf.* 16). In the present study when PBL were exposed both to simple sugars and to T11 antibody, no additive effect could be detected. In tests with PBL from 6 individuals exposure to D-ribose, at a concentration of 110 mM resulted in a 37.75 ± 4.31% NK-inhibition while T11 inhibited the NK activity by 49.00 ± 3.51%. Exposure of PBL to both T11 antibody and D-ribose inhibited the NK activity by 45.37 ± 3.05%.

IV. DISCUSSION

It has recently become clear that T-lymphocytes utilize a number of cell surface structures in order to discriminate "self" from "non-self" determinants. In addition to the specific T-cell receptor, T-lymphocytes utilize either the OKT 4 or the OKT 8 cell surface components in the recognition of alien determinants (17). It has been suggested that the OKT 4 molecules on helper T-cells enable the cells to recognize class II determinants of major histocompatibility antigens, while the OKT 8 molecules recognize class I MHC determinants. In contrast to the OKT 4 and OKT 8 molecules, which enable T-lymphocytes to recognize alien MHC structures, the cell surface E-receptors seem to be involved in the recognition of determinants shared by all individuals and which are also present on autologous cells (5-7). Thus, E-receptors seem to be vital for human T-cells not for the recognition of SRBC but rather for the recognition of "self" determinants, in a broad sense.

The cell surface receptors used by NK-cells for the recognition of target cells is unclear. *A priori* it seems that E-receptors may play a role in the recognition of "self" structures by NK cells, similar to the role of these receptors in T-lymphocytes. Fast *et al.* (13) have shown that the 9.6 monoclonal E-receptor antibody inhibits NK activity by about 40%. Later, Martin *et al.* (14) found that the 9.6 antibody, at a concentration of 10 μg/ml or higher blocked NK lysis almost completely, while the 35.1 monoclonal antibody to another epitope of the E-receptor had no effect on NK lysis. In the present study rabbit antisera against the E-receptor and the OKT 11 and T11 monoclonal antibodies inhibited NK activity by about 40%. NK cells from adults belonging to various age groups, and from the cord blood displayed a similar sensitivity to inhibition by monoclonal antibodies to the E-receptor. The fact that antibodies against the E-receptor consistently inhibit about half of the NK activity may indicate that only a distinct subset of human NK cells is endowed with the E-receptor (9). This subset seems to be characterized not only by its cell surface markers but also by a relatively high bouyant density (12).

The inhibitory effect of antibodies against the E-receptor on NK activity could reflect two, not mutually exclusive mechanisms: 1. The antibodies could block specific sites for the recognition of target cells. 2. Antibody-coating of the NK cells could exert an inhibitory effect on their cytotoxic function. It has previously been shown that OKT 11A antibody inhibits a number of T-cell functions (18), and possibly this antibody may also be capable of regulating the activity of NK cells. The present observations on the effects of simple sugars and of α interferon on the activity of antibody coated NK cells do not allow a discrimination between the two alternative mechanisms by which E-receptor antibodies may affect NK activity. It seems possible, however, that like in T-lymphocytes, E-receptors may be involved both in the recognition of target structures by NK-cells and in regulation of their activity.

REFERENCES

1. Jondal, M., Holm, G., and Wigzell, H. (1972). J. Exp. Med. 136, 207.
2. Wybran, J., Carr, M.C., and Fudenberg, H.H. (1972). J. Clin. Invest. 51, 2537.
3. Kamoun, M., Martin, P.J., Hansen, J.A., Brown, M.A., Siadek, A.W., and Nowinski, R.C. (1981). J. Exp. Med. 153, 207.

4. Verbi, W., Greaves, M.F., Koubek, K., Janossy, G., Stein, H., Kung, P., and Goldstein, G. (1982). Eur. J. Immunol. 12, 81.
5. Schlesinger, M., Levy, J., Laskov, R., Hadar, R., Weinstock, J., Ben-Bassat, H., and Rabinowitz, R. (1983). Clin. Immunol. Immunopathol. 29, 349.
6. Scheffel, J.W., and Swartz, S.J. (1982). J. Immunol. 128, 1930.
7. Schlesinger, M., Haubenstock, H.S., and Bekesi, G.J. (1984). J. Clin. Lab. Immunol. (In press.)
8. West, W.H., Cannon, G.B., Kay, H.D., Bonnard, G.D., and Herberman, R.B. (1977). J. Immunol. 118, 355.
9. Caraux, J, Serrou, B., and Weigle, W.O. (1982). Cell Immunol. 74, 404.
10. Ortaldo, J.R., Sharrow, S.O., Timonen, T., and Herberman, R.B. (1981). J. Immunol. 127, 2401.
11. Zarling, J.M., Clouse, K.A., Biddison, W.E., and Kung, P.C. (1981). J. Immunol. 127, 2575.
12. Schlesinger, M., Lew, F., and Bekesi, J.G. (1984). J. Clin. Lab. Immunol. 13, 195.
13. Fast, L.D., Hansen, J.A., and Newman, W. (1981). J. Immunol. 127, 448.
14. Martin, J., Longton, G., Ledbetter, J.A., Newman, W., Braun, M.P., Beatty, P.G., and Hansen, J.A. (1983). J. Immunol. 131, 180.
15. Mendes, N.F., Bernd, L.A.G., Cintra, H.H.A., Mendes, C.M.F., Praca, C.L., and Moura, N.C. (1982). Cell Immunol. 72, 143.
16. Schlesinger, M., and Bekesi, J.G. (1984). J. Clin. Lab. Immunol. (In press.)
17. Reinherz, E.L., Meuer, S.C., and Schlossman, S.F. (1983). Immunol. Rev. 74, 83.
18. Wauwe, J. van, Goossens, J., Decock, W., Kung, P., and Goldstein, G. (1981). Immunology 44, 865.

Analysis of NK Activity Against Virus-Infected Target Cells[1]

Carlos Lopez
Patricia A. Fitzgerald
Thomas E. Schindler[2]

Laboratory of Herpesvirus Infections
Sloan-Kettering Institute for Cancer Research
New York, New York

1. Supported by NIH Grants CA 08748, CA 34989, CA 35793, CA 23766 and American Cancer Society FRA # 193.

2. Present Address: Dr. Thomas Schindler
Xytronyx, Inc.
10 West 35th Street
Chicago, IL 60616

ISBN 0-12-341370-2

I. Introduction

Most studies of natural killer (NK) cell function have been carried out using cultured tumor cells as targets (1,2). However, ample evidence has been presented which indicates that these effectors can also lyse certain normal cells (3), as well as virus-infected targets (4-6). Virus infection of primary cell strains (4,6), spontaneously transformed cell lines (5,7), or cultured tumor cells (8) with any ofa number of RNA or DNA viruses has usually been associated with markedly increased lysis by spontaneously cytotoxic effectors. In general, NK cells which lyse any of these targets share three basic characteristics; they are not mature T-cells, B-cells or macrophages, they are found in Percoll gradient fractions enriched for large granular lymphocytes, and they require no presensitization in order to be active.

Studies have been undertaken to determine the mechanisms involved in NK cytolysis of target cells. Lysis of K562 erythroleukemia tumor targets can be dissected into four phases; 1. conjugate formation between effector cells and targets, 2. trigger, 3. calcium dependent programing for lysis, and 4. killer cell independent lysis of targets (9). Different experimental approaches have been used to attempt to determine the surface structures on tumor targets which participate in conjugate formation and/or triggering (8) lysis. For example, Roder _et al_ (10) showed that certain membrane glycoproteins from murine YAC-1 tumor targets inhibited conjugate formation but not lysis of mouse effectors. On the other hand, Newman (11) showed that monclonal antibody 13.1, which recognized a T-200 molecule, failed to block conjugate formation but blocked NK lysis of K562 targets by inhibiting the trigger. These studies suggest that binding structures may be different from those strutures responsible for trigging the lytic mechanisms.

Our laboratory has been interested in NK cells as mediators of natural defense against herpes and, perhaps, other virus infections (12,13). For our studies, we have used herpes simplex virus (HSV) infected fibroblasts (HSV-Fs) as targets to evaluate NK activity [(NK(HSV-Fs))] (6). Although the effectors which lyse these cells have the general characteristics of NK cells, evidence has been provided which indicates that NK cells are heterogeneous and that the effectors which lyse the commonly used K562 tumor targets can be differentiated from the effectors which lyse HSV-Fs (14).

Primary human skin fibroblasts are poor targets for NK effectors but infection with HSV-1 converts them into targets efficiently lysed by these cells (6). There are at least three ways in which HSV-1 infection of targets might result in increased lysis: First, interferon might be induced and this cytokine might non-specifically augment lysis of targets. Second, the virus infection might make the target cell less stable and more easily lysed. Third, infection of fibroblasts may alter cell surface determinents leading to recognition of target cells as foreign and lysis by NK effectors. In this report, we summarize results of investigations into these possibilities and suggest mechanisms which might account for induced lysis.

II. Role of Interferon In NK(HSV-Fs)

In early studies, Trinchieri *et al* (15) suggested that the greater levels of lysis of virus-infected or virus transformed cells versus uninfected cells may be dependent on the generation of interferon-alpha during the assay and the non-specific augmentation of NK activity by this cytokine. Studies from our laboratory showed that interferon-alpha was made during the NK(HSV-Fs) assay but that this cytokine was not responsible for greater lysis of infected over uninfected cells (16). Specifically, neutralizing all detectable anti-viral activity by adding antibody against interferon-alpha failed to significantly reduce ^{51}Cr-release with HSV-Fs. Furthermore, we were unable to correlate the level of ^{51}Cr-release and the amount of interferon-alpha generated during the assay. This was especially evident with peripheral blood mononuclear cells from several patients with

AIDS. Effector cells from these patients demonstrated normal NK(HSV-Fs) and NK(K562) even though they failed to generate detectable levels of interferon-alpha in the same wells (17). Therefore, even though interferon-alpha is capable of augmenting lysis of target cells which may or may not normally be lysed by NK effectors, this cytokine is not responsible for the selective lysis of HSV-infected over uninfected fibroblasts.

III. Are HSV-Fs More Easily Lysed Than Fs Targets?

Inoculation of human fibroblasts with HSV-1 results in a lytic infection and the death of the cells. Host protein and DNA synthesis is turned off and replication of virus continues (18). Since HSV-1 is a lytic virus, infected fibroblasts may be less stable and more vulnerable to lysis than uninfected cells. Brooks *et al* (19) found a correlation between osmotic sensitivity of rat NK targets and their susceptibility to NK lysis; i.e., those targets more susceptible to disruption when placed in hypotonic solution tended to be those which were most easily lysed by NK effectors. However, incubation of targets in media containing 0 to 67% water demonstrated no significant difference in ^{51}Cr-release between HSV-infected and uninfected fibroblasts (Fitzgerald *et al*, unpublished data).

Another way to evaluate the relative sensitivity of HSV-infected versus uninfected cells to lysis was to compare their susceptibility to lysis using HLA-directed cytotoxic T-cells. In such an experiment, we found that HSV-1 infected fibroblasts were no more easily lysed by cytotoxic T cells than were the uninfected targets (Fitzgerald *et al*, unpublished data).

These data suggest that an inherent instability of the infected fibroblasts does not account for their preferential lysis. Rather, it is more likely that virus infection causes changes in the targets which result in induction of NK cell lysis.

IV. Kinetic Analysis of NK(HSV-Fs) and NK(Fs)

In most studies, NK activity has been quantitated using fixed numbers of target cells and different effector: target cell ratios. However, since effector-target cell interactions are analogous to enzyme-substrate reactions, enzyme kinetic analysis has been found to be suitable for the comparison of NK activities against different targets (20). Such an analysis, when combined with the single cell assay, can be used to approximate the number of NK effector cells active against a given target as well as their ability to recycle and kill again. We have used these assays to show that target cell lysis increased with increases in target cell concentration, approaching a cytotoxic maximum. Plots of target cell concentration versus the number of target cells lysed fit a hyperbolic curve characteristic of enzyme-substrate interaction and the Michaelis-Menton equation could be applied to the data (20).

We have analysed lysis of HSV-Fs and have found that lysis of these targets increased with increasing target cell concentration, approaching a cytotoxic maximum (Schindler and Lopez, in preparation). Plots of target cell concentration versus number of target cells lysed fit a hyperbolic curve characteristic of enzyme-substrate interaction. Thus, the Michaelis-Menton equation could be applied to the analysis of the data. Kinetic analysis of NK(HSV-Fs) demonstrated that maximum lytic capacity (V_{max}) was dependent on the number of lytic effector cells. In experiments using different concentrations of effector cells, there was a strong correlation ($r \geq 0.979$) between cell concentration and V_{max} (Schindler and Lopez in preparation.) Enrichment of NK(HSV-Fs) effectors by Percoll density fractionation was also associated with increased V_{max}. In addition, the efficiency of lysis also appeared to be reflected by V_{max} since preincubation of effectors with interferon resulted in higher lytic activity and greater V_{max} (Schindler and Lopez, in preparation). Data obtained from the study of eight normal controls over a period of six months was consistent for each individual but significant variation was found between individuals.

Parallel experiments were carried out using uninfected fibroblasts as targets. Using these targets, increasing target cell concentration failed to yield increasing target cell lysis; in fact, the curve was flat with very little or no significant chromium release. The data did not fit a hyperbolic curve and could not be analysed with the Michaelis-Menton equation. Furthermore, there was no increase in V_{max} with increasing concentration of effector cells nor when large granular lymphocyte enriched preparations were evaluated. Thus, the very low level of lysis obtained with Fs targets cannot be analysed in this manner. Chromium release with Fs cells may not be the result of lysis of targets by NK effectors and may be an artifact of the culture conditions used.

Using the single cell assay in agarose, we found that there were twice as many stable (30°C) conjugates between effectors and HSV-Fs than effectors with Fs cells ($p<0.005$). These results suggest that HSV-infection of targets results in their binding avidly to effector cells, leading to triggering of lysis.

V. In Theory

Using data generated in this laboratory as well as others, we would like to propose mechanisms which might be responsible for NK(HSV-Fs). Although there appears to be a small subpopulation of effectors capable of lysing both HSV-Fs and K562 targets, most lysis of each of these targets is by non-overlapping effector cells populations. The cells which only lyse HSV-Fs appear to recognize a structure on HSV-Fs and Fs targets while the effectors which only lyse K562 targets appear to interact via another. Cold target inhibition studies showed that HSV-Fs and Fs were capable of efficiently inhibiting lysis of labeled HSV-Fs while only partially inhibiting lysis of K562 cells. Conversely, K562 erytholeukemia targets inhibited NK(K562) lysis but only partially competed NK(HSV-Fs). It thus appears that both HSV-Fs and Fs express a structure which mediate cold competition of NK(HSV-Fs). Comparison of Fs and HSV-Fs as targets in the single cell assay in agarose indicated that stable binding to depended on HSV-infection of target. Thus, Fs targets have a structure which cold compete HSV-Fs but do not have what is required to generate stable conjugates. There are not least two ways in which HSV infection of targets might lead to these reuslts: First, Fs and HSV-Fs cells may have the same binding structure and HSV-infection may modify that structure such that conjugate formation is stable. Secondly, initial (perhaps unstable) binding and cold competition may be mediated by a structure found on Fs and HSV-Fs targets. Virus infection may then result in a second structure which, when pesent, causes that conjugate to become stable. Such a stable conjugate then might induce triggering and lysis of targets.

Our results with HSV-Fs are different than the results obtained by yasukawa and Zarling <u>et al</u> (4) using HSV-infected lymphoblastoid cell lines (LCLs). These workers found that the uninfected LCLs failed to cold compete with HSV-infected LCLs whereas HSV-LCLs were effective cold competitors. In addition, they found cross competition between HSV-LCL and K562 tumor targets. These results suggest that LCLs do not regularly express a target structure recognized by NK effectors but that HSV-infection results in such a structure. In this case, viral antigen (by itself), a viral antigen modified normal cell surface determinent, or a virus antigen-independent, HSV-infection induced modification of the target cell might be responsible for binding and induction of lysis of these cells. These results suggest that different subpopulations of NK effectors may interact differently with targets, even when each of the targets are infected with the same virus.

Our studies indicate that HSV-infection does not increase lysis of fibroblasts by making them less stable or more easily lysed. Instead, infection may change some normal cell surface determinant or add a surface structure which acts as a trigger for NK cell lysis. The studies of Bishop *et al* (22) suggest that HSV viral glycoproteins may be required for induction of NK lysis. Since there is no requirement for presensitization, virus antigens would not be recognized in an antigen-specific manner similar to that of humoral or cell-mediated immunity. Instead, viral antigens might act as lectins for a receptor on a subpopulation of NK effectors. An interaction of receptor structure and "viral lectin" might result in binding and activation of NK effectors and lead to the destruction of virus infected cells.

It is also possible that virus infection might alter some cell surface determinants so that the virus infected cells are recognized as foreign. Karre *et al* (23) have proposed that alterations of cell surface MHC might trigger NK. Since virus infections can markedly alter expression of these cell surface determinants (23), such a change might make cells recognizable as foreign and thus lysed by NK effectors. Another possibility is that virus infection may augment the expression of transferrin receptors which have also been suggested as possible NK target structures (24).

VI. Conclusions

Studies of the mechanisms of lysis of target cells by NK effectors indicate that lysis can be dissected into at least four phases; binding, trigger, programming for lysis and killer cell independent lysis. Definition of the recognition structure and/or trigger will be difficult using tumor targets because of the large number of cell surface determinent which might be responsible. However, the use of exogenous agents such as viruses, should allow the definition of these important determinants of NK lysis because relatively fewer viral gene functions may be involved. Furthermore, the use of viral mutants and gene transfer technology should allow the specific definition of the viral genes and gene products responsible for induction of NK lysis.

REFERENCES

1. Herberman, R.B., Nunn, M.E. and Lavrin, D.H. (1975). Int. J. Cancer 16, 216.
2. Kiessling, R., Klein, E. and Wigzell, H. (1975). Eur. J. Immunol. 5, 112.
3. Hansson, M., Kiessling, R., Andersson, E., Karre, K. and Roder, J. (1979). Nature 278, 174.
4. Trinchieri, G., Santoli, D., Dee, R.R. and Knowles, B.B. (1978). J. Exp. Med. 147, 1299.
5. Welsh, R.M. and Hallenbeck, L.A. (1980). J. Immunol. 124, 2491.
6. Ching, C. and Lopez, C. (1979). Inf. Imm. 26, 49.
7. Bishop, G.A., Glorioso, J.C. and Schwartz, S.A. (1983). J. Exp. Med. 157, 1544.
8. Minato, N., Bloom, B.R., Jones, C., Holland, J. and Reid, L.M. (1979). J. Exp. Med. 149,1117.
9. Hisexodt, J.C., Britian, L.S. and Targan, S. (1982). J. Immunol. 129, 2266.
10. Roder, J.C., Rosen, A., Fenyo, E.M. and Troy, F.A. (1979). Proc. Natl. Acad. Sci. 76, 1405.
11. Newman, W., Fast, L.D. and Ross, L.M. (1983). J. Immunol. 131, 1742.
12. Lopez, C. (1975). Nature 258, 152.
13. Lopez, C., Kirkpatrick, D., Read, S.E., Fitzgerald, P.A., Pitt, J., Pahwa, S., Ching, C.Y. and Smithwick, E.M. (1983). J. Inf. Dis. 147, 1030.
14. Fitzgerald, P.A., Evans, R., Kirkpatrick, D. and Lopez, C. (1983). J. Immunol. 130, 1663.
15. Trinchieri, G., Santoli, D. and Koprowaki, H. (1978). J. Immunol. 120, 1819.
16. Fitzgerald, P.A., von Wussow, P. and Lopez, C. (1982). J. Immunol. 129, 819.
17. Lopez, C., Fitzgerald, P.A. and Siegal, F.P. (1983). J. Inf. Dis. 148, 962.
18. Nahmias, A.J. and Roizman, B. (1973). N. Engl. J. Med. 289, 667.
19. Brooks, C.G., Wayner, E.A., Webb, P.J., Gray, J.D., Kenwick, S. and Baldwin, R.W. (1981). J. Immunol. 126, 2477.
20. Callaewaert, D.M., Smeekens, S.P. and Mahle, N.H. (1982). J. Immunol. Methods 49, 25.

21. Yasukawa, M. and Zarling, J.M. (1983). J. Immunol. 131, 2011.
22. Bishop, G.A., Blorioso, J.C. and Schwartz, C.A. (1983). J. exp. Med. 157, 1544.
23. Karre, K., Pointek, G., Ljunggren, H., Gronberg, A. and Kiessling, R. (1984). Second Internations Workshop on Natural Killer Cells (In Press).
24. Kay, H.D., Bierman, P.J. and Klasser, L.W. (1984). Second International Workshop on Natural Killer Cells (In Press).

Part III.

Mechanisms of Postrecognition Interactions Between NK Cells and Targets

BIOCHEMISTRY OF NK CYTOTOXICITY

Ronald H. Goldfarb

Department of Immunology and Infectious Disease,
Central Research Division
Pfizer, Inc.
Groton, Connecticut

This session reviewed the substantial and rapid progress that has recently been made in our understanding of the biochemistry and molecular interactions involved in the recognition and lysis of tumor cells by NK cells. The discrete steps involved in the molecular mechanism of NK cytotoxicity were discussed from a variety of perspectives including: the identification and characterization of soluble NK cytolytic factor(s), NKCF; the isolation and examination of cytolytic granules derived from large granular lymphocytes (LGL); the elucidation of target damage and the ultrastructure of NK cell-target cell interactions; and the exploration of biochemical pathways which appear to contribute to NK cytotoxicity. These diverse lines of research addressed key questions and themes dealing with the potential relationship of NKCF and LGL granules; the exact role of LGL granules and/or NKCF in tumor cell lysis by intact NK cells; biochemical similarities and differences among various NKCFs; and the potential synergy among biochemical pathways involved in the NK lytic mechanism.

TARGAN (UCLA) introduced the subject by defining the sequential functional substages of NK-target cell interactions in binding, triggering, programming, and lethal hit stages of NK cytotoxicity. The binding step appears to be magnesium dependent, and requires intact receptors on both the NK cell and target cell. Chemical crosslinking agents and various proteolytic enzymes inhibit the binding stage. Triggering was reported to be both calcium and temperature dependent. Phospholipase A_2 activation may be required to alter target cell phospholipids for activation of the NK trigger mechanism and it appears that an epitope of the T-200 glycoprotein, associated with the NK binding receptor, may be a NK trigger for some target cells. TARGAN reported that the programming for lysis stage has an early calcium and temperature dependence. Calcium

ISBN 0-12-341370-2

channel blockers interfere with the early portion of programming for lysis suggesting that calcium dependency requires calcium flux into the NK cell. During the latter portion of the programming stage, NK cell-derived structures are secreted and placed onto the target cell. TARGAN reported that such structures may involve NKCF subunits and a focusing molecule which leads to the formation of a lytic complex through disulfide bond formation.

The terminal lethal hit stage of NK cytotoxicity was found to be temperature dependent during early substages and calcium independent. TARGAN reported that the killer cell independent NK lethal hit can be subdivided into at least four substages: an NK lytic complex/NKCF binding stage; an assembly/activation stage; an NK lethal complex/NKCF effector stage; and a lytic stage. The lytic factor binding substage is inhibited by cross-linking reagents. The assembly/activation stage takes place immediately following the binding stage and is inhibited by agents that disrupt disulfide bonds. Since cross-linking agents enhance lysis within this stage, subunit polymerization or complex formation may be required for completion of the lethal hit. Inhibitors of metabolism and microtubules inhibit later in this substage and suggests a role for membrane movement and endocytosis. The effector substage is interfered with by inhibitors of serine proteolytic enzymes suggesting that proteases may play an enzymatic role during the later substages of the lethal hit. Effector and lytic substages are inhibited by benzyl alcohol which may inhibit membrane fluididty and disrupt NK lytic factor-mediated transmembrane phenomena. The definable sequentially coupled events on the target membrane described by TARGAN that occur during the NK lethal hit appear to be rquired for completion of target cell lysis. The identification of functional substages within the NK lytic mechanism provides an important frame of reference for analysis of distinct biochemical events during NK killing.

HENKART (NIH) reviewed his findings examining the hypothesis that LGL granules play a role in the NK lethal hit. Earlier EM studies by HENKART on NK-target conjugates indicated that granule-contents undergo morphologic changes and are secreted within the micro-environment of target cell membranes during NK killing. The cytolytic properties of purified cytoplasmic granules from rat LGL tumors showing NK and/or ADCC activity were described. Following ultracentrifugation of cell homogenates on Percoll gradients a high density granule peak, pure upon analysis by EM, was obtained. The morphology of the granules in fractions containing lytic activity was similar to that found in rat LGL tumors. Fractions assayed for lytic activity against SRBC showed potent lytic activity with less than 1 ug/ml protein of the most active fractions. The lytic activity was associated with a peak in the activity of four different lysosomal enzymes distinct from fractions exhibiting marker enzymes for mitochondria, plasma membrane, and cytosol. HENKART reported that analysis of Percoll gradient fractions by SDS PAGE showed that the granule fraction was free of contamination from other gradient fractions. The granules contained major protein bands of 62, 58, 3

29, and 28 kilodaltons. The granules lyse a large variety of nucleated cells, but the nucleated cells are approximately 100 fold more resistant than red cells. Chromium release from target cells is noted within a few minutes at 37°C. Lysis mediated by LGL granules requires calcium at 10^{-4}M. However, preincubation of granules with calcium prior to the addition of target cells leads to a loss of lytic activity which has been interpreted by HENKART to suggest an activation event followed by an unstable intermediate. Granule preparations derived from normal spleen cells, noncytotoxic tumors, and peripheral T cells do not demonstrate the cytotoxicity of the LGL tumors, whereas granules from purified normal peripheral blood LGL do exhibit comprable cytotoxicity. Upon exposure of LGL tumor granules to calcium and EM examination by negative staining, ring shape structures with an approxmate diameter of 15nm are noted. Ring structures, previously reported by HENKART on NK and ADCC target cells, are not observed in preparations not exposed to calcium. HENKART has suggested that the correlation between the generation of ring structures and lysis supports the hypothesis that pore formation is a principal component of the NK lethal hit. HENKART also noted the apparent similarity and appearence of the rings to complement-mediated lesions and hypothesized that lytic activity might take place by a complement-like insertion of ring structures into the target cell membrane.

REYNOLDS (NCI-FCRF) described the functional characterization of rabbit antibodies against purified cytoplasmic granules of rat LGL tumors. Two precipitin lines with granule extracts was observed when antisera was employed in Ouchterlony analysis. The antisera also stained a dimer of granule proteins when reduced SDS PAGE gels are Western blotted (60/58 kilodaltons). In addition, when non-reduced gels were blotted, two additonal proteins were reactive. REYNOLDS reported that in immunofluorescence studies the antibodies reacted with cytoplasmic granules in permeabilized rat LGL as well as normal rat or human LGL but not with permeabilized Blood T cells. $F(ab')_2$ preparations of anti-LGL granule antibodies completely blocked granule-mediated lysis of both nucleated targets and SRBC. $F(ab')_2$ preparations were also found to effectively inhibit both rat and human NK and ADCC reactivity but not the activity of mouse cytolytic T-cell clones. The antibodies had to be present during the NK or ADCC assays and pretreatment of either target cells or effecor cells had no effect. Preincubation of human NKCF supernatants with anti-granule $F(ab')_2$ preparations, but not control preparations, completely inhibited the cytolytic activity of the NKCF supernatants. REYNOLDS suggested that a granule component(s) may be required for the lytic activity of LGL in both NK and ADCC. These studies therefore further support a secretory process involving granules in the lytic event.

GOLDFARB and HISERODT questioned HENKART about the significance of calcium blockers to inhibit at the calcium pulse phase of programming for lysis in intact LGL, which suggests a need for calcium influx into LGL for subsequent secretion. HENKART discussed calcium requirements for NK killing within, and potentially beyond, the calcium requirment of cytolysis. TARGAN raised the question of why LGL are not lysed by its own potent cytolysin upon release. HENKART suggested that LGL might have a temporal period of resistance. In probing the potential interrelationship between NKCF and granule cytolytic activity, SAKSELA raised the question of whether the antisera described by REYNOLDS binds to NKCF, which has not yet been fully resolved. In response to further questions by SAKSELA, HENKART noted that the granule-derived lytic activity that he observed and described is more striking in differences from NKCF than in any similarity. HENKART and REYNOLDS have not detected NKCF from isolated granules to date. Whereas the antibody experiments show some partial inhibition of human NKCF activity suggesting that NKCF may come from granules, HENKART expresssed the view that the polyclonal antisera employed can not definitely lead to this conclusion. In response to queries from GOLDFARB and HISERODT, REYNOLDS noted that upon further purification of lytic proteins isolated from LGL granules, cytolytic activity appears to be coincident with Mr bands of approximately 60 kilodaltons. SAKSELA noted that the staining pattern of LGL observed with anti-granule antibodies appeared to be similar to a Golgi distribution suggesting that NKCF may be secreted in the Golgi and subsequently be deposited in the granules in a non-lytic form. HENKART noted, however, that the light microscopy staining pattern does not appear to be as compatible with a Golgi distribution as a granule. In response to questions by WIGZELL, HENKART noted that cloned CTL react with the anti-LGL granule antibodies in approximately the same way as LGL. The antisera does not inhibit CTL killing, although it does inhibit CTL cytolysin activity. BONVAVIDA raised the issue of whether the antibody also inhibited ADCC, and whether the use of the antibody could determine mechanistic similarities or differences between NK killing and ADCC. REYNOLDS concluded that monoclonal rather than polyclonal antibodies will be required to further sort this issue. In response to questions by BONAVIDA, REYNOLDS noted that antibody cross reactivity with the mouse is much less than that observed for human, and it has been very difficult to show inhibition of mouse NK activity by using anti-rat tumor LGL antibodies. HERBERMAN raised the issue that since the antibodies do not inhibit mouse NK killing very well, the point concerning the inability of the antibody to effectively block mouse CTL killing shold be considered an ambiguous result.

Following the discussions of the stages of the LGL lytic mechanism, the role of granules in the NK lytic mechanism, and the potential relationship between granule cytolysins and NKCF, attention was then directed towards questions concerning the nature, size, and relevance of NKCF. BONAVIDA (UCLA) presented a preliminary overview of the NKCF area in which he reviewed the original observation that there is release from NK cells in soluble form, factors that can mediate the lysis of NK-sensitive target cells. BONAVIDA then described the correlation established by his laboratory between the role of NKCF and direct NK cell-mediated cytotoxicity: NKCF is produced by mouse NK cells, human peripheral blood-derived NK cells and LGL enriched for NK cells; murine lymphoid cells that release NKCF are non-adherent, Thy 1.2 negative and Asialo GM1 positive cells; the stimulation of lymphoid cells from NK-deficient Beige/Beige mice with NK-sensitive target cells led to poor release of NKCF. BONAVIDA also described studies by WRIGHT directed towards examination of the role of proteolytic enzymes in NKCF, due to their implication in NK cell-mediated cytotoxicity. It was found that one site at which protease inhibitors play a role is at the induction of NKCF from NK cells. Pre-treatment of effector cells with IFN, followed by stimulation for release of NKCF, resulted in augmented NKCF production. Target cells pre-treated with IFN, which are known to be resistant to NK cell-mediated lysis, can be stimulated to release only poor yields of NKCF...even though the target cell is susceptible to lysis mediated by NKCF. As for killing by intact NK cells, NKCF activity can be blocked by the addition of simple sugars into the assay. BONAVIDA then described a model for what constitutes an NK sensitive target cell, and what is required for NK lytic activity. Putative interaction between an NK cell and an NK-sensitive target cell can include a binding step and the stimulation of the release of NKCF. IFN-treated target cells, described above, were used to define the action of NKCF. The first step was described as the release of NKCF in supernatants, followed by the binding of NKCF to target cells to allow for target cell lysis. Binding by itself, however, is not sufficient since several target cells can bind NKCF, but remain resistant to the activity of NKCF. BONAVIDA therefore postulated that following binding of NKCF an additional step must be involved in making the target more susceptible to NKCF. This step might be either a processing or internalization step. Converseley, additonal factors might also play an important role. BONAVIDA concluded that further insights into the role of potential receptors and various interactions can now be further analyzed independently of the whole phenomenon by careful disection of the discrete steps involved in the lytic mechanism.

WRIGHT (UCLA) discussed the biochemical characterization of NK cytolytic factors, as well as the release of soluble cytotoxic factors from peripheral blood leukocytes stimulated by ADCC target cells. Methods were described for the generation of NKCF from murine or rat spleen cells or human peripheral blood lymphocytes *in vitro* when stimulated with either lectin or various types of tumor cells. After 24 hours supernatants are titred for cytotoxicity against appropriate target cells in a 20 hour chromium release test for which it was essential to prepare target cells by overnight chromium labeling procedures. WRIGHT noted that cultures are prepared under serum-free conditions, and that stimulator cells and target cells employed are tested for, and free of, mycoplasma. Rat and murine NKCF are partially inactivated at 56°C and completely inactivated at 63°C. While insensitive to neuraminidase, NKCFs are trypsin sensitive and unstable at pH 2. Reduction and alkylation were found to eliminate lytic activity. Exposure to 2-mercaptoethanol results in a 50% reduction of lytic activity. Protease inhibitors had no effect on the lytic activity of NKCF, whereas oxidation with sodium periodate led to complete inhibition of lytic activity suggesting that carbohydrate -containing structures may be involved in the lytic activity of NKCF. Evidence derived from gel filtration by HPLC suggests that NKCFs have an apparent Mr of approximately 20 - 30 kilodaltons. NKCF preparations derived from rat tumor LGL display similar gel elution profiles. WRIGHT therefore concluded that NKCF are heat sensitive proteins or glycoproteins which require disulfide bonding for lytic activity. In addition, it was suggested that carbohydrate-containing structures, on NKCF or on the target cell membrane, play a role in the binding of these factors to the target cell. NKCF, with apparent Mr of 20,000 -30,000 appear to bind to target cell membranes, and mediate target cell lysis in the lymphocyte-independent phase of the NK lytic mechanism.

In response to questions from OELTMANN, WRIGHT noted that the increase in activity from NKCF recovered from peak gel filtration fractions is approximately 4 fold. It was also reported that concentration of NKCF can lead to an enhancement in the kinetics of lysis. In response to HUDIG's question concerning biochemical characteristics of rat LGL -derived NKCF, WRIGHT noted that NKCF production is variable when compared to normal rat spleen cells. Upon dialysis of rat tumor LGL NKCF, lytic activity often increases, in contrast to WRIGHT's experience with normal rat or mouse factors. The possibility was therefore raised that the rat NK tumor cells might produce a low Mr inhibitor. ORTALDO questioned the use of U937 target cells for human NKCF rather than K562 cells; WRIGHT noted that U937 cells are more sensitive than K562 cells to NKCF. In addition, U937 cells were found to maintain a very low spontaneous release in a 24 hour chromium release assay. In response to a question by KOREN, it was ascertained that antibody to rat tumor LGL granules had not been tested against WRIGHT's NKCF. KOREN asked for an explanation of the slow kinetics observed for NKCF as opposed to intact NK cells. WRIGHT argued that NKCF might actually be high within the local microenvironment of stimulated cells which might allow for rapid cell lysis. The possibility of inhibitors or degrading molecules were also considered. KOREN suggested experiments to directly probe these possibilities including pretreament of target cells with NKCF to attempt to accelerate kinetics of lysis.

ORTALDO (NCI-FCRF) described studies analyzing the mode of activity of NKCF with various agents previously demonstrated to inhibit NK cell-mediated lysis of target cells including: phosphorylated sugars; protease inhibitors; antibodies; cations; lipomodulin; nucleotides; prostagalandins; and inhibitors of lysosomal enzymes. These agents were tested for their effect on various steps of NKCF generation or function: production during stimulation with targets or target cell membranes;in the binding of NKCF to target cell structures; and in NKCF-mediated lysis following adsorption with or without washing of target cells. Most of the agents tested by ORTALDO have had no effect on the production of NKCF. It is unclear as to whether anti-LGL granule antibodies effect production since it is difficult to eliminate the antibodies after using them. In contrast to the observations of WRIGHT, ORTALDO found that protease inhibitors had no effect on production, or any phase,of NKCF function

ORTALDO noted that phosphorylated sugars of mannose, which inhibit NK lysis, significantly inhibits absorption of the binding of the NKCF to the target cell but have no effect on lysis once on the target cell. Ca++/Mg++ inhibit NKCF prodcution but have not been fully assessed for binding and lysis. ATP and PGE_2, as well as ammonium chloride and strontium chloride which have been reported to effect NK cell degranulation, interfere with NKCF production but have no effect on NKCF binidng or lysis of target cells. ORTALDO reported that monensin severely interfered with the production of NKCF, and also effects the lysis of target cells once NKCF has been absorbed.

ORTALDO also summarized progress dealing with the biochemical analysis of internally labeled NKCF derived from human LGL in serum free media supplemented with serum albumin and additional defined media-supplements. The internal label employed by ORTALDO is ^3H-arginine or ^{35}S methionine. Upon HPLC gel filtration the Mr of NKCF appears to be between 18 and 37 kilodaltons, and the lytic activity indicates a 10-20 fold enrichment.

A number of questions were raised dealing with the incorporation of NKCF into liposomes (HUDIG), and whether EM analysis had explored the potential of NKCF to generate ring-like structures within the target cell membrane (GOLDFARB). SAKSELA noted that the monensin effect described by ORTALDO might imply some internalization of NKCF, since monensin is known to be specific for inhibition of vesicle fusion dependent traffic as well as endocytotic events. SAKSELA noted that this might explain why it appears that NKCF is not internalized in a cytotoxic manner. DJEU inquired as to whether anti-LGL granule antibody bind to radio-labeled NKCF. ORTALDO noted that such experiments are on-going and results are dependent upon the long time of autoradiographic analysis required by the low specific activity of radiolabeled NKCF. LOPEZ inquired as to whether the HPLC purified material still maintained selective activity against good versus poor NK cell targets. ORTALDO indicated that in preliminary experiments, with K562 and MOLT targets, this is the case.

OELTMANN (Vanderbilt U.) also presented evidence for a soluble lytic factor and its involvement in the NK lytic mechanism. Based on the ability of certain hexose phosphates to inhibit lysis of target cells, but not binding, OELTMANN developed an affinity column as a probe to examine the possible role of hexose-containing macromolecules in the NK lethal hit. ^{35}S labeled NKCF which exhibits lytic activity against K562 cells was described. G200 gel filtration chromatography indicates an Mr of approximately 60 kilodaltons. This NKCF is sensitive to low pH, proteases, chloroquine, and ammonia. Direct binding experiments with the ^{35}S-labeled NKCF indicates that both binding to and lysis of target cells can be inhibited by several hexose phosphates, particularly mannose 6-phosphate. OELTMANN suggested that simple hexose phosphates may inhibit at a receptor site which mediates the uptake of NKCF, distinct from recognition and binding site(s).

JOHNSON (U. Nebraska) presented data characterizing soluble cytotoxic factors produced by marmoset monkey lymphoid cells transformed by lymphotropic Herpesvirus, which exhibit human NK cell-like cytolytic activity towards specific target cells. NKCF produced by these cells lyse NK sensitive target cells but fail to lyse NK-insensitive targets, nor can the latter cells stimulate NKCF production. Epstein-Barr virus-transformed marmoset B cells, which have no NK activity, cannot be induced to produce NKCF. This NKCF is insensitive to UV irradiation, is inactivated by proteases, and is inhibited by certain mono- and disaccharides. Following gel permeation chromatography, the NKCF appears to have an Mr of approximately 40-50 kilodaltons. A second peak activity was noted at 20-25 kilodaltons. It was noted that this system has the potential to produce large amounts of cytotoxic supernatants, required for biochemical analysis.

In response to questions by GOLDFARB, JOHNSON noted that the marmoset lines are large lymphocytes which contain granules, but that the granules are smaller than those noted in human LGL. HENKART raised the general question of the differences between the various forms of NKCF and lymphotoxins. HISERODT, in response, suggested clear involvement of classical lymphotoxin in various forms of cell-mediated cytotoxicity including natural killing. Antibody directed against highly purified alpha lymphotoxin (Mr of 60-70 kilodaltons) can block lymphotoxin-mediated lysis of K562 cells. HISERODT made analogies between the different Mr forms of lymphotoxin with the current heterogeneity of Mrs reported for NKCFs, and also stressed the need for comparison of classical lymphotoxins wit NKCFs. ORTALDO noted that the lymphotoxin purified and sequenced by GRANGER does not kill K562 cells, whereas the lymphotoxin of EVANS seems very similar to NKCF with respect to its lytic profile. HISERODT had used some forms of lymphotoxin to kill K562 cells. ORTALDO noted that the recDNA-derived material produced by Genentech failed to lyse K562 cells in both his hands as well as their hands. HERBERMAN noted that natural material derived from the 1788 cell line also fails to kill K562 cells. Nevertheless, the situation is confusing since since a monoclonal antibody directed agains the latter lymphotoxin also partially inhibits NKCF. WRIGHT reported that supernatants with high NKCF activity, sent to GRANGER for lymphotoxin assay on L929 cells, had no detectable cytotoxic activity in their lymphotoxin assay. WRIGHT therefore concluded that NKCF may be a molecule related to lymphotoxin which is inactive in the assay, or may be of an entirely different class.

HERBERMAN raised the issue with OELTMANN of whether he was abl to show that the NKCF activity will bind to the mannose receptor-column, and be eluted off with sugars. Although insufficient data to date does not allow for a clear answer, HERBERMAN re-emphasized the significance of the question: it is important to distinguish whether the receptor for mannose-phosphate is on the NKCF or whether it is on the target cells.

Further lively and controversial discussion ensued from a presentation by BROOKS doubting the significance and relevance of NKCF. BROOKS reported that he was completely unable to find any evidence of NKCF production in systems where target cells (eg. YAC-1 cells) are free of mycoplasma. BROOKS has noted NKCF production when he has used cells obtained from WRIGHT, which he claims were mycoplasma contaminated. Upon elimination of mycoplasma, BROOKS claims that all ability to generate NKCF was lost. BROOKS also claimed that upon exchange of reagents with WRIGHT, she too was unable to find NKCF production. BROOKS asserted that putative correlation between NKCF "phenomenology" and NK killing was obtained using preparations of mycoplasma-contaminated cells, and therefore questions the significance of the NKCF phenomena. BROOKS extended his concerns to questions of correlations between NKCF, NK specificity, and the bining of NK cells to targets. BROOKS was concerned that there is no correlation between the ability of tumor cells to induce NKCF under NK sensitivity. BROOKS also voiced concerns with the concept that cells which fail to bind to NK cells are able to induce a mediator of NK lytic activity. BROOKS also suggested that the evidence that NK cells themselves produce NKCF is weak. An additional concern of BROOKS was the differences in the kinetics between the factor system and the direct lytic assay. BROOKS also reported the inability in his laboratory to find any evidence of stable factors from cloned cell lines expressing high levels of NK lytic activity.

WRIGHT reported the investigation of a very large number of _in vitro_ cultured cell lines for both the presence of mycoplasma, and for their ability to stimulate NKCF release. WRIGHT noted that under the appropriate culture conditions, it is of general occurrence among the tumor lines tested that they are able to stimulate factor release. WRIGHT also noted that there are indeed many NK resistant lines that are still resistant but can bind NK cells (eg., RAJI). WRIGHT noted that U937 cells, in a mycoplasma-free state, is a very good stimulator of human NKCF release and is a good stimulator of murine NKCF release since it is NK resistant in the rodent system. WRIGHT noted the need for following strict and critical technical parameters including overnight labelling procedures. WRIGHT also reviewed her findings with NK cell lines, in which she reported the ability to derive highly cytotoxic NKCF. WRIGHT reported that NKCF is released into the culture supernatants of BROOKS'cloned cell lines in her hands. She noted, however, that as for rat tumor LGLs, the supernatants must be dialyzed to detect the cytotoxic activity at high levels. Further issues of debate included the potential variability among YAC-1 cells in different laboratories and issues of NK specificity and "promiscuity" of cell lines. BOLHUIS asked BROOKS by what criterion the clones he employed to check for NKCF was an NK clone. BROOKS noted that the clones are CTL clones with specificity for NK targets. BOLHUIS noted that the ability of CTL to kill NK targets does not qualify a clone to be an NK clone. GOLDFARB noted that the best way to solve this type of issue is to have exchange visits to each laboratory and to perform side by side experiments. BONAVIDA reiterated that BROOKS points are well taken and that WRIGHT had indeed tested a large battery of mycoplasma-free lines. BONAVIDA also extended an invitation to BROOKS to visit his laboratory to settle the issue. ORTALDO agreed with the concept that mycoplasma infected target cells would probably stimulate more NKCF than non-infected cells. ORTALDO reported that the use of mycoplasma-free target cells, as judged by multiple methods and criteria, has led to good production of NKCF by human cells. ORTALDO noted that good NK targets tend to stimulate higher levels of NKCF than poor targets. Considerable discussion was then had concerning issues of mycoplasma culture, assay for mycoplasma under aerobic and anaerobic conditions, and the selection process that accompanies curing of mycoplasma from cell lines, and the need to reinfect cured lines with mycoplasma to complete the cycle of examination. ORTALDO pointed out that the original way in which NKCF was stimulated was with lectin and not with any effector cell type, and therefore NKCF can be stimulated by a mycoplasma-free source.

The next session dealt with the role of proteolytic enzymes in NK killing. HUDIG discussed proteinases as a trigger mechanism in human natural cytotoxicity. HUDIG stressed the problems of using an indirect approach in the absence of isolated components: proteases and natural substrates. Probes need to be both highly specific as well as non-toxic to both lymphocytes and targets. HUDIG's data suggests a role for a serine dependent proteinase in the post-binding events of human natural cytotoxicity. Human NK activity was found to be abrogated by irreversible inhibitors specific for the serine dependent class of proteinases: PMSF and alpha-1 antichymotrypsin. HUDIG observed that NK activity is inhibited by alternate proteinase substrates containing aromatic amino acids, but not by similar compounds containing basic amino acids. HUDIG's data suggests that the NK-associated proteinase resembles chymotrypsin more than trypsin in its specificity of cleavage. Compounds that inhibited NK activity of reshly isolated peripheral blood lymphocytes also inhibited the NK-like activity of the cloned IL-2 dependent cell line 6G12. Alternate substrates did not inhibit the binding of this effector cell line to K562 cells, and do not affect the lysis of targets when introduced after the lethal hit has been initiated. HUDIG interpreted her results as an indication that continued activity of the NK-associated proteinase is not required for final target cell disruption. HUDIG claimed evidence for a role of a chymotrypsin-like enzyme during the initiation of the lethal hit stage, but not for the final stages of target cell death. HUDIG argued that the enzyme is more likely to have a catalytic rather than a direct cytotoxic function in NK killing since incubation of K562 cells with high concentrations of proteinases, including chymotrypsin, had no effect on either ^{51}Cr or disruption of K562 cells. HUDIG noted the same class and substrate specificity of protease requirements as that noted for mast cell degranulation and other degranulation systems. HUDIG therefore hypothesized that the role of proteases in NK killing may function as a control mechanism for degranulation and may not play any role in the final assembly of the final lytic substance. HUDIG has also argued that there may not be a high concentration of substrate within the granules and is consistent with lack of direct protease-mediated digestion of target cells.

GOLDFARB (PFIZER, INC.) discussed the role of neutral serine proteases in the NK lytic mechanism of highly purified human LGL (>95%). One of the approaches taken by GOLDFARB was to determine whether NK cells produce proteolytic activity. It was shown that highly purified populations of human LGL, as well as IL-2 cultured LGL, produce both cell-associated as well as extracellular proteolytic activity. Upon subcellular fractionation it was found that the subcellular distribution of the cell associated enzyme is associated with a cell-surface enriched fraction that copurified with the plasma membrane markers 5'nucleotidase and Na^+, K^+ ATPase. The neutral serine protease proteolytically converted the serum zymogen plasminogen to the active enzyme, plasmin, and is therefore a plasminogen activator. GOLDFARB also reviewed data showing that a number of synthetic protease inhibitors with specificity for both tryptic and chymotryptic-like neutral serine proteases, inhibited the lysis of K562 cells by LGL when assayed under serum-free conditions. The inhibitors of proteases inhibited both spontaneous NK activity as well as IFN-boosted activity of LGL in a dose-dependent manner. In these experiments effector cells were not washed following preincubation with inhibitors since it was found that the inhibitors must be present in the assay. GOLDFARB also reported the results of reciprocal experiments in which exogenously added proteases were examined for their effect on LGL-mediated lysis of K562 cells. At low concentrations, chymotrypsin and trypsin enhanced the level of the killing of K562 cells by LGL. GOLDFARB noted that the use of ng levels of homogeneously purified proteases under serum-free conditions is stimulatory, unlike the higher concentrations of proteases routinely used for the cleavage of cellular receptors that lead to diminished NK activity.

These observations were extended by the examination of several LGL-derived samples for proteolytic activity: NKCF derived from LGL, supplied by ORTALDO; and rat LGL granules obtained from HENKART and REYNOLDS. As a point of comparison these samples were compared to the cytolytic protease of activated murine macrophages of ADAMS, which has been partially purified by GOLDFARB. GOLDFARB reported that HUMAN LGL-derived NKCF is unable to cleave specific fluorescent substrates for plasminogen activator, trypsin, plasmin, elastase, or chymotrypsin. In contrast, GOLDFARB observed that rat LGL granules contain proteolytic activity that cleaves substrates for plasminogen activator, plasmin, and trypsin. Rat LGL granules failed to cleave substrates for chymotrypsin or elastase. It was also noted that rat LGL granular proteolytic activity is inhibited by TLCK, leupeptin, NPGB, Gly-gly-arg-CMK, and to a lesser extent, trasylol, when a number of enzymes are tested on a substrate for plasminogen activators.

TARGAN provided evidence for the role of serine proteases during the NK lethal hit. TARGAN examined TLCK and TPCK as probes to define the mechanism of protease action in NK cytolysis. Pretreatment of either effector cells or target cells had no effect on NK-K562 conjugate formation. Pretreatment of effector cells, but not targets, completely inhibited cytolysis in a standard chromium release assay. TARGAN noted that TLCK blocked beyond calcium dependent programming for lysis in a calcium pulse assay at time points following initial "triggering" events. To investigate the effects of TLCK on killer cell independent lysis, TLCK was added to programmed and dispersed targets in a KCIL assay, and was found to inhibit lysis. TARGAN examined the actions of TLCK in an NKCF assay with L929 cells, and noted that TLCK was able to ihibit lysis. TARGAN suggested that trypsin-like proteases participate in the initial trigger mechanism as well as the terminal events in NK cytolysis. TARGAN also suggested that the transferred NK lytic complex, as well as NKCF, may be dependent upon protease actions for completion of its lytic effect on target cells. TARGAN discussed the interaction between NKCF and KCIL, particularly of the complex and the target cell.

HENKART inquired whether the activity described by GOLDFARB in his assay of granules, basically lysosomal in character, could be due to known lysosomal protease. GOLDFARB noted that it is possible. Nevertheless, in the case of LGL plasminogen activator, the subcellular distribution of the enzyme did not associate with lysosomal markers. It was considered possible that other enzymes within the granules could yield similar activities. GOLDFARB discussed the findings of ZUCKER-FRANKLIN, who has suggested that NK cells carry several enzymes with different substrates thay may be involved in disparate cellular functions. For example, enzymes that mediate cytolysis are not the same as those responsible for proteolysis of specific substrates; LGL elastase was found not to play a role in cytotoxicity. REYNOLDS, in agreement with HUDIG and TARGAN, noted that TPCK does inhibit a post-binding step. HUDIG expressed caution in interpreting experiments done with TLCK and TPCK. GOLDFARB noted the studies of RISTOW, who has also employed TLCK and TPCK and had been concerned with its broad spectrum of activity influencing both proteolytic and non-proteolytic events. GOLDFARB also mentioned the work of PASTERNACK, who has employed TLCK as an affinity probe for functionally important CTL surface molecules, when studied with anti-tosyl antibodies and immunoprecipitation, and who have shown TLCK binding to T 200 in CTL. HUDIG raised the issue of the potential autolytic role of the target cells in contributing to cell death, and noted that effects studied can be on the very last stages, and have little to do with initial stages of the NK lytic mechanism. BONAVIDA expressed caution for those who investigate NKCF or NKCF-like preparations from the viewpoint of the use of various reagents and inhibitors with these preparations. BONAVIDA noted supernatants derived from cells can have many components, and until homogeneously purified NKCF, or lytic proteases are employed, data must be interpreted with caution.

The next topic covered in the Biochemistry of NK cytotoxicity session was the role of oxidative metabolism. RODER revised an earlier hypothesis which had suggested that NK cells generate a chemiluminescence response when mixed with NK sensitive target cells, that is important in the NK cytolytic pathway. RODER now suggests chemiluminescence noted following mixing of NK cells and target cells is not generated by the NK cell, but is ultimately generated by the monocytes. RODER claims that this reaction is NK dependent. In agreement with KAY, RODER has noted that NK cells from CGD patients, which fail to generate chemiluminescence, continue to kill targets normally suggesting that the reaction is not essential in the NK lytic pathway. Another revision suggested by RODER, was that superoxide is not the radical species invovled, but hydroxyl radical might play a necessary role in the NK lytic pathway. RODER reported that monocytes are also involved in the generation of chemiluminescence since treatment of Percoll-enriched LGL with anti-Mo2 and complement abrogated the response and passage of PBL over nylon wool prior to Percoll fractionation also abrogated chemiluminescence. RODER suggested that monocytes alone do not respond to NK sensitive target cells, but synergize with NK ells in the chemiluminescence response. RODER reported that a panel of 13 non-toxic hydroxyl radical scavengers strongly inhibited NK ctyolysis, and that the degree of inhibition of lysis correlated with their rate constants for reacion with hydroxyl radicals. RODER discussed STENZEL's work and suggested that the hydroxyl radical is generated from the lypoxygenase pathway. RODER suggested that the hydroxyl radical is working at the target level, although a role on the NK cell has not been excluded.

GOLDFARB discussed studies examining the potential role of oxidative metabolism by NK cells, and reported that highly purified human LGL (>95%) failed to produce a detectable oxidative burst in response to number of stimuli including PMA, FMLP, digitonin, CON A, A23187, and phospholipase C, under conditions that elicited an appreciable oxidative burst in monocytes. Superoxide production was measured by the reduction of ferricytochrome C and by the reduction of nitroblue tetrazolium. GOLDFARB noted that purified populations of LGL (95%) also failed to produce a detectable oxidative burst in response to K562 target cells. In contrast, K562 cells were found to inhibit the oxidative burst produced by monocytes. GOLDFARB reported that enriched LGL (90%) with few monocytes (1.5%) lacked a detectable oxidative burst, whereas enriched LGL (90%) with a substantial monocytic contamination (8.5%) showed an appreciable oxidative burst in response to PMA.

KAY (U. NEBRASKA) reported that the in vitro cytolytic activity of human NK lymphocytes is independent of oxidative metabolites. KAY examined the NK CMC of lymphocytes from 4 patients with CGD whose neutrophils and monocytes lack oxidative burst capacity, and who showed good NK activity. Kay also studied LGL isolated from 14 healthy donors. Luminol-enhanced chemiluminescence was used to determine production of several oxidative metabolites including superoxide anion, hydroxyl radical, and singlet oxygen. Superoxide anions were detected by reduction of cytochrome C and of nitroblue tetrazolium dye. H_2O_2 production was monitored with the fluorescent dye leucodiacetyl-2-7-dichlorofluorescein. KAY noted that LGL from normal donors with potent NK activity, did not yield a detectable chemiluminescence response when stimulated with K562 cells, zymosan, particles, PMA, endotoxin, or A23187. Stimulated LGLs also failed to reduce NBT dye or cytochrome C. In contrast KAY noted that control monocytes and neutrophils produced oxygen metabolites in all assays. Fluorescent LDDF positive granules were noted in non-stimulated LGLs and K562 cells, but were never triggered by NK cell-K562 interaction. With respect to hydroxyl radical scavengers, KAY has shown that catalase and mannitol failed to inhibit NK CMC. KAY concluded that the human NK lytic mechanism is independent of oxygen intermediates.

BROOKS has also evaluated the role of products of oxygen metabolism in NK lysis. BROOKS reported on studies with human NK activity mediated by PBL, mouse splenic NK activity, and NK activity expressed by cloned cell lines induced by IFN. BROOKS reported that neither catalase or superoxide dismutase, used at concentrations that strongly inhibit the generation of chemiluminescence responses by monocytes or macrophages, caused an inhibition of NK activity in any of the systems evaluated. BROOKS reported on his studies dealing with the issue of whether hydroxyl radicals may be involved in NK lysis. BROOKS noted that the correlation between inhibition of NK activity with rate constants for reactions with some hydroxyl radical scavengers is not observed for all classes of scavangers. BROOKS reported that some of the best hydroxyl scavengers that he studied were totally incabable of inhibiting NK, or CTL, lysis. BROOKS stressed the need for using these inhibitors under isotonic conditions since NK killing is very susceptible to changes in ionic strength. In the NK clone system, BROOKS noted that IFN-induced cells had high NK activity but displayed no sign of chemiluminescence burst during any part of the cytotoxic reaction when assayed by a highly sensitive assay. BROOKS concluded that reactive oxygen intermediates do not play a key role in NK lysis by either uncloned or cloned NK cells.

In response to questions by KOREN dealing with the role of target cell specificity, RODER noted experiments by GREENBERG in which NK cells, upon interaction with NK sensitive target cells, release factor(s) that can trigger a chemiluminescence response by monocytes. Discussions between RODER and BROOKS further explored the issue of scavengers. Discussion on DMSO as a scavenger, and its effect on killing by LGL and in target binding was reviewed by ORTALDO, GOLDFARB, and HISERODT. SEAMAN noted that non-adherent cells not only fail to produce superoxide anion but also fail to produce hydrogen peroxide.

The remaining areas discussed in the biochemistry of NK killing session dealt with additional biochemical pathways, and the ultrastructure of NK cell-target interaction, and target cell damage. HISERODT reviewed the role of phospholipase A_2 and arachidonic acid in the cytolytic activity of human NK cells. Addition of phospholipase A_2 inhibitors BPB, mepacrine, or quercetin, or the lipoxegenase inhibitor NDGA,showed dose dependent inhibition of cytolysis. The cyclooxygenase inhibitor, indomethacin, was found to enhance NK activty. HISERODT concluded that phospholipase A_2 has an important role in the establishment of a strong conjugate between NK cells and their targets, but does not appear to function as a cytolytic effector molecule. HISERODT concluded that lipoxygenase products may play an important role in the early stages of the activation of the NK cytolytic response.

SEAMAN reviewed data dealing with inhibition of NK activity by reactive oxygen intermediates. Upon stimulation by PMA, monocytes can release reactive oxygen intermediates and thereby inhibit NK activity. The major effector species was hydrogen peroxide. In patients with CGD, PMA activated NK activity. SEAMAN also reviewed recent data dealing with the effect of lipoxygenase inhibitors on NK cell activity by fresh or cultured mouse spleen cells. SEAMAN reported that human NK activity is inhibited by antagonists of arachidonate lipoxygenation: NDGA and BW755C. Similar results were noted for inhibition of murine NK killing, including cultured NK cells. SEAMAN concluded that cloned cells can be used to examine the mechanism of NK inhibition by lipoxygenase antagonists and to determine the metabolism of arachidonic acid.

PROSS discussed studies with a patient with a rare disease: mucolipidosis III, with an enzymatic defect in phosphorylation of mannose residues needed for lysosomal incorporation of an acid hydrolase. PROSS concluded that the observed NK cytotoxicity is independent of the phosphorylation of the lysosomal enzyme's mannose residues, and subsequent mannose-6-phosphate dependent transport of these enzymes into lysosomes.

DENNERT described the steps leading to target cell destruction by NK cells. After binding of killer cells to target cells, it was noted that a rapid and coordinate repositioning of two organelles within the killer cell is observed. The microtubule organizing center and the perinuclear Golgi apparatus within the NK cell reposition towards the contact area between killer and target cells. DENNERT proposed that the repositioning serves the function of directing secretory vesicles derived from the Golgi to the binding site. DENNERT noted that the Golgi apparatus might be a packaging plant for cytolytic granules. It appears that the reorientation of organelles requires polymerized microtubules and the presence of calcium. This orientation is noted in the NK cell but not in the target cell. Following the reorientation of granules, granules migrate towards the target binding site and DENNERT suggests that microtubules play a role in the migration of granules to the target binding site.

HISERODT reviewed ultrastructural studies of human NK-target cell interacitons leading to target cell death. Highly purified human LGL and K562 cells were used in the analysis of cytolytic interactions by TEM and SEM electron microscopy. Within minutes following binding HISERODT noted complex interactions with numerous interdigitations that appear to go from the effector cell to the target, and from the target to the NK cell. NK cells participating in the cytolytic interactions developed large uropods with microvillus tufts. By SEM, the target cell changes leading to lysis could be ordered: first, rapid loss of surface microvilli, followed by the appearence of surface blebs. Numerous small holes appear on the target surface which then coalesce into larger defects. Subsequently, HISERODT noted that the target cell begins to disintegrate, with a ruffling of the membrane with large portions of surface membrane appearing stripped or dissolved from the target leading to complete denudation of its plasma membrane. The LGL is left bound to the target to a microvillus structure denuded of plasma membrane and left only with cytoskeletal remnants. HISERODT discussed how LGL, when conjugated with more than one target cell, polarize in one direction and kills one cell before subsequent polarization and killing of additional targets. HISERODT also noted a need for calcium in polarizing LGL towards target cells to inflict a lethal hit.

BONAVIDA reviewed studies dealing with the effect of membrane fluidity on NK cytotoxic function. The effect of cholesterol and phosphatidyl choline were examined for their effect on NK cells, target cells, and on the cytotoxic function of NK cells. Fluorescence polarization was used to determine target cell rigidity. Pretreatment of LGL with cholesterol led to an increase in the frequency of conjugates and a decrease in the frequency of killers. Pretreatment with phosphatidylcholine resulted in a decrease of both conjugates and killers. Treatment of target cells with cholesterol resulted in no decrease in conjugates, but decrease in the frequency of killers. Treatment with phosphatidylcholine resulted in no inhibition. BONAVIDA suggested that the membrane fluidity of both NK effector cells, and target cells, play a role in both target cell recognition and lysis.

The remainder of the session was devoted to discussion. LATTIME discussed similarities between cloned CTL lines and NK cells with respect to their inhibition by sugars and mechanism of action. SAKSELA noted that the mechanism of polarization directly linked to secretion, described for the murine system, was similar to that described by his laboratory in human LGL a few years earlier. SAKSELA pointed out that microtubules in human LGL and in NK like cells are not necessary for the secretory process, and was pleased to note analogies with the murine system described by DENNERT. SEAMAN questioned HISERODT about the holes noted by EM over the target cell surface and the data presented for sequential killing of individual targets when effector cells are bound to 2-3 targets. SEAMAN suggested that holes in only the lysis target argues against a soluble factor engulfing all the targets and generating holes, and an internal event may take place instead. HISERODT speculated that a protective mechanism, or selective focus, may occur.

HERBERMAN raised the issue of whether compounds that inhibit phospholipase A_2 might also inhibit conjugate formation through additional mechanisms as well. HERBERMAN questioned PROSS as to whether the NK cells (LGL) that he described from patients with mucolipidosis III were defective in the enzyme that he described. With respect to the presentations dealing with reorientations, HERBERMAN asked whether DENNERT, or others, had any formal evidence to show that the granules themselves are reorienting. DENNERT noted that there is no evidence other than that provided by EM sections that always show granules concentrated by the side. With respect to holes over large areas of the target cells, DENNERT suggested that following tight binding over a large area between killer cells and targets, the target breaks away but the cytolytic molecule may already be transplanted with the target membrane. HISERODT noted that at the site of conjugation coated pits are seen on the target cells, suggesting active membrane movement on the target which is being injured. HISERODT raised the possibility of a great deal of membrane movement from that cell with resultant global distribution on that target which would protect, or not affect, another target. PROSS, in response to HERBERMAN, noted that LGL have not been examined to date, and most of the enzyme data has been derived from studies with fibroblasts. In response to GOLDFARB, PROSS noted that there were no observable morphological alterations in LGL. SAKSELA noted that one child in Finland with the same disease displayed completely normal NK activity. HENKART questioned the calculations of HISERODT in the determination of "collarette" holes since they appeared to HENKART to be of different size than the ring structure he has described. REYNOLDS asked for clarification, from DENNERT, of the need for calcium dependence for reorientation in conditions of killer cell dependent and killer cell independent stages of lysis. HENKART explored issues of rearrangement in the presence or absence of calcium, and optimal methods for removal of calcium.

It is clear from the above accounts that important advances have been achieved in our understanding of the biochemistry of NK cytotoxicity. With the availability of highly purified LGL, which account for NK cell-mediated cytotoxicity, rapid progress has been made in elucidating biochemical aspects of NK cell function. The identification of cloned cells with NK properties, and the identification of tumor LGLs allow for sufficient material for biochemical studies. The delineation of the discrete steps in the NK lytic mechanism provides a framework for understanding the significance of biochemical pathways, granules, and cytotoxic factors from the perspective of: target cell recognition; effector-target cell interaction; triggering; programing for lysis; secretion; and the substages of the lethal hit including effector phase lytic complex formation, activation, and the lytic phase leading to target damage and destruction.

NK-TARGET CELL INTERACTIONS IN BINDING, TRIGGERING, PROGRAMMING, AND LETHAL HIT STAGES OF NK CYTOTOXICITY[1]

Stephan R. Targan
Richard L. Deem

Geriatric Research, Education and Clinical Center
Wadsworth VA Medical Center
Los Angeles, California
and Departments of Medicine, and Microbiology and Immunology
UCLA School of Medicine
Los Angeles, California

I. INTRODUCTION

Studies have shown that NK cytotoxicity can be resolved into several stages: NK-target cell binding, triggering, programming and killer cell independent lysis (KCIL), or the NK lethal hit (1, 2). The early stages of NK cytolysis can be studied kinetically by using the calcium pulse technique (1). By comparing the kinetics of lysis to the kinetics of inhibition of EDTA, one can obtain a measure of the calcium-independent events, or KCIL. KCIL can also be determined directly by using the EDTA-dextran dispersion technique (1). NK cells produce soluble cytolytic factors (NKCF) upon effective interactions with appropriate target cells (3, 4) and have been proposed to be part of the NK lethal hit of direct cell-mediated lysis (5). These factors can lyse NK-sensitive targets and have been used as another measure of the NK lethal hit (3, 4). These studies were designed to examine the mechanisms of NK cytotoxicity with primary emphasis on the terminal stage, the lethal hit.

---

[1]Supported by United States Public Health Service Grants CA-37205 and AI-15332 and Veteran's Administration research funds. Stephan R. Targan is recipient of a VA Clinical Investigator Award.

ISBN 0-12-341370-2

II. RESULTS AND DISCUSSION

A. NK-Target Cell Binding

NK-target cell binding is temperature independent, since it has been shown that it will occur rapidly either at 4°C or 37°C (1). It has also been shown that conjugate formation requires the presence of magnesium but not calcium. Thus, EDTA, which chelates magnesium, but not EGTA, which does not chelate magnesium, blocked formation of target-effector conjugates (Table I) (1). Conjugate formation was prevented in calcium-magnesium-free HBSS, but upon addition of magnesium, binding occurred at control levels (1).

Conjugate formation requires intact receptors on both the NK cell and the target cell, as well as an intact NK microtubule system. Thus, monoclonal antibody with specificity for NK cells, RH2 7.2, was able to inhibit conjugate formation (Table I) (6). Protein cross-linking agents, glutaraldehyde (7), crotonaldehyde, and dimethyl suberimidate (DMS), as well as the membrane-modifying agents, dimethyl sulfoxide (DMSO) and 2-mercaptoethanol (2-ME) (8), were also able to modify the NK cell to prevent conjugate formation (Table I). Modification of the target cell binding receptor and NK binding receptor can be achieved with the proteolytic enzymes, trypsin, chymotrypsin, papain (5), and bromelain (Table I). Thus, pretreatment of K562 target cells, and pretreatment of NK cells, resulted in reduced conjugate formation and inhibition of cytotoxicity. Glutaraldehyde has also been shown to

TABLE I. Effect of pharmacologic agents upon NK-K562 binding

Reagent	% Binding[a]	Control Binding[b]
2.5 mM EGTA	9 (18)	11 (1)
10 mM EDTA	1 (91)	11 (1)
100 mM DMSO	2 (83)	12 (8)
5.0 mM 2-Mercaptoethanol	1 (92)	12 (8)
1:2 RH2 7.2 Ab	4 (64)	11 (6)
0.1 % Trypsin	2 (78)	9 (5)
0.1 % Chymotrypsin	2 (78)	9 (5)
2.5 u/ml Papain	1 (89)	9 (5)
2.5 mg/ml Bromelain	1 (86)	7
.05 ug/ml Glycopeptide A	8 (50)[c]	16 (10)
1.0 mM Glutaraldehyde	1 (92)	12 (7)
3.0 mM Crotonaldehyde	2 (83)	12
2.0 mM Dimethyl Suberimidate	1 (91)	11

[a]Percent inhibition in parentheses.
[b]Reference in parentheses.
[c]Percent binding 45 min after K562 added to PBL.

inhibit NK-target cell binding by altering the target cell (7). The target cell binding receptors are probably glycoproteins (9), and it has been shown that a glycoprotein isolated from K562 targets, glycopeptide A, dissociated NK-K562 conjugates (Table I) (10). Conjugate formation also requires intact microtubules, since cytochalasin B, a reversible microtubule disrupter, inhibited conjugate formation (11).

B. NK Triggering

The first event to occur following the initial NK-target cell interaction has been shown to be NK triggering (2). The NK cell membrane structure involved in this K562 target cell interaction has been functionally defined, by a monoclonal antibody termed 13.3, as an epitope of the T-200 molecule (12). It appears as if the function of this structure is closely linked to calcium, in that two minutes following addition of calcium to NK-target cell conjugates, 13.3 can no longer block lysis. The question arises as to what this calcium link is and what subsequent events in the NK cell or target cell results from this interaction prior to programming for lysis.

Other investigators have demonstrated that one early event of interferon activation of the lytic process is the increase of phospholipase A_2 (PA_2) production (13). Since interferon enhances the kinetics of initiation or completion of programming (1), PA_2 may be important in early lytic events. In addition, inhibitors of PA_2 function have been shown to inhibit NK cytolysis only early post binding (J.C. Hiserodt, personal communication). Because of these findings, we felt it would be advantageous to determine what effect exogenous PA_2 has upon NK lysis. Table II summarizes our recent findings on the effect of added PA_2 in various assays, which are capable of measuring the precise stages of NK lysis affected by this enzyme (1, 4, 8). PA_2 enhances at the level of the target cell (Table II, line 3). It not only enhances binding, but post-binding events (Table II, lines 4-6). In a calcium pulse

TABLE II. Effect of PA_2 on the NK lytic reaction sequence

Assay	PA_2 Treatment	Stage of NKCC	Effect on Lysis
NKCC	In Assay	- - -	increases
NKCC	LGL	- - -	none
NKCC	K562	- - -	increases
Single Cell	K562	Binding	increases
Single Cell	K562	Post-Binding	increases
Ca^{++} Pulse	In Assay	Programming	increases (very early)
NKCF Release	K562	Triggering	increases

TABLE III. Effect of 13.3 antibody upon NK cytolysis of PA_2-pretreated and untreated K562 target cells

K562 Treatment	Fold Inhibition of Lytic Units
None	40
PA_2	40

assay, it enhances only early (Table II, line 6), suggesting that PA_2 affects only early post-binding events in direct cytolysis. Thus, in direct killing, PA_2 appears to enhance expression of a target cell structure capable of initiating calcium-dependent programming[2].

Further evidence[2] that PA_2 uncovers a structure similar to the target trigger present on untreated K562 is shown in Table III. These results show similar blocking of cytolysis by 13.3 monoclonal antibody using both untreated and PA_2-pretreated K562 target cells (Table III). Since 13.3 blocks only the T-200 epitope which interacts with K562, these results suggest that the trigger exposed by PA_2 treatment goes through this same K562 trigger process on the NK cell.

Some recent results have indicated that NKCF may be one of the molecules involved in the putative lethal hit in direct NK-mediated lysis (5). Thus, release of this factor is dependent upon the completion of the lethal hit (4), and would be a relative measure of lytic capability. Results comparing PA_2-treated K562 targets to untreated targets (Table II, line 7) demonstrated that there was an enhanced release of NKCF with PA_2-treated targets. These results suggest that there was enhanced stimulation of triggering by NK effector cells interacting with PA_2-treated targets.[2]

All of these results[2] suggest that effective lytic triggering results from more than a single event. In addition, they suggest that the initial interaction of the NK cell with the target cell causes a NK cell-associated event (i.e., activation of PA_2) to interact and alter the target cell membrane (i.e., phospholipid alteration) such that this now allows a functional interaction between the altered target and NK effector to create a calcium flux into the NK cell and initiate subsequent programming for lysis.

---

[2]Targan, S.R. and L. Britvan. 1984. The use of phospholipase A_2 to further analyze the NK lytic triggering mechanism. In submission.

C. Programming for Lysis

Immediately after the trigger stage is calcium-dependent programming. Although NK-target cell conjugates can be held in media containing EGTA plus magnesium for up to 3 hr at 37°C with no decline in target cell binding, no target cell lysis occurs (1). However, lytic events proceed as soon as calcium is added. EDTA is added kinetically to the calcium pulse to determine the length of calcium-dependent programming. Thus, at early time points in a calcium pulse assay, EDTA completely inhibits lysis, and at later time points EDTA has no effect upon subsequent lysis, even though lysis is not yet completed (i.e., calcium-independent events) (1). By comparing the kinetics of inhibition of EDTA with the kinetics of inhibition of various pharmacologic agents, insights can be obtained concerning the biochemical events occurring during NK cell-associated processes which program a target cell for subsequent lysis.

The calcium dependence of programming for lysis was confirmed by the action of lidocaine and the calcium channel blocker, vermapamil, in calcium pulse assays (8). Both reagents failed to affect binding and KCIL, but did inhibit early stages of programming (8). Both reagents affect ion gradients by interfering with calcium flux, strongly suggesting the requirement for a calcium flux into the NK cell for completion of programming.

Interferon-pretreatment of PBL significantly enhances not only the level of NK lysis, but also the rate of lysis (1). By adding EDTA to interferon-activated and control PBL in calcium pulse assays, it was shown that the increase in kinetics of lysis was due solely to increases in the rate of calcium-dependent programming (1). The rate of lysis after addition of EDTA (during calcium-independent events) was the same with interferon-treated and control NK effectors. Thus, interferon enhances NK lysis by recruiting pre-NK cells to become active NK cells (14, 15), increases the rate of NK cell recycling (14, 15), and increases the kinetics of lysis by increasing the rate of calcium-dependent programming (1).

Glutaraldehyde (7), crotonaldehyde, dimethyl suberimidate, osmium tetroxide, and other chemical cross-linking reagents (Deem and Targan, unpublished data) inhibit in calcium pulse assays at the calcium-dependent stage of programming, although they do not dissociate conjugates (7). Although this inhibition could be due to the inhibition of release of lytic factors, it has been shown that glutaraldehyde (16), crotonaldehyde, and osmium tetroxide[3] alter the NKCF receptor to prevent binding of the lytic factors to

---

[3] Deem, R.L. and S.R. Targan. 1984. Analysis of the mechanisms of chemical cross-linking-mediated enhancement of the NK lethal hit. In submission.

the target cell. Thus, there is a late stage of programming or early stage of the lethal hit that involves binding of NK cell-derived lethal factors to the target cell.

The reducing agent 2-ME inhibits in calcium pulse assays beyond the calcium-dependent stage of programming but does not inhibit KCIL effectively (8). Thus, 2-ME probably inhibits too early after lytic factor binding to be detected in KCIL assays.

Monoclonal antibodies (RH2) made against human cytotoxic lymphocytes have been shown to inhibit NK cytolysis (6). Calcium pulse assays demonstrated that RH2 7.2 and RH2 17.2 antibodies inhibited calcium-dependent and calcium-independent events in the absence of complement. In addition, these antibodies did not dissociate conjugates, suggesting that they affected a NK-derived protein that was involved in cytolysis. KCIL assays showed that these antibodies inhibited the lethal hit, suggesting that this NK-derived protein was transferred to the target cell.

D. NK Lethal Hit

The NK lethal hit is that portion of the cytotoxic mechanism that occurs after the target cell has been programmed to lyse and is completely independent of the NK cell. Three types of assays have been used to examine the lethal hit in this study. The calcium pulse not only provides information on programming, but can be used as a screening assay to test the effect of reagents upon KCIL. The EDTA-dextran dispersion technique measures KCIL directly, but has the disadvantage of removing metal ions, which may be required for the activity of certain enzymes or lectins being tested. Because of this, NKCF assays have been used as another measure of the NK lethal hit (3, 4). Although NKCF acts much more slowly than direct NK cytotoxicity, results from KCIL assays and NKCF assays have been in agreement (5, 7, 8, 16). In addition, NKCF assays are much more easily manipulated and the kinetics are such that analysis of the sequential events are more easily studied.

1. Evidence for the Transfer of Protease Sensitive Structures During the NK Lethal Hit

Preliminary evidence indicated that NK cytotoxicity could be inhibited by certain proteolytic enzymes, trypsin, chymotrypsin, and papain. The kinetics of inhibition by these enzymes during a calcium pulse assay indicated that they inhibited the terminal stages of NK cytotoxicity (17). KCIL assays confirmed that all three enzymes inhibited the lethal hit (5). These findings indicated that protease-sensitive structures from the NK cell were transferred to the target cell during NK cytotoxicity.

To further analyze the mechanism of inhibition of these enzymes, NKCF assays were preformed. Only trypsin and chymotrypsin inhibited NKCF-mediated cytolysis (5), indicating that papain was probably affecting a component of the direct NK lethal hit mechanism not present in NKCF-mediated lysis. Cold target absorption and cold target competition assays indicated that trypsin, chymotrypsin, and papain pretreatment of K562 target cells had no effect upon the ability of those cells to absorb NKCF (5). These results suggested that these enzymes did not inhibit the lethal hit by removing NKCF receptors. Pretreatment of NKCF with trypsin and chymotrypsin and subsequent neutralization of the enzymes with soybean trypsin inhibitor (SBTI) dramatically reduced the lytic ability of the NKCF, although SBTI itself had no effect upon NKCF activity (5). Papain pretreatment of NKCF had no effect upon NKCF-mediated lysis (5).

These results suggest that the various structures involved in the lethal hit have differential susceptibilities to proteolytic enzymes. Although KCIL is inhibited by all three enzymes tested, NKCF subunits are only sensitive to trypsin and chymotrypsin, suggesting that papain affects a component not present in NKCF. This hypothesized "focusing molecule" is thought to be involved in the assembly of NKCF subunits on the target cell membrane during the lethal hit. The existance of this focusing molecule is further supported by the ability of RH2 antibody to inhibit KCIL and yet to be removed by absorption on purified NK cells (6). The lack of this focusing molecule in NKCF preparations may account for the slow kinetics of lysis observed during NKCF-mediated cytolysis.

2. Kinetic Model for the NK Lethal Hit

Early studies on the NK lethal hit indicated that KCIL is temperature sensitive (1). Thus, when preprogrammed K562 target cells were dispersed into EDTA and dextran and incubated at 4°C, no KCIL was observed (1). But, if after this incubation period, these cells were warmed to 37°C, the amount of KCIL returned to control levels (1). It was also shown that PGE_2 inhibited KCIL (8). Three hypotheses were consistant with these findings. Since low temperatures are known to reduce or eliminate enzyme activities, it was proposed that the lytic molecules may have enzymatic activity. Both low temerature and PGE_2 have effects on membrane fluidity/movement, which may play a role during the lethal hit. Also consistant with these results may be the requirement for active metabolic processes of the target cell for completion of the lethal hit.

Glutaraldehyde, a bifunctional protein cross-linking reagent, was chosen to examine the role of membrane movement/fluidity during the NK lethal hit. When 100 μM glutaraldehyde was added into an NKCF assay, cytolysis of L929 target cells was inhibited 64% (16). Pretreatment of L929 targets with 100 μM glutaraldehyde inhibited NKCF-mediated cytolysis 100%, but pretreatment of NKCF had little or no effect (16), indicating that glutaraldehyde was affecting the target cell directly. The cold target competition assay (5) was performed to determine if glutaraldehyde was altering the NKCF receptor. Two to four fold more glutaraldehyde-pretreated K562 target cells were required to achieve the same inhibition observed with untreated cold targets (16). Thus, glutaraldehyde-pretreated K562 and L929 cold target cells were not able to compete as well as untreated cold target cells to inhibit lysis of labeled L929 targets (16). To comfirm these results, cold target absorption of NKCF was performed at 4°C, to eliminate the possibility that glutaraldehyde was inhibiting endocytic uptake of NKCF. Again, glutaraldehyde-pretreated K562 target cells were not able to absorb NKCF activity as well as untreated target cells (16). Thus, the earliest substage of the NK lethal hit, lytic complex/NKCF binding, was inhibited by glutaraldehyde. This inhibition was due to its direct cross-linking effect upon the NKCF receptor.

Although glutaraldehyde inhibited binding of lytic factors to target cells, once these factors became bound, lysis was enhanced by the addition of glutaraldehyde (7, 16). It was shown that KCIL was enhanced by 13 ± 3% lysis (7). In addition, enhancement of lysis was directly proportional to the amount of KCIL exhibited (7). However, this enhancement was not nonspecific, but occurred only during an early substage of KCIL (within 5 min after dispersion of conjugates) (7). Likewise, glutaraldehyde enhanced NKCF-mediated cytolysis, but only when added during a critical time period (1 to 4 hr after addition of NKCF), after NKCF binding, but well before initiation of lysis (16). Thus, there is a second substage of the lethal hit which occurs after NK lytic complex/NKCF binding, which is enhanced by the addition of glutaraldehyde. This substage probably represents assembly or activation of the lytic complex, which is facilitated or stabilized by cross-linking with glutaraldehyde.

Two possible hypotheses are proposed to explain glutaraldehyde-mediated enhancement of NKCF-mediated cytolysis. First, glutaraldehyde may enhance lysis by stabilizing or preventing repair of the NKCF-produced lesion. This hypothesis seems unlikely, due to the nature of the critical time period for glutaraldehyde enhancement. If glutaraldehyde were affecting the NKCF-produced lesion, it would be expected that the critical time period would be much later than that observed (i.e., well after NKCF binding and closer to the beginning of lysis). Additionally, it has been observed that when NKCF is bound to L929 target cells at 4°C, then removed and replaced with glutaraldehyde that lysis is enhanced maximally (Deem and Targan, unpublished data). It is unlikely that any lesion would be produced at 4°C, since the NK lethal hit is temperature dependent (1).

It was originally proposed that glutaraldehyde might enhance NKCF activity directly, as has been shown for lymphotoxin (17). However, it has been shown that glutaraldehyde does not affect the lytic capacity of nonmembrane-bound NKCF (16). Thus, the second hypothesis is that glutaraldehyde stabilizes or enhances the lytic capacity of membrane-bound NKCF. This model would more closely fit the observed critical time point for glutaraldehyde enhancement of NKCF-mediated cytolysis. It is possible that glutaraldehyde affects membrane-bound NKCF only before the NKCF becomes "activated" for lysis. Alternatively, glutaraldehyde may stabilize the membrane-bound NKCF to prevent target cell-mediated inactivation or spontaneous disintegration of the membrane-bound NKCF.

Following the glutaraldehyde enhanceable stage is a stage during which little or no lysis occurs, and yet glutaraldehyde is unable to enhance lysis. In KCIL, this stage occurs 5 to 15 min after dispersion of target-effector conjugates (7). In an L929 assay, this "effector" stage occurs 5 to 10 hr after addition of NKCF (16).

The terminal stage of the NK lethal hit, the lysis stage, is also unaffected by glutaraldehyde. In KCIL, the majority of lysis occurs between 15 and 30 min after conjugate dispersion and is completed by 45 min (1). When glutaraldehyde was added to preprogrammed K562 target cells, 50% of total KCIL occurred between 15 and 30 min after conjugate dispersion (7). Thus, the kinetics of lysis of glutaraldehyde-enhanced KCIL are nearly identical to the kinetics of KCIL without the addition of glutaraldehyde. NKCF-mediated lysis is much slower, with nearly all lysis occurring 11 to 14 hr after addition of NKCF (S.R. Targan, unpublished data). However, glutaraldehyde can significantly accelerate the rate of NKCF-mediated lysis (16), indicating that its effects may represent more than simple stabilization of membrane-bound NKCF.

In summary, the NK lethal hit can be divided into 4 substages based upon modulation by glutaraldehyde (Table IV). The first substage, the lytic factor binding stage, is inhibited by glutaraldehyde. Although programming for lysis was inhibited by addition of glutaraldehyde (7), it has not be determined if this was due to inhibition of triggering, inhibition of lytic factor release or alteration of the lytic factor binding receptor. It is clear that glutaraldehyde blocks NKCF-mediated lysis early

Table IV. Functional Substages of the NK Lethal Hit

	NKCF-mediated Cytolysis		Killer Cell Independent Lysis	
Substage	Time (hr)	Glut. Effect	Time (min)	Glut. Effect
1. Binding	0 - 2	inhibits	n/a	inhibits?
2. Assembly/ Activation	2 - 5	enhances	0 - 5	enhances
3. Effector	5 - 10	none	5 - 15	none
4. Lysis	10 - 15	none	15 - 45	none

cross-linking the NKCF receptor to prevent NKCF binding. The second substage, assembly/activation, occurs immediately after lytic factor binding (the beginning of KCIL or 2 hr after addition of NKCF) and is enhanced by glutaraldehyde. This enhancement of lysis is specific and is directly related to the cross-linking time of various concentrations of glutaraldehyde (7). Following this stage is the effector substage, during which little or no lysis occurs, but is not affected by glutaraldehyde. The final substage, lysis, is also unaffected by glutaraldehyde.

3. Analysis of Chemical Crosslinking-Mediated Enhancement of the NK Lethal Hit

In order to begin to analyze the mechanism by which glutaraldehyde enhanced lysis of the assembly/activation substage of the NK lethal hit, and define what types of molecules may be involved in this modulation, several specific chemical cross-linking agents were chosen for study. Malondialdehyde (MDA), a bifunctional cross-linking aldehyde, enhanced NKCF-mediated cytolysis 177%, but had no effect upon KCIL (18). MDA is a smaller molecule than glutaraldehyde and cross-links over shorter distances. In addition, it cross-links more slowly and less effectively than glutaraldehyde (19). This probably accounts for the inability of MDA to enhance KCIL, since the critical time period for enhancement of KCIL is only 5 min (7). Monofunctional aldehydes also cross-link slowly and incompletely (19). Thus, formaldehyde and acetaldehyde had little or no effect upon KCIL or NKCF-mediated cytolysis (18). Crotonaldehyde, a monofunctional unsaturated molecule, is more reactive than other monofunctional aldehydes (20), but had no effect upon KCIL, although it did inhibit NKCF-mediated lysis 42%, probably by altering the NKCF receptor (18). The noncross-linking aldehydes, butryaldehyde and valeraldehyde, had no effect upon KCIL or NKCF-mediated cytolysis (18). Thus, it appears that cross-linking and not just the presence of an aldehyde functional group is required for enhancement of lysis during the NK lethal hit. It is also likely that the speed of cross-linking is important, especially in KCIL, and that a certain minimum cross-linking distance may be required for enhancement of the lethal hit, suggesting that intermolecular cross-linking may be required. (The bifunctional reagents cross-link more effectively and over a longer distance than monofunctional reagents and enhance lysis more effectively.)

Although glutaraldehyde cross-links primary amino groups, particularly lysine, it can also cross-link other functional groups, such as tyrosyl, histidyl, and sulfhydryl groups (21). Because of this lack of specificity by aldehyde cross-linking reagents, more specific reagents were tested. The bis-imidates are very specific protein cross-linking reagents which react only with primary amino groups (22). All the bis-imidates tested, DMS, DMP, and DMA, enhanced KCIL slightly but inconsistantly (25 to 44%) (18). This is probably due to the reaction rate of the imidates with proteins. Although the imidates react within a few minutes, the intermediate products which are formed, have a half life of 5 to 30 min (22). Thus, it is likely that most cross-links are not formed during the first critical 5 minutes of KCIL. However, all the bis-imidates tested enhanced NKCF-mediated cytolysis nearly as well as glutaraldehyde (139 to 190%) (18).

Osmium tetroxide enhanced KCIL, although variably, and inhibited NKCF-mediated cytolysis 71%, probably by affecting the NKCF receptor (18). Although osmium tetroxide primarily cross-links unsaturated lipids, it can cross-link or cleave (by oxidation) proteins (23), and is especially reactive toward the amino acids tryptophan, cysteine, and histidine (24). Therefore, whether or not lipid cross-linking is involved during enhancement of the lethal hit remains to be determined.

Lectins can cross-link biological membranes by binding to glycoproteins. Concanavalin A (con A, specificity: D-mannose, D-glucose) did not enhance KCIL (18), but this would be expected, since its cross-linking activity requires calcium and manganese (25), which are removed during the EDTA dispersion technique. However, NKCF-mediated cytolysis was enhanced significantly (168%) (18). This enhancement of NKCF-mediated cytolysis appears to be specifically related to the cross-linking ability of con A, since a-methyl-D-mannoside can inhibit con A enhancement, whether added concurrently with con A or added to L929 target cells that have been pretreated with con A (Deem and Targan, unpublished data). Likewise, wheat germ agglutinin (WGA, specificity: N-acetyl-D-glucosamine) enhanced NKCF-mediated lysis 135%, whereas lectins with specificty for D-galactose, peanut agglutinin and soybean agglutinin had no effect (18). Thus, it appears that cross-linking of the target cell membrane with specificity for glucose residues may be important during enhancement of the lethal hit.

The cross-linkers glutaraldehyde, DMS, and DMA demonstrated an early critical time period (0 to 4 hr after NKCF binding), only during which time, they were effective at enhancing NKCF-mediated cytolysis[3]. The other enhancing cross-linkers, DMP, con A, WGA, and MDA, did not show this same early critical time period, since enhancement of lysis was still approximately 60% of the maximum value at 6 hr[3]. Thus, it appears that these reagents may enhance NKCF-mediated lysis at different or multiple substages of the lethal hit.

Three of the four cross-linking agents (con A, WGA, and DMP) that failed to demonstrate an early critical time period for enhancement of NKCF-mediated cytolysis were also able to modulate the target cell through pretreatment to make them more sensitive to lysis by NKCF (61 to 135% enhancement of lysis) (18). These results suggest that these agents may enhance NKCF-mediated lysis at more than one substage. Cold target competition assays were performed to determine if these reagents were increasing binding or uptake of NKCF by the target cell to increase NKCF cytolysis. The results indicated that this was not the mechanism of enhancement and that although the target cells were pretreated with the agents, they were affecting later substages of NKCF-mediated cytolysis[3].

There are several hypotheses that can explain these results. Cross-linking of cell membranes causes patching and capping of the affected areas. If NKCF receptors are involved in this movement, it is possible that they may aggregate to form a focus, at which point NKCF may polymerize more easily to become a more effective complex. By speeding up this polymerization process, even the kinetics of lysis would be increased, as observed in this study. There is recent evidence that suggests that NKCF-mediated cytolysis may involve a polymerization process, since 2-mercaptoethanol and dithiothreitol inhibit NKCF cytolysis, but only when added soon after addition of NKCF (18).

Once cell membranes patch and cap, these areas of the membrane are often internalized by the cell. If internalization of NKCF is a substage of NKCF-mediated cytolysis, then cross-linking may enhance this substage of the process. Further, it has been suggested that protein cross-linking by transglutaminase may be required for receptor-mediated endocytosis of some protein and polypeptide hormones (26). Several lines of evidence suggest that endocytosis or internalization of NKCF may be involved in completion of NKCF-mediated cytolysis. Inhibitors of energy metabolism, cyanide, 2-deoxyglucose, and dinitrophenol, which inhibit endocytosis, inhibit NKCF-mediated cytolysis (18). Likewise, cytocholasin B, an inhibitor of microtubules also inhibits NKCF-mediated lysis[4]. These reagents also inhibit glutaraldehyde-enhanced NKCF-mediated cytolysis, but must be added prior to the addition of glutaraldehyde to be completely effective[4].

Cross-linkers may also enhance NKCF-mediated cytolysis by cross-linking NKCF subunits into a more effective lytic complex on the target cell membrane. Although this is not the only possible mechanism for enhancement by cross-linking reagents such as DMP, Con A and WGA, it may represent a portion of NKCF-mediated lysis

---

[4]Targan, S.R. and A. Niederlehner. 1984. Energy metabolism and intact microtubules in target cells are involved in the NK lethal hit. In submission.

that can be enhanced by chemical cross-linking agents. To this end, it has also been shown that some cross-linkers can by pre-treatment make NKCF a more effective lytic compound[3]. Although this treatment enhances NKCF activity only a fraction of the amount that occurs when cross-linkers are added directly into an NKCF assay, it does suggest that a polymeric NKCF structure is a more effective compound. Thus, NKCF may represent subunits of the lytic molecule secreted during interaction of NK cells and susceptible targets.

These results suggest that the assembly/activation substage may involve an association of cell-free proteins or glycoprotein subunits with multiple protein and/or glycoprotein structures on the target cell membrane. The aggregation of these structures by chemical cross-linking agents facilitates this stage of lysis and implies that formation of a "complex" on the target cell membrane may be critical for optimal lysis to occur.

4. Effect of Pharmacologic Agents Upon Substages of the NK Lethal Hit.

As noted previously, the lytic factor binding stage is inhibited by cross-linking agents, which can modify the NKCF receptor. These reagents include glutaraldehyde (16), osmium tetroxide, and crotonaldehyde (18).

Reagents that break disulfide bonds, such as 2-mercaptoethanol (2-ME) and dithiothietol (DTT), inhibited NKCF-mediated lysis nearly completely (95 to 97%) and KCIL partially (42 to 45%)(18). In addition, 2-ME inhibits beyond calcium-dependent events in a calcium pulse assay (8), suggesting that 2-ME inhibits at a very early stage of KCIL. Kinetic addition of 2-ME into an L929 assay showed that inhibition of lysis decreased almost immediately after addition of NKCF, indicating that these reagents affect an early process of NKCF-mediated lysis (18). This early inhibition is even more evident during glutaraldehyde-enhanced NKCF-mediated lysis, where inhibition decreased from nearly 100% to less than 10% in 5 hr (18). Thus, it appears that 2-ME and DTT inhibit during an early phase of the assembly/activation substage. Since KCIL is so rapid, it seems likely that these reagents do not totally inhibit KCIL because this stage occurs within minutes after conjugate dispersion.

To test whether intact energy metabolic processes are required in the target cell during the NK lethal hit, several inhibitors of glycolysis, oxidative phosphorylation, and electron transport were chosen for study. These inhibitors had little or no effect upon KCIL (18), probably beacause depletion of intracellular ATP levels requires several hours (27), whereas KCIL is completed within 45 min. All the reagents tested inhibited NKCF-mediated cytolysis 30 to 80% (18). The effects of pretreatment of L929 target cells with dinitrophenol and fluoride were totally reversible, whereas the effects of 2-deoxyglucose and cyanide were only partially reversible (18). Analysis of the kinetics of inhibition of these metabolic inhibitors has shown that they inhibit maximally through 4 hr[4]. Thus, they probably inhibit until late in the assembly/activation substage of the lethal hit, although this is uncertain because of the delay of action of these agents. Thus, it seems likely that some energy requiring process, possibly involving membrane movement or endocytosis of the lytic complex, may be necessary for completion of the lethal hit.

It has been shown that inhibitors of serine protease activity can inhibit CTL cytotoxicity (28) and ADCC (29), but the stage at which these inhibitors act has not been adequately defined. All inhibitors of serine protease activity tested, TLCK, TPCK, and TLME inhibited KCIL nearly completely (18). TLCK and TPCK (irreversible alkylating agents) inhibited NKCF-mediated lysis, whereas the competitive trypsin inhibitor, TLME, had little effect (18). Pretreatment of L929 target cells with these inhibitors partially inhibited NKCF-mediated cytotoxicity (18), indicating that a target cell originating protease may be involved in the lethal hit. In addition, it has been found that pretreatment of NKCF with these inhibitors and dialysis had no effect upon its ability to lyse L929 target cells[5]. Kinetic analysis of inhibition of NKCF-mediated lysis has shown that TLCK inhibits maximally through 6 hr and continues to inhibit until the beginning of lysis[5]. These data suggest that protease activity is required late in the NK lethal hit, probably during the effector substage.

Since cross-linking of the target cell membrane resulted in enhancement of the lethal hit, it was postulated that the opposite effect, increasing membrane fluidty, might inhibit the NK lethal hit. The neutral anesthetic, benzyl alcohol was chosen because of its specific fluidizing effect upon the lipid bilayer of biological membranes (30). This reagent inhibited direct NK cytotoxicity

[5]Brogan, M. and S.R. Targan. 1984. The role of proteases in the terminal stages of NK lysis: comparison of direct cell-mediated and NKCF-mediated lysis. In submission.

completely, although pretreatment of target or effector cells had no effect[6]. Likewise, both KCIL and NKCF-mediated lysis were inhibited, and pretreatment of L929 target cells had no effect upon NKCF-mediated lysis (18). Analysis of the kinetics of inhibition of benzyl alcohol showed that it completely inhibited NKCF-mediated cytolysis through 6 hr[6]. In addition, a benzyl alcohol pulse experiment has shown that addition (for up to 5 hours) and removal of benzyl alcohol to L929 targets prebound 2 hr with NKCF had little effect upon subsequent lysis[6]. Thus, benzyl alcohol effects only the late substages of the NK lethal hit (effector and lysis substages). Since NK lysis may involve formation of transmembrane channels (31), it is possible that benzyl alcohol's fluidizing effect may disrupt these channels to prevent lysis. However, benzyl alcohol's membrane fluidizing effects have no effect upon complement-mediated lysis[6]. Alternatively, benzyl alcohol may inhibit some required enzymatic activity by changing conformations of integral membrane proteins, as has been shown to occur for the enzyme adenylate cyclase (32). Benzyl alcohol has little effect upon free NKCF[6], but may alter the target cell membrane once NKCF becomes bound to prevent its further activity. In addition, benzyl alcohol has been shown to inhibit fusion of endocytic vesicles (33), which may be required for completion of the NK lethal hit.

III. SUMMARY

NK cytotoxicity consists of 4 stages: binding, triggering, programming, and lethal hit (Figure 1). Binding ① is magnesium dependent, and calcium and temperature independent. Binding requires intact receptors on both the NK cell and target cell, since both proteolytic enzymes and chemical crosslinking agents can inhibit this stage. In addition, inhibitors of microtubules can block conjugate formation, demonstrating the need for an intact NK microtubule system.

Triggering ② is temperature dependent and calcium dependent, and is probably completed within minutes after addition of calcium. In addition, it appears that the NK trigger against some target cells is an epitope of the T-200 glycoprotein, which is probably closely associated with the NK binding receptor ①. It appears that phospholipase A_2 activation may be required to alter the target phospholipids for activation of the NK trigger mechanism ① & ②.

---

[6]Deem, R.L. and S.R. Targan. 1984. Specific inhibition of the terminal substages of the NK lethal hit by the membrane fluidizing agent benzyl alcohol. In submission.

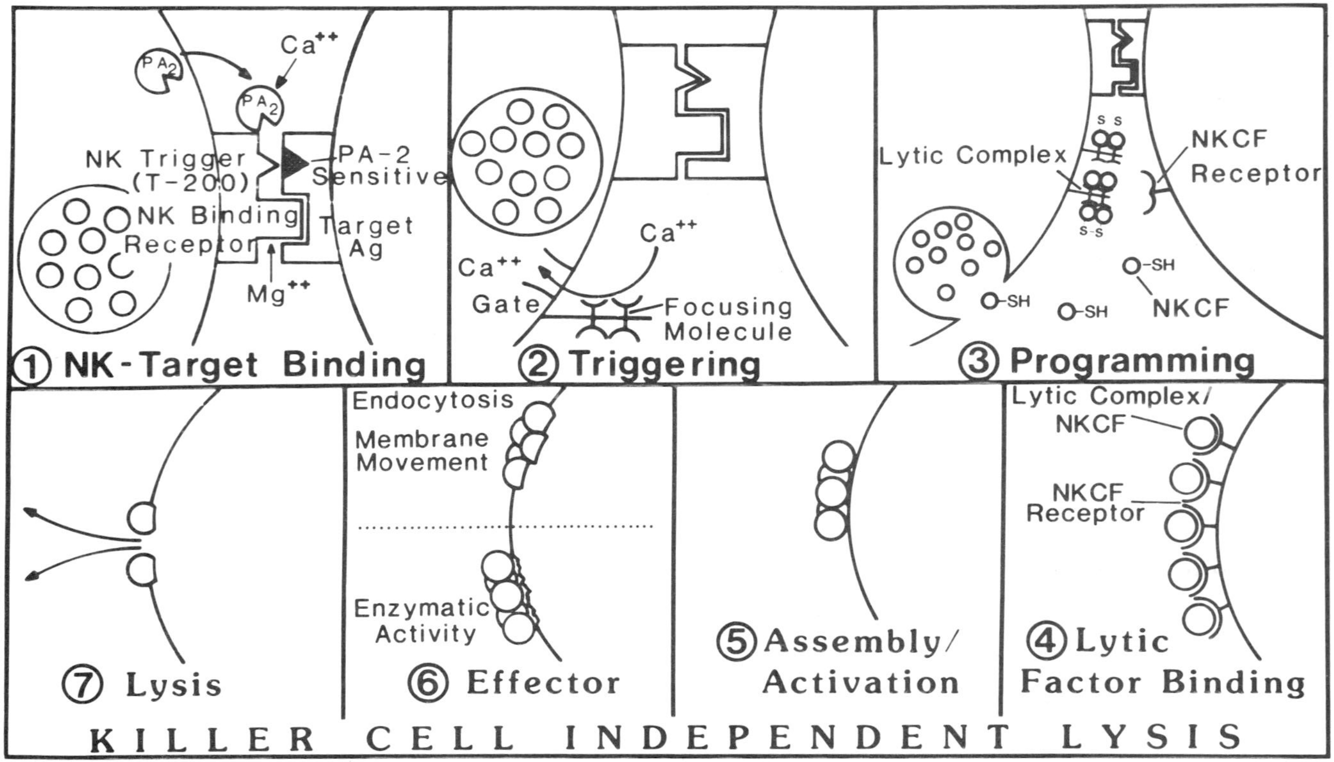

Figure 1. Sequential stages of NK cytotoxicity.

Programming for lysis ③ is temperature dependent, and calcium dependent only early during this stage. This calcium dependence probably requires a calcium flux into the NK cell, since calcium channel blockers can block the early portion of programming. During the late portion of programming, NK cell-derived structures are secreted and placed onto the target cell. These structures may involve a focusing molecule and NKCF subunits, which may form a lytic complex through formation of disulfide bonds.

The terminal stage of the NK cytolysis, the lethal hit, is calcium independent, and temperature dependent only during early substages ④ & ⑤. It has been studied kinetically using KCIL and NKCF-mediated cytolysis. There appear to be at least 4 substages of the lethal hit based upon modulation by pharmacologic agents. The lytic factor binding substage ④ is inhibited only early by certain cross-linking reagents. Immediately after binding is the assembly/activation substage ⑤, which is inhibited early by agents that break disulfide bonds. In addition, chemical cross-linking agents enhance lysis at this substage, suggesting that complex formation or polymerization of subunits may be required for completion of the lethal hit. Inhibitors of energy metabolism and microtubules inhibit late in this substage, suggesting a role for membrane movement/endocytosis. The effector substage ⑥ is inhibited by inhibitors of serine protease activity, suggesting a role for enzymatic activity during the later substages of the lethal hit. The effector and lysis substages ⑥ & ⑦ are inhibited by benzyl alcohol, which may inhibit by fluidizing the membrane to disrupt NK lytic factor-induced transmembrane channels or by inhibiting endocytosis of the assembled lytic complex.

REFERENCES

1. Hiserodt, J.C., Britvan, L.J., and Targan, S.R. (1982). J. Immunol. 129: 1782.
2. Targan, S.R., and Newman, W. (1983). J. Immunol. 131: 1149.
3. Wright, S.C., and Bonavida, B. (1982). J. Immunol. 129: 433.
4. Farrum, E., and Targan, S.R. (1983). J. Immunol. 130: 1252.
5. Hiserodt, J.C., Britvan, L., and Targan, S.R. (1983). J. Immunol. 131: 2705.
6. Hiserodt, J.C., Britvan, L., and Targan, S.R. (1982). J. Immunol. 129: 2248.
7. Deem, R.L., and Targan, S.R. (1984). J. Immunol. 133: 72.
8. Hiserodt, J.C., Britvan, L., and Targan, S.R. (1982). J. Immunol. 129: 2266.
9. Roder, J.C., Rosen, A., Fenyo, E.M., and Troy, F.A. (1979). Proc. Nat. Acad. Sci. (U.S.A.). 76: 1405.

10. Targan, S., Decker, J.M., and Ades, E.W. (1984). Nat. Immun. Cell Growth Regul. In press.
11. Roder, J.C., Argov, S., Klein, M., Petersson, C., Kiessling, R., Andersson, K., and Hansson, M. (1980). Immunology 40: 107.
12. Newman, W. (1982). Proc. Natl. Acad. Sci. (U.S.A.) 79: 3858.
13. Hoffman, T., Hirata, L.F., Bougnoux, P., Fraser, B.A., Goldfarb, R.H., and Axelrod, J. (1981). Proc. Natl. Acad. Sci. (U.S.A.) 78: 3839.
14. Ullberg, M., Merrill, J., and Jondal, M. (1981). Scand. J. Immunol. 14:285.
15. Targan, S. (1981). J. Immunol. 127: 1424.
16. Deem, R.L., and Targan, S.R. (1984). J. Immunol. 133: in press.
17. Devlin, J.J., Yamamoto, R.S., and Granger, G.A. (1981). Cell. Immunol. 61: 22.
18. Deem, R.L., and Targan, S.R. (1984). In "Mechanisms of Cell-Mediated Cytotoxicity" (P.A. Henkart, ed.). Plenum Publishing Corp., New York. In press.
19. Bowes, J.H., and Cater, C.W. (1965). J. Roy. Micro. Soc. 85: 193.
20. Flitney, F.W. (1965). J. Roy. Micro. Soc. 85:353.
21. Richards, F.M., and Knowles, J.R. (1968). J. Mol. Biol. 37: 231.
22. Peters, K., and Knowles, F.M. (1977). Ann. Rev. Biochem. 46: 523.
23. Hopwood, D. (1969). Histochimie 18: 250.
24. Bahr, G.F. (1954). Exptl. Cell Res. 7: 457.
25. Goldstein, I.J., and Hayes, C.E. (1976). Arch. Biochem. Biophys. 173: 127.
26. Davies, P.J.A., Davies, D.R., Levitxki, A., Maxfield, F.R., Milhaud, P., Willingham, M.C., and Pastan, I.H. (1980). Nature 283: 162.
27. Silverstein, S.C., Steinman, R.M., and Cohn, Z.A. (1977). Ann. Rev. Biochem. 46: 669.
28. Chang, T.W., and Eisen, H.E. (1980). J. Immunol. 124: 1028.
29. Trinchieri, G., and DeMarchi, M. (1976). J. Immunol. 116: 885.
30. Shibata T., Sugiura, Y., and Iwayanagi, S. (1982). Chem. Phys. Lipids 31: 105.
31. Podack, E.R., and Dennert, G. (1983). Nature 302: 442.
32. Needham, L., Whetton, A.D., and Houslay, M.D. (1982). FEBS Lett. 140: 85.
33. Tolleshaug, H., and Berg, T. (1982). Biochem. Pharmac. 31: 593.

NK MEDIATED CELL LYSIS: A DIRECTED SECRETORY EVENT OF CYTOLYTIC TUBULAR STRUCTURES

Gunther Dennert

Comprehensive Cancer Center
University of Southern Californnia, School of Medicine
Los Angeles, CA 90033

Abraham Kupfer
J.S. Singer

Department of Biology
University of California, San Diego
La Jolla, CA 92037

I. Introduction

Natural killer (NK) cells are thought to constitute an important defense mechanism against neoplasia, viral infections and to be responsible for the rejection of bone marrow transplants. The mechanisms by which NK cells perform this function are largely unknown. One possibility that has received recent support is that NK cells secrete cytolytic components that lyse bound targets. Several studies have indicated that an intact secretory system in the effector cells is important for target lysis to occur (1,2). Generally cell secretion is accomplished via vesicles which derive from the Golgi apparatus (GA), these vesicles fuse with the cell membrane and thus release their contents (3). Several years ago Goldstein (4) using specific cytotoxic T cells of different specificities such that one killer cell recognized the other but not visa versa, observed that cell lysis is unidirectional and proceeds in the direction of cell recognition. If therefore cell lysis is a secretory event then it should be directed towards the target and could be manifested in a polarization of secretory cell organelles in the effector cell. To explore this possibility we observed the GA and the microtubule organizing center (MTOC) in conjugates of

ISBN 0-12-341370-2

cloned NK cells with their targets (5,6). This was accomplished by using immunofluorescense microscopy with specific anti GA antibodies and anti tubulin antibodies which serve to visualize the MTOC. The MTOC is closely associated with the GA and is that part of the interphase cell including the centriole pair from which the microtubules emanate. Results showed that following NK target cell binding there is a rapid and coordinate reorientation of the GA and the MTOC in the effector but not in the target cell toward the binding site (5,6). We proposed that this event may serve to direct secretion of vesicles towards the bound target cell. In another approach electronmicroscopic evidence was obtained which showed that presumably GA derived granules and vesicles contain two types of tubular structures called polyperfovin I and II (7,8). These complexes are secreted by cytolytic effector cells and can be visualised on the target cell membrane by negative staining electronmicroscopy. We proposed that these tubular structures are cytolytic effector molecules that lead to target cell death. In this chapter we want to summarize our present knowledge on the mechanism of NK mediated target cell lysis and present evidence for the hypothesis that NK mediated lysis is due to a directed secretory event which presumably culminates in the insertion of tubular structures into target cell membranes

II. RESULTS

A. Repositioning of the GA/MTOC in the NK cell toward the target binding site

In the interphase cell the GA and MTOC appear to be spacially related, due probably to a close physical association of the microtubules with elements of the GA (9,10). At the level of light microscopy therefore both organelles appear to be invariably colocalized (9), an observation that was verified in NK cells by double labeling immunofluorescense microscopy using specific antibody reagents (5, and Fig. 1). Therefore to monitor the localization of the GA, specific labeling for either the GA or the MTOC can be employed. A number of different NK clones (11,12) with high cytolytic activity were conjugated to the NK sensitive targets S194 or YAC-1 (5). It was observed that at early times after conjugation i.e. up to 20 minutes, less than 70% of the effector cells have their GA and their MTOC oriented towards the target binding site. At later times i.e. 30 minutes or later essentially all the effector cells have their GA and their MTOC oriented towards the target contact areas. These results suggested that GA orientation in the NK target conjugate is initially random but is followed by a rapid reorientation prior to target lysis. This result raises the question whether GA reorientation is obligatory

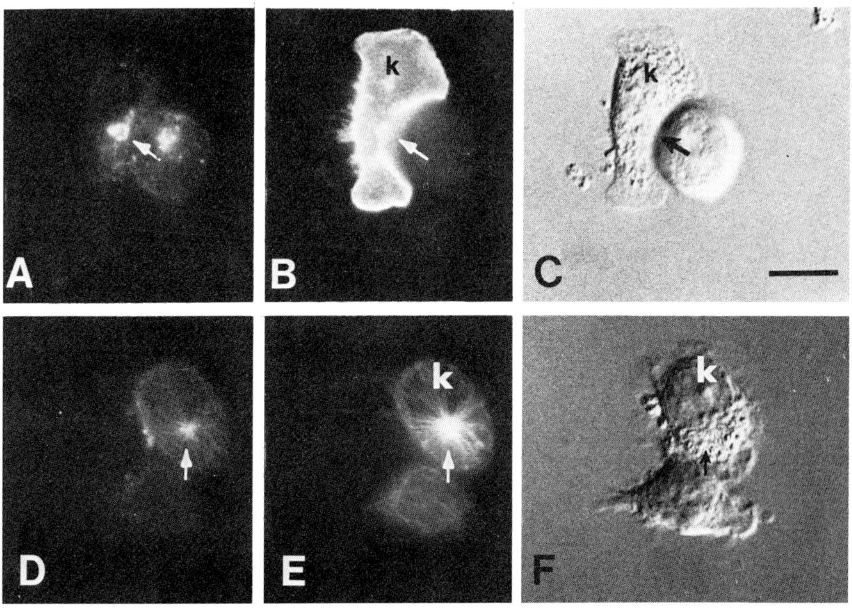

Fig. 1: Light microscopy of couples of cloned NK cells (K) and S194 (A-C) or YAC-1 (D-F) targets, 20 minutes after mixing the cells. The Nomarjki optics images of these cell couples are shown (C and F). A-C. The cells were double immunofluorescently labeled with anti Thy-1 antibodies (B) to identify the NK cell and with antibodies specific for the GA (A). D-F. The cells were double immunofluorescently labeled to reveal their GA (D) and their MTOC (E). The GA (arrows in A and D) and the MTOC (arrow in E) are facing towards the target cell. Bar in C indicates 10 μm.

for cell lysis to occur. We therefore studied various parameters known to interfere with either NK cell lysis or GA/MTOC reorientation.

B. The assembly status of microtubules influences both GA/MTOC repositioning and cell lysis.

As discussed above the MTOC and GA are organelles that are colocalized and reorient together. It is known that microtubule depolymerising drugs like nocodazole cause a complete dispersion of the elements of the GA into the cell periphery, without however significantly impairing its secretory function (10). It was therefore interesting to explore whether treatment of NK cells with nocodazole inhibits both MTOC reorientation and target cell lysis. Treatment of cells with nocodazole results in elimination of the MTOC structure detectable by immunofluorescent labeling for tubulin. If however cells are treated with nocodazole and then with a mixture of nocodazole and taxol an abortive MTOC is formed which can be visualized by immunofluorescence microscopy. This abortive MTOC is close to the nucleus and consists of short strands of microtubules, whereas the bulk of the tubules remain disassembled (13) and importantly the elements of the GA remain dispersed in the cell periphery (Kupfer and Singer, unpublished results). In NK cells that were treated in this way i.e. nocodazole, followed by nocodazole/taxol and then mixed with target cells, binding appeared to be normal, but the MTOC was randomly oriented with respect to the target binding site even after long incubation times (5). Assay of cytolytic activity of such cells revealed that cytotoxicity is either completely or severely inhibited depending on the particular NK clone employed (5). Both effects were readily reversible when the NK target conjugates were transferred into normal tissue culture medium (5). There was a rapid reappearance of microtubules and reorientation of the GA/MTOC towards the target binding site. It had been shown in other studies that drug removal which allows microtubule reassembly, invariably results in a recompaction of the previously dispersed elements of the GA into a compact perinuclear configuration that is again associated with the MTOC (10). These results show that an intact microtubule MTOC complex is obligatory for MTOC/GA reorientation to occur and that the signal from a viable target cell surface to an effector that induces reorientation can persist for a long period of time.

C. The absence of CA^{++} inhibits both NK cytotoxicity and GA/MTOC reorientation.

Since the assembly status of the GA/MTOC complex does not only influence its ability to reorient to the target binding site but also the ability of the effector to lyse the target, we next examined whether conditions known to inhibit target lysis but not binding, interfere with GA/MTOC reorientation. Cloned NK cells were mixed with S194 targets in the absence of Ca^{+2} but presence of 2.5 mM Mg^{+2} and 5 mM EGTA. Under these conditions target cytolysis was completely inhibited as expected (14,15). Fluorescence microscopy revealed that effector target binding was not inhibited but there was no GA/MTOC reorientation towards the target binding site (15). Addition of 5 mM Ca^{+2} to NK target conjugates resulted in a rapid and coordinate reorientation of the GA/MTOC and onset of target cell lysis. This result showed that the presence of Ca^{+2} is important for both target cell lysis and reorientation of the GA/MTOC complex.

D. Nonlysable targets do not emit the reorientation signal and nonfunctional NK cells do not respond to the signal.

One could argue that the withdrawal of Ca^{+2} or addition of nocodazole may result in several effects some of which indirectly inhibit reorientation and cell lysis. Therefore more physiological means were used to explore the relationship of reorientation and cytolysis. In one approach NK cells were conjugated to nonlysable targets (11) and in the other, an NK cell line that had lost its cytolytic activity (G. Dennert unpublished results) was reacted with lysable targets. The non lysable targets were either YAC-8 an NK resistant derivative of YAC-1, or 10 ME, a derivative of BCN (1,12). In both cases no GA/MTOC reorientation was observed (6,16). This suggested that these two targets cannot emit the signal to the effector necessary for reorientation of the two organelles. In the reverse situation the NK cell line NK 3 that had lost its cytolytic activity was reacted with the lysable target S194 and no MTOC/GA reorientation was observed (15). This showed that NK 3 lost its ability to lyse targets, not because of its inability to recognize and to bind to the target but presumably because of its failure to respond with GA/MTOC reorientation to the signal emitted by a lysable target.

E. What is the purpose of GA MTOC reorientation?

The experiments discussed so far demonstrate a close relationship of target cytolysis and reorientation of the GA/MTOC complex. This brings up the question whether reorientation is an obligatory step in the sequence of events leading to cytolysis. In this sense reorientation is an active process in the effector cell which most likely results in more than just the observed repositioning of the secretory machinery. For instance it is possible that stimulation of secretion goes hand in hand with the process that lead to GA/MTOC reorientation. It is of importance to note that in the absence of reorientation i.e. in use of a random orientation of the NK-organelles about 50% of the GA/MTOC complexes are oriented towards the target binding site at any point in time. However as discussed above, in such cases target cell lysis is not observed, which would suggest that yet other processes might play a role in the mechanisms of cellular cytotoxicity. In this regard our experiments in which nocodazole was used to inhibit reorientation are of relevance. From all we know treatment of NK cells with nocodazole completely inhibits reorientation and leads to a dispersion of the GA to the cell periphery. In this situation secretion is not focused to any part of the cell surface, as previously shown by the expression of an integral membrane protein from the rough endoplasmic reticulum on the plasma membrane (17). Results discussed above have shown that treatment of NK cells with nocodazole does not always completely inhibit cytolysis. Moreover in other experiments it was seen that addition of the lectin Concanavalin A to NK target cell mixtures may overcome, at least in part, the inhibition by nocodazole (16). Therefore, one could hypothesize that a reoriented GA primarily serves the function of directing the secretion of toxic components to a confined area of the plasma membrane a mechanism by which lysis would become more efficient. Under this hypothesis the enhancing effects of Con A in nocodazole treated effector cells could be either due to stimulation of secretion or due to a closer contact between effector and target caused by the agglutinating activity of the lectin. It would seem that although the phenomenon of reorientation of the NK-organelles (but not the target-organelles) would suffice to impose unidirectionally of killing in the direction of recognition, the fact that killer cells are not lysed by their own cytolytic components requires the participation of additional mechanisms.

F. Cytolytic components appear to originate from the dense granules

One of the characteristic features of NK cells are the electron dense granules, which have led to calling human NK cells large granular lymphocytes. Using thin sectioning electron-microscopy it was observed in pairs of NK cells and target cells that the dense granules tend to collect in the area of the killer target binding site (7). Moreover, it was seen that the granules give rise to vesicles that carry tubular structures which are reminiscent of the membrane attack complex of complement (7). Subsequently experiments by Henkart and collaborators (18) showed that isolated granules from a rat NK type tumor are cytolytic to erythrocytes or nucleated targets (18). Essentially identical results were seen with isolated granules from our cloned NK cell lines. Negative staining electronmicroscopy of granules isolated by percoll gradient centrifugation revealed the presence of tubular structures in the granules (19). There is therefore compelling evidence for the presence of cytolytic components in granules of NK cells. Assay of cytolytic activity in granules derived from a number of lymphoid tumors was negative (19). There is therefore a correlation between expression of cytolytic activity on the cellular level and on the level of the granule. The cytolytic specificity of the NK derived granule however appears to be non-specific since isolated granules from distinct NK cell clones lyse erythrocytes as well as the NK target YAC-1. Specificity of NK mediated cytolysis therefore does not reside in the granule, but presumably in the recognition of target antigens by the NK cell receptor.

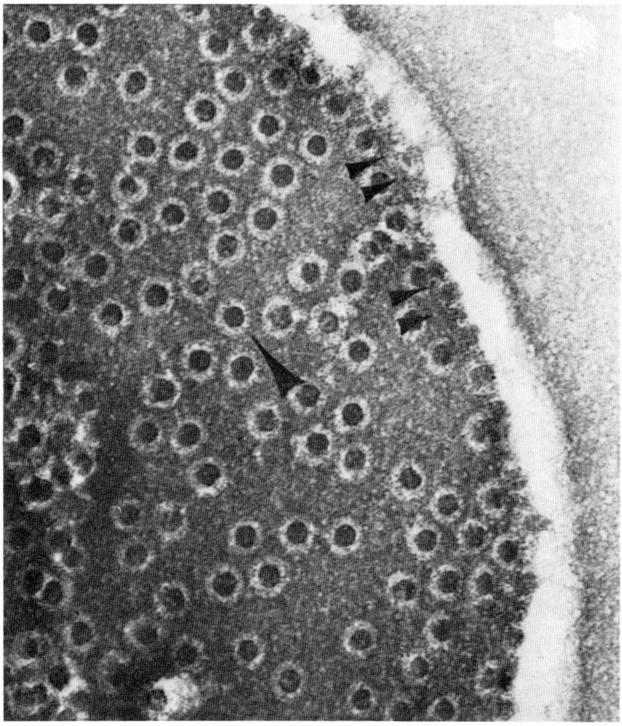

Fig. 2: P_1 complexes generated by cloned NK cells on target cells imaged by negative staining electron microscopy (7). The complexes have an outer diameter of 26nm.

G. The NK derived cytolytic components appear to be tubular structures which are inserted into the target membrane

Incubation of cloned NK cells with YAC-1 targets followed by immunolabeling of target membranes and negative staining electromicroscopy revealed that two kinds of ring type structures can be detected on target membranes (7). The larger structure was called Polyperforin I (P_1) and has an inner and outer diameter of 16 and 26 nm respectively. The smaller ring structure was called Polyperforin 2((P_2) and has an inner and outer diameter of 5 nm and 11 nm respectively (7). Side views of these structures reveal that both are tubules with a small collar that have a length of 16 nm (P_1) and 12 nm (P_2) respectively. The overall appearance of P_1 and P_2 is reminiscent of the membrane attack complex of complement but both P_1 and P_2 have different dimensions than both human and mouse complement lesions (7). Moreover antibodies specific for human C9 do not react with P_1

(P_2 was not tested) (7). Taken together these results show that NK cells incubated with targets may give rise to membrane lesions that are similar to those caused by complement but the dimensions of the lesions are different from complement lesions. We therefore proposed that the P_1 and P_2 lesions observed on target membranes may be the cytolytic effector molecules responsible for target cell lysis. Support for this hypothesis comes from the finding that cytolytic granules appear to contain the precursors of P_1 and P_2, that target lysis and appearance of P_1 on target membranes follow the same time course, and that cells which are not cytolytic do not generate P_1 and P_2 complexes on target membranes (7). Moreover drugs that inhibit NK cell lysis like the carboxylic ionophore monensin also inhibit the appearance of P_1 and P_2 lesions (7,8). The mechanism by which P_1 and P_2 are inserted into the target cell membrane is presently not understood. It is possible that granule derived vesicles are secreted by the NK cell and that the vesicles carry P_1 and P_2 complexes. Transfer of P_1 and P_2 into target membranes from the vesicles could occur via fusion of the vesicle with the target membrane. This mechanism would not explain why NK cells are not lysed by their own vesicles. But one possibility namely that vesicles are specifically cytolytic to their target does not seem to be verified since NK granules are not specifically cytolytic for NK targets (18,19).

III. CONCLUSION

The events which we have studied with light microscopy and that precede target cell lysis are schematically summarized in Fig. 3. NK cells and targets collide and bind randomly. After the initial binding the GA/MTOC complex in the NK cell reorients towards the target binding site. This event requires recognition of specific cell surface structures on the target and is Ca^{+2}-dependent. Reorientation of the GA/MTOC complex only takes place in the killer cell and not in the bound target cell. Reorientation was not observed in a situation in which the killer cell lost its cytolytic activity. This and the finding that non lysable targets do not induce reorientation in the effector cell suggests that reorientation is an important and integral part of the events leading to target lysis. The integrity of the microtubules appears to be important for GA/MTOC reorientation and for NK-dependent lysis. Interestingly, one of the conditions that inhibits reorientation, i.e., treatment of cells with nocodazole does not completely inhibit cytolysis in particular when the lectin Con A is added to NK target mixtures. We therefore propose that reorientation serves to direct the secretion of cytolytic components rather than enables it. Support for this comes from

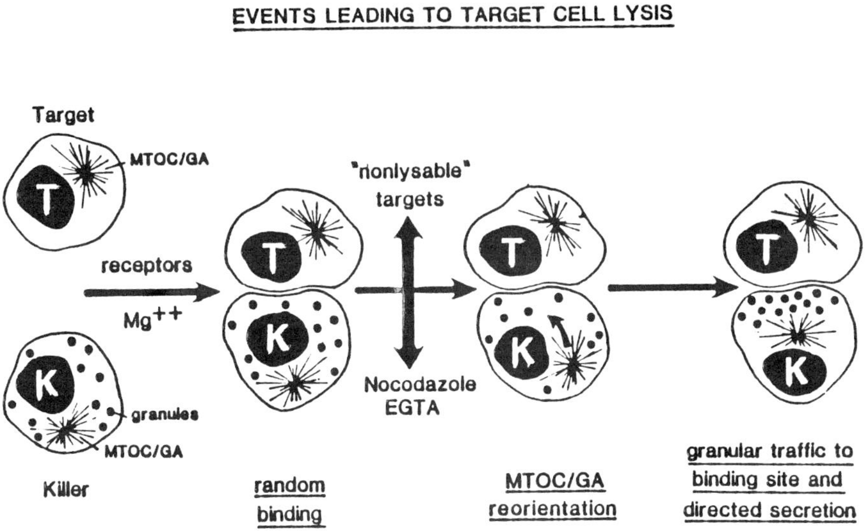

Figure 3

the observation that the dense granules tend to collect in the area of target cell binding. A scheme depicting the next sequence of events is shown in Fig. 4. The recognition of target structures by the NK cell leads to secretion of vesicles that originate from the granules. The vesicles carry P_1 and P_2 perhaps in a preassembled way and either fuse with the target membrane or transfer P_1 and P_2 by an as yet unknown mechanism. Since vesicle mediated cytolysis appears to be nonspecific the effector cell must have a protective mechanism to avoid autolysis. Future experiments will have to be done in order to elucidate what these protective mechanism may be and whether the vesicles carry yet other lymphocytotoxic substances besides P_1 and P_2.

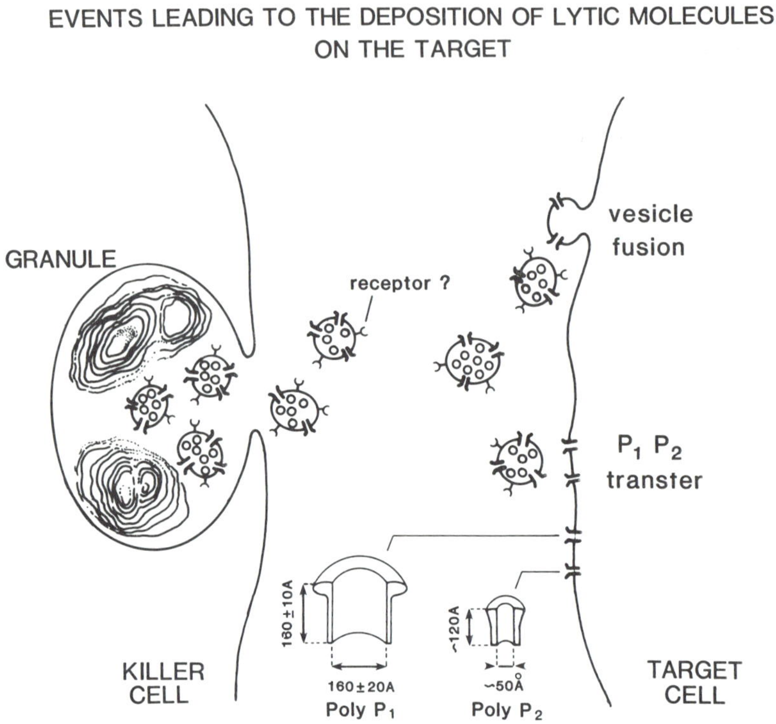

Figure 4

ACKNOWLEDGMENTS

We are grateful to Mrs. Margie Adams and Carol Gay Anderson for their excellent technical assistance. This work was supported by U.S. Public Health Service grants CA-39501, CA-37706 and CA-19334, and American Cancer Society grant IM-284, to G.D., and by U.S. Public Health Serivce grants AI-06659 and unit 3 of HL-14197 to S.J.S., who is an American Cencer Society Research Professor.

REFERENCES

1. O. Carpen, I. Virtanen, and E. Saksela, Ultrastructure of human natural killer cells: nature of the cytolytic contacts in relation to cellular secretion, J. Immunol., 128:2691 (1982).

2. O. Carpen, I. Virtanen, and E. Saksela. The cytotoxic activity of human natural killer cells requires the intact secretory apparatus. Cell. Immunol. 58:97 (1981).

3. M.G. Farquhar and G.E. Palade. The Golgi apparatus (complex) -(1954-1981) -from artifact to center stage. J. Cell. Biol 91:77a (1981).

4. P. Goldstein. Sensitivity of cytotoxic T cells to T cell mediated cytotoxicity. Nature 252:81 (1974).

5. A. Kupfer, G. Dennert, and S.J. Singer. Polarization of the Golgi apparatus and the microtubule organizing center within cloned natural killer cells bound to their targets. Proc. Natl. Acad. Sci. USA 80:7224 (1983).

6. A. Kupfer and G. Dennert. Reorientation of the microtubule organizing center and the Golgi appartus in cloned cytotoxic lymphocytes triggered by binding to lysable target cells. J. Immunol., (in press).

7. E.R. Podack and G. Dennert. Cell-mediated cytolysis: assembly of two types of tubules with putative cytolytic function by cloned natural killer cells. Nature (Lond) 302:442 (1983).

8. G. Dennert and E.R. Podack. Cytolysis by H-2 specific T killer cells. Assembly of tubular complexes on target membranes. J. Exp. Med. 157:1483 (1983).

9. A. Kupfer, D. Louvard, and S.J. Singer. Polarization of the Golgi apparatus and the microtubule-organizing center in cultured fibroblasts at the edge of an experimental wound. Proc. Natl. Acad. Sci. USA 79:2603 (1982).

10. A.A. Rogalski and S.J. Singer. Association of elements of the Golgi apparatus with microtubules. J. Cell Biol., in press.

11. G. Dennert. Cloned lines of natural killer cells. Nature (Lond), 287:47 (1980).

12. G. Dennert, G. Yogeeswaran, and S. Yamagata. Cloned cell lines with natural killer activity. Specificity, function and cell surface markers. J. Exp. Med, 153:545 (1981).

13. M. De Brabander, G. Geuens, R. Nuydens, R. Willebrords, and J. De Mey. Microtubule stability and assembly in living cells: the influence of metabolic inhibitors, taxol and pH. Cold Spring Harbor Symp. Ouant. Biol., 46:227 (1981).

14. J. C. Hiserodt, L.J. Britwan and S.R. Targan. Characterization of the cytolytic reaction mechanism of the human natural killer (NK) lymphocytes: resolution into binding, programming, and killer cell-independent steps. J. Immunol., 129:1782 (1982).

15. A. Kupfer, G. Dennert, and S.J. Singer. The reorientation of the Golgi apparatus and the microtubule organizing center in the cytotoxic effector cell in an obligatory early event in the lysis of bound target cells (submitted) (1984).

16. G. Dennert, A. Kupfer, C.G. Anderson, S.J. Singer. Reorientation of the Golgi apparatus and the microtubule organizing center: is it a means to polarize cell mediated cytotoxicity? 2nd International Workshop on Cell Mediated Cytotoxicity, Ed: P. Henkart. Academic Press (in press) (1984).

17. A.A. Rogalski, J.E. Bergmann, and S.J. Singer. The effect of microtubule assembly status on the intracellular processing and surface expression of an integral protein of the plasma membrane. J. Cell Biol., in press.

18. P.A. Henkart, P.J. Milland, C.W. Reynolds, M.P. Henkart. Cytolytic activity of purified cytoplasmic granules from cytotoxic rat large granular lympocyte tumors. J. Exp. Med., (1984).

19. M. Criada, J. Lindstrom, R. Baker, G. Dennert. Mechanism of cell mediated lysis: properties of polyperforin membrane channels (submitted) (1984).

INVOLVEMENT OF THE CYTOSKELETON AND AN ASSOCIATED 95 KD PROTEIN IN RECOGNITION AND LYSIS BY NK CELLS[1]

Eero Saksela
Olli Carpén
Ismo Virtanen

Department of Pathology
University of Helsinki
Finland

I. INTRODUCTION

Human NK mediated killing is a multistage process. Following an initial step of target cell recognition and binding of the effector cells the contact area widens and the effector cell Golgi apparatus is directed towards the effector-target contact site. This polarization of the cytoplasmic organelles of the effector cell is followed by a directed secretion of cytotoxic substances, and the target cell lysis can be inhibited with drugs affecting cellular secretion, specifically with monensin which inhibits the fusion of Golgi vesicles with the cellular membranes (1). More recently, cytotoxic material has been demonstrated in the supernatants of NK cell/target cell co-cultures and isolated from the granules of rat LGL tumor cells (2,3). Similar secretory mechanisms of killing have also been implicated in cytotoxic T cell lysis (4) and antibody dependent cellular cytotoxicity (5).

The phase following the effector/target conjugate formation has been termed "programming for lysis" which lasts approximately 30 to 120 minutes (6). The lethal hit phase follows after which the lysis of the target cell proceeds

---

[1]Supported by grant no Ca 23809-06 from the National Cancer Institute, NIH, Bethesda, Md. USA.

ISBN 0-12-341370-2

independently of the effector cells. The events taking place during the "programming for lysis" have been largely unknown. Timewise, this period coincides with the widening of the initial effector-target cell contat area and polarization of the effector cell Golgi apparatus (7).

The cellular cytoskeleton is known to be involved in such basic functions as regulation of the cellular shape, mitosis, redistribution of cellular organelles and cellular motility (for review see 8). A functional linkage has also been demonstrated between cell surface receptors and actin-containing microfilaments in lymphocytes, and both spectrin and vimentin have been implicated in the cell surface receptor capping phenomena (9-11). Cell-to-cell adhesions as well as cell-to-matrix adhesions seem to be mediated by the cytoskeletal components. Against this background it seemed reasonable to expect that the cytoskeleton might play a role in the steps taking place during the "programming for lysis".

II. POLARIZATION OF NK CELL CYTOSKELETON IN LYTIC CONJUCATES

To analyze the post-binding steps leading to lysis of the target cells we selected two target cell types, K-562 and Raji. At 1:1 ratio with highly enriched LGL effector cells both target cell types bind 30-35 % of the effector cells, but practically only the LGL-K562 conjugates lead to target cell lysis indicating that the differences are at the post-binding steps. The LGL/target cell conjugates were stained at various times after conjugate formation with monospecific antibodies against actin, vinculin, myosin, vimentin and p230, an immunoanalogue of erythroid alpha-spectrin (for refererence of antibodies see 12). The conjugates of LGLs with K-562 but not with Raji were associated with a polarization of the LGL actin containing microfilaments and vinculin at the effector-target contact area within 60 minutes (Figure 1). Myosin, spectrin and vimentin showed a diffuse staining pattern on both effector and target cells. These latter cytoskeleton components have been found, as well as actin, to be involved in the ligand-induced capping on lymphoid cells and their behaviour in the NK-target conjugate formation is thus different from these phenomena. The focal concentration of actin and vinculin at the contact sites of NK cells and sensitive target cells is similar to the so called focal adhesions found on cell-to-cell contacts of epithelial cells in vivo and at adhesion sites of adherent cells, such as fibroblasts or macrophages on their growth substrata in vitro (13). The precise molecular organization of the components of

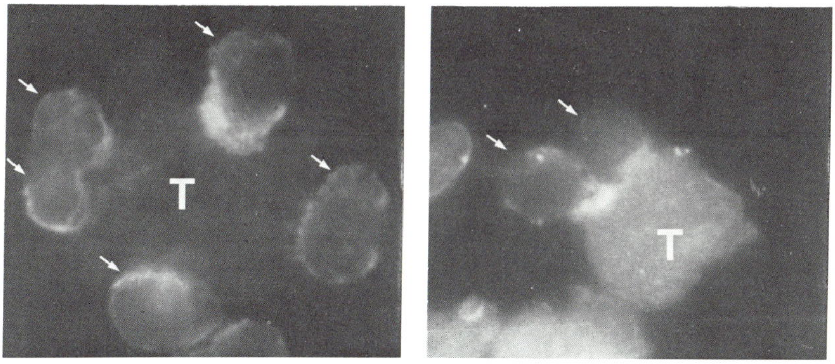

Fig. 1. Orientation of NK cell actin and vinculin during conjugation with K562 target cells. Conjugating cells were incubated for 60 min and fixed with 3.5 % paraformaldehyde. Actin was visualized with NBD (7-nitrobenz-2-oxa-1,3-diazole phallacidin) (a). For visualization of vinculin fixed cells were briefly permeabilized with 0.5 % Triton X-100, and stained with monospecific rabbit antivinculin antibody. As a secondary antibody rhodamine-conjugated swine anti-rabbit IgG was used (b). The arrows show NK cells and T indicates K562 targets.

focal adhesions is not known. Geiger (8) has postulated that their formation is preceded by clustering or exposure of specific cell surface attachment sites for vinculin. Subsequent patching of vinculin could then lead to formation of actin fibers and function as anchorage sites between cell surface proteins and microfilaments. Via this cascade of events the cytoskeleton could thus play a critical role in the polarization of the secretory apparatus towards the contact area of sensitive but not insensitive target cells. Analogously, a change in the shape of effector cells to form a wider contact area and polarization of the effector cell Golgi complex seems to preferentially take place in contacts with sensitive target cells (7).

III. EFFECT OF DETERGENTS ON NK/K-562 CONJUGATES

The focal adhesions with the characteristic concentration of actin and vinculin microfilaments described in other cellular systems are known to be detergent resistant (8,14). In order to test this aspect of the NK/target cell conjugates

Table 1. The effect of detergent extraction on NK/target conjugates

% Triton X-100[a]	0	0.5	1.0	2.0
% conjugating NK cells	29	33	39	23

[a]Highly enriched LGLs and K-562 target cells were allowed to conjugate and thereafter extracted with various concentrations of Triton X-100. After 20 min the number of conjugating cells were counted. A representative of 5 experiments.

we used various concentrations of the non-ionic detergent Triton X-100. The conjugates of NK cells and K-562 proved to be highly detergent resistant. As shown in Table I the number of the NK/target conjugates showed only a small decrease during the extraction and most of the conjugates were not affected by treatment with up to 2 % Triton X-100 for 20 minutes. The cells lost their refractorines in phase contrast microscopy already after a few minutes extraction with 0.5 % Triton X-100 but the conjugate formation and polarization of the effector cells was maintained even with the prolonged extraction procedure above (Figure 2). In this respect the

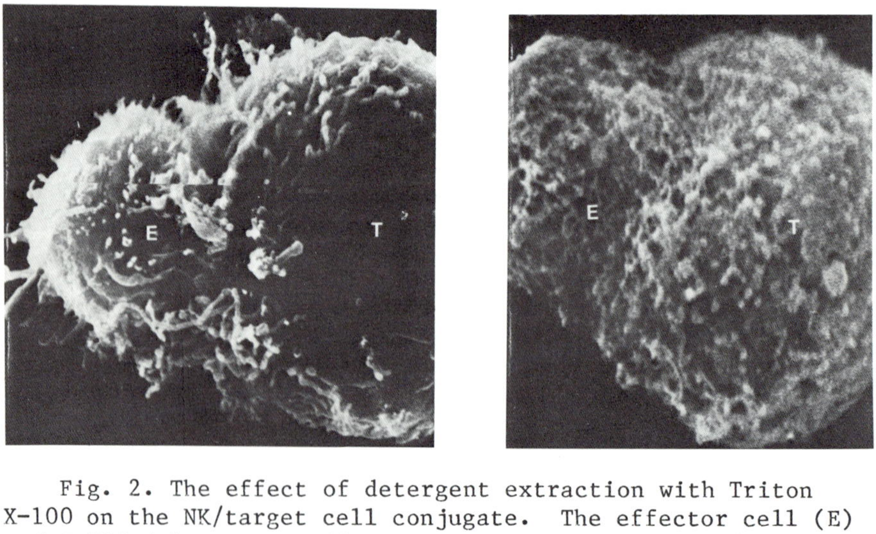

Fig. 2. The effect of detergent extraction with Triton X-100 on the NK/target cell conjugate. The effector cell (E) and K-562 (T) target cell conjugate is shown before (a) and after (b) treatment with 0.5 % Triton X-100 for 20 minutes.

NK/target conjugates thus shared the physical properties of characteristic focal adhesions.

IV. ANALYSIS OF SURFACE STRUCTURES RESPONSIBLE FOR CONJUGATION

Detergent extraction removes most of the cellular proteins and lipids, and leaves behind a bare cytoskeleton together with associated proteins firmly anchored to it (8). When metabolically prelabelled NK cells were used in the target cell conjugation and the enriched conjugates were extracted with Triton X-100 over 90 % of the labelled protein was lost into the supernatant (Table II). As, however, the integrity of the conjugates was maintained as shown above, it seemed reasonable to assume that also the surface structures responsible for the conjugate formation were resistant to extraction. In order to analyse such putative recognition

Table 2. Effect of detergent extraction on the cellular proteins of NK/target conjugates.

Experiment no[a]	C	NC
1	13	19
2	4	6
3	12	2
4	5	3
Mean	8.5	7.5

[a]The loss of ^{35}S-Met. label in conjugating (C) and non-conjugating (NC) NK cells during extraction with 0.5 % Triton X-100 for 20 min. Results are expressed as percentage of the label present in non-extractable portion of NK cells.

structures on the NK cells we devised a strategy of pre-labelling the NK cells, allowing them to conjugate with sensitive target cells and look for preferentially retained labelled proteins after extraction of the conjugates. Figure 3 shows a SDS-PAGE fluorograph of a representative experiment where a protein band of an approximate molecular weight of 95 kD is greatly enriched in the Triton resistant fraction of conjugating effector cells. Conversely, this band was not

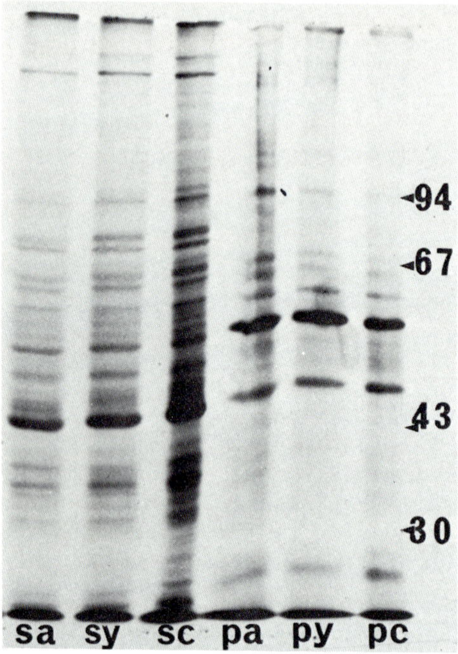

Fig. 3. Fluorography of Triton X-100 extracted conjugating and non-conjugating NK cells prelabeled with ^{35}S-Met. In sa (extractable portion) and pa (non-extractable portion) 73 % of Nk cells were binders, in sy (extr.) and sp (non-extr.) 13 % were binders and in sc (extr.) and pc (non-extr.) 0 % were binders. The enrichment of a 95kd protein in the detergent resistant fraction of binding NK cells (pa) was a consistent finding in several experiments.

visible in the extractable fraction of conjugating cells although it could be seen in the extractable fraction of non-conjugating control effectors. An alternative method utilizing the lactoperoxidase-^{125} mediated labelling confirmed that the 95kD band remaining in the residues of the conjugating effector cells was indeed a cell surface exposed protein.

V. THE EFFECT OF THE 95 KD PROTEIN ON CONJUGATE FORMATION

In order to test whether the 95kD protein described above could competitively inhibit conjugate formation, we used an electrophoretic method to isolate polypeptides from gels

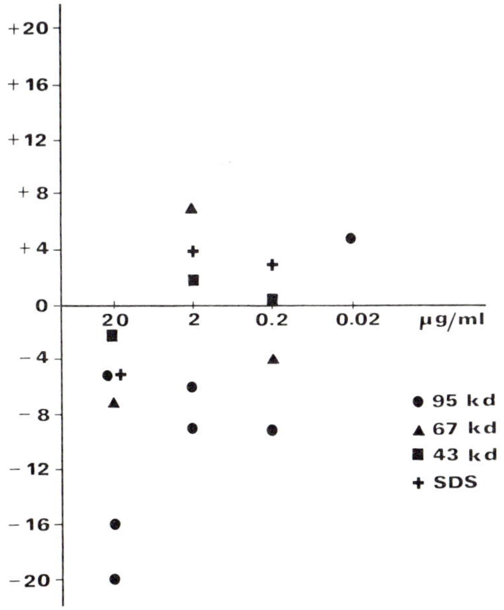

Fig. 4. Inhibition of the conjugate formation of highly enriched LGLs and K-562 target cells with various concentrations of 95 kD protein and two unrelated proteins eluted from SDS gels of human LGLs. The target cells were pretreated with the putative competitive eluted protein or SDS and washed prior to addition of the effector cells. The percentual difference to the control is indicated.

containing LGL proteins. Polypeptides from the 95kD regions as well as two unrelated areas were purified and extensively dialysed against water to remove free SDS. As seen in Figure 4 0.2 ug of the eluted 95 kd protein caused a significant reduction in conjugate formation when the target cells were incubated with the eluted material prior to addition of the effector cells.

In contrast, similar concentrations of eluted proteins from unrelated areas, free SDS or BSA did not competitively inhibit the conjugate formation suggesting that the eluted 95kD protein could be functionally associated with the NK/target conjugation.

VI. DISCUSSION

The "programming for lysis" stage of human NK mediated killing involves polarization of the cellular secretory apparatus towards the target cell contact area (7). This polarization is associated with concentration of actin containing microfilaments and vinculin in the effector/target contact region (15). Several examples are known in other cellular systems showing that cytoskeletal elements attach to cell surface structures via transmembrane proteins, and that surface patching and capping phenomena result in unmasking of filament binding structures and subsequent cytoskeletal polarization (8). It is conceivable that in the NK/target cell conjugation the putative recognition molecules undergo modifications resulting in unmasking of vinculin binding sites which again lead to organization of actin bundles and subsequent movement of the cellular organelles required to deliver the secretory lethal hit. The vinculin anchorage sites responsible for these phenomena are now known. The present experiments suggest, however, that such structures form a transmembrane continum to the target cell attachment molecules since the conjugates of NK cells with sensitive target cells were highly resistant to detergent in analogy to adhesion sites known in other cellular systems. The cellular structures attached to cytoskeleton are known to be resistant to detergent treatment although most of the cell surface proteins and lipids are removed (8).

We utilized the detergent resistance of NK/target cell conjugates to analyze more closely the putative recognition molecules on human NK cell surface which could be held responsible for the retained integrity of the conjugates. In such experiments a 95kD cell surface protein was identified which was retained in NK/target cell conjugates after extraction but was freely extractable from the surface of non-conjugating effector cells. The results thus suggest that a post-binding modification of this protein either directly or indirectly led to attachment of vinculin and actin filaments thus anchoring the structures into a detergent resistant complex. Analogous experiments have been reported by Woda and McFalden (15) who showed that rat lymphocyte surface Ig became detergent insoluble following capping induction. Painter et al. (17) have studied the effect of Con A on three different platelet surface membrane proteins. They showed that after ConA binding 85-90 % of the proteins could be collected in residual material remaining after the Triton extraction although before ConA binding the percentage was less than 10 %.

The 95kD protein when eluted from the SDS-PAGE gels of enriched LGLs caused a moderate inhibition of NK cell binding on target cells pretreated with the protein. Such competitive inhibition could not be seen with other unrelated proteins from the same gels nor with free SDS or BSA. The inhibitory capacity of about 20-30 % was of the same order of magnitude seen in analogous experiments where a similar approach of competitive inhibition has been used (18,19). The experiments above have thus led us to suggest that the 95kD protein on human LGL surface is functionally associated with the recognition and subsequent polarization of the NK cells upon target cell contact (29).

NK cells and CTLs use different structures for target cell recognition. However later stages of the NK and CTL killing are quite similar involving polarization of the secretory apparatus and possibly a directed delivery of similar cytotoxic factors (3, 4). It is thus interesting to note that the 95kD protein has the approximate molecular weight of the beta chain of the recently described lymphocyte functional protein, LFA-1 (21). Anti LFA-1 antibodies inhibit the binding of CTLs to appropriate targets by about 50 % and also inhibit the cytotoxic action of cloned NK cells (22). In the CTL system LFA-1 has been viewed as a molecule strengthening the target cell contact and it seems to be an adhesion molecule apparently involved in many other leukocyte functions as well, such as lymphocyte transformation by soluble antigens, mitogens or by allogeneic cells and in neutrophil migration (23). LFA-1 may thus be related to a membrane dependent activation pathway in several leukocyte functions known to involve cytoskeleton modifications. Whether the 95kD protein described here and LFA-1 are actually the same structure is not known at the present but they could at least be closely related. The previously described Rat* antibody which inhibits post-binding steps in the cytolysis by CTLs and both binding and post-binding stages in NK cells also detects, among others, a 95kD protein (24).

The 95kD protein appears to be the major component induced to become anchored to the cytoskeleton at the NK/target cell contact. It may not, however, be the only one (25). Other components may be present in the complex whose modifications may be important in the cascade of lytic conjugate formation. The experiments have also been performed so far with only one sensitive target cell type and the results must be expanded to other NK lytic components in order to test the generality of the conclusions.

REFERENCES

1. Carpén, O., Virtanen, I., and Saksela, E. 1981. Cell. Immunol. 58, 97.
2. Wright, S.C., and Bonavida, B. 1982. J. Immunol. 129, 433.
3. Millard, P., Henkart, M.P., Reynolds, C.W., and Henkart, P.A. 1984. J. Immunol. 132, 3197.
4. Eckhard, E., Podack, R., and Konigsberg, P.J. 1984. J. Exp. Med. 160, 695.
5. Pedersen, B., Norrild, B., and Krebs, H.J. 1982. Cell. Immunol. 40, 164.
6. Hiserodt, J.C., Britvan, L.J., and Targan, S.R. 1982. J. Immunol. 129, 1782.
7. Carpén, O., Virtanen, I., and Saksela, E. 1982. J. Immunol. 128, 2691.
8. Geiger, B. 1983. Biochim. Biophys. Acta 737, 305.
9. Health, J.P. 1983. Nature 302, 532.
10. Nelson, W.J., Colaco, C.A.L.S., and Lazarides, E. Proc. Natl. Acad. Sci. USA 80, 162.
11. Dellagi, K., and Brouet, J.-C. 1982. Nature 298, 284.
12. Virtanen, I., Badley, R.A., Paasivuo, R., and Lehto, V.-P. 1984. J. Cell Biol. 99, 1083.
13. Chen, W.-T., and Singer, S.J. 1982. J. Cell Biol. 95, 205.
14. Lehto, V.-P., 1983. Exp. Cell. Res. 143, 271.
15. Carpén, O., Virtanen, I., Lehto, V.-P., and Saksela, E. 1983. J. Immunol. 131, 2695.
16. Woda, B., and McFadden, M.L. 1983. J. Immunol. 131, 1917.
17. Painter, R.G., and M. Ginsberg. 1982. J. Cell. Biol. 92, 565.
18. Ades, E.W., Hinson, A., and Decker, J.M. 1981. Immunobiol. 160, 248.
19. Vodinelich, L., Sutherland, R., Schneider, C., Newman, R., and Greaves. M. 1983. Proc. Natl. Acad. Sci. USA 80, 835.
20. Carpén, O., Virtanen, I., and Saksela, E. 1984. J. Immunol. submitted.
21. Springer, T.A., Davignon, D., Ho, M.-K., Kürzinger, K., Martz, E., and Sanchez-Madrid, F. 1982. Immunol. Rev. 68, 171.
22. Krensky, A.M., Robbins, E., Springer, T.A., and Burakoff, S.J. 1981. J. Immunol. 132, 180.
23. Beatty, P.G., Ledbetter, A.J., Martin, P.J., Price, T.H., and Hansen, J.A. 1983. J. Immunol. 131, 2913.
24. Kahle, R., Hiserodt, J., and Bonavida, B. 1983. Cell. Immunol. 80, 97.
25. Fast, L., Beatty, P., Hansen, J., and Newman, W. 1983. J. Immunol. 131, 2404.

ULTRASTRUCTURAL ANALYSIS OF HUMAN NATURAL KILLER CELL - TARGET CELL INTERACTIONS LEADING TO TARGET CELL LYSIS

John C. Hiserodt and Theodore F. Beals

Department of Pathology
University of Michigan
Veterans Administration Medical Center
Ann Arbor, Michigan 48109

INTRODUCTION

Natural Killer (NK) Cells are a subpopulation of Fc receptor positive, non adherent, nonphagocytic cytotoxic effect or cells having light microscopic features of large granular lympocytes (1). NK cells have the capacity to spontaneously lyse a wide variety of tumor or virally infected target cells as well as certain normal cells *in vitro*. Because of these observations it has been postulated that NK cells may play important roles in primary host defense mechanisms and/or immunoregulation (2,3).

The mechanism(s) by which NK cells mediate the lysis of tumor target cells remains an elusive problem. While complex, recent studies have resolved the cytolytic reaction into a series of discrete stages (4,5). In the most simplistic scheme, the NK cell must first recognize and bind to the target. Once bound, the killer cell undergoes a series of physiological responses collectively termed activation. It is during this phase that the killer cell initiates infliction of the lethal hit on the target. This involves transfer of protease sensitive cytolytic mediators from the killer cell to the target via a secretory process (6,7). It is thought these mediators may be derived from the cytoplasmic granules characteristic of NK cells (8). Once the lethal hit has been completed the killer cell can detach and recycle to other targets. Target cells having received a lethal hit are "programmed to lyse" and will rapidly do so during the final

ISBN 0-12-341370-2

stage of the lytic reaction known as the killer cell independent stage (5).

While several studies have analyzed the ultrastructural characteristics of purified NK cells (9,10), few studies have analyzed NK-target cell interactions leading to target cell lysis. Employing high resolution scanning and transmission electron microscopy, highly purified NK cells and methods developed to study the terminal stages of the NK cytolytic mechanism, we analyzed in a kinetic study NK-target cell interactions leading to target cell lysis.

MATERIALS AND METHODS

Target Cells and Culture Media:

The erythroleukemia cell line, K562, was grown in medium RPM1-1640 supplemented with 10% heat inactivated fetal calf serum and 20 µg/ml gentamcin.

Isolation of Natural Killer Lymphocytes

Mononuclear leukocytes were obtained from the peripheral blood of normal healthy donors after Ficoll-Hypaque density gradient separation. These cells were routinely plated onto plastic petri dishes for 1 hour at 37°C to remove adherent cells. Purified NK cells were obtained by centrifugation of the nonadherent lymphocytes through Percoll density gradients (11). Cells obtained from the 43 to 45% percoll interface were employed. Sixty to 80% of these cells had large granular lymphocyte (LGL) morphology. LGL were also obtained by flow cytometric sorting of non adherent PBL stained with Leu 11a antibody. These cells had approximately 80% LGL morphology.

Cytotoxicity Assays

The ^{51}Cr release microcytotoxicity assay was employed using 96-well round bottom microplates (Lindbro). 5×10^3 ^{51}Cr labeled target cells (in 50 ul) were mixed with various numbers of NK cells to give E:T ratios between 1:1 and 50:1 in a final volume of 200 ul. these cells were allowed to interact at 37°C for 3 to 4 hours at 37°C after which the percentage of cytoxicity was determined by standard formulas. The Ca^{++} pulse method was employed exactly as described previously (5).

Scanning (SEM) and Transmission (TEM) Electron Microscopy

Specimens for TEM were fixed by adding equal volumes of 5% glutaraldehyde to the cell suspensions, post-fixed with OsO_4, en bloc stained with uranyl nitrate, dehydrated in ethanol and embedded in an EPON/Araldide mixture. Thin sections were stained with uranyl nitrate and lead citrate. Specimens for SEM were collected on Nuclepore® filters, fixed with 2.5% glutaraldehyde post-fixed with OsO_4, dehydrated in ethanol, critical point dried in liquid CO_2 and gold sputter coated.

RESULTS AND DISCUSSION

SEM and TEM Ultrastructural analysis of NK-target cell interactions leading to target cell lysis were studied employing highly purified human LGL, K562 targets and a Ca^{++} pulse cytotoxicity assay. The Ca^{++} pulse technique was used for two purposes: a) to separate ultrastructural alterations induced only during the binding events compared to those occurring during cytolysis and b) this technique synchronizes initiation of the cytolytic response upon the addition of Ca^{++} ions.

Figure 1 shows a representative SEM and TEM of an LGL-K562 conjugate established in the presence of EGTA/Mg^{++} (thus representing a conjugate in which killing cannot occur). The SEM micrograph (Fig. 1a) shows that under these conditions the binding interactions are complex consisting of broad surface areas and numerous long microvillus projections extending from the NK cell to the target as well as from the target to the NK cell. In many instances the bound NK cell rapidly developed a "terminal tuft" (i.e. a highly compact area of microvilli) at one pole of the cell (not shown). The TEM micrograph (Fig. 1b) shows that the NK microvillus projections interdigitate intimately with the target cell membrane but rarely approach the target cell nucleus. Analysis of many NK-target conjugates did not reveal any specialized junctions between the two membranes as has been suggested for CTL-target cell conjugates (12). Also, under conditions of EGTA/Mg^{++} note the random arrangement of the characteristic granules (arrows) within the NK cytoplasm. Finally, multivesicular bodies (arrowheads), which also appear to be a feature of NK cells (10) are indicated. These are also randomly oriented. Similar binding interactions were observed with conjugates prepared in standard medium containing both Ca^{++} and Mg^{++}. Thus, no ultrastructural differences could be detected in the initial binding

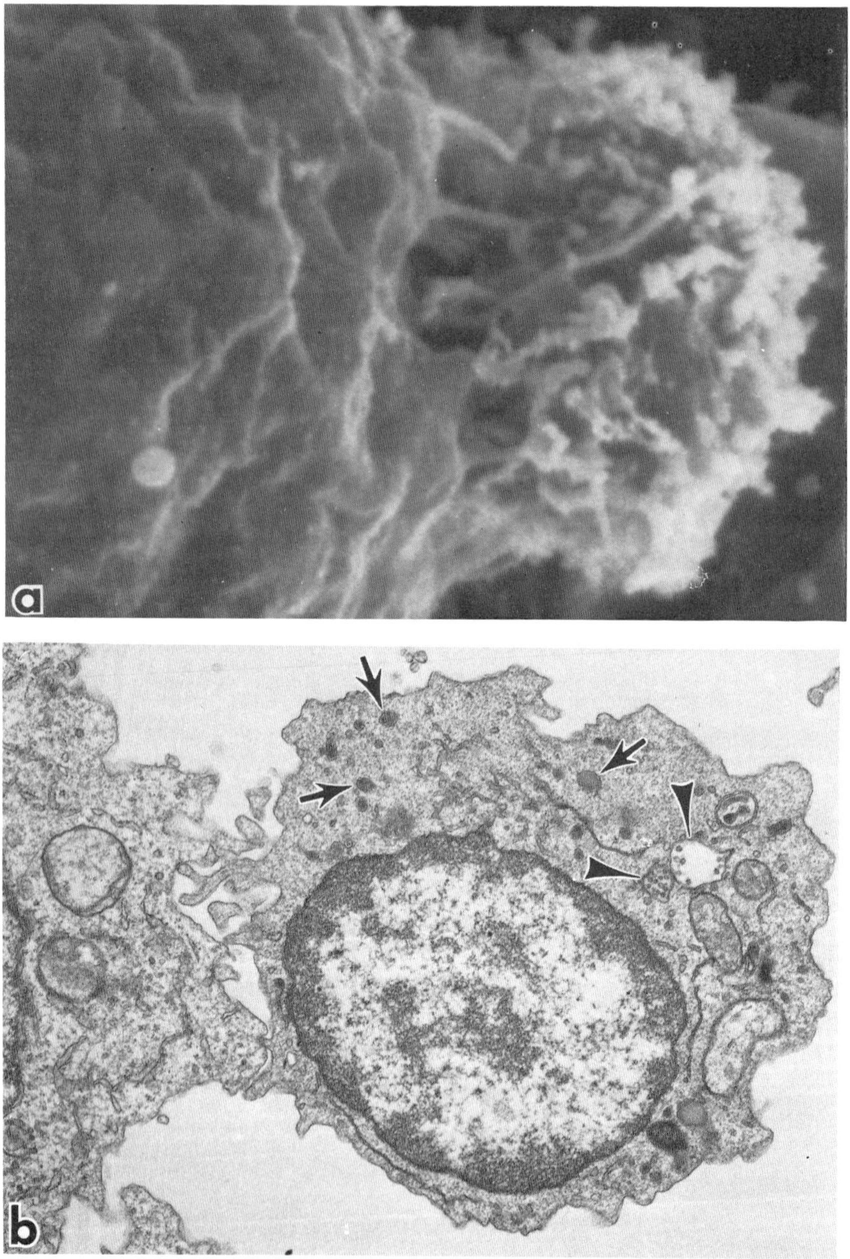

Figure 1: SEM and TEM of NK-K562 conjugate prepared in the presence of EGTA/Mg^{++}. a) SEM (width of field, 10μ) b) TEM (X 15,500). Note complex interdigitations and random orientation of LGL granules.

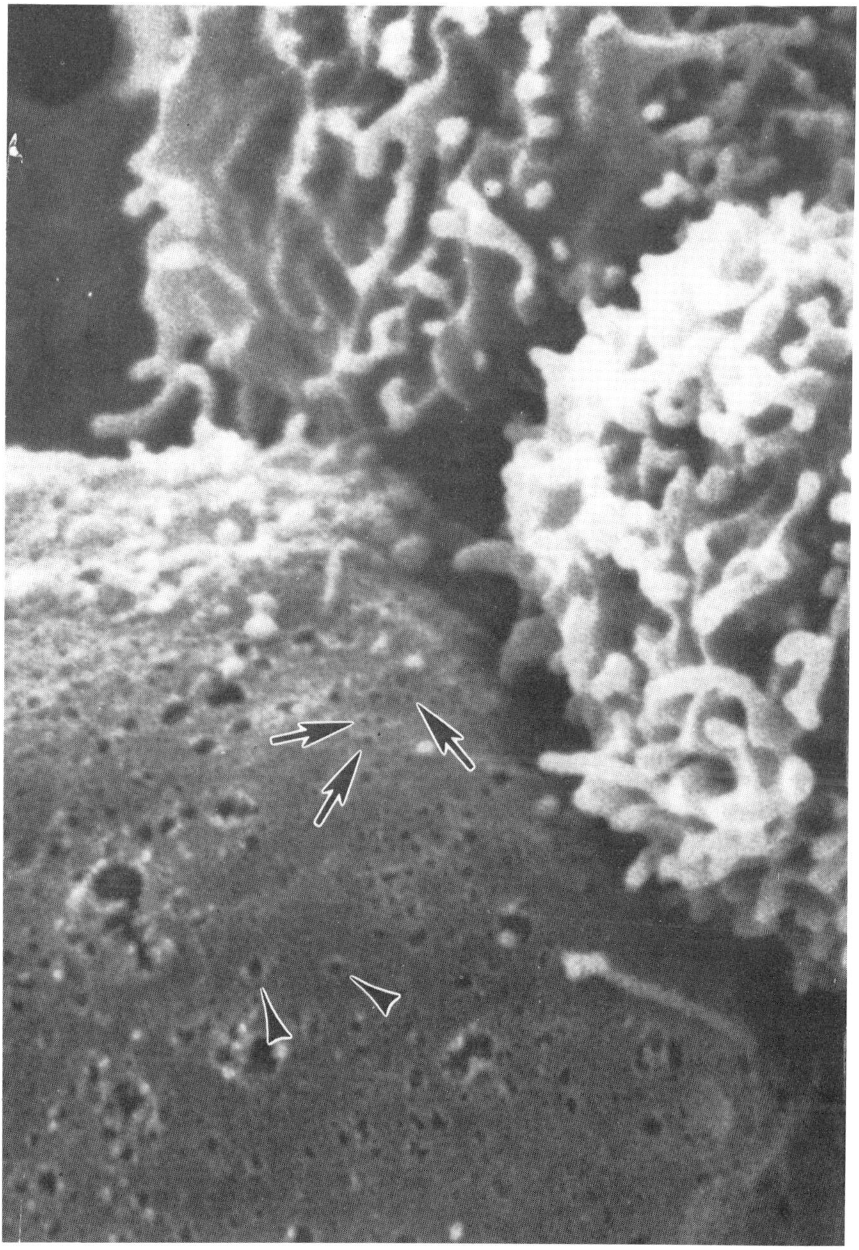

Figure 2: SEM (width of field, 6 μ) of NK-K562 conjugate 15 minutes after addition of Ca^{++}. Note numerous surface "pore" like lesions with peripheral collarettes (arrows) and heterogenous sizes.

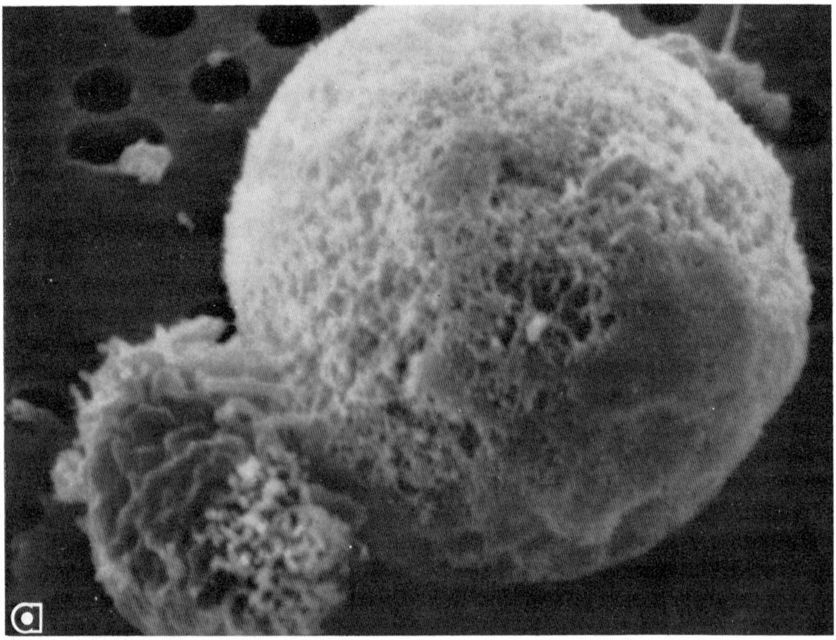

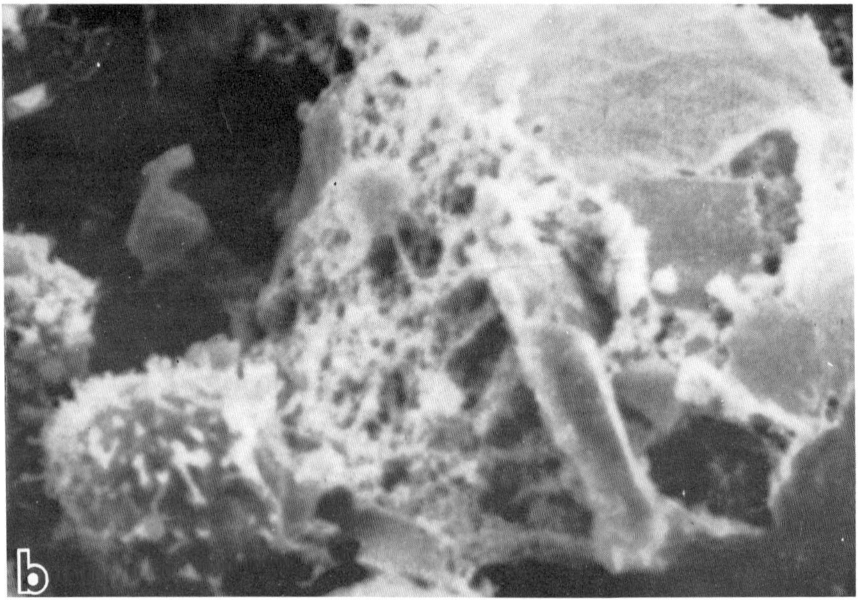

Figure 3: SEM of NK-K562 conjugates 30 minutes after addition of Ca^{++}. Two different K562 targets showing characteristic surface membrane changes: a) width of field, 12.7μ, b) width of field 18.8μ.

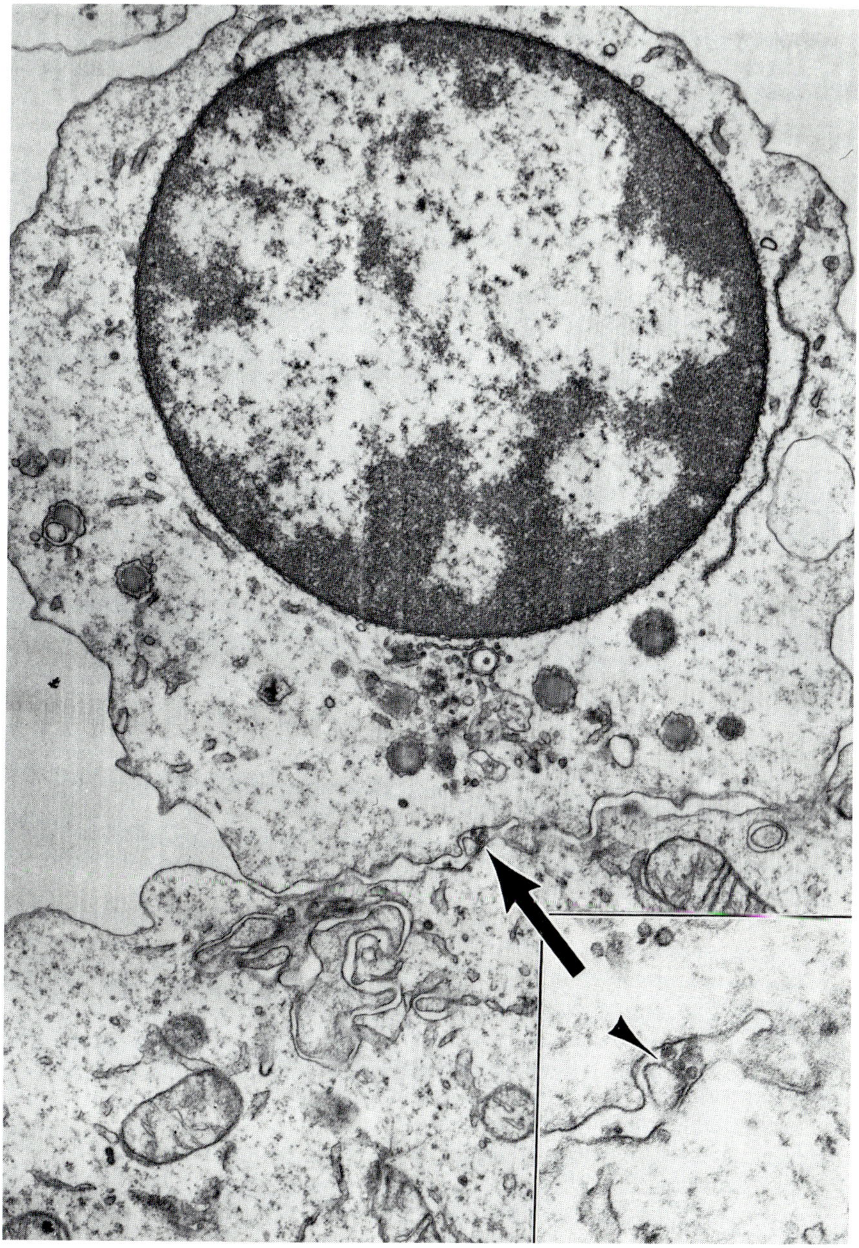

Figure 4: TEM (X 25,300) of NK-K562 conjugate 15 minutes after addition of Ca^{++}. Note densely staining ring-shaped material in the intercellular space (Arrow). Inset shows close up (X 39,700) of this material.

interactions in the presence of Mg^{++} with or without Ca^{++}. Furthermore, in the absence of added Ca^{++} ions these binding interactions persisted without significant changes in NK cell or target cell morphology or cytotoxicity.

Upon the addition of calcium (Ca^{++} pulse) ultrastructural changes occurred rapidly in the target cell membrane. Within 5 minutes the target cell lost its surface microvilli and became smooth. Large surface blebs (zeiosis) also began to occur (data not shown). By 15 minutes target cells began to show numerous small surface membranes defects with the appearance of "pores" having a peripheral collarette (Figure 2). The sizes of these surface lesions could be measured by comparison to standardized grids. The smallest of these surface lesions (arrow) measured approximately 40nm outside diameter (O.D) and 20nm inside diameter (I.D.) Larger membrane defects with irregular indurated borders are also seen (arrowhead) possibly representing coalescence or polymerization of the smaller lesions. Interestingly, these alterations were observed globally over the entire surface of the target and appeared to be unique to targets lysed by NK cells since targets killed by other methods (i.e. 56°C, 10 min.) did not show these types of membrane lesions.

When analyzed by a ^{51}Cr release assay (CRA), cytotoxicity was low at time points in which only target cell surface blebbing was the major ultrastructural change (i.e. less than 5-10 minutes after Ca^{++} pulsing). By 15 minutes after Ca^{++} pulsing (when surface lesions were present) cytotoxicity had reached significant levels (~25%). Cytotoxicity generally plateaued at around 60% by 30 to 45 minutes. However, it is difficult to compare SEM morphology to CRA since CRA data is obtained with killer to target ratios ten times higher (10:1 E:T ratio) than that used for ultrastructural analysis of NK-target interactions (1:1 E:T ratio).

Figure 3 shows target cell damage thirty minutes after initiation of cytolysis. Some targets showed breaking up and sloughing of their plasma membranes into large sheets (Fig. 3b). Parts of these membranes remained attached to the NK cell. In other instances, however, the NK cells remained attached to the target via long irregular microvillus projections (Fig. 3a). These targets appeared to be nearly completely devoid of their outer plasma membrane with only cytoskeletal elements present in the cytosol keeping the target intact.

Upon initiation of cytolysis ultrastructural alterations were also noted in the NK cell. Generally, upon the addition of calcium most NK-target conjugates demonstrated a rapid polarization of NK granules toward the interface of the target. This did not occur in the absence of added Ca^{++}. This would suggest that if the NK granules are involved in the cytolytic response their mobilization and/or release would be a clacium dependent process (5,13). This does not preclude the possibility that the activity of the granular contents is also Ca^{++} dependent (8). By serially sectioning a single NK-target conjugate we noted secreted material within the intercellular space between killer and target (Fig. 4). This material had the appearance of densely staining ring shaped structures having dimensions of approximately 15nm I.D. by 50nm O.D. (Arrow, inset shows magnified view of this material). Such material was not seen in NK-target conjugates prepared in the absence of calcium. A few of these structures appeared to be fusing with cell membranes via short "stalk-like" projections (arrowhead in inset).

It is interesting that the size of the ring shaped structures observed in this study is nearly identical to those previously noted in target cells lysed in an ADCC reaction (14) and targets lysed by cloned murine NK cells (15) and CTL (16). Taken collectively, these findings indicate that <u>secretion</u> of material which can polymerize into such structures in a target membrane may represent a general phenomenon common to all killer lymphocytes. This may, in fact represent the cellular analog of the C' system in which inactive cytolytic precursors (carried within a cell) are secreted into the extracellular space where in the presence of activating surfaces (target membrane and Ca^{++}) polymerize into ring shaped complexes with potent cytolytic capacity. While somewhat speculative at present these hypothesis are in complete agreement with such concepts previously proposed for a general mechanism of cell mediated cytotoxicity (17,18). Certainly, further ultrastructural analysis of NK-target cell interactions leading to target cell lysis will be exciting.

REFERENCES

1. Saksala, E., Timonen, T., Ranki, A., and P. Hayrig. 1979. Immunol. Rev. 44:71.
2. Stutman, O., Lattine, E., and E. Figarella. 1981. Fed. Proceedings. <u>40</u>:2699.
3. Herberman, R.B. 1982. Molec. Immunol. <u>19</u>:1313.
4. Roder, J. and T. Haliotis. 1980. In Natural cell mediated Immunity against tumors. Vol. I. R.B. Herberman, ed. Academic Press, New York. p. 379.
5. Hiserodt, J., Britvan, L. and S. Targan. 1982. J. Immunol. <u>129</u>:1782.

6. Hiserodt, J. Britvan, L. and S. Targan. 1983. J. Immunol. 131:2710.
7. Carpen, O., Virtanen, I., and E. Saksala. 1981. Cell. Immunol. 58:97.
8. Henkart, P., Millard, P., Reynolds, C., and M. Henkart. 1984. J. Exp. Med. 160:75.
9. Carpen, O., Virtanen, I., and E. Saksela. 1982. J. Immunol. 128:2691.
10 Grossi, C., Zicca, A., Cadoni, A., Mingari, M., Moretta, A., and L. Moretta. 1983. Eur. J. Immunol. 13:670.
11. Timonen, T., Ortaldo, J., and R. Herberman. 1981. J. Exp. Med. 153:569.
12. Grimm, E., Price, Z. and B. Bonavida. 1979. Cell. Immunol. 46:77.
13. Quan, P., Ishizaka, T. and B. Bloom. 1982. J. Immunol. 128:1786.
14. Dourmashkin, R., Deteix, P., Simone, C. and P. Henkart. 1980. Clin. Exp. Immunol. 43:554.
15. Podack, E. and G. Dennert. 1983. Nature (Lond.) 302:442.
16. Dennert, G. and E. Podack. 1983. J. Exp. Med. 157:1483.
17. Hiserodt, J. and G. Granger. 1978. J. Reticuloendothelial Soc. 24:472.
18. Ware, C. and G. Granger. 1981. J. Immunol. 126:1919.

ROLE OF PROTEASES IN NK ACTIVITY

Ronald H. Goldfarb

Department of Immunology and Infectious Disease,
Central Research Division
Pfizer, Inc.
Groton, Connecticut

I. INTRODUCTION

Proteases are degradative enzymes with regulatory capacity in diverse physiological and pathological processes that function through selective and limited enzymatic cleavage of peptide bonds. The specificity of peptide bond cleavage by proteases, coupled with the irreversible consequences of limited proteolysis, are key factors involved in the regulatory action of proteases in multiple biological processes including: blood coagulation; fibrinolysis; digestion; and hormone activation (1). Limited proteolysis may be restricted to independent events or may trigger secondary, sequential, cascading steps by rapid and specific irreversible zymogen activation(s). Proteolytic enzymes are often associated with cellular migration, invasiveness, and degradative tissue remodeling of intracellular substrates, cytoskeletal components, and cell surface molecules (2).

Neutral serine proteases have been examined for their role in cell-mediated tumor cell lysis by several lymphoid effector cells including T cells, activated macrophages, and NK cells (3-14).

II. INHIBITION OF NK KILLING BY PROTEASE INHIBITORS

Several reports have suggested that neutral serine proteases may play a role in the NK lytic mechanism (8-12). Evidence for this hypothesis has come from the examination of the action of protease inhibitors on NK killing, as well as the examination of NK-containing lymphoid cells and large granular lymphocytes

ISBN 0-12-341370-2

(LGL) that account for NK cell activity for the production of proteolytic activity. A number of studies have demonstrated that NK lysis can be blocked by a variety of both low and high Mr inhibitors of proteolytic activity(8-23). We have previously described the inhibition of highly purified LGL (>95%) by inhibitors of neutral serine proteases (10,13,14). Active site titrants and inhibitors of neutral serine proteases, including DFP and NPGB, inhibited both spontaneous, and IFN-boosted LGL activity in a dose-dependent manner, as did chloromethyl ketones of tryptic (TLCK) and chymotryptic (TPCK) specificity (10,13, 14). In addition, the NK activity of LGL was inhibited by benzamidine, p-aminobenzamidine, chymostatin, and leupeptin, but only slightly inhibited by elastatinal or trasylol (10,13,14). Hirudin, the selective inhibitor of alpha thrombin was also found to have no effect on LGL reactivity. Our results have therefore indicated that inhibitors of tryptic enzymes (p-aminobenzamidine, benzamidine, leupeptin, and TLCK) and of chymotryptic enzymes (chymostastin and TPCK), inhibit the NK activity of highly purified LGL against K562 cells in a serum-free chromium release assay. Additional studies with LGL have also indicated that TLCK and DFP can inhibit NK cytotoxicity, whereas inhibitors of elastase do not appear to play a role in cytotoxicity(15). Evidence has accumulated that NK cells produce several proteolytic enzymes, all of which are not involved in mediating cytolysis (10,12,15-16). A number of additional studies have also demonstrated that small protease inhibitors can inhibit NK killing of target cells in chromium release assays (12,17-19). t is important to note that results obtained with some inhibitors, particularly the chloromethyl ketones (eg. TLCK and TPCK) must be interpreted with some caution. It has been demonstrated that TLCK, for example, can react with sulfhydryl or imidazole groups of membrane components in addition to proteolytic enzymes (20). Although TLCK is an active site titrant for tryptic enzymes, TLCK can react with additional proteolytic and non-proteolytic enzymes and components (18,20). For example, TLCK can bind to the T200 CTL glycoprotein (21)

Macromolecular anti-proteinases have also been used as probes to determine whether a proteinase plays a role in the NK lytic mechanism (22-24). Recent studies have indicated that at least one serine-dependent proteinase with aromatic amino acid cleavage specificity is involved in the NK lytic mechanism; in addition to inhibition of NK lysis of K562 cells by DFP, and PMSF, it was also observed that alpha-1-antichymotrypsin and chymostatin blocked NK cytotoxicity (17).

III. DIRECT EVIDENCE FOR PROTEASE PRODUCTION BY NK CELLS

A number of studies have demonstrated that NK cells produce proteolytic activity (12-16). In studies with ^{3}H DFP- labeled NP-40 lysed preparations of peripheral blood lymphocytes, and SDS PAGE analysis, major components with Mr of 200 and 35 kilodaltons, and minor components with Mr of 58 and 45 kilodaltons were observed (12). The 35 kilodalton Mr band was found to be a TLCK-sensitive tryptic-like protein, perhaps related to a previously described lymphocyte-derived cytolytic protease (5, 12). The 58 and 45 kilodalton Mr bands were found to bind to ^{3}H-acetyl-ala-ala-pro-val chloromethylketone, which is known to bind to elastase (12).

In studies with highly purified preparations of human LGL (>95%), and IL-2 cultured LGL, it was demonstrated that NK cells produce the specific neutral serine protease, plasminogen activator (16). The enzyme was found in both an extracelluar form as well as in a cell associated form, which upon subcellular distribution, associated with a cell-surface membrane enriched fraction. The extracellular proteolytic activity, upon concentration and analysis on SDS PAGE, consisted of multiple Mr species of 100, 78, 52, 45, 28, and 26 kilodaltons (16). IFN-treament of LGL led to a substantial increase in cell-associated, but not extracellular proteolytic activity (16). LGL isolated from patients with Chediak-Higashi syndrome, who are known to be defective in NK activity, displayed low plasminogen activator activity when compared to LGL from normal donors (16).

Additional studies with LGL have provided evidence for the production of several enzymes including aryl sulfatase and more than one neutral proteinase including an elastase-like activity capable of degrading serum amyloid A (15). In contrast to TLCK, which blocked cytotoxicity but not degradation of serum amyloid A, inhibitors of elastase prevented degradation of serum amyloid A, but not cytotoxicity (15). The results therefore suggest that elastase does not participate in NK killing, and that NK cells produce a variety of enzyme classes with different substrate specificities and functions (15). Additional studies have also suggested that elastase does not play a direct role in NK lysis of target cells (10,12).

An additional study suggesting a role for proteolytic enzymes in the NK lytic mechanisms has been the observed correlation between NK activity and a growth-inhibitory serum protein with protease like activity (25). In addition, a number of proteases, including ng quantitites of trypsin and chymotrypsin, as well as granulocyte-derived medullasin, enhances NK killing (10,26). The enhancement of LGL killing of K562 cells with ng quantities of trypsin and chymotrypsin, is reminiscent of studies that have shown protease-enhanced ADCC and CMC against tumor target cells (27).

IV. EXAMINATION OF NKCF AND LGL GRANULES FOR PROTEASE ACTIVITY

To further explore the role of proteolytic enzymes in the NK lytic mechanism, we have examined both soluble cytotoxic factors derived from human LGL (NKCF) and cytolytic granules derived from rat tumor LGL. Both NKCF and LGL granules have been studied for thier role in NK cell-mediated cytotoxicity of tumor cells (28-30). The human LGL-derived NKCF that we have studies was provided by Drs. John Ortaldo and Isaac Blanca (NCI-FCRF), and the rat tumor LGL granules were obtained from Drs. Pierre Henkart and Craig Reynolds (NCI, NIH and NIC-FCRF). The proteolytic activities of these samples were compared to a partially purified protease from murine activated macrophages for their enzymatic selectivity in cleaving a series of synthetic fluorogenic substrates, selective for a number of neutral, serine proteases, which utilize the fluorescent leaving group 7-amino-4-methylcoumarin (AMC) (31). The results of these studies are shown in Table I. The results indicate that NKCF is devoid of proteolytic activity when assayed on a number of substrates with selectivity for plasminogen activator, plasmin, trypsin, chymotrypsin and elastase. In contrast, the results in Table I demonstrate that the rat LGL granules contain neutral serine proteolytic activity with the capacity to cleave substrates for plasminogen activator, plasmin, and trypsin, but not for chymotrypsin and elastase. Our results also demonstrate that rat LGL granular proteolytic activity is inhibited by leupeptin, NPGB, Gly-Gly-Arg-CK, TLCK, and to a lesser extent trasylol, when examined on a CBZ-Gly-Gly-Arg-MCA substrate (31). In addition, leupeptin inhibited rat granular proteolytic activity assayed on a Boc-Phe-Ser-Arg-MCA tryptic substrate and trasylol more efficiently inhibited plasmin activity on the plasmin Suc-Ala-Phe-Lys-MCA substrate. The partially purified 38 kilodalton Mr protease, derived from activated macrophages, cleaved the CBZ-Gly-Gly-Arg-MCA substrate, but failed to cleave substrates for trypsin, chymotrypsin, plasmin, or elastase. When assayed on the CBZ-Gly-Gly-Arg-MCA substrate, proteolytic activity of the macrophage protease is inhibited by leupeptin, TLCK, and benzamidine, and to a lesser extent, by Soybean trypsin inhibitor. It should be kept in mind that this substrate, although designed for cleavage by plasminogen activators, may also be cleaved by trypsin, plasmin, alpha thrombin, chymotrypsin, and elastase. In contrast plasminogen activators fail to cleave the substrate for chymotrypsin or elastase substrates. The results therefore suggest that rat LGL granules contain enzymatic activity capable of cleaving selective substrates for several neutral serine proteases: plasminogen activator, plasmin, and trypsin.

It has prevously been reported that rat LGL granules, when assayed on fluorogenic substrates for chymotrypsin, trypsin, and elastase(N-succinoyl-MCA derivatives: N-Suc-Ala-Ala-Phe-MCA, Ala-Phe-Lys-MCA, and Ala-Ala-Ala-MCA), did not reveal substantial granule-associated proteolytic activities (29). It was concluded that several proteolytic enzymes were not major components of NK granules (29). Although in agreement with the findings with chymotrypsin and elastase, our results differ with re-

Table I. Proteolytic activity of LGL granules and NKCF[§]

Enzyme	Inhibitor	Substrate	△ Fluorescence/min.	Per Cent Inhibition
Urokinase[a]	-----	CBZ-Gly-Gly-Arg-MCA	22	---
"	benzamidine(1.4M)	"	0	100
Plasmin[b]	-----	SUC-Ala-Phe-Lys-MCA	40	---
"	trasylol(.82M)	"	0	100
Trypsin[c]	-----	BOC-Phe-Ser-Arg-MCA	48	---
"	leupeptin(10^{-3}M)	"	0	100
Chymotrypsin[d]	-----	SUC-Ala-Ala-Phe-MCA	64	---
"	chymostatin(10^{-3}M)	"	0	100
Elastase[e]	-----	SUC-Ala-Phe-Lys-MCA[f]	20	---
"	elastatinal(10^{-3}M)	"	0	100
NKCF(Human LGL)	-----	CBZ-Gly-Gly-Arg-MCA	0	---
"	-----	SUC-Ala-Phe-Lys-MCA	0	---
"	-----	BOC-Phe-Ser-Arg-MCA	0	---
"	-----	SUC-Ala-Ala-Phe-MCA	0	---
"	-----	SUC-Ala-Phe-Lys-MCA	0	---
Granules(Rat LGL)	---	CBZ-Gly-Gly-Arg-MCA	9	---
"	leupeptin(10^{-3}M)	"	0	100
"	NPGB (10^{-3}M)	"	0	100
"	Gly-Gly-Arg-CK(10^{-3}M)	"	0	100
"	TLCK(10^{-3}M)	"	0	100
"	trasylol(.82M)	"	4	56
"	-----	SUC-Ala-Phe-Lys-MCA	18	---
"	trasylol(.82M)	"	3	83
"	-----	BOC-Phe-Ser-Arg-MCA	16	---
"	leupeptin(10^{-3}M)	"	5	69
"	-----	SUC-Ala-Ala-Phe-MCA	0	---
"	-----	SUC-Ala-Phe-Lys-MCA	0	---
MacroØ-Protease	-----	CBZ-Gly-Gly-Arg-MCA	9	---
"	leupeptin(10^{-3}M)	"	0	100
"	TLCK (10^{-2}M)	"	0	100
"	benzamidine(1.4M)	"	0	100
"	SBTI (500 BAEE U.)	"	6	33
"	-----	SUC-Ala-Phe-Lys-MCA	1	---
"	-----	BOC-Phe-Ser-Arg-MCA	0	---
"	-----	SUC-Ala-Ala-Phe-MCA	0	---
"	-----	SUC-Ala-Phe-Lys-MCA	0	---

[a] plasminogen activator, 2 Ploug units
[b] 2 X 10^{-2} caseinolytic units
[c] 6 X 10^{-3} ng
[d] 2 X 10^{-3} mg
[e] 1 X 10^{-2} elastinolytic units
[f] Similar results were found on Meth-Suc-Ala-Ala-Pro-Val-MCA
[§] Purified reference proteases, LGL-derived NKCF, LGL-granules, and activated macrophage derived protease were assayed as previously described (7,31).

pect to plasmin-like activities and/or plasminogen activator-like enzymatic activities (32).

V. ROLE OF PROTEASES IN STAGES OF THE MECHANISM OF NK KILLING

Studies with the inhibitors TLCK, TPCK, and ATEE have been used as probes in the analysis of proteases in different stages of the NK lytic mechanism (33). It has been reported that TLCK inhibited killing beyond Ca++ dependent programming events when tested in a Ca++ pulse assay. When TLCK was added to pre-programmed target cells in a killer cell independent assay, lysis was completely inhibited, suggesting that TLCK affected events occuring on the target cell during the NK lethal hit (33). Additional findings have suggested that serine proteases may be involved not only in an NK cell associated early triggering stage of NK lysis, but also during the terminal lethal hit stage on the target cell as part of a transferred lytic complex (33). It has been proposed that proteases involved in NK cytotoxicity may be involved in either the initiation of cytolysis, as part of a stimulus and secretion of cytolytic molecules, or in the final series of events that leads to the formation of the final lytic substances that participate in lytic events (11b, 17,19). It has been suggested that cell-surface associated NK proteases become exposed and activated upon interaction with target cells during cytolysis and that expression of the cytotoxic protease may require the action of the contractile apparatus of the cell (12).

VI. ENZYMATIC SPECIFICITIES OF PROTEASES ASSOCIATED WITH NK CELLS

A number of studies have suggested that tryptic or chymotryptic proteolytic enzymes may be involved in NK cell mediated lysis of tumor cells (8-18). A number of reports have suggested that chymotryptic, rather than tryptic, activity is involved in NK killing (11, 17). Several reports have suggested that elastase activity does not play a direct role in NK cell-mediated cytotoxicity (10,12,15). One report has suggested that metalloproteases, as well as serine proteases, may contribute to NK killing (19). One report has demonstrated that LGL produce plasminogen activator activity and suggests that the plasminogen activator/plasmin system may contribute, either directly or indirectly, to the lysis of tumor cells by NK cells (16). The latter finding is of interest since plasminogen activator-like activity has also been examined as a protease involved in the abrogation of lymphocyte cytotoxicity (34).

VII. CONCLUSIONS

The exact role and scope of action of proteolytic enzymes in the killing of tumor cells by NK cells remains undefined. It is clear, however, that a number of inhibitors of proteolytic enzymes can bloc NK killing, and that highly purified populations of NK cells produce proteolytic activity. Mechanistic studies have suggested that serine proteases may be involved in both the early triggering stage of NK lysis as well as during the terminal lethal hit stage. The relationship between proteolytic activity, NKCF, and LGL cytolytic granules remain to be fully explored within the substages of the NK lytic mechanism. Similarly, the mechanism of the processing and export of LGL proteases, their potential interaction with cytoskeletal elements, their modulation by ion fluxes, and their potential role in cleavage, processing, and export of cytolytic molecules from LGL granules, through the LGL granular membrane, and ultimate release and insertion into target cells remain open issues. It is likely that the critical analysis of these issues will require the employment of highly purified LGL, purified proteolytic enzymes from LGL, and the identification of the natural substrates of protease action. It is likely that a thorough understanding of LGL proteolytic activity will contribute to our understanding of the mechanism of NK killing, as well as to aspects of NK cell migration through extracellular barriers to allow for NK cells to reach and lyse primary and metastatic tumor foci _in vivo_ (2,15).

ACKNOWLEDGEMENTS

We thank Drs. Pierre Henkart and Craig Reynolds for their gift of rat tumor LGL cytolytic granules, Drs. John R. Ortaldo and Isaac Blanca for their gift of human LGL-derived NKCF, and Drs. William Johnson and Dolph O. Adams for their gift of murine macrophage cytolytic-protease-containing supernatants. We thank Ms. Valerie Van Dyke-Phillips for her expert technical assistance.

REFERENCES

1. Reich,E., Rifkin, D.B., and Shaw, E. (1975) "Proteases and Biological Control", Cold Spring Harbor Press, New York.
2. Goldfarb, R.H. (1984) _In_ "Mechanisms of Metastasis: Therapeutic Implications"(K. Honn and B. Sloane, eds.) Martinus Nijhoff, Amsterdam, In Press.
3. Trinchieri,G., and De Marchi, M. (1976). J. Immunol. _116_, 885.
4. Redelman, D. and Hudig, D. (1980). J. Immunol. _124_: 870.
5. Hatcher, V.B., Oberman, M.S., Lazarus, G.S., and Grayzel, A.I. (1978). J. Immunol. _120_, 665.
6. Adams, D.O., Kao,K.J., Farb, R., and Pizzo, S.V. (1980). J. Immunol. _124_: 293.
7. Johnson,W.J., Goldfarb, R.H., Van Dyke-Phillips, V., and Adams, D.O. (1984) _In_ "Lymphokines" (S.B. Mizel, ed.) Academic Press, New York. In Press.
8. Roder, J.C., Kiessling, R., Bibberfeld, P., and Anderson, B. (1978). J. Immunol. _121_: 2509.
9. Hudig,D., Redelman, D., and Minning,L. (1982) _In_ "NK Cells and Other Natural Effector Cells" (R.B. Herberman, ed.),p. 939, Academic Press, New York.
10. Goldfarb, R.H., Timonen,T.,and Herberman, R.B.(1982) _In_ "NK Cells and Other Natural Effector Cells" (R.B. Herberman, ed.), p. 931, Academic Press, New York.

11. Quan,P., Ishizaka, T., and Bloom, B.R. (1982) In "NK Cells and Other Natural Effector Cells" (R.B. Herberman, ed) p. 989, Academic Press, New York.
11b. Quan,P., Ishiazka, T, and Bloom, B.R. (1982) 128, 1786.
12. Lavie,G.(1982) In "NK Cells and Other Natural Effector Cells" (R.B. Herberman, ed) p. 939, Academic Press, New York.
13. Goldfarb, R.H., and Herberman, R.B. (1982) Adv. Inflamm. Res. 4, 45.
14. Goldfarb, R.H., Timonen, T., and Herberman, R.B. (1982) Adv. Exp. Med. Biol., 146, 403.
15. Zucker-Franklin,D., Yang, J., and Fuks, A. (1984) Proc. Natl. Acad. Sci. (USA), 132, 1451.
16. Goldfarb, R.H., and Herberman, R.B. (1984) J. Exp. Med., 159, 935.
17. Hudig,D., Redelman,D., and Minning, L. (1984) J. Immunol., 133, 2647.
18. Ristow, S.S., Starkey,J.R., and Hass, G.M. (1983) Immunol., 48, 1.
19. Floyd, R.A., Kuo, C.Y., Yoo, T.J., Song, A., Postlethwaite, A.E., and Mainardi, C. (1984) J. Lab. Clin. Med., 103, 215.
20. Chang, T.W., and Eisen, H.N.,(1980) J. Immunol., 124: 1028.
21. Pasternack, M.S., Sitkovsky, M.V., and Eisen, H.N. (1983), J. Immunol., 131: 2477.
22. Hudig, D., Haverty, T., Fulcher, C., Redelman, D., and Mendelsohn, J. (1981) J. Immunol., 126, 1569.
23. Ades, E.W., Hinson, A., Chapuis-Cellier, C., and Arnaud,P. (1982) Scand. J. Immunol., 15, 109.
24. Gravagna, P., Gianazza, E., Arnaud, P., Neels, M., and Ades, E.W. (1982) Scand J. Immunol., 15, 115.
25. Gundersen, S., Funderud, S., Bloom, B.R., and Godal, T. (1983) Acta. Path. Microbiol. Scand. C., 91: 137.
26. Aoki,Y., Sumiya,M., and Oshimi, K. (1982) J. Clin. Invest., 69, 1223.
27. Kedar, E., De Landazuri, M.O., and Fahey, J.L. (1974), J. Immunol., 112, 26.
28. Wright, S.C. and Bonavida, B. (1983), Proc. Natl. Acad. Sci. (USA), 80: 1688.
29. Millard, P.J., Henkart, M.P., Reynolds, C.W., and Henkart,P. A. (1984), J. Immunol., 132, 3197.
30. Henkart, P.A., Millard, P.J., Reynolds, C.W., and Henkart, M.P. (1984), J. Exp. Med., 160, 75.
31. Zimmerman, M., Quigley, J.P., Ashe, B., Dorn, C., Goldfarb, R.H., and Troll, W. (1978), Proc. Natl. Acad. Sci. (USA), 75, 750.
32. Pierzchala, P.A., Dorn, C.P.,and Zimmerman, M., (1979), Biochem. J., 183, 555.
33. Brogan, M., and Targan, S., (1984) Fed. Proc., 43: 1443.
34. Kundar, S.K., Bergeron,J., and Menezes, J. (1984), Clin. Exp. Immunol., 56, 701.

ENZYMES INVOLVED IN NATURAL KILLER CELL ACTIVITY: EVIDENCE FOR PHOSPHOLIPASES, LIPOXYGENASES, AND SERINE DEPENDENT PROTEINASES

Dorothy Hudig
Doug Redelman

Department of Microbiology
School of Medicine
University of Nevada, Reno
Reno, Nevada

Kathleen Carine
Lory Minning

UC San Diego Cancer Center
University of California, San Diego
La Jolla, California

I. INTRODUCTION

The mechanism by which lymphocytes kill other cells is poorly understood. Somehow cytotoxic lymphocytes deliver a lethal lesion to their targets, yet are themselves undamaged and are able to lyse additional targets. Despite considerable effort, no real understanding of the nature of the lesion, its generation and/or release has accrued. However, new data from Henkart et al. (1,2,3) provide direct evidence that there is a lytic substance found in killer cell granules which can mediate cytotoxicity. It is likely that there must be a release mechanism for the cytotoxic granules and also a specific "trigger" to control this release. Assembly may also be needed before the lytic substance is fully active.

We have used indirect approaches to the mechanism of lymphocyte-mediated cytotoxicity by assessing the effects of specific inhibitors on the activity of live cells. Initial

ISBN 0-12-341370-2

experiments to evaluate the role of phospholipases and lipoxygenases in natural cytotoxicity (NK) were inspired by proposals that lysolecithin and leukotriene products could be fusogens and Ca++ ionophores, respectively (4,5). The similarities between complement-mediated and lymphocyte-mediated cytotoxic lesions (6,7), and the incompleteness of experiments designed to implicate complement components in lymphocyte-mediated killing, led us to determine if any serine dependent proteinases, complement or other, could be implicated in the NK process. Our approaches, described below, indicate that phospholipases, lipoxygenases, and serine dependent proteinases are all likely to participate in the cellular process of killing.

II. MATERIALS AND METHODS

A. NK Assays

We used standard ^{51}Cr release assays with freshly prepared human peripheral blood mononuclear cells as effectors (8), and K562 cells obtained directly from the Drs. Lozzio (9) in all assays unless otherwise indicated. These assays were always performed at several effector lymphocyte to K562 target (E:T) ratios. Details may be found in reference 10.

B. Assays for Lymphocyte-K562 Conjugate Formation

Conjugates were formed by centrifuging 2:1 ratios of peripheral blood mononuclear cells to K562 cells for 3 minutes at 40g at room temperature. The resultant conjugates were kept on ice and gently resuspended prior to direct counts with a hemocytometer. The larger K562 cells could be readily distinguished from the normal mononuclear cells.

C. Assays for Effects of Antiproteinases on Killer Cell Independent Lysis

It is possible that a proteinase or proteinases might digest K562 cells or selectively destroy critical K562 proteins. In such situations, the proteinase effects might continue after the killer lymphocytes detach from their targets, during the stage of killing marked by a rate of ^{51}Cr release. If a critical post lymphocyte-mediated event were blocked, one might observe slower rates and/or less final

^{51}Cr release. To evaluate these possibilities, lymphocytes were permitted to kill for 1 hour in routine ^{51}Cr assays, and then stopped killing by the concurrent addition of 10 mM NaN3 and 25 mM 2-deoxyglucose (final concentrations); see (10) for details. (The combination of these two metabolic inhibitors leaves the cells with no source of energy for cytolysis, and was selected instead of the Ca^{++} pulse with EDTA because many serine dependent proteinases require divalent cations for activity). For these experiments, either alternate substrates or control medium was added concurrently with the metabolic inhibitors and then the ^{51}Cr release measured.

D. Reagents

Dexamethasone, diisopropylfluorophosphate (DFP), p-bromophenacyl bromide (BPB), hydrocortisone, nordihydroguaiaretic acid (NDGA), phenylmethylsulfonyl fluoride (PMSF) and quinacrine were obtained from Sigma Chemical Co, St. Louis, MO. Eicosatetraynoic acid (ETYA) and hydroxyphenylretinamide were generously provided by Drs. W.E. Scott and P. Sorter of Hoffmann-LaRoche, Inc., Nutley, NJ. Rosenthal's inhibitor was obtained from Calbiochem, La Jolla, CA. All the synthetic alternate substrates were obtained from Vega Biochemicals, Tucson, AZ. Alpha-1-antichymotrypsin (a-1-X) was a gift from Dr. James Travis, University of Georgia, Athens, GA. Hydrophobic reagents, such as AcTyrOEt were dissolved first in 0.1 ml 100% ethanol and then diluted to appropriate concentrations in medium. Further dilutions were then made in medium containing identical concentrations of ethanol. Concurrent controls of medium with these ethanol concentrations were always included.

III. RESULTS

A. Lipoxygenases Are Implicated in NK, but the Role for Phospholipase A2 Cannot Be Unequivocally Affirmed

Many reagents can affect phospholipase A2 activity of cells or tissues. These reagents include glucocorticoids such as natural hydrocortisone and synthetic dexamethaxone, quinacrine, Rosenthal's inhibitor and p-bromophenacyl bromide (BPB). Not all of these reagents affect phospholipase A2 directly. The glucocorticoids are thought to induce a protein termed lipomodulin which in turn reduces phospholipase A2 activity. The mechanism by which quinacrine can also reduce phospholipase A2 activity is unclear. Rosenthal's inhibitor acts as a competitive nonhydrolyzable substrate. BPB alone directly interacts with the phospholipase A2 active site and irreversibly inactivates the enzyme. When we evaluated the effects of these reagents on NK (11) we found that only BPB was a suitable reagent for the reasons summarized in Table 1.

Additional studies indicated that only BPB affected radiolabeled arachidonate metabolism of lymphocytes (11), an activity that would result from phospholipase A2 activity or the combined activity of phospholipase C and diglyceride lipase. The troubling observation is that BPB also affected conjugate formation, an effect that could be ascribed to the ability of BPB to react with sulfhydryl groups (12). Since we have demonstrated that alkylation of cell surface thiol

Table 1. Summary of the Effects of Phospholipase "Inhibitors" on Human Natural Killer Activity

Reagent	Effect on NK	Comments
Dexamethasone, Hydrocortisone	None at physiological concentrations	Effects at extremely high concentrations could not be ascribed to lipomodulin induction
Rosenthal's reagent	Inhibited	Toxic to lymphocytes
Quinacrine	Inhibited Effects of 37° incubation not reversible	Effects should have been fully reversible. Suspect quinacrine affects NK as a lysosomotropic reagent altering intragranule pH.
BPB	Inhibited by pretreatment of lymphocytes	Effects were with predicted activity. However, also affected conjugate formation.

groups compromises NK (13), these experiments mean that BPB could have dual effects on NK cells, either of which would inhibit killing. Thus, there are no reagents which provide unequivocal evidence that phospholipase A2 is required for NK.

The inhibitors that can affect the lipoxygenase pathway of arachidonate metabolism, ETYA, NDGA and hydroxyphenyl-retinamide all inhibited NK activity when included in the assays but not when lymphocytes were pretreated (11), as would be expected if the reagents were affecting production of leukotrienes. The activity of hydroxyphenylretinamide, which selectively affects the 5-lipoxygenase pathway, could block formation of the leukotriene B4 which can act as a Ca^{++} ionophore (5). It is possible that formation of this leukotriene is one regulatory mechanism that controls the Ca^{++} required for NK.

B. A Proteinase of the Class Characterized by Serine in the Active Site is Implicated in the Lethal Events of NK Cytotoxicity, but not in Target Binding or in Killer Cell Independent Lysis

Table 2 summarizes the results which are presented in greater detail in references 14 and 15. The major points are that the proteinase is of the serine dependent class and that only those reagents which would affect proteinases with cleavage specificity at aromatic amino acids were highly effective in blocking NK activity.

The inhibitions with DFP, PMSF and the plasma antiproteinase α-1-antichymotrypsin assign the activity to the serine dependent class of proteinases, as opposed to other classes such as the cysteine dependent (e.g., papain), the aspartic dependent (e.g., pepsin), or the metallo dependent (e.g., collagenase). However, because of the thiol-reactive contaminants in commercial DFP, and the alkylating properties of PMSF, the most compelling data are the experiments with α-1-antichymotrypsin. This plasma antiproteinase reacts only with serine dependent proteinases with aromatic specificity of cleavage. Therefore, this reagent provides independent confirmation of the implications of the experiments with both the class specific inhibitors DFP and PMSF and with the alternate substrates.

Table 2. Proteinase Inhibitors that Block NK Activity

Reagent	Concentration (mM) (ID_{50})	Implications	Comments
DFP	4	Serine class	DFP also has thiol reactive contaminants
PMSF	1	Serine class	Can also alkylate thiols
A-1-X	Approx. 0.025	Serine class & aromatic specifity	No other reactivity; too big to readily enter cells
AcTyrOEt, AcTrpOEt	0.4 0.1	Aromatic specificity	Nonaromatic amino acid derivatives had little effect

Only the synthetic substrates containing aromatic amino acids effectively inhibited NK. The overall ability of these synthetic substrates and inhibitors to block NK activity correlates well with their reported interactions with purified chymotrypsin. I.e., L-enantiomers competed NK 2-fold better than D-enantiomers, and ester derivatives competed better than amides. For chymotrypsin, the K_m of L forms is about twofold lower than the K_i of the nonhydrolyzable D forms. (In situations in which an alternate substrate is added with the substrate for which hydrolysis will be measured, the alternate substrate acts as a competitive inhibitor. Esters are sensitive substrates for the serine proteinase activity because these proteinases form ester intermediates between the serine of the proteinase active site and the carboxy terminus of the protein natural substrate and then hydrolyze this ester bond.) Representative data are provided in Table 2. The ID_{50} concentrations are those that inhibited NK sufficiently so that it required twice as many inhibited lymphocytes for lysis equivalent to control, untreated cells.

Table 3. **Inhibition of NK by Synthetic Alternate Substrates**

Reagent	ID_{50} (mM)
Ac-L-ArgOMe	None at 20
Ac-L-Tyr	20.0
AcL-IleOMe	10.0
Ac-L-TyrNH2	3.5
Ac-D-TyrOEt	0.9
Ac-L-TyrOEt	0.4
Ac-L-TrpOMe	0.1

The aromatic alternate substrates also did not affect conjugate formation of either freshly isolated lymphocytes to K562 cells (10) or of cloned human lymphocytes with NK-like activity (Hudig and Minning, unpublished results). It can always be argued that with fresh lymphocytes, only a fraction of the binding cells are actually killers, and that one could easily miss a low percentage of inhibition because the sample numbers are always relatively small when cells are hand counted. However, these objections become less valid with cloned cells where all the cells represent the binding potential of killer cells. Furthermore, the cloned killer cells remained attached to their targets even after 15 hours of culture in medium containing AcTyrOEt. Nonetheless, abrogation of net binding affinity of killer cells to targets could be undetected by these assays.

Our experiments with the addition of AcTyrOEt at the killer cell independent stage, i.e., addition concurrently with the metabolic inhibitors NaN_3 and 2-deoxyglucose, indicate that the presence of the alternate substrates had no effect on subsequent ^{51}Cr release (not illustrated). We have also observed that K562 cells are not lysed in the presence of mg/ml concentrations of pronase, papain, chymotrypsin or trypsin, although these enzymes hydrolyze K562 surface structures necessary for recognition by killer cells (10).

On the basis of these two series of experiments, we consider it extremely unlikely that an NK-associated proteinase has any direct effects on targets cells.

There is one anomaly in the correlation between inhibition of NK and inhibition of chymotrypsin which may be significant: amino acid derivatives which are N-acylated are better substrates for serine dependent proteinases, whereas the nonacylated derivates were better competitiors for NK activity (Table 3). These nonacylated forms would be expected to be lysosomotropic because they would have charged amino groups in the acidic pH of lysosomes, and thus be unable to leave the lysosomes. Thus, nonacylated forms could accumulate in lysosomes at a preferential rate in comparison to acylated forms which could not be charged at the amino group of the amino acids.

Table 4. Nonacylated Alternate Substrates Block NK More Effectively than Corresponding Acylated Substrates

Reagent	ID_{50} (mM)
TyrNH2	3.5
AcTyrNH2	8.0
TyrOEt	0.06
AcTyrOEt	0.4

IV. DISCUSSION

The experiments outlined above represent a strong case for a role for phospholipases, lipoxygenases and serine dependent proteinases in NK. However, there are inherent problems in these experimental approaches which are not easy to overcome. "Specific" inhibitors may have more than one reactivity when applied to whole cells. The difficulties of interpretation of the experiments with the phospholipase A2 inhibitor BPB are a good example. While BPB will selectively inactivate phospholipase A2 molecules in their active site, it can have additional effects which could influence the killing by intact lymphocytes. In the case of BPB, we could at least demonstrate that it did have the effect on cellular lipase activity consistent with its effect on isolated lipases. However, we could not exclude the additional effects.

What one would like to observe is that a population containing only killer lymphocytes mobilizes arachidonate selectively during cytotoxicity. One would also like to recover particular radiolabeled arachidonate products from these cytotoxic situations in which only the killer cells have and can mobilize arachidonate. An ultimate resolution would be a subcellular system in which one could reconstitute the enzymes and their substrates, and generate the situation that was postulated on the basis of whole cell experiments. In this particular situation, one might like to see Ca^{++} flux linked to phospholipase A2 and lipoxygenase activity, and fusion of phospholipid vesicles.

Similar limitations affect the interpretation of the cellular experiments with the proteinase inhibitors. Although we have not isolated or characterized specific proteinases from lymphocytes, others have (16). However, one has no real assurance that a particular lymphocyte proteinase, even with the right class and substrate specificity, has a role in NK until one has identified the natural substrate of the proteinase and its role in cytotoxicity. The current state of knowledge is analogous to a hypothetical study of blood coagulation with the knowledge that serine proteinases were important but without knowledge of fibrinogen. In such a study, once fibrinogen and fibrin were identified as the components of the "clot", it would be possible to identify the coagulation cascade.

It is to be hoped that identity of the final lytic substance (1,2,3) will be the critical breakthrough to allow the delineation of the NK process. For now, we can follow the guidelines from the indirect evidence to design experiments to test whether this NK-associated proteinase has a role in assembly of a final lytic substance, whether the apparent NK-associated proteinase is a ubiquitous "trigger" for degranulation which will be shared with mast cell degranulation (4), or whether the proteinase has another currently unpostulated role in the NK cellular process.

V. ACKNOWLEDGEMENTS

This work was supported by NIH CA 38942 (D.H.) and NIH CA38396 (D.R.) and NIH T32 09290 (K.C.). The helpful discussions of Dr. Alan Barrett are gratefully acknowledged.

REFERENCES

1. Millard, P.J., M.P. Henkart, C.W. Reynolds, and P.A. Henkart. (1984) J. Immunol. 132:3197.

2. P.A. Henkart (1984) J. Exp. Med. 160:75.

3. P.A. Henkart (1984) Proc. Natl. Acad. Sci. USA 81:5551.

4. Ishizaka, T. (1982) Fed. Proc. 14:17.

5. Serhan, C.H., J. Fridovich, E.J. Goetzl, P.B. Dunham, and G. Weissmann. (1982) J. Biol. Chem. 257:4746.

6. Dourmashkin, R.R., P. Deteix, C.B. Simone and P.A. Henkart. (1980) Clin. Exp. Immunol. 42:554.

7. Simone, C.B., and P.A. Henkart. (1980) J. Immunol. 124:952.

8. Boyum, A. (1968) Scand. J. Clin. Lab. Invest. 21 (Suppl. 97):77.

9. Lozzio, C.B., and B.B. Lozzio. (1975) Blood 45:321.

10. Hudig, D., D. Redelman, and L. Minning. (1984) J. Immunol. 133:408.

11. Carine, K., and D. Hudig. (1984) Cell. Immunol 87:270.

12. Kyger, E.M., and T.C. Franson. (1984) Biochim. Biophys. Acta 794:96.

13. Hudig, D., D. Redelman, and L. Minning. (1982) In, "NK Cells and Other Natural Effector Cells", edited by R.B. Herberman. P 1021. Academic Press, N.Y.

14. Hudig, D., T. Haverty, C. Fulcher, D. Redelman, and J. Mendelsohn. (1981) J. Immunol. 126:1569.

15. Hudig, D., D. Redelman, and L. Minning. (1984) J. Immunol. 133 (No. 6), in press.

16. Hatcher, V.B., M.S. Oberman, G.S. Lazarus and A.I. Grayzel. (1978) J. Immunol. 120:665.

THE ROLE OF PHOSPHOLIPID METHYLATION AND CHROMOSOMALLY REGULATED TARGET STRUCTURES IN PRE- AND POST-RECOGNITION EVENTS IN THE NK CYTOLYTIC PATHWAY[1]

Robert J. Lauzon
Tatsuo Kiyohara
John C. Roder

Department of Microbiology and Immunology
Queen's University
Kingston, Ontario, Canada

INTRODUCTION

In recent years, there has been a considerable amount of interest in attempting to elucidate the mechanism by which Natural Killer (NK) cells recognize and destroy their targets. A stimulus-secretion model was first proposed (1) and later elaborated upon (2,3). Briefly, NK cytolysis involves a series of discrete stages including: (i) target cell recognition and binding, (ii) triggering and activation of the effector cell (4), (iii) a Ca++-dependent programming for lysis (5), (iv) the reposition of the microtubule organizing center and the perinuclear Golgi apparatus (6) followed by the release of cytotoxic moieties (perforins) capable of polymerization on the target cell surface, and (v) target cell death resulting from the membrane insertion and transmembrane channel formation of polyperforins(7).

Our laboratory has been actively involved with many

----

[1]This research was supported by the Medical Research Council of Canada.

ISBN 0-12-341370-2

facets of the NK cytolytic pathway over the years. Our most recent aim has been to investigate the nature of recognition and post recognition determinants on both the effector and tumor cell surfaces.

RESULTS

A. An alpha-mannosidase sensitive triggering site on the NK cell surface is intimately coupled to phospholipid methylation

In many receptor-ligand interactions at the cell surface, two methyl transferase enzymes are activated within minutes of receptor aggregation (8). These enzymes cause a transfer of methyl groups from phosphatidyl-ethanolamine to phosphatidylcholine, which is then transferred to the outer side of the lipid bilayer with a resultant increase in membrane fluidity. Depending on the system, this reaction is followed by either a calcium influx or cAMP generation, which leads to responses as diverse as chemotaxis, histamine release and hormonal triggering (9). Early work using inhibitory compounds suggested that this pathway might be employed in NK mediated cytolysis (10). Our work confirms this hypothesis and extends it to reveal a mannose containing molecule on the surface of NK cells which is coupled to transmethylation of membrane phospholipids and which serves to trigger the cytolytic pathway in NK cells following binding to tumor cell targets (11).

In the first part of our study, pretreatment of nylon wool-enriched CBA/J spleen cells or human peripheral blood lymphocytes (PBL) with an exoglycosidase, alpha-mannosidase led to a marked dose-dependent inhibition of NK lytic activity against NK^s YAC-1.2 or K-562 tumor cells, although target cell binding was not affected. NK cytolytic activity of a murine NK clone was also inhibited. Maximum inhibition occured after a 60 min. pretreatment of murine effectors at 37^oC, and the kinetics of NK inhibition by this enzyme was similar to the reported kinetics for enzymatic activity. Inactivation of enzymatic function of alpha-mannosidase with EDTA reversed the NK inhibitory effect of this enzyme. Released hexose was detected chemically in the supernatant of mouse spleen cells treated with NK inhibitory doses of alpha-mannosidase. These results suggest that alpha-mannosidase inhibited NK function by virtue of its enzymatic action. Culture of human PBL for 20 hours after treatment with this enzyme led to a greater than 70% recovery in NK lytic function. Such recovery was however blocked by the incorporation of tunicamycin, a glycosylation inhibitor of asparagine-linked oligosaccharides, into the culture medium. These findings indicate that the alpha-mannosidase sensitive site is a de novo synthesized glycoprotein. Neuraminidase, β-galactosidase, endo-β-N-acetylglucosaminidase-D and -H, peptide N-glycosidase treatments did not inhibit human NK cell lysis of K-562 cells. Calcium-pulse experiments revealed that the mannosidase sensitive site was involved following effector:target binding, but prior to the Ca++-influx.

In the second part of our study, we asked if the mannosidasesensitive NK structure was coupled to transmethylation of membrane phospholipids. NK enriched spleen cells, target cells or combinations of both, were incubated one hour at 37° with ^{3}H-methyl-methionine followed by chloroform/methanol extraction, and thin-layer chromatography (TLC). The phospholipid fraction was scraped and counted in a scintillation counter. As shown in table I, the addition of small numbers of target cells stimulated a 40 fold increase in methyl transfer into the phospholipid pool of NK

TABLE I. The effect of alpha-mannosidase on phospholipid methylation

Cells	Alpha-mannosidase Pre-treatment of effectors[b]	^{3}H-[methyl] incorporation (CPM)
YAC-1.2	-	5
NWP spleen	-	220
NWP spleen +YAC-1.2[a]	-	8450
NWP spleen	+	351
NWP spleen +YAC-1.2	+	924

[a] 10^{6} nylon wool passed (NWP) CBA/J spleen cells were incubated 1 hour at 37°C with 10^{4} YAC-1.2 cells in the presence of 20μCi of L-[methyl-^{3}H] methionine. Chloroform/-methanol extracts were run on TLC and the phospholipid fraction was counted in a liquid scintillation counter. The data is expressed as counts per minute (cpm)

[b] NWP spleen cells were pre-treated 1 hour at 37°C with 1 unit of alpha-mannosidase per 10^{6} cells.

cells. Experiments using two-dimensional TLC have shown methyl transfer to both phosphatidyl-ethanolamine (PE) and phosphatidylcholine (PC). NK cells pretreated with alpha-mannosidase failed to incorporate the label. Therefore, we hypothesize that the alpha-mannosidase sensitive site on NK cells is a triggering receptor which may be coupled to the transmethylation of membrane phospholipids as a trans-membrane signalling device.

B. Human x Mouse somatic cell hybrids retain an intermediate level of sensitivity to human NK cells

In an attempt to investigate the nature of NK recognition or post-recognition determinants on human NK sensitive target cells, we fused two highly NK sensitive (NK^s) human tumor lines (K-562 and Y-79) with murine $LMTK^-$ cells (12). The rationale for generating interspecies hybrids resides in their preferential loss of human chromosomes, over the course of time. One can therefore assign a characteristic phenotype (ie. NK sensitivity) to a specific chromosome. As shown in table II, several K-562 x $LMTK^-$ (KL-8, KL-12 and KL-13), and Y-79 x $LMTK^-$ (YL-2, YL-40 and YL-43) hybrids retain an NK sensitive phenotype that is intermediate between that of the human NK^s and the murine NK^r parents. The YL-22 hybrid was originally highly NK^s in the first stages of this study , after which, it consistently gave an intermediate level of sensitivity to cytolysis. Other lines (KL-1, KL-9 and YL-44) were consistently NK^r. These findings suggest that chromosomal segregation is occuring amongst these hybrids. Further support to this hypothesis emanates from data obtained with cloned lines derived from several NK^s somatic hybrids. Clones derived from the KL-12 hybrid clearly demonstrate segregation at the level of NK sensitivity (12). Analogous findings were obtained with cloned lines derived from YL-2, YL-22 and YL-43. Interestingly, three somatic hybrids generated from the fusion of the NK^r human Raji line with $LMTK^-$ fibroblasts were completely NK^r. Preliminary evidence also suggests that NK sensitivity correlates with the presence of a limited number of human chromosomes.

As the Transferrin receptor (TfR) has been postulated as a possible target structure (TS) in human NK cell mediated recognition (13,14), the construction of somatic cell hybrids retaining an NK^s phenotype prompted us to investigate the level of TfR expression in our hybrid clones by flow cytometry with the B3/25 monoclonal antibody (12). We could not find evidence of any correlation between competitive ability and TfR expression or NK sensitivity and TfR expression. Several of the NK^{s} cloned lines derived from KL-12 were as potent cold targets as parental K-562 cells, yet were either weakly B3/25+ or B3/25-. In addition, NK^{s} YL clones were B3/25-. These data clearly indicate that other as yet uncharacterized determinants are involved in human NK cell recognition.

TABLE II. Human x mouse somatic cell hybrids segregate at the level of the NK sensitive phenotype

Target Cells	# Lytic units/10^{6} effectors at 20% lysis[a]
K-562	37.7
LMTK^{-}	<1
KL-2[b]	<1
KL-8	18.0
KL-9	<1
KL-12	25.8
KL-13	10.8
Y-79	55.8
YL-2[c]	12.3
YL-22	66.1
YL-40	15.5
YL-43	15.6
YL-44	<1
RAJI	<1
RL-1[d]	<1
RL-2	<1
RL-3	<1

[a] The number of lytic units were derived according to the exponential fit equation as previously described by Pross et al (15). The LU value takes into account all experimental points of the cytotoxic curve, which range from effector/target ratios of 20/1 to 0.63/1.

[b] K-562 x LMTK^{-}

[c] Y-79 x LMTK^{-}

[d] Raji x LMTK^{-}

C. Involvement of the hydroxyl radical in human NK cell mediated cytolysis

Recent findings indicate that human NK cells require O_2 to efectively kill tumor cells (16). NK lysis was inhibited by approximately 52% under a non-toxic N_2 atmosphere in 11/12 individuals tested. Consistent with the hypothesis that oxygen intermediates are involved in the cytolytic pathway of NK cells, a panel of 13 non-toxic hydroxyl radical scavengers were found to strongly impede NK lysis of K-562 cells. The scavengers differed widely in structure, but the degree of inhibition of lysis correlated highly with their reactivity (k) for the hydroxyl radicals (k vs log inhibitor concentration required to inhibit NK lysis by 50%: $r= -0.9202$; $p<0.001$). In addition, it was found that the majority of hydroxyl radicals impeded NK cytolysis without affecting conjugate formation. It would appear however, that the chemiluminescence generated following binding to NK sensitive target cells is not a required event for cytolysis to occur, as (i) cytolysis occurs in the absence of any detectable chemiluminescence in Chronic Granulomatous Disease (CGD) patients, (ii) chemiluminescence occurs in the absence of cytolysis in Chediak-Higashi patients, and (iii) chemiluminescence appears to be dependent on the presence of monocytes (synergy between NK and monocytes) which are not required in cytolysis.

D. Expression of the T cell receptor gene in NK cells

Until recently, it has been extremely difficult to ascertain the lineage of NK cells as they were found to exhibit surface features characteristic of both T cells and monocytes. The molecular cloning of the Beta chain of the T cell receptor (17,18) has enabled us to verify the hypothesis wether NK cells express, reaarange and utilize T cell receptor genes for cytolytic function.

First, DNA from five cloned cell lines with cytolytic function (HY-1, HY-3, Balb/c NK, NKB61B10 and NKB61A2) were individually digested with a variety of restriction endonucleases and analysed by Southern Blot analysis with ^{32}P-labelled probes J1 or C1 of the mouse genome (19) for the presence of DNA rearrangements. It became apparent that (i) all six lines were rearranged, and that (ii) a limited number of rearrangements existed (20). Second, mRNA homologous to the T cells receptor B chain gene were found to be expressed in these NK cell lines. In all cases examined, a 1.3 Kb transcript was detected, presumably encoding the complete rearranged (V-D-J-C) B chain gene. Th was no preferential use of the constant regions cloned cell lines. Fourth, RNA isolated from h riched poly:IC boosted splenic NK cells hybridi cell receptor gene, wheras RNA from non-boosted cells hybridized weakly (20). These data collectively support the hypothesis that NK cells are of the T cell lineage, and raise the possibility that they may utilize the T cell receptor in target cell recognition and binding.

SUMMARY AND CONCLUSIONS

The data can be summarized in the following hypothetical model: our results with somatic cell hybrids suggest that the NK target structure (NK-TS) on human NK^s target cells is distinct from the transferrin receptor, and is regulated by a limited number of human chromosomes. These target structures then interact with a recognition receptor on the NK cell that is encoded in part by the Beta gene of the T cell receptor. This gene undergoes a limited number of rearrangements in NK lines, and is expressed at the RNA level in fresh NK cells as well as NK clones. Following target: effector binding, a mannosidase sensitive site on the NK cell surface delivers a triggering signal via the transmethylation of membrane phospholipids. Transmethylation is presumably followed by a calcium influx that culminates in the secretion of the lytic moiety, which subsequently binds to receptors on the target cell membrane. At some undefined point in this pathway, hydroxyl radicals are generated and play an essential role in NK cytolysis. A greater understanding of the NK cytolytic mechanism may lead to more rational modalities for therapeutic intervention in human malignancies.

REFERENCES

1. Roder, J.C., Kiessling, R., Biberfield, P. and Anderson, B., (1978). J. Immunol., 121, 2509.
2. Roder, J.C., Karre, K. and Kiessling, R., (1981). Prog. Allergy (Karger, Basel), 28, 66.
3. Roder, J.C. and Pross, H.F., (1982). J. Clin. Immunol., 2, 249.

4. Newman, W., Fast, L.D. and Rose, L.M., (1983). J. Immunol., 131, 1742.
5. Hiserodt, J., Britvan, L.J. and Targan, S.R., (1982). J. Immunol., 129, 1782.
6. Kupfer, A., Dennert, G. and Singer, S.J., (1983). Proc. Natl. Acad. Sci. U.S.A., 80, 7224.
7. Podack, E.R. and Dennert, G., (1983). Nature, 302, 442.
8. Hirata, F. and Axelrod, J., (1980). Science, 209, 1082.
9. Traill, K.N. and Wick, G., (1984). Immunol. Today, 5, 70.
10. Hoffman, T., Hirata, F., Bougnoux, A., Fraser, B.A., Goldfarb, R.H., Herberman, R.B. and Axelrod, J., (1981). Proc. Natl. Acad. Sci. U.S.A., 78, 3839.
11. Kiyohara, T., Dennis, J., Boegman, R.J. and Roder, J.C., Manuscript submitted for publication.
12. Lauzon, R.J. and Roder, J.C., Manuscript submitted for publication.
13. Vodinelich, J., Sutherland, R., Schneider, C., Newman, R. and Greaves, M., (1983). Proc. Natl. Acad. Sci. U.S.A., 80, 835.
14. Newman, R.A., Warner, J.F. and Dennert, G., (1984), J. Immunol., 133, 1841.
15. Pross, H.F., Baines, M.G., Rubin, P., Shragge, P. and Patterson, M., (1981). J. Clin. Immunol., 1, 51.
16. Duwe, A., Werkmeister, J., Lauzon, R.J. and Roder, J.C., Manuscript submitted for publication.
17. Yanagi, Y., Yoshikai, Y., Leggett, K., Clark, S.P., Aleksander, I. and Mak, T.W., (1984). Nature, 308, 145.
18. Hedrick, S.M., Cohen, D.I., Nielsen, E.A. and Davis, M.M., (1984). Nature, 308, 149.
19. Mallissen, M., et al., (1984). Cell, 37, 1101.
20. Yanagi, Y., Caccia, N., Kronenberg, M., Chin, B., Roder, J.C., Rohel, D., Kiyohara, T., Lauzon R.J., Rosenthal, K., Dennert, G., Hengartener H., Hood, L. and Mak, T.W., Manuscript submitted for publication.

MONOSACCHARIDE CONCENTRATION DEPENDENT INHIBITION OF HUMAN NATURAL CYTOTOXICITY

E. W. Ades
M. Culwell
A. Hinson

Department of Immunology Research
Lilly Research Laboratories
Indianapolis, Indiana

I. INTRODUCTION

The phenomenon of natural killing and the ability of natural killer cells (NK) to recognize and kill a wide variety of targets suggests that NK activity is directed against multiple antigenic determinants, some of which may be cross-reactive with other target cells (1,2). No information is yet available about the nature of the recognition structures on either the effector or the target cells although a large body of evidence is accumulating which suggests that carbohydrates on cell surfaces are of considerable importance in the immune response through their involvement in cell-cell recognition and/or communication (3,4,5).

We (6,7) and others (8,9) have previously shown that NK mediated ^{51}Cr release could be blocked by certain monosaccharides and glycoproteins (10). Our work with human peripheral blood NK cells and the erythroleukemic target cell K562 indicated that N-acetyl-galactosamine, galactose, N-acetyl-glucosamine and arabinose were particularly effective in preventing cytolysis (6). Washing of NK cells preincubated with saccharides before assay did not abrogate blocking while K562 targets washed after sugar incubation were still susceptible to lysis. Both glycosidase treatment of target cells (7) and growth of target cells in tunicamycin (11), an antibiotic which blocks protein glycosylation (12), rendered the K562 cells resistant to NK-mediated lysis. These results are consistent with our hypothesis that a receptor on NK cells is responsible for recognizing oligosaccharide determinants on the K562 targets as a

ISBN 0-12-341370-2

preliminary to lysis. In addition, we have recently shown that a glycopeptide isolated from the plasma membranes of K562 cells and fractionated on lectin affinity adsorbents inhibits NK lysis whereas control glycopeptides do not inhibit (13). Furthermore, this inhibitory glycopeptide does not inhibit the initial binding of effector to target but rather dissociates conjugates following initial effector target interactions as well as inhibiting lysis induced by natural killer cell derived soluble cytotoxic factor(s) (NKCF) in a soluble factor assay (13).

In this study, we examine the effect of monosaccharides shown to inhibit NK activity on Interleukin-2 (IL-2) treated effector cells against non-virally infected and virally-infected tumor target cells. We report that certain monosaccharides significantly inhibit the enhancing effect of IL-2.

II. MATERIALS AND METHODS

Mononuclear cells (MC) from normal individuals were first separated from heparinized blood on a Ficoll-Hypaque gradient as described previously (11). In some instances, the mononuclear cells were fractionated on nylon wool columns to obtain the null cell population. Previously, we have shown that between 85-90% of this null cell population were large granular lymphocytes (14) and reacted with Leu 7 (BD Monoclonals, Sunnyvale, CA). Cell viability was at least 98% as determined by trypan blue dye exclusion.

Target cells (K562, erythroleukemic cells; Raji, EBV-infected human B lymphoblastoid cells) were grown in culture, diluted to the desired concentration, and labeled with ^{51}Cr as previously described (11). Excess ^{51}Cr was removed by centrifuging labeled target cells over 100% fetal calf serum. Ratios of 60:1, 30:1 and 10:1 effector (NK) cells to target cells were used. Controls included effector cells alone and effector cells plus detergent (2% Triton) to obtain 100% lysis of targets. All samples were tested in triplicate and results are shown as mean values. At least three experiments were performed for each assay using cells from different individuals.

The percent specific release was calculated as [(C-D)/(M-D) x 100] where C = percent release from target cells (K562) in the presence of MC, D = percent release from target cells alone, and M = maximum percent release (from targets cells + 2% Triton). The percent inhibition was calculated as

$$100 - \frac{\text{percent specific release (experimental)}}{\text{percent specific release (control)}}$$

A. Sugar Treatment of Target or Effector Cells

Sugars (L-Fucose, Calbiochem; N-acetyl-D-galactosamine, Calbiochem; N-acetyl-D-glucosamine, Calbiochem; D-xylose, Calbiochem; L-rhamnose, Calbiochem; D-Arabinose, Calbiochem; D-Mannose-6-phosphate, Sigma; rhamnose, Difco; galactose, Calbiochem; and D-mannose, Calbiochem) were dissolved in RPMI 1640 (Gibco) and prepared as 200 mM, 100 mM, 50 mM and 10 mM solutions. The final pH of each preparation was between 7.0 and 7.2. The final concentration of sugar used when examined was 100 mM, 50 mM, 25 mM and 5 mM (i.e., cells were resuspended to 1 x 10^6 cells/ml and 0.5 ml of cells [5 x 10^5] was mixed with 0.5 ml of monosaccharide at the desired concentration). Each sugar preparation was added to the target or effector cells and not removed (Fig. 1). In no case did these sugars affect spontaneous release from the targets. Viability of either the targets or effectors in the presence of sugar was determined after 24 hour incubation. Cell viability was at least 97% as determined by trypan blue dye exclusion. Optimal sugar concentrations were employed after being established by dose-response curves in preliminary experiments.

B. IL-2

Human interleukin-2 (Lot 1464-64) was purchased from Cell Science Laboratory (Electronucleonics, Inc., Silver Spring, Maryland 20904). This highly enriched IL-2 preparation averaged between 8,000-16,000 half-maximal units per ml when tested over a series of weeks. This material was devoid of detectable gamma interferon and macrophage activating factor activity (15).

C. Interleukin-2 Treatment of Effector Cells

Effector cells were resuspended to 10^6 cells/ml in RPMI with 10% fetal calf serum and incubated with approximately 50 μl of human Interleukin-2 per 10^6 cells. After treatment, the cells were washed three times in RPMI with 10% FCS and resuspended to appropriate concentration for testing. Addition of IL-2 directly to the assay without pre-treatment of effector cells made no difference in enhanced NK cytolytic activity. The sequence of addition of treated or untreated cells to the assay had no significant effect on the inhibition.

III. RESULTS

Figure 1 shows a schematic representation of our previous results (6). These data support the concept that human NK cells are polyvalent and can recognize a wide variety of target cell surface components. When either the effector or target cells are treated with sugars and the sugar is not removed, NK activity is inhibited (Fig. 1A, B). Our interpretation of this observation is that excess sugar blocks the sugar recognition site on the effector cell, similar to cold target inhibition. When the target cells are treated with sugar and then washed, lysis occurs (Fig. 1C). We interpret this to indicate a lack of cold target inhibition. In contrast, when treated effector cells are washed to remove excess sugar, NK lysis of target is completely abolished (Fig. 1D). We interpret these results as suggesting the sugar bound to the receptor on the effector cell blocks the recognition of antigenic sites on the target cell.

In Table I, we report the effects of human natural cytolytic activity with and without interleukin-2 on two different target cells (K562 and Raji). The cytotoxic activity was observed to be enhanced with IL-2 against both targets; enhancement was greater (% change) for the EBV infected target than for K562.

In Fig. 2, monosaccharide concentration dependent inhibition of cytotoxicity is demonstrated with galactosamine and glucosamine against the target, K562. Fig. 3 similarly demonstrates concentration dependent inhibition only with the sugar glucosamine although when effector cells are activated with IL-2, galactosamine inhibits the lysis of the Raji target cell [i.e., 39% cytotoxicity with IL-2 to 20% ($p = 0.1$) with galactosamine].

Fig. 4 demonstrates a broad range of sugars with inhibitory properties against the target, K562. As with the Raji target, when effector cells were activated with IL-2, several sugars previously unable to significantly inhibit cytolytic activity were now capable of inhibiting this activity. When a concentration dependent inhibition curve was examined, Fig. 5, it was noted that greater amounts of glucosamine were required for the inhibition of IL-2 treated effector cells in order to obtain equivalent killing.

TABLE I. Human Natural Cytotoxicity

	% Lysis		(Target:K562)[a]
E:T	60:1	30:1	10:1
Without IL2	54 ± 4[c]	44 ± 4	25 ± 3
With IL2[b]	69 ± 2	61 ± 3	45 ± 5
% change	+ 28%	+ 39%	+ 80%
P value[d]	P = .01	P = .01	P = .01

			(Target:Raji)[a]
	60:1	30:1	10:1
Without IL2	15 ± 4[c]	11 ± 3	11 ± 3
With IL2[b]	44 ± 7	39 ± 6	25 ± 5
% change	+ 193%	+ 255%	+ 127%
P value[d]	P = .01	P = .001	P = .05

[a]$N \geq 9$.

[b]5 µl of IL2 in 0.2 ml (10,000 half-maximal units/ml as determined against CTLL).

[c]Mean ± S.E.M.

[d]Students T test.

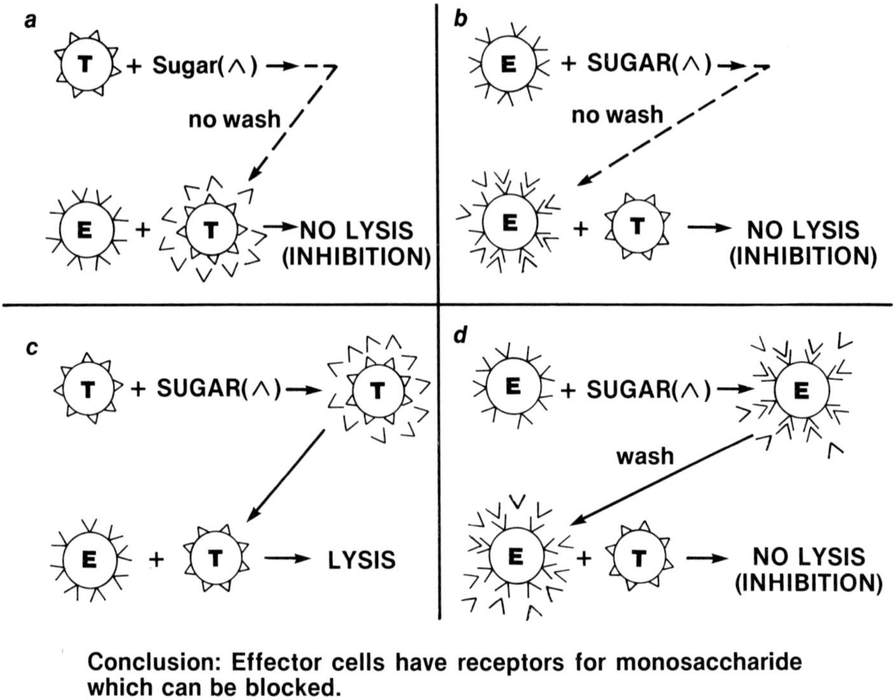

Fig. 1. Schematic representation of inhibition of NK activity by monosaccharide.

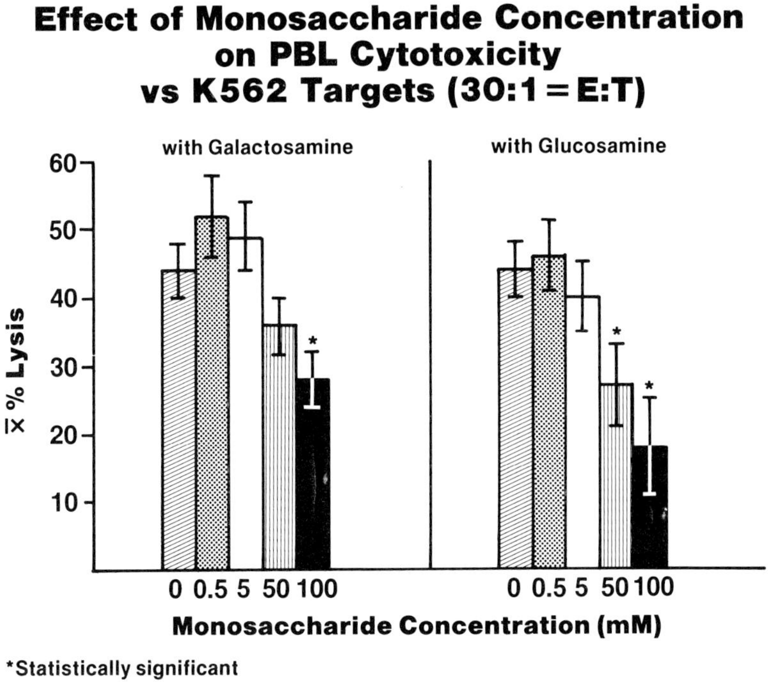

Fig. 2. Monosaccharide concentration dependent inhibition of cytotoxicity (K562).

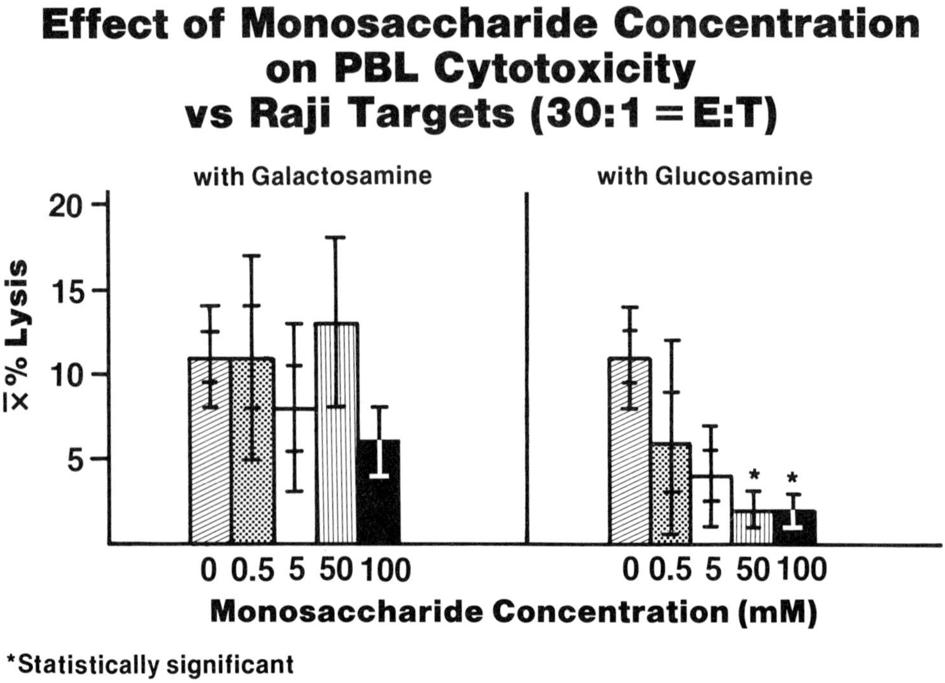

Fig. 3. Monosaccharide concentration dependent inhibition of cytotoxicity (Raji).

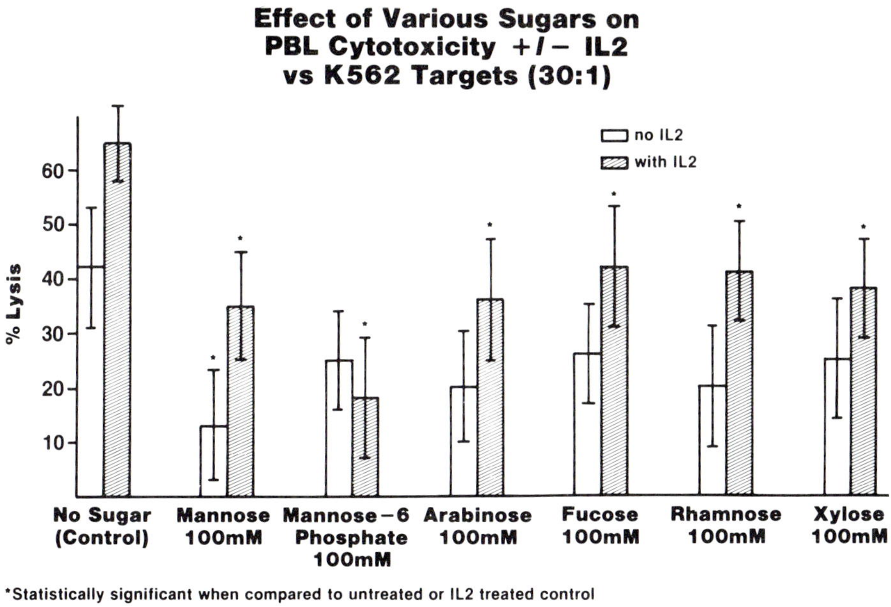

Fig. 4. Broad range of sugars with inhibitory properties for cytolytic activity against the target, K562.

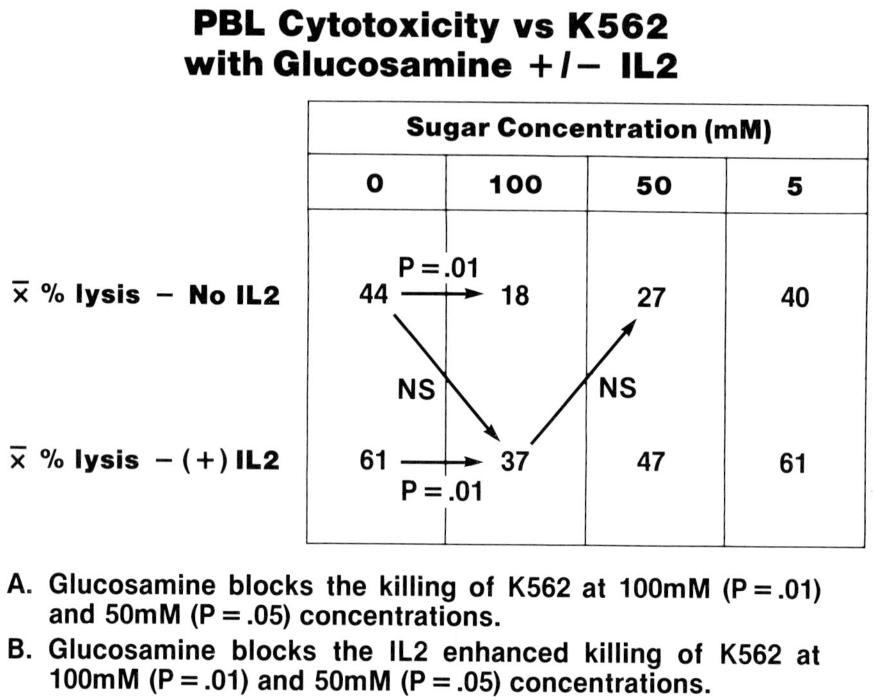

Fig. 5. IL-2 treated cells: comparison of sugar concentration required for inhibition of effector cells to obtain equivalent killing.

ACKNOWLEDGEMENTS

We would like to thank Mrs. Mary Bolton for excellent assistance in the typing of this manuscript.

REFERENCES

1. Takasugi, M., Koide, Y., Akira, D., and Ramseyer, A. (1977). Int. J. Cancer, 19, 291.
2. Herberman, R. B., Djeu, J. Y., Kay, H. D., Ortaldo, J. R., Riccardi, C., Bonnard, G. D., Holden, H. T., Fagnani, R., Santoni, A., and Puccetti, P. (1979). Immunol. Rev. 44, 143.
3. MacDermott, R. P., Kienker, L. J., Bertovich, M. J., and Muchmore, A. V. (1981). Immunology, 44, 143.
4. Werkmeister, J. A., Roder, J. C., Curry, C., and Pross, H. F. (1983). Cell. Immunol., 80, 172.
5. Ulrich, F. (1983). Cell. Immunol., 80, 241.
6. Ades, E. W., Hinson, A., and Decker, J. M. (1981). Immunobiol., 160, 248.
7. Ades, E. W., and Hinson, A. (1983). J. Clin. Lab. Immunol., 11, 173.
8. Boldt, D. H., MacDermott, R. P., and Jorolan, E. P. (1981). J. Immunol., 114, 1532.
9. Stutman, O., Dien, P., Wisun, R. F., and Lattime, E. C. (1980). Proc. Natl. Acad. Sci., U.S.A., 77, 2895.
10. Decker, J. M., Hinson, A., and Ades, E. W. (1984). J. Clin. Lab. Immunol., in press.
11. Ades, E. W., and Hinson, A. (1982). Cell. Immunol., 72, 326.
12. Takatsuki, A., Kohno, K., and Tamura, G. (1975). Agric. Biol. Chem., 39, 2089.
13. Targan, S., Decker, J. M., and Ades, E. W. (1984). Nat. Immunity and Cell Growth Reg., in press.
14. Stewart, L. D., and Ades, E. W. (1984). Clin. Immunol. and Immunopath., 31, 78.
15. Ades, E. W., Hinson, A., Marder, P., and Butler, L. D. (1984). Clin. Immunol. and Immunopath., 32, 119.
16. Brunda, M. J., Wiltrout, R. H., Holden, H. T., and Varesio, L. (1983). Int. J. Cancer, 31, 373.
17. Weitzen, M. L., Innins, E., Yamamoto, R. S., and Granger, G. A. (1983). Cell. Immunol, 77, 42.
18. Zaunders, J., Werkmeister, J., McCarthy, W. H., and Hersey, P. (1981). Brit. J. Cancer, 43, 5.
19. Millard, P. J., Henkart, M. P., Reynolds, C. W., and Henkart, P. A. (1984). J. Immunol., 132, 3197.

20. Sly, W., Kaplan, A., Achord, D., Brot, F., and Bell, C. (1978). In Cell Surface Carbohydrates and Biological Recognition, (V. T. Marchsi, V. Ginslourg, P. W. Robbins, and C. F. Fox, eds.), Alan R. Liss, Inc., NY, p. 547.
21. Phillips, W. H., Ortaldo, J. R., and Herberman, R. B. (1980). J. Immunol., 125, 2322.

DISCUSSION

The experiments presented in this paper were designed to examine whether monosaccharides can inhibit human IL-2 enhanced cytolytic activity against tumor target cells. Previous results suggest that a receptor on the NK effector cell recognizes oligosaccharide determinants on the K562 or Raji tumor target cells as a preliminary to lysis. It is our contention that incubation with modulating agents such as IL-2 increases the number of susceptible target binding cells by inducing increased expression of endogenous monosaccharide binding receptors on the effector cell. This is best demonstrated in Fig. 4. Here it is shown that monosaccharides previously having no inhibitory effect on effector cell lysis of K562 do inhibit lysis by the effector cell after stimulation by IL-2. Additionally, the expression of these target binding structure on the effector cell appears to be increased upon incubation with IL-2 (Fig. 5); greater amounts of glucosamine are required to bring the level of cytolysis of IL-2 treated cells down to that of untreated cells (37 ± 5%).

Several studies (3,5,16) have implicated a role for different carboyhydrate moieties in the effector:target cell interaction of natural cytolysis. This hypothesis has been based on simple sugar inhibition studies (3,6,9), inhibition of soluble cell-lytic molecules with various saccharides (17) and isolation and characterization of putative recognition structures from susceptible target cells (10,13,18). The site of inhibition can be at any point in the cycle of complex events that begins with effector:target cell recognition and ends with target cell lysis. Werkmeister et al. (4) indicated that some sugars (N-acetyl-D-glucosamine and β-gentiobiose) function by inhibiting cytolysis at the recognition stage while others (D-Mannose) do not appear to work solely at this stage. Phosphorylated sugars were speculatd to be signals which regulate cell-mediated cytolytic mechanisms (4). Attachment of saccharides to other sites on the cell surface which are associatied with cytotoxic expression or a decrease by sugars in the mobility of surface receptors needed for cytolysis are possibilities (4) while non-specific effects related to an increase in the osmolarity of medium containing the monosaccharides cannot be ruled out. Brunda et al. (16) demonstrated no correlation between inhibition of cytotoxicity and increased osmolarity of monosaccharide-containing medium at the concentrations tested. Artificial effects, according to Ulrich, (5) (nonspecific cytotoxicity, changes in thymidine transport or changes in isotonicity of the culture medium) cannot explain the inhibition of thymocyte proliferation by monosaccharides. Additionally, he points out that monosaccharide inhibition (D-ribose) does not appear to act by generating suppressor cells, by blocking the absorption of IL-2 or by acting as a general metabolic inhibitor. Whatever the mechanism of inhibition may be, it is apparent that monosaccharides are capable of inhibiting NK cytolytic activity. Currently, we are examing the inhibition of lysosomal enzyme activity by sugars since lysosomal enzymes have been shown to be involved in target cell lysis through recognition by saccharide structures on the cell membrane (19,20).

Disparity in the sugar(s) shown to inhibit lysis (Figs. 3 and 4) might be analogous to lytic susceptibility. Phillips et al. (21) have demonstrated that there is variation among donors in recognition of any one target cell line and that multiple sugar specificities are responsible for these differences in susceptibility to lysis. Certainly our data further reinforces the growing belief that effector cells express receptors with different sugar specificities and suggests differences in sugar specificities in the effector population as one explanation for the heterogeneous nature of the human NK cell population.

THE POTENTIAL ROLE OF ARACHIDONIC ACID METABOLISM IN NATURAL KILLER CELL ACTIVITY[1]

William E. Seaman[2]
Janet Woodcock

Department of Medicine
VA Medical Center and University of California
San Francisco, California

Colin G. Brooks

Fred Hutchinson Cancer Research Center
Seattle, Washington

I. INTRODUCTION

Arachidonic acid (AA; 5,8,11,14-eicosatetraenoic acid; 20:4) is an unsaturated fatty acid containing 20 carbons. It is derived from linoleic acid and is stored in cell membranes and/or in cytoplasmic lipid bodies where it is esterified within phospholipids and triglycerides (1,2). Mechanisms for the release of AA from most cells appear to utilize phospholipase A_2, acting preferentially on methylated phospholipids (1,3). Platelets, utilize phospholipase C to release precursors of AA, and mast cells may also possess this pathway (1).

---

[1]Supported by Public Health Service grants Nos. CA34529 and AI5384 and by the Veterans Administration.

[2]Address correspondence to Dr. Seaman at VAMC 111R, 4150 Clement St., San Francisco, CA. 94121

ISBN 0-12-341370-2

Free AA may be metabolized by either of two major pathways. In the first, cyclooxygenase promotes the formation of an endoperoxide that is the precursor of prostaglandins, thromboxanes, and prostacyclins, all of which have important biologic activities (1,4). In the second, different lipoxygenases promote oxygenation of AA by addition of a hydroperoxy group (-OOH) at a carbon adjacent to a double bond. The products are hydroperoxyeicosatetraenoic acids (HPETEs) (1,4-6). Examples of lipoxygenation at the 5 and 12 positions are shown in Figure 1. The site of oxygenation is enzyme-dependent and varies with the type of cell, the tissue, and the species. In general, platelet lipoxygenase(s) act predominantly at C_{12} (forming 12-HPETE). A variety of leukocytes can form 5-HPETE. These include neutrophils, macrophages, mast cells, and eosinophils. Some polymorphonuclear cells also can produce 15-HPETE.

The biologic importance of lipoxygenase products of AA is rapidly being established. Among the most important products are those of 5-lipoxygenation. 5-HPETE is the precursor of the leukotrienes, agents with a variety of biologic activities. Leukotriene B_4 (LTB_4) has potent activity as a chemoattractant and as an activator of polymorphonuclear cells (1,4-7). It is also an immunosuppressant (7,8). LTC_4, LTD_4, and LTE_4 contract smooth muscle, including bronchial and vascular muscle; together they comprise slow-reacting substance of anaphylaxis (SRS-A) (9).

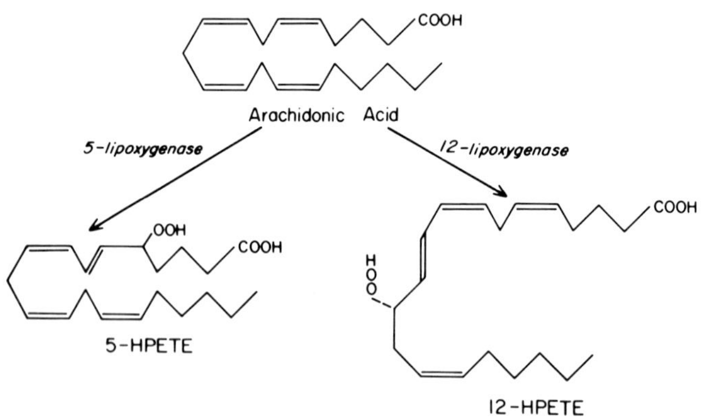

Figure 1. The structure of hydroperoxyacids formed from arachidonic acid by the action of 5-lipoxygenase or 12-lipoxygenase.

II. EVIDENCE THAT AA METABOLISM IS REQUIRED FOR NK CELL ACTIVITY

A possible role for AA metabolism in the activity of NK cells was suggested by the finding of Hoffman et al. that exposure of human blood leukocytes to NK targets stimulated phospholipase A_2 activity and the methylation of membrane phospholipids (10). This group also demonstrated that inhibitors of phospholipase A_2, including lipomodulin, inhibited NK cell activity (10,11).

If AA metabolism contributes to NK cell activity, it appears unlikely that cyclooxygenase activity is important; indomethacin and other inhibitors of this pathway do not inhibit NK cell activity (12). This suggests that lipoxygenation of AA may be important.

In support of this, we and others have found that inhibitors of lipoxygenation can reversibly inhibit human or murine NK cell activity (13-16). Our studies utilized the antioxidants BW755C and nordihydroguaiaretic acid (NDGA) (13,16). The arachidonic acid analogue ETYA (5,8,11,14-eicosatetraynoic acid) is also inhibitory (15). Rola-Pleszczynski and his colleagues have moreover shown that LTB_4 stimulates NK cell activity against HSV-1 infected tumor targets (13). LTB_4 reversed inhibition of NK cell activity by NDGA (13). In our studies, inhibition of human NK cell activity was independent of monocytes and was not accompanied by an appreciable change in binding to K562 targets (12).

III. AREAS OF CURRENT RESEARCH INTEREST

The above evidence supports the contention that lipoxygenation of AA is important in NK cell activity. However, there remain uncertainties that must be addressed before this is established.

First, our original reports and those from other laboratories used mixed effector cell populations, either from human blood or mouse spleen (13-16). LTB_4 and/or lipoxygenase inhibitors might therefore be acting indirectly on NK cells through other cells. Against the latter possibility, we have recently demonstrated that cytotoxicity by cloned murine NK cells against YAC-1 is inhibited by the lipoxygenase antagonists BW755C or NDGA in the same concentrations that inhibit cytotoxicity by unseparated mouse spleen cells or by human peripheral blood cells. These studies used two cloned lines, both grown without feeder cells. The first, M4G8, was cloned from spleen cells from mice that had been pretreated with poly I C to stimulate NK cell activity. Spleen cells were then preselected for the expression of Mac-1 antigen before cloning. Cytotoxicity by this cell was inhibited by BW755C or NDGA (Figure 2). Similar results were obtained with clone F5A/MB, a cytotoxic T cell clone that had been induced to express broader (NK-like) activity by exposure to 20% conditioned medium (17). These studies exclude the contribution of other cell types to suppression of natural killing by lipoxygenase inhibitors.

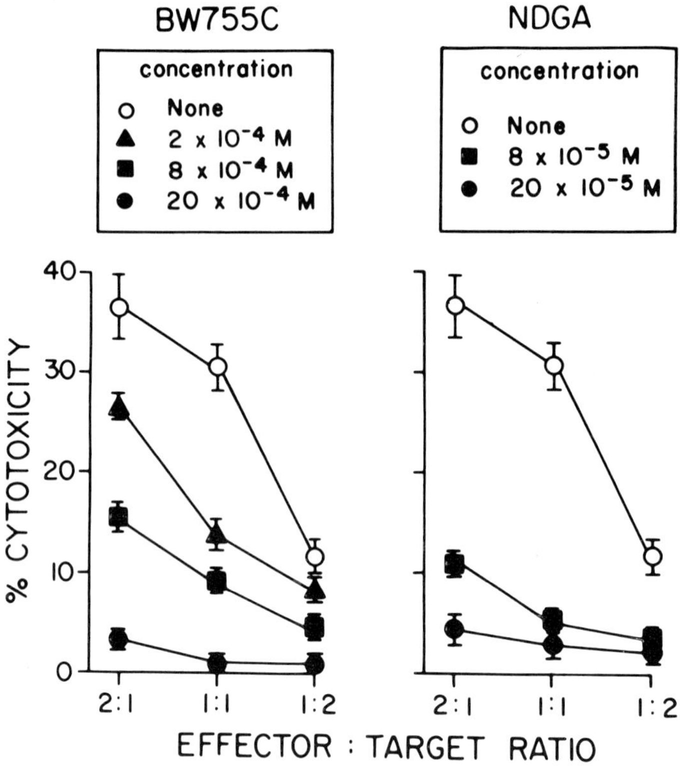

Figure 2. The lipoxygenase inhibitors BW755C and NDGA inhibit cytotoxicity by the cloned NK cell line, M4G8, against YAC-1 targets. Killing was assessed in a standard 5-hr chromium-release assay in round-bottom microtiter plates with 10^4 targets per well and triplicate samples for each determination. Inhibitors were added at the start and were present throughout the assay.

A second problem is that no available agent exclusively inhibits lipoxygenase activity, and different agents have varying effects depending on the tissue or cell type. Most available lipoxygenase inhibitors may be expected to interrupt oxygenation reactions in general. In particular, they are known to inhibit cyclooxygenase activity, despite a relative selectivity for lipoxygenase. It does not appear, however, that lipoxygenase inhibitors interrupt NK cell activity by blocking cyclooxygenase, because cyclooxygenase inhibitors that are relatively specific have little or no effect on natural killing (12). We have demonstrated also that the lipoxygenase inhibitors BW755C and NDGA do not act by preferentially shunting AA metabolism from the lipoxygenase pathway to the cyclooxygenase pathway because indomethacin does not alter the effect of these lipoxygenase antagonists on natural killing (13).

The lipoxygenase inhibitors, then, do not block NK activity by inhibiting cyclooxygenase activity, but it remains to be established that their effect is specific for inhibition of the lipoxygenase pathway. To examine this, the most important question to be answered is whether NK cells can metabolize AA acid and, if so, whether changes in AA metabolism correlate with changes in NK cell activity. The answer to this is not known, but several clues exist regarding the possibility that NK cells can metabolize AA. First, the removal of adherent cells from human peripheral blood mononuclear cells removes most (some would say all) of the detectable AA metabolism in response to a variety of stimuli (18). The remaining nonadherent cells should include NK cells, so it is probable that if NK cells can metabolize AA they must do so at a rate that is very much lower than the metabolism of AA in polymorphonuclear cells or macrophages. It has been argued that lymphocytes cannot metabolize AA at all (18). There is, however, evidence for the metabolism of AA by human and murine lymphocytes, including T lymphocytes (19-21). Moreover, the T cell clone HT2 expresses phospholipase A_2 activity, and can metabolize AA by lipoxygenation and cyclooxygenation (22, and Handwerger, BS, personal communication). It has also been noted that NK cells share many histologic and biologic features with mast cells, which are potent metabolizers of AA. They may thus have enzymes for the metabolism of AA that are not shared by other lymphocytes. Thus, several lines of indirect evidence support the possibility that NK cells may metabolize AA but at relatively low levels.

The elucidation of the role of AA in NK cells will require the use of cloned NK cells. Attempts to purify NK cell populations in order to examine the metabolism of AA will not likely succeed in excluding the contribution of small numbers of platelets or other cells that are potent producers of AA metabolites. The detection of AA metabolism in cloned NK cells may also require the use of methods that can detect products that are several orders of magnitude less than in macrophages, polymorphonuclear cells, or platelets. On the other hand, it may also be necessary to specifically add certain AA metabolites in order to stimulate AA metabolism by cloned NK cells; the formation of 5-HETE and LTB_4 by human leukocytes, for example, is promoted by 12-HPETE from platelets (23).

A second issue that is related to, but distinct from, the metabolism of AA by NK cells is their response to exogenous products of AA metabolism. The effects of LTB_4 have been noted. Payan, et al, have recently described specific binding of LTB_4 by human T cells (24), and it will be of interest to look for this receptor on cloned NK cells and to examine the correlation between binding and biologic effects of this leukotriene, as well as other products of AA metabolism.

We have previously demonstrated that NK cell activity is suppressed by reactive metabolites of oxygen (RO), released either by monocytes or by polymorphonuclear cells (25,26). It is possible that suppression of natural killing by RO is secondary to effects of RO on AA metabolites. Leukotrienes B_4, C_4, D_4, and E_4 can be inactivated extracellularly by polymorphonuclear cells through the production of RO (27-29). Inactivation is dependent on H_2O_2 which, in high concentration, may act alone or, in low concentration, requires peroxidase. Whether the effects of RO on leukotrienes pertain to the effects of RO on NK cell activity is unknown and the question, as with many others, awaits the delineation of AA metabolism within NK cells.

REFERENCES

1. Weissman, G. (1983). Seminars Arthritis Rheum. 13 (Suppl 1), 123.
2. Dvorak, A.M., Dvorack, H.F., Peters, S.P., et al. (1983). J. Immunol. 131, 2965.
3. Hirata, F., and Axelrod, J. (1980). Science 209, 1082.
4. Kuehl, F.A., Jr., and Egan, R.W. (1980). Science 210, 28.
5. Hammarstrom, S. (1983). Ann. Rev. Biochem. 52, 355.

6. Lewis, R.A., and Austen, K.E. (1984). J. Clin. Invest. 73, 889.
7. Goetzl, E.J., Payan, D.G., and Goldman, D.W. (1984). J. Clin. Immunol. 4, 79.
8. Rola-Pleszczynski, M., Borgeat, P., and Sirois, P. (1982). Biochem. Biophys. Res. Comm. 108, 1531.
9. Parker, C.W. (1982). In "Advances in Inflammation Research" (G. Weissman, ed.), Vol. 4, p. 1. Raven Press, New York.
10. Hoffman, T., Hirata, F., Bougnoux, P., et al. (1981). Proc. Natl. Acad. Sci. U.S.A. 78, 3839.
11. Hattori, T., Hirata, F., Hoffman, T., Hizuta, A., and Herberman, R.B. (1983). J. Immunol. 131, 662.
12. Koren, H.S., Anderson, S.J., Fischer, D.G., Copeland, C.S., and Jensen, P. (1981). J. Immunol. 127, 2007.
13. Seaman, W.E. (1983). J. Immunol. 131, 2953.
14. Rola-Pleszcyznski, M., Gagnon, L., and Sirois, P. (1983). Biochem. Biophys. Res. Comm. 113, 531.
15. Suthanthiran, M., Solomon, S.D., Williams, P.S., Rubin, A.L., Novogrodsky, A., and Stenzel, K.H. (1984). Nature 307, 276.
16. Seaman, W.E., and Woodcock, J. (1984) J. Allergy Clin. Immunol. In press.
17. Brooks, C.G., Urdal, D.L., and Henney, C.S. (1983). Immunol. Rev. 72, 43.
18. Goldyne, M.E., and Stobo, J.D. (1981). CRC Crit. Rev. Immunol., p. 189.
19. Parker, C.W., Stenson, W.F., Huber, M.G., and Kelly, J.P. (1979). J. Immunol. 122, 1572.
20. Goetzl, E.J. (1981). Biochem. Biophys. Res. Comm. 101, 344
21. Kato, K., Koshihara, Y., Fujiwara, M., and Murota, S. (1983). Prostaglandins, Leukotrienes and Medicine 12, 273.
22. Abraham, R.T., Shipley, G.D., and Handwerger, B.S. (1984). Arthritis Rheum. (No. 4, Supplement), S22.
23. Maclouf, J., Fruteau-de-Laclos, B., Borgeat, P. (1982). Proc. Natl. Acad. Sci. U.S.A. 79, 6042.
24. Payan, D.G., Missirian-Bastian, A., and Goetzl, E.J. (1984). Proc. Natl. Acad. Sci. U.S.A. 81, 3501.
25. Seaman W.E., Gindhart T.D., Blackman, M.A., Dalal, B., Talal, N., and Werb, Z. (1981). J. Clin. Invest. 67, 1324.
26. Seaman, W.E., Gindhart, T.D., Blackman, M.A., Dalal, B., Talal, N., and Werb, Z. (1982). J. Clin. Invest. 69, 876.
27. Henderson, W.R., Jorg, A., and Klebanoff, S.J. (1982). J. Immunol. 1289, 2609.

28. Lee, C.W., Lewis, R.A., Corey, E.J., et al., (1982). Proc. Natl. Acad. Sci. U.S.A. 791, 4166.
29. Lee, C.W., Lewis, R.A., Tauber, A. I., Mehrotra, M.M., Corey, E. J., and Austen, K.F. (1982). J. Biol. Chem. 2582, 15004.
30. Boeynaems, J.M., Brash, A.R., Oates, J.A., and Hubbard, W.C. (1980). Analyt. Biochem. 104, 259.
31. Henderson, W.R., and Klebanoff, S.J. (1983). J. Biol. Chem. 258. 13522.

THE ROLE OF THE LIPOXYGENASE IN REGULATION OF NK ACTIVITY[1]

Kam H. Leung[2]
Margot M. Ip

Grace Cancer Drug Center
Roswell Park Memorial Institute
Buffalo, New York

Hillel S. Koren

Division of Immunology
Duke University Medical Center
Durham, North Carolina

I. Introduction

Recent data suggest that NK (natural killer) activity is under the regulating influences of a variety of substances, most prominently those of interferons (IFN) (1) and PGE_2(2,3). Arachidonic acid (AA) metabolites, such as PGE_2 are now recognized as highly active substances that play a central role in modulating a variety of immune cell functions (4), including NK activity (5). The role of lipoxygenase products in NK regulation is not well understood. Lipoxygenase converts AA into hydroperoxyeicosatetraenoic acids (HPETE) which are

---

[1]This work was supported by NCI grants CA35641, and CA23354 and Core grant CA24538.
[2]Present address: Central Research and Development Department, Glenolden Laboratory, E. I. du Pont de Nemours and Company, Glenolden, Pennsylvania

ISBN 0-12-341370-2

unstable and are reduced to hydroxy fatty acids (HETE). 5-HETE, 12-HETE, and 15-HETE have been isolated from various cell types (6). 5-lipoxygenase activity also leads to formation of a variety of leukotrienes (LTA_4, LTB_4, LTD_4, LTE_4, LTF_4). The transformation of LTA_4 to LTC_4 is carried out by glutathione-S-transferase (7).

It has been demonstrated that inhibiton of phospholipase A_2 activity (8) and lipoxygenase activity (9-11) suppressed human NK cytotoxicity. These findings suggest that lipoxygenase metabolites may be required for effective NK cytolytic activity. It has also been reported that lipoxygenase products may play a role in IFN production by mouse spleen cells (12). It therefore seemed important to investigate whether lipoxygenase products are involved in the activation of NK activity by IFN. We report here that four inhibitors of lipoxygenase activity, nordihydroguaiaretic acid (NDGA), α-phenanthroline, esculetin and 6, 9-deepoxy-6,9-(phenylimine)-6,8-prostaglandin I (U-60257) (7, 14-16), inhibit not only NK cytotoxicity but also inhibit the maintenance of endogenous NK activity and the activation by IFN. However, IFN-activated NK cells are less sensitive to suppression mediated by these inhibitors of lipoxygenase.

II. Materials and Methods.

Human peripheral blood lymphocytes (13) depleted of adherent cells (NA) by plastic and nylon wool adherence were used routinely.

To test the effect of various reagents on NK activity, effector cells ($2x10^6$/ml) were incubated in RPMI-1640 medium supplemented with 10% FCS in 12 x 75 mm polypropylene tubes at 37°C in a 5% CO_2 atmosphere. Poly I:C, NDGA, esculetin, α-phenanthroline (Sigma), U-60257 (Upjohn), human fibroblast interferon (IFN_β) (HEM), human lymphoblastoid interferon (IFN) (Burroughs Wellcome) and recombinant IFN_γ (Biogen) were added to the cell suspensions. After 18 hr the cultured cells were collected, washed and resuspended in medium. NK activity was measured in a 2 hr ^{51}Cr release assay against K562 target cells in the presence or absence of the lipoxygenase inhibitors. All data presented are for triplicate determinations.

III. Results

To investigate a role of leukotrienes in NK lysis, two specific inhibitors, U-60257 and esculetin along with two general inhibitors of lipoxygenase, α-phenanthroline and NDGA were tested. As shown in Table I, NK activity against ^{51}Cr labeled K562 target cells was suppressed by all four inhibitors.

Table I. Effect of lipoxygenase inhibitors on NK lysis

Agents in Assay	0	1	5	10	25	50	100
	Concentration, µg/ml						
	Percent Specific Cytotoxicity						
NDGA	30	25	15	9	5		
U-60257	48				45	43	33
α-Phenanthroline	30	30		29	21	15	
Esculetin	30				21	15	8

Freshly isolated NA cells were tested for NK cytolytic activity against ^{51}Cr-labeled K562 target cells. NDGA, U-60257, α-phenanthroline and esculetin were present during the two hour assay period. E:T ratio was 10:1.

We have previously shown that IFN-activated NK cells became less sensitive to suppression by PGE_2 (2). To examine if U-60257 and NDGA would inhibit the NK killing of IFN-activated cells, NA cells were cultured with IFN or poly I:C for 18 hr, washed and assayed for the NK activity in the presence or absence of U-60257, NDGA or PGE_2. As shown in Table II, U-60257 (100 µg/ml), NDGA (25µg/ml) and PGE_2 (1 µg/ml) inhibited killing by untreated cells by 61-67%. In contrast, poly I:C and IFN_α-activated cells became less susceptable to suppression by U-60257, NDGA and PGE_2, when compared to the untreated controls.

Furthermore, we determined whether pretreatment of NA cells with the inhibitors of lipoxygenase at 37°C for 18 hr will affect the maintenance of endogenous NK activity. NA cells were cultured with medium, NDGA, esculetin, α-phenanthroline or U-60257 for 18 hr at 37°C, washed and assayed for the NK cytotoxicity (Table III). The data indicate that pretreatment of NA cells with the inhibitors decreased their NK activity in a dose-dependent manner. To further investigate whether these inhibitors interfere with the actual process of NK activation by IFN or poly I:C, NA cells were stimulated with IFN_α (10^3 U/ml) or poly I:C

Table II. Loss of Sensitivity of IFN-activated human NK cells to Suppression by Lipoxygenase Inhibitors

Pretreatment	Medium	Agents in NK assay NDGA (10μg/ml)	U-60257 (100μg/ml)	PGE_2 (1μg/ml)
	Percent Cytotoxicity against K562			
Medium	20	7(65)	8(60)	7(65)
IFN_α(100 μ /ml)	43	50(-25)	34(21)	28(37)
Poly I:C(100μg/ml)	49	48(2)	30(39)	33(33)

NA cells were cultured with medium alone, IFNα or poly I:C for 18 hr at 37°C. NK activity was assayed in the presence or absence of NDGA, U-60257 or PGE_2. E:T ratio was 10:1 in a 2 hr assay. Numbers in parentheses represent percent of inhibition by each agent.

Table III. Effect of Pretreatment of NA cells with Lipoxygenase Inhibitors on Endogenous NK activity

Pretreatment	0	1	5	10	25	50	100
	Concentration, μg/ml; Percent Cytoxicity Against K562						
NAGA	43	41 (5)	31 (28)	25 (58)	5 (88)		
Esculetin	56				50 (11)	49 (12)	20 (64)
α-phenanthroline	50	49 (2)	35 (30)	30 (40)	5 (90)		
U-60257	43				32 (26)	15 (65)	2 (95)

NA cells were cultured with medium alone, NDGA, esculetin, α-phenanthroline as U-60257 for 18 hr at 37°C. The cultured cells were washed and assayed for NK activity. E:T ratio was 20:1 in a 2 hr assay. Numbers in parentheses represent percent of inhibition by each agent.

(100μg/ml) for 18 hr at 37°C in the presence or absence of the lipoxygenase inhibitors. The findings demonstrate that IFN and poly I:C boosted the NK activity but to a lesser extent in the presence of these lipoxygenase inhibitors (Figure 1), suggesting that inhibitors of lipoxygenase interfere with optimal activation of NK cells.

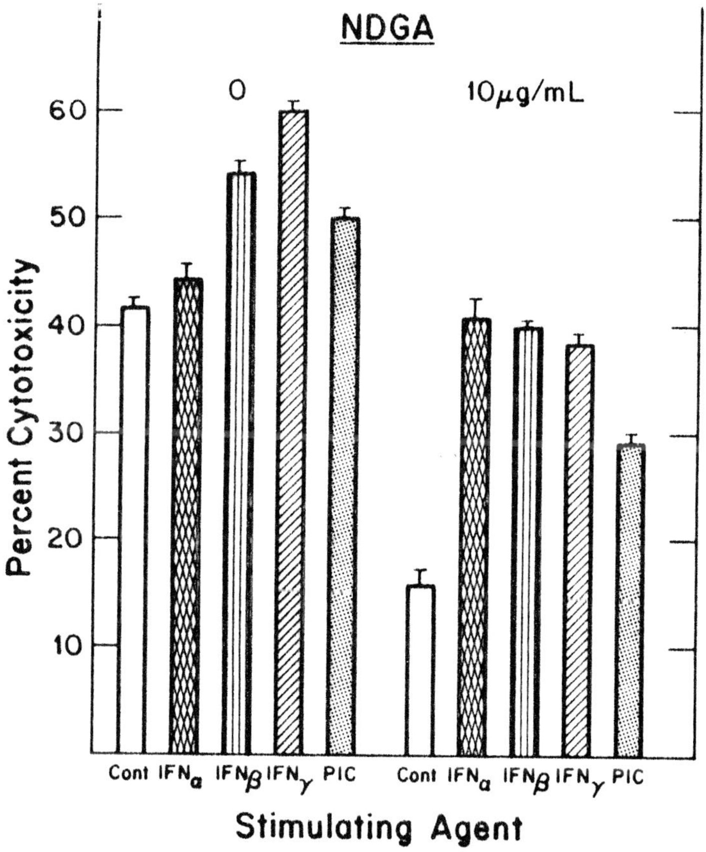

Figure 1. Effect of NDGA on NK activation by IFN. NA cells were cultured with NDGA, IFNα (10^3 U/ml), IFNβ (10^3 U/ml), IFNγ (10^3 U/ml), poly I:C (100μg/ml) or their combinations for 18 hr at 37°C. The cells were washed and assayed against K562 target cells at E:T ratio of 10:1 in a 2 hr assay.

IV. Discussion

In this chapter we have shown that NDGA, U-60257, α-phenanthroline and esculetin, agents that inhibit the lipoxygenase pathways of AA metabolism inhibit NK mediated cytotolysis and the maintenance of endogenous NK activity. U-60257 (a specific inhibitor of 5-lipoxygenase and glutathione-S-transferase) (7), esculetin (a specific inhibitor of 5- and 12-lipoxygenase) (16), and α-phenanthroline as well as NDGA (general inhibitors of lipoxygenation) (14,15) inhibited NK lysis. The involvement of lipoxygenase in NK lysis explains the early observations that inhibition of phospholipase A_2 resulted in diminished NK lysis (8). Others have reported that NDGA, BW755C and ETYA inhibit NK lysis (9-11). The observation that pretreatment of NK cells with the lipoxygenase inhibitors for 18 hrs resulted in inhibition of the endogenous NK activity is novel and intriguing. The decrease in endogenous NK activity is unlikely due to a direct toxicity of the drugs because NA cells cultured with those inhibitors showed comparable recovery and viability with that of untreated control cultures as measured by trypan blue dye exclusion. A more likely explanation for the effect of lipoxygenase inhibitors on endogenous NK activity may be their effect on the synthesis and release of IFN and other lymphokines, known to be produced by these cells (17). These lymphokines are important for the maintenance of NK activity and may depend on lipoxygenase products.

We have previously shown that indomethacin has little effect on the augmentation of NK cell activity by IFN or poly I:C (18). In this study, we showed that lipoxygenase inhibitors block the activation induced by IFN or poly I:C. It is possible that IFN may induce a greater pool of lipoxygenases which are involved in NK lysis. Alternatively, IFN may enhance a production of lipoxygenase metabolites which in turn may affect the events leading to activation of NK cell activity. The former possibility is intriguing since IFN-activated NK cells become partially resistant to suppression by NDGA and U-602567. Therefore, induction of lipoxygenase enzymes may be one of the mechanisms by which IFN activates NK activity.

Recent data from our laboratory (2,3,18) and others (1,5,8-11) suggest that NK activity is under the influences of IFN, and AA metabolites. Possible regulatory mechanisms of NK cells are summarized in Figure 2. When mononuclear cells are stimulated with poly I:C NK augmentation is less pronounced than that by stimulated nonadherent cells (3) suggesting negative influence by monocytes. In fact, the addition of indomethecin during the stimulation of mononuclear cells with poly I:C resulted in further augmentation of NK activity (3), confirming that the monocytes contribution was primarily in producing PGE_2 while the NA cells were induced to produce IFN in response to poly I:C stimulation (3). We have also shown that pretreatment of NA cells with PGE_2 lead to a decrease in NK activity (18), however, PGE_2 did not interfere with the activation of NK activity induced by IFN. Furthermore, we have shown that IFN-activated cells are partially

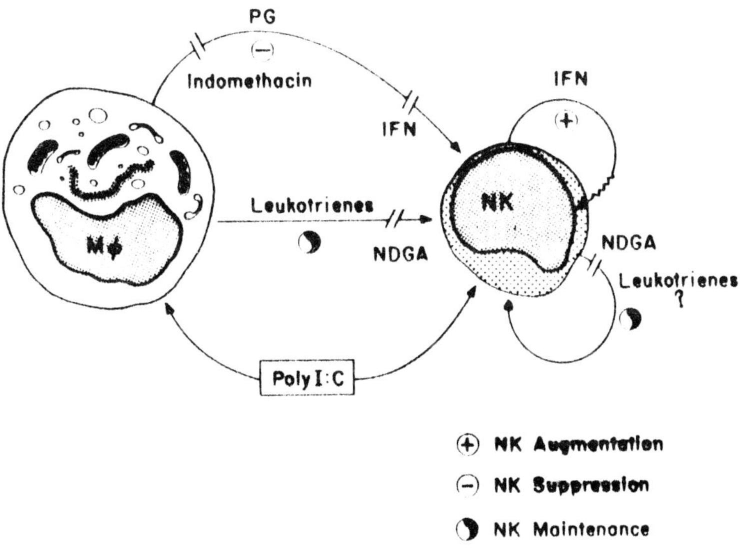

Figure 2. Possible regulatory mechanisms of NK cells.

resistant to the inhibitory effect of PGE_2 (2,18). Thus, IFN not only activates NK cells but also protects them from suppression by PGE_2 which may be of tumor and/or monocyte origin. In this chapter, we further extend our findings and show that lipoxygenase activity is involved in NK lysis and maintenance of NK activity. We submit that lipoxygenase activity is necessary for NK activation induced by IFN, as lipoxygenase inhibitors block IFN activation. It seems that PGE_2 and lipoxygenase products exert opposing effect on NK cells: PGE_2 and lipoxygenase inhibitors inhibit NK lysis and maintenance of endogenous NK activity. However, leukotrienes and 5-HETE on our hands have little effect in these two parameters and they did not reverse the inhibitory effects of NDGA (data not shown). We have also established that IFN-activated NK cells are relatively less sensitive to suppression by inhibitors of lipoxygenase as compared to the sensitivity of unstimulated cells. Taken together, the regulation of NK activity is complex and seems to be regulated by multiple factors. The precise nature of their action and the effects of the interactions of different mediators as part of the complex regulatory circuit will require further investigation.

Acknowledgement

We thank Dr. M. Bach of the Upjohn Company for kindly providing us with U-60257 and Ms. E. Freas for the preparation of the manuscript.

REFERENCES

1. Herberman, R.B., Ortaldo, J.R., and Bonnand, G.D. (1979) Nature 277, 221.
2. Leung, K.H. and Koren, H.S. (1982). J. Immunol. 129, 1742.
3. Koren, H.S., Anderson, S.J., Fischer, D.G., Copeland, C.S., and Jensen, P.J. (1981). J. Immunol. 127, 2007.
4. Goodwin, J.S. and Ceuppens, J. (1983). J. Clin. Immunol. 3, 295.
5. Droller, J.J., Schneider, M.U., Perlmann, P. (1978). Cell Immunol. 39, 165.
6. Samuelsson, B. (1983). Science 220, 568.
7. Sun, F.F. and McQuire, J.C. (1983). Prostaglandins 26, 211.
8. Hoffman, T., Hirata, F., Bougnoux, P., Fraser, B.A., Goldfarb, R.H., Herberman, R.B., and Axelrod, J. (1981). Proc. Natl. Acad. Sci. USA 78, 3839.

9. Rola-Pleszcynski, M., Gagnon, L. and Sirois, P. (1983). Biochem. Biophys. Res. Comm. 113, 531.
10. Seaman, W.E., (1983). J. Immunol. 131, 2953.
11. Suthanthiran, M., Solomin, S.D., Williams, P.S., Rubin, A.L. Novogrodsky, A., and Stengel, K.H. (1984). Nature 307, 276.
12. Johnson, H.M. and Torres, B.A. (1984) J. Immunol. 132, 413.
13. Boyum, A. (1968) Scand. J. Clin. Lab Invest. 21 (Suppl. 97), 9.
14. Sircar, J.C., Schwender, C.F., and Johnson, E.A. (1983). Prostaglandins 25, 393.
15. Toppel, A.L., Lundberg, W., Boyer, P.D. (1953). Archo. Biochem. Biophys. 42, 293.
16. Sekiya, K., Okuda, H. and Arichi, S. (1982). Biochem. Biophys. Acta 713, 68.
17. Kasahana, T., Djeu, J.Y., Dougherty, S.F. and Oppenheim, J.J. (1983). J. Immunol. 131, 2379.
18. Leung, K.H. and Koren, H.S. (1984). J. Immunol. 132, 1445.

NO CONFIRMED ROLE FOR REACTIVE OXYGEN INTERMEDIATES IN NATURAL KILLER (NK) CELL-MEDIATED CYTOLYSIS

H. David Kay*

Experimental Immunology Laboratory,
Department of Internal Medicine
University of Nebraska Medical Center
Omaha, Nebraska

Ronald H. Goldfarb**

Cancer Metastasis Research Group
Department of Immunology and Infectious Disease
Central Research Division
Pfizer, Inc.
Groton, Connecticut

Elizabeth A. Wayner
Colin G. Brooks

Basic Immunology
Fred Hutchinson Cancer Research Center
Seattle, Washington

---

*Coworkers with HDK in these studies were Deirdre L. Smith; Gerald R. Donowitz, M.D.; Gerald L. Mandell, M.D.; and Gail Sullivan; Department of Internal Medicine, University of Virginia School of Medicine, Charlottesville, Virginia.

**Coworkers with RHG in these studies were Tuomo Timonen, M.D.; Edgar Pick, M.D., Ph.D.; Thomas Hoffman, M.D.; John R. Ortaldo, Ph.D.; Ezio Bonvini, Ph.D.; and Ronald B. Herberman, M.D.; Laboratory of Immunodiagnosis, NCI, NIH, Bethesda, Maryland.

ISBN 0-12-341370-2

It has been proposed by Roder and colleagues that human natural killer (NK) cells exhibit a burst of oxygen radical production during the early stages of the lytic attack on K562 erythroleukemia target cells in vitro (1-4). A role for such products in NK killing was attractive at the time (5) because it mechanistically linked the functions of NK cells with those of other peripheral blood-derived effector cells (monocytes, granulocytes) known to use reactive oxygen intermediates (ROI) in carrying out their part of normal host defense mechanisms (6,7). Other investigators, however, reported evidence which showed that reactive oxygen radicals were not involved in the NK cytolytic event (8-14). Thus began a controversy which has not yet been clearly resolved.

It is the aim of this article, therefore, to examine in detail the question of whether ROI participate in NK-mediated cytolysis. In particular:

1. Can monocytes/macrophages "assist in" the NK-cell-mediated lysis of NK-sensitive targets?

2. Do studies with lymphocytes from patients with chronic granulomatous disease (CGD) provide unique insight into the role of oxygen radicals in the NK cytolytic mechanism?

3. Do highly purified NK cells from normal donors release detectable amounts of reactive oxygen radicals following interaction with suitable NK target cells?

4. How helpful are those studies which purport to show inhibition of NK cytotoxicity by a variety of hydroxyl free radical scavengers?

Answers to these questions should provide an appropriate framework from which to determine if reactive oxygen radicals are of significance in the NK cytolytic process.

I. IS THE SHORT-TERM NK CELL-MEDIATED LYSIS OF TARGET CELLS ENHANCED IN THE PRESENCE OF MONOCYTES?

Roder and colleagues have recently modified their original proposal, having determined that the oxidative intermediates detected in their earlier studies were produced by monocytes which remained in their NK cell preparations (4, note added in proof; 15). They state, however, that the release of the ROI detected early in their NK assays requires the simultaneous presence of both NK cells and monocytes with the target cells in the assay. Either cell type alone will not release ROI in the presence of target cells. Such a concept of "synergy" (i.e., the amount of ROI detectable in an assay is greater than the sum of the ROI released by either NK cells or monocytes alone) has been supported by the recent reports of Ernst _et al_ (16), and Pohajdak _et al_ (17). In contrast, no synergistic effect in the production of ROI was observed by Ramstedt _et al_ (18) when purified monocytes were mixed back with NK cells.

Because a number of early reports demonstrated that the presence of monocytes clearly is not required for the cytolytic activity of NK cells (19-22), the relevance to the NK lytic event of a purported NK cell/monocyte synergism for ROI production must be questioned. It would be helpful, therefore, to review some of the published reports which have examined the relationship of monocytes to the cytolytic activity of NK cells.

A. Depletion Studies

Using K562 erythroleukemia cells as targets, West _et al_ (21), and Kay _et al_ (22) observed that cytotoxicity was unaffected or even slightly enhanced when monocytes/ macrophages were removed from effector cell preparations either by adherence, by magnet following ingestion of carbonyl iron, or by passage over Sephadex G10 columns. Tanaka _et al_ (23), Ottenhof _et al_ (20), Abrams _et al_ (24), and Seaman (25), found that removal of monocytes from blood lymphocytes often led to higher levels of NK activity being expressed by the remaining non-adherent lymphocytes. This phenomenon was explored in depth by Koren _et al_ (19) who reported that the presence of normal (nonactivated) monocytes in effector cell preparations in vitro did not influence endogenous NK levels. While they, too, found that monocyte-depleted mononuclear cells (<1% monocytes) usually exhibited enhanced NK activity, they concluded that this was not a result of removal of cells with inhibitory activity but rather that removal of extraneous cells led to an inevitable enrichment of NK effector cells. Recently, Pohajdak _et al_ (17) reported that removal of monocytes from NK cell suspensions with the monocyte-specific monoclonal MO2 antibody and complement had either no effect on NK lysis or slightly enhanced it.

B. Addback Studies

From another perspective, if NK effector cells were to act synergistically with monocytes in the release of a product required in the NK cytolytic event, it might be assumed that addback of monocytes to an NK assay would lead to an observable increase in the NK levels detected in the assay. This, however, has not been the experience. Kay and Smith (26), for example, found that NK activity against K562 was usually unaffected by adding back monocytes, and, in fact, was occasionally inhibited (up to 30%) in some assays. Likewise, Koren et al (19) showed that monocytes added back to nonadherent cells never augmented NK activity, most commonly had no effect on the assay (even when added back to a final concentration of 20%), and occasionally inhibited cytotoxicity (by 46% in one experiment). Ernst et al (16) reported that addback of monocytes to phagocyte-depleted mononuclear cells resulted in decreased NK cell activity.

From a third perspective, were reactive oxygen intermediates (ROI) required early in the NK lytic event, then one might suppose that adding more of these oxidative species (derived from monocytes or even from granulocytes) to an NK assay would enhance target cell lysis. This is, however, not the case. Seaman and colleagues (12,27) have shown that cells which are capable of an oxidative burst (e.g., monocytes, granulocytes) significantly inhibit NK activity when activated by appropriate agents (such as zymosan or phorbol diesters) to release ROI in the presence of NK cells. Similarly, Kay and Smith (unpublished data) found that granulocytes which had been stimulated to release ROI by activation with soluble immune complexes (tetanus antigen plus human IgG antitetanus antibodies), and, then thoroughly washed to remove free, excess immune complexes, were far more potent as inhibitors of autologous NK cell activity than were nonactivated granulocytes tested in an identical manner. Thus, under conditions where ROI release during cytotoxicity assays is enhanced, a corresponding enhancement of NK activity does not occur.

Overall, the data obtained in several laboratories

argue strongly against the participation of monocytes in NK-mediated cytolysis.

II. USING LYMPHOCYTES FROM PATIENTS WITH CHRONIC GRANULOMATOUS DISEASE (CGD) CAN ADDRESS THE ROLE OF REACTIVE OXYGEN RADICALS IN THE NK LYTIC EVENT

CGD comprises a group of rare pathological conditions, usually detected in children. This disease is characterized by a defect in bactericidal activity of phagocytic cells directed mainly against catalase-positive microorganisms (28-31). The defect stems from the complete inability of the phagocytic cells (monocytes, polymorphonuclear leukocytes) to mount a normal respiratory, oxidative burst. Therefore, they fail to reduce molecular oxygen to the free radical superoxide (O_2^-), derived from the activation of the hexose monophosphate shunt required for the generation of other ROI, such as hydrogen peroxide (H_2O_2) and hydroxyl radical (OH•). The inability of cells from these patients to reduce nitroblue tetrazolium dye or ferric cytochrome C, or to produce chemiluminescence (all ROI-dependent reactions) are important tests used to establish the diagnosis of CGD (30,31).

Studies of NK cell activity by peripheral blood mononuclear cells of patients with CGD have, therefore, been very useful in determining the importance of oxidative mechanisms in target cell lysis by NK cells. Tanaka <u>et al</u> (23) were among the first to report that NK activity of lymphocytes from the blood of CGD patients was entirely normal, when tested against K562 cells. The studies of Lopez <u>et al</u> (34), Seaman <u>et al</u> (12), Kay <u>et al</u> (11,35), El-Hag and Clark (36), Ramstedt <u>et al</u> (18), and Duwe and Roder (15), have confirmed Tanaka's observations, thus establishing that NK cytolytic activity by mononuclear cells from CGD patients (all of whom exhibited classical defects in oxidative metabolism) is identical to that exhibited by cells from healthy controls tested simultaneously (Figure 1). In addition, Katz <u>et al</u> (37) showed that lymphocytes, monocytes, and neutrophils of CGD patients all exhibited normal levels of antibody-dependent cellular cytotoxicity (ADCC) against IgG-sensitized targets. These results, taken together, demonstrate that both NK and ADCC activities are expressed normally in patients whose blood cells show

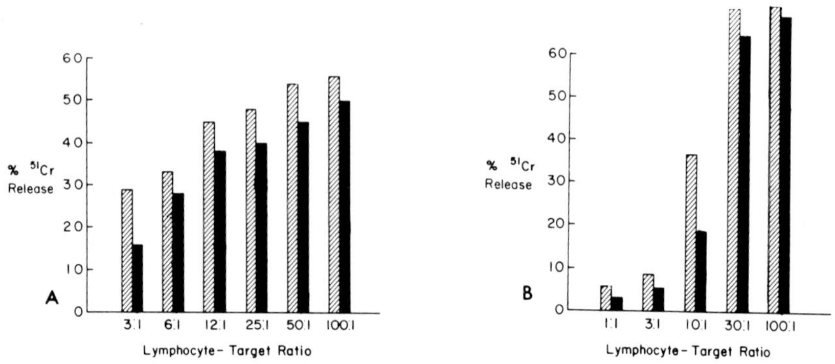

Figure 1. Percentage of ^{51}Cr released from radiolabeled K562 erythrleukemia target cells after in vitro incubation for 4 hr with mononuclear cells from the blood of patients with CGD (hatched bars) or from normal, healthy volunteers (open bars). A and B, representative assays of NK cell activity of two different patients with CGD, each tested on at least two separate occasions. (Data from Kay et al (35), reproduced with the permission of the Journal of Immunology.)

a severe defect in their ability to generate ROI, and provide a further powerful argument against the participation of monocyte-derived ROI in NK cell lysis. The possibility that NK cells themselves might generate ROI by enzymatic pathways distinct from those which are deleted in CGD patients remains a formal possibility. The question of whether purified NK cells can produce ROI has therefore been examined in great detail.

III. ARE NK CELLS CAPABLE OF ELICITING AN OXIDATIVE BURST IF APPROPRIATELY STIMULATED IN VITRO?

A. Studies measuring release of ROI

Several laboratories have been investigating the role of the oxidative burst in the NK lytic event by using highly purified preparations of NK cells (large granular lymphocytes, or LGL) from normal donors. Goldfarb *et al* (9,10) noted in 1981 that highly purified preparations of LGL did not display a detectable oxidative burst in response to challenge with a number of well-characterized inducers, which included phorbol myristate acetate, the calcium ionophore A23187, Concanavalin A, sodium fluoride, digitonin, the synthetic oligopeptide FMLP, and phospholipase C. In contrast, autologous monocytes tested in parallel under identical conditions were stimulated to release high levels of detectable oxidative intermediates, as expected. These investigators measured superoxide production by the reduction of ferric cytochrome C and by the reduction of nitroblue tetrazolium dye. More recently they reported that purified (>95%) populations of LGL also failed to produce a detectable oxidative burst in response to K562 target cells (38) (Table I). With a 1.5% contamination with add-back monocytes, enriched LGL (90%) failed to produce a detectable oxidative burst. When monocyte contamination was deliberately raised to 8.5%, an appreciable oxidative burst was observed in response to stimulation by PMA, confirming the earlier-mentioned conclusion that the presence of monocytes can significantly alter the results of an oxidative burst assay.

Kay *et al* (11,35) used purified LGL from 12 healthy controls, and likewise found no evidence to support a hypothesis that LGL could generate superoxide (as measured by reduction of nitroblue tetrazolium dye or ferric cytochrome C) or reactive oxygen species capable of inducing luminol-dependent chemiluminescence (superoxide, hydrogen peroxide, hydroxyl radicals, singlet oxygen) when exposed to zymosan, PMA, or K562 cells (Table II). Yet, monocytes and neutrophils were readily stimulated in the same experiments to produce a strong oxidative burst. In further experiments, increasing the number of LGL per assay to 5×10^6, adding K562 cells in ratios up to 30:1 (instead of 10:1), and prolonging the assay to 60 min (instead of 10 or 30 min)

TABLE I. Oxidative burst of human lymphoid subpopulations in response to various stimuli[a]

Lymphoid Sub-population	LGL (%)	Phagocytes (%)	Cytochrome C Test:[b] O_2^- (nMoles/10^6 cells/hour) Resting	Stimulated	(Stimulus)	+K562[c]	+Rlo1[c]
NK	>98	<2	0.7 ± 0.5	0.6 ± 0.5	(PMA, 20 nM)	ND[d]	ND
"	"	"	0.7 ± 0.5	0.7 ± 0.6	(PMA, 200 nM)	ND	ND
"	"	"	0.7 ± 0.5	0.6 ± 0.5	(FMLP, 100 nM)	ND	ND
"	"	"	0.7 ± 0.5	0.7 ± 0.5	(Digitonin, 2.5 µg/ml)	ND	ND
"	90	8.5	0.7 ± 1.2	2.2 ± 1.0	(PMA, 200 nM)	0.7±0.7	ND
"	90	1.5	-0.2 ± 0.7	-0.5 ± 0.7	(PMA, 200 nM)	0.0±0.4	-0.1±0.6
Monocytes	0	95	4.0 ± 0.6	15.0 ± 1.5	(PMA, 200 nM)	ND	ND
"	20	50	2.5 ± 0.6	11.0 ± 1.5	(PMA, 200 nM)	1.7±0.8	2.5±0.6
"	0	95	4.0 ± 0.5	6.4 ± 0.6	(FMLP, 100 nM)	ND	ND
"	0	95	4.0 ± 0.6	3.9 ± 0.5	(Digitonin, 2.5 µg/ml)	ND	ND
T cells	<1	<0.1	1.0 ± 0.5	0.9 ± 0.6	(PMA, 20 nM)	ND	ND
"	"	"	1.0 ± 0.5	1.2 ± 0.7	(PMA, 200 nM)	ND	ND
"	"	"	1.0 ± 0.5	0.9 ± 0.5	(FMLP, 100 nM)	ND	ND
"	"	"	1.0 ± 0.5	0.8 ± 0.5	(Digitonin, 2.5 µg/ml)	ND	ND
"	3	"	0.1 ± 0.9	0.1 ± 0.5	(PMA, 200 nM)	-0.7±1.0	-0.2±0.6

[a]Human lymphoid subpopulations were isolated as previously described (39).
[b]Superoxide production was measured by the cytochrome C test as previously described (40). Cell concentrations were 0.5 x 10^6/well. Data shows mean of quadruplicates ± SD. (n=5).
[c]E:T = 20:1. Results depict Resting Cells + Tumor Cells without additional stimulation.
[d]ND = Not Determined.

still failed to generate an oxidative burst detectable by any of these procedures. The recent results of Ernst *et al* (16), Pohajdak *et al* (17), and Ramstedt *et al* (18) are all consistent with these observations In each of these reports it was shown that NK cell suspensions lost all capacity to produce ROI when monocytes were removed but maintained (or improved) their NK cytolytic capacity.

In the mouse, a stringent test of whether ROI are generated during NK reactions has been reported by Wayner and Brooks (41). These investigators took advantage of the fact that cloned cytotoxic T cells (CTL) express potent NK activity after treatment with interferon (IFN). Such IFN-treated CTL (or control untreated CTL) were incubated with YAC-1 cells (NK target) or EL4 cells (CTL target) in a cytotoxicity assay in the presence of luminol. As noted above, luminol-dependent chemiluminescence is an extremely sensitive means of detecting ROI. Although almost 60% lysis of YAC-1 targets occurred, there was no detectable chemiluminescence generated at any time during the 4-hr assay (Figure 2). By contrast, peritoneal macrophages generated a strong burst of chemiluminescence when incubated under the same conditions with mycoplasma-infected tumor cells. The fact that chemiluminescence responses even between macrophages and tumor cells are only generated if the latter are infected with mycoplasma is discussed in greater detail below in Section V.

B. Studies using scavengers of ROI

The conclusions of the previous section have more recently been confirmed by Abrams and Brahmi (42), who found that the addition of catalase (up to 1000 μg/ml) or superoxide dismutase (SOD; up to 1000 μg/ml) to LGL-mediated cytotoxic assays had no effect on the NK cytotoxicity observed. Furthermore, by chemiluminescence and SOD-inhibitable reduction of cytochrome C, they failed to detect an oxidative burst associated with K562-stimulated or PMA-stimulated NK cells. Nor did they find that pretreatment of LGL with interferon-B resulted in detectable production of oxygen radicals by K562-stimulated NK cells. Polymorphonuclear leukocytes, on the other hand, rapidly responded with an easily-detected oxidative burst, when appropriately stimulated in each of these assays. In support of these results, Ramstedt *et al* (18) also found no NK inhibition when the ROI scavengers SOD, cytochrome c, and catalase were added to assays at concentrations sufficient to block

Table II. Responses in assays (means ± SEM) for production of oxidative metabolites by NK cells (=LGL), monocytes, or neutrophils obtained from normal, healthy donors

Assay used to monitor production of oxidative metabolites	Units of measure	Stimulant	Responder Cells		
			NK cells[a]	Monocytes	Neutrophils
A. Reduction of nitroblue tetrazolium dye (n = 3)	Percentage of cells which contained insoluble blue formazan	Opsonized zymosan (K562 used for LGL)[b]	0	92 ± 3%	97 ± 3%
B. Reduction of ferric cytochrome-C (n = 5)	nmoles ferric cytochrome C reduced/10^6 cells/30 min	Opsonized zymosan (K562 used for LGL)	0	9.8 ± 2.5	14.9 ± 3.4
C. Luminol-enhanced chemiluminescence (n = 5)	Fold increase above resting chemiluminescent baseline	Opsonized zymosan (K562 used for LGL)	0	420 ± 75	441 ± 84
D. Luminol-enhanced chemiluminescence (n = 1)	Fold increase above resting chemiluminescent baseline	Phorbol myristate acetate (100 ng/well)	0	375	660

[a]NK cells used were purified large granular lymphocytes (LGL), 1×10^6 per assay.
[b]Purified LGL tested with zymosan as stimulant were always totally negative.

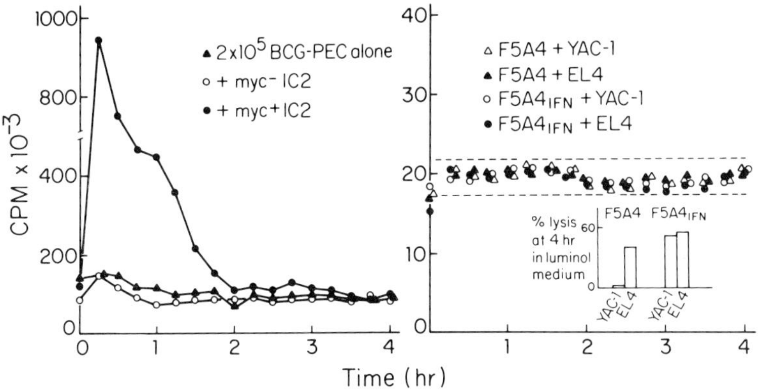

Figure 2. Luminol-dependent chemiluminescence produced by BCG-induced peritoneal macrophages (BCG-PEC), resting F5A4 cells, or IFN-induced F5A4 cells, when incubated with various timuli over a 4-hr period. The straight lines in the right hand panel indicate the range of chemiluminescence activity in control vials containing effector cells alone. Inset shows the cytotoxic activity of resting and IFN-treated F5A4 cells incubated in parallel vials under identical conditions but with ^{51}Cr-labeled target cells (Data from Wayner and Brooks (41), reproduced with permission of Plenum Press).

chemiluminescence responses of monocytes by 70 to 95%.

The recent report by Wayner and Brooks (41) expands these observations to include murine effector cells as well. They found that catalase and SOD had no inhibitory effect either in the human NK system, using blood mononuclear cells (MNC) as effectors, or in the murine NK system, using as effectors either fresh spleen cells or cloned IFN-induced CTL. The only significant change they reported was an occasional enhancement by SOD of lysis of K562 targets by human NK cells, an enhancement which was probably due to an abrogation of the suppression of human NK function caused by monocyte-generated ROI (12).

Together with the earlier report by Kay and Smith (26) in which it was shown that neither catalase (up to 3000 units/ml) nor mannitol (up to 125 mM) has any inhibitory effect on the human NK assay, it seems clear that the earlier reports by Roder and colleagues of scavenger (SOD)-mediated inhibition of NK activity be substantiated. There always remains the argument that macromolecular ROI scavengers fail to penetrate into the sites of interaction between NK cells and target cells. However, it is difficult to understand how the accessibility of scavengers to points of effector cell-target cell interaction would vary from laboratory to laboratory.

IV. SCAVENGERS OF FREE HYDROXYL RADICALS.

Duwe and Roder (15) have recently modified their original hypothesis, and now agree that superoxide radicals are not involved in the NK cytolytic sequence. This conclusion was based on their observations that 1) when using highly-purified LGL, NK cytolysis was not affected by SOD and ferric cytochrome C at levels which were sufficient to block O_2 production by granulocytes; and also that 2) NK cells from CGD patients, whose blood cells are incapable of producing detectable quantities of ROI, expressed normal cytolytic activity in vitro.

Recently, it has been reported that O_2 is required in order for NK cells to express cytolysis (3, "note added in proof"; 15). Such findings that hypoxic conditions reversibly inhibited NK lysis in vitro are consistent with the findings of Nathan _et al_ (43) who have shown that hypoxic conditions reversibly inhibited T cell-mediated lysis of specific targets. It has been proposed that oxygen consumed by the NK cell might be used to produce hydroxyl free radicals (OH•), an extremely potent ROI which has been implicated amongst the bactericidal substances released by granulocytes and monocytes (7). In support of this hypothesis, Duwe and Roder investigated a panel of 13 hydroxyl radical scavengers, and reported that they inhibited NK cytolysis in vitro (15). While these scavengers differed widely in structure, the degree of inhibition of lysis correlated well with their known rate constants for reaction with hydroxyl radical. Accordingly, it seemed appropriate to conclude that lysis by NK cells requires the generation of reactive hydroxyl radicals which then conceivably initiate target cell lysis by membrane peroxidation. This hypothesis, however, is still in apparent conflict with the many observations (including their own) that NK cells from CGD patients express normal NK potential, in spite of their lack of ability to release ROI.

Some inhibition of NK cell lysis was seen by Ramstedt et al (18) when 4,5-dihydroxy-m-benzenedisulfonic acid (Tiron) or 2,3-dihydroxybenzoate were added to the assay. These compounds, however, required significantly higher concentrations to achieve inhibition of NK cell lysis than were required to inhibit chemiluminescence, suggesting an alternative action for these compounds.

A recent report by Suthanthiran et al (44) is in apparent accord with the role for hydroxyl radicals proposed by Duwe and Roder. Investigating the effects of various scavengers of ROI on human NK activity, Suthanthiran and colleagues found that several scavengers of free hydroxyl radical (DMSO, thiourea, dimethylurea, tetramethylurea, benzoic acid, ethanol, methanol, and ethylene glycol) inhibited NK-mediated cytotoxicity against both MOLT-4 and K562 target cells. Like Duwe and Roder (15), Kay and Smith (26), Kay et al (35), Wayner and Brooks (41), and Abrams and Brahmi (42), they too found that catalase (a scavenger of H_2O_2) and SOD (a scavenger of superoxide anion) had no effect on NK cell activity. Because inhibitors of the lipoxygenase pathway of arachidonic acid metabolism also inhibited NK activity [in agreement with similar observations by Seaman (25)], Suthanthiran and coworkers concluded that hydroxyl free radicals -- possibly generated via the lipoxygenase pathway -- are critical for NK cell cytotoxicity. They did not, however, report whether these chemical scavengers had any affects on effector or target cell morphology (although they did find that the target cell-binding assay was not altered). Nor did they report whether other cellular functions were altered by these chemical scavengers, such as protein synthesis, RNA synthesis, or O_2 consumption This is certainly important, especially since such "non-specific" toxic effects were earlier noted in the CTL system by Nathan et al (43). Nor did they report whether these hydroxyl radical scavengers affected the ionic strength of the culture medium in which the assays were performed.

This last point is an insidious one, and has been

clearly addressed by the recent work of Wayner and Brooks (41), who tested for NK inhibitory action a large number of hydroxyl radical scavengers of low molecular weight, each differing in their rate constants for reaction with the hydroxyl free radical. When dissolved directly in tissue culture medium, it was found that all of these scavengers could inhibit NK cell function. It soon became clear, however, that the inhibitory effects of at least some of these scavengers could be explained by an unexpected technical "artifact"; that is, by their capacity to significantly alter the ionic strength of the support medium. Figure 3 illustrates this problem. When the OH• scavenger mannitol, or the sugar mannose, was dissolved in medium, significant abrogation of NK functions occurred at concentrations as low as 33mM. However, if these solutions were first made isotonic and then diluted to appropriate concentrations with isotonic NaCl before being used in the NK assay, then no significant inhibition of NK functions was observed, even at scavenger concentrations as high as 135mM. In further experiments, it was noted that NK cytolysis could be inhibited by as little as 9mM NaCl when added directly to the culture medium of the assay. This exquisite sensitivity of NK cell function to changes in ionic strength is of great import, especially since similar changes in tonicity seem to have no telltale effects on cellular viability (as, for example, when tissue culture cells grow quite happily under such non-isotonic conditions). Thus, extreme caution must be exercised to maintain isotonic conditions when utilizing pharmacological reagents to probe the subtle mechanisms of NK-mediated lysis.

When tonicity precautions were taken, the results shown in Table III were obtained. Among the amino acid and alcohol scavengers tested, there was an inverse correlation between the rate constant for reaction with OH• radical, and the concentration of scavenger needed to cause 30% inhibition of NK activity. The data show that human NK activity against K562, mouse splenic NK activity against YAC-1, and NK activity of a cloned mouse CTL line (F5A4) against YAC-1 were all equally susceptible to inhibition (as was specific CTL lysis of EL4 targets). While at first glance these data seem to be in agreement with those of Duwe and Roder (15) and Suthanthrian et al (44), the fact that a number of other hydroxyl radical scavengers (under isotonic conditions) completely failed to inhibit cell-mediated lysis brings the mechanism of this type of inhibition into serious question. For

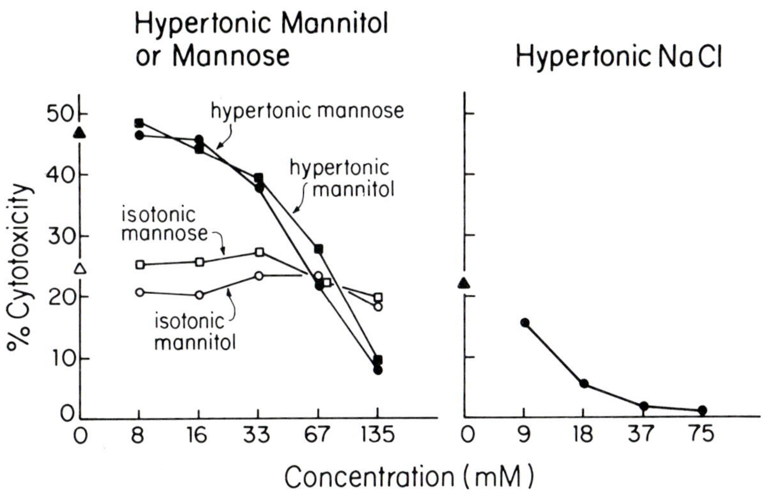

Figure 3. Effect of hypertonicity on lysis of K562 cells by human MNC (E:T ratio = 25:1) (Data from Wayner and Brooks (41), reproduced by permission of Plenum Press).

example, while benzoate and thiourea are equally reactive with OH•, benzoate was a significantly better inhibitor of cytolysis, especially in the murine system. DMSO, on the other hand — one of the most avid of OH• scavengers — had little or no inhibitory activity in any cytotoxicity assay.

The question of isotonicity is again brought directly into focus by this last point, since Suthanthiran _et al_ (44) reported that NK activity against MOLT-4 and K562 targets was inhibited by 75% and 85% respectively, in the presence of DMSO, but only when this potent OH• scavenger was used at concentrations 3 to 6 times that of the other scavengers tested in the same assays.

Ethanol, on the other hand, appears to be a relatively less efficient scavenger of OH•, and its effectiveness at inhibiting the NK system has been variable. For example, Ramstedt _et al_ (18) reported some inhibition of NK cell lysis by ethanol at 0.5% to 0.06%

TABLE III. CONCENTRATION OF SCAVENGERS REQUIRED TO INHIBIT CELL-MEDIATED CYTOTOXICITY AT THE E:T RATIOS SHOWN

Scavenger	Rate constants at neutral pH $(M^{-1}s^{-1})$[a]	Human NK v K562 (12:1)	Mouse spleen v YAC-1 (50:1)	Resting F5A4 v EL4 (0.5:1)	IFN-ind. F5A4 v YAC-1 (0.5:1)
Alcohols					
Methanol	$3.9 - 8.4x10^{8}$	>75[b]	>75	>75	>75
Ethanol	$.72 - 1.8x10^{9}$	40 ± 20	72	>75	72
n-Butanol	$2.2 - 3.4x10^{9}$	6 ± 1	9 ± 2	14 ± 6	12 ± 8
Amino acids					
Glycine	$.6 - 1x10^{7}$	>75	>75	>75	>75
Serine	$1.9 - 3.2x10^{8}$	>75	>75	>75	>75
Tryptophan	$.85 - 1.4x10^{10}$	9 ± 2	7 ± 3	12 ± 2	10 ± 2

TABLE III (Continued):

Others					
Urea	$<7\times10^5$	>75	>75	>75	>75
Mannitol	$>1\times10^9$	>75	>75	>75	>75
Benzoate	$3.5 - 5.4\times10^9$	7 ± 1	11 ± 8	8 ± 2	11 ± 5
Thiourea	4.7×10^9	17 ± 3	75	28 ± 15	>75
DMSO	7×10^9	>75	>75	>75	>75

[a]Rate constants for the reaction of scavengers (except mannitol) with hydroxyl radicals in aqueous solution were taken from Anbar and Neta (45) and Dorfman and Adams (46). The rate constant for mannitol was from Cederbaum *et al*. (47).

[b]Figures show the concentration of scavenger (mM) required for 30% inhibition of cytotoxicity. Percent lysis in control cultures lacking scavengers was always in the range of 20-40%. Results are mean ± SD from 4 experiments.

final concentration in the assay. Suthanthiran et al observed at least a 50% inhibition of NK activity by LGL against MOLT-4 and K562 targets when ethanol (80mM) was added to the assay. Wayner and Brooks (41) found that human NK against K562 was inhibited 30% by ethanol at 40 ± 20mM, but found almost no effect by ethanol at similar concentrations on murine NK cells. In contrast, Kendall and Targan (48) found that 25 mM of ethanol in an NK cytotoxicity assay for 3 hr seemed to significantly augment NK lysis compared to untreated controls ($p < 0.01$). Brunda et al (49) reported that ethanol at from 0.2% to 0.5% final concentration (approximately 350 to 850 mM) had no effect on murine NK cell activity.

Further, the studies of Saxena et al (50) revealed significantly higher than normal levels of NK activity in human alcoholics (mean alcohol intake 218 g/day) as well as in mice drinking alcohol. When added to in vitro NK assays, however, ethanol induced a dose-dependent inhibition of anti-K562 NK activity by human effectors, and a dose-dependent inhibition of anti-YAC NK activity by murine spleen cells (51). Trypan-blue viability and target cell binding did not seem to be affected, even at ethanol concentrations of 2%, whereas activation of mouse spleen NK activity by interferon or IL-2 was very significantly blocked by ethanol at the same doses (2%) which inhibited NK activity.

It seems evident, therefore, that the effects of ethanol in an NK assay are dose dependent and highly variable, making any interpretation of its mechanism of action somewhat hazardous. Because ethanol, and some of the other chemical scavengers, are known to have broad chemical reactivity and multiple membrane actions, the precise interpretation of their effects solely as OH• radical scavengers must be questioned.

An important insight into the actions of these chemical agents might be gained by examining their potential "toxicity", such as by looking for an inhibition of leucine uptake by human lymphocytes or cloned CTL lines in the presence of these "scavengers". Wayner and Brooks (41) found that compounds which inhibited NK lysis also inhibited, at similar concentrations, protein synthesis in ConA blasts prepared from human MNC, and in mouse F5A4 cells (Table IV). Thus, rather than inhibiting NK- (and also CTL-) mediated cytotoxicity by direct interaction with OH•, it seems more likely that the purported inhibitory effects of

TABLE IV. INHIBITION OF PROTEIN SYNTHESIS BY HYDROXYL RADICAL SCAVENGERS

Scavenger	ConA blasts	F5A4
Methanol	>75[a]	>75
Ethanol	>75	>75
n-Butanol	12	50
Glycine	>75	>75
Serine	70	60
Tryptophan	1	1
Urea	70	64
Thiourea	9	27
Benzoate	12	10
DMSO	>75	>75

[a]Figures show the concentration of scavenger required for 30% inhibition of leucine uptake. Data shown are representative of 3 separate experiments(41).

these hydroxyl radical scavengers can be attributed to metabolic modulation of the effector cells, a conclusion which is in agreement with that of Nathan et al (43), who observed morphological deterioration and loss of target cell binding capacity when cytotoxic T cells were treated with benzoate and thiourea.

When the effects of hydroxyl radical scavengers are examined in other cytotoxicity systems, the results again seem to say that ROI are not part of the cytolytic effector mechanisms. For example, Mavier and Edgington (52) recently reported that the hydroxyl radical scavengers histidine (20 mM) and ethanol (40 mM) did not affect K562 killing by phorbol-activated human monocytes. The potent OH• scavenger, mannitol (50 mM), had only a slight inhibitory effect (23%), leading these investigators to conclude that hydroxyl radicals played no significant role in this type of cytotoxic event.

Similarly, Weinberg and Haney (53) examined spontaneous monocyte-mediated killing of a large variety of cultured tumor cell lines, including HeLa, HL60, and K562, and they found that the hydroxyl radical scavengers mannitol (up to 50 mM) and L-histidine (up to 5 mM) did not diminish the killing in any of their assays. Devlin et al (54) reported that mannitol had no effect on murine CTL-mediated killing of P815 target cells, even when used at concentrations (1 to 10 mM) known to protect PMN's from OH• (55). They concluded that these data, while not proof, were strong evidence that hydroxyl radicals were not involved in murine CTL killing, a notion entirely consistent with the conclusions of Nathan et al (43).

At this time, then, in examining the available data for a participation by hydroxyl radicals in the lytic effector mechanism of NK cells (and, for that matter, in CTL or monocyte-mediated cytotoxicity) it is hard to find an unambiguous case to champion its involvement. Rather, there seems to be a consensus among most investigators that the available data support an oxygen-radical independent mechanism as the means whereby natural killer cells interact with and lyse their target cells.

V. WHAT PRICE MYCOPLASMA?

Koppel et al (56) have reported that in vitro chemiluminescence could be produced by interaction of cultured tumor cells with rat NK cells, polymorphonuclear leukocytes, or activated macrophages only when tumor cells were contaminated with mycoplasma. Yet the NK targets were equally susceptible to cytolysis whether or not mycoplasma was present. Wayner and Brooks (41) have reported data in complete agreement with these conclusions (see Figure 2).

These results further support the conclusion that ROI as detected by chemiluminescence do not play a role in the NK lytic event. They also provide yet another example of the importance of using only cell lines demonstrably free of mycoplasma for biological research. The availability of a simple and sensitive technique for preliminary screening of cell lines for mycoplasma contamination (57), as well as effective methods of decontamination (see for example reference 58), makes it both unnecessary and unacceptable that infected lines should be used.

IV. CONCLUSIONS

Taken together, the evidence discussed in the previous sections argues strongly against any participation of ROI in NK-mediated lysis of cultured tumor cell lines. Because NK cell activity is normal in CGD patients who have major oxidative defects, and because no ROI production could be detected in highly purified NK cells which display potent cytolytic activity, and because no inhibition of K-mediated killing is found using a variety of scavengers of ROI under non-toxic conditions, we conclude that NK activity by mononuclear effector cells involves a lytic mechanism(s) independent of oxygen metabolites.

ACKNOWLEDGEMENTS

We wish to thank Ms. Lucie Case for expert assistance in preparation of this manuscript. We thank Dr. A.H. Greenberg for his kindness in providing (to RHG) a preprint of his study.

REFERENCES

1. Helfand SL, J Werkmeister, and JC Roder. 1982. J. Exp. Med. 156: 492.
2. Helfand SL, J Werkmeister, and JC Roder. 1982. In: NK Cells and Other Natural Effector Cells (R.B. Herberman, editor). Academic Press, New York, p. 1011.
3. Roder JC, SL Helfand, J Werkmeister, R McGarry, TJ Beaumont, and A Duwe. 1982. Nature 298: 569.
4. Werkmeister J, S Helfand, J Roder, and H Pross. 1983. Eur. J. Immunol. 13: 514.
5. Babior BM, and DW Parkinson. 1982. Nature 298: 511.
6. Babior BM, RS Kipnes, and JT Curnutte. 1973. J. Clin. Invest. 52: 741.
7. Klebanoff SJ. 1981. In: Advances in Immunopharmacology (J Hadden, L Chedid, P Mullen, and F Spreafico, editors). Pergamon Press, New York, p. 207.
8. Brahmi Z, K Lazarus, A Hubbard, R Bray, C Butterick, and R Baehner. 1982. Fed. Proc. 41: 601.
9. Goldfarb RH, T Timonen, and RB Herberman. 1982. Adv. Exp. Med. Biol. 146: 403.
10. Goldfarb RH, and RB Herberman. 1982. Adv. Inflamm. Res. 4: 45.
11. Kay HD, DL Smith, GR Donowitz, and GL Mandell. 1982. Clin. Res. 30: 850A.
12. Seaman WE, TD Gindhart, MA Blackman, B Dalal, N Talal, and Z Werb. 1982. J. Clin. Invest. 69: 876.
13. Bloom B. 1982. Nature 300: 214.
14. Koren H, and RB Herberman. 1983. Immunol. Today 4: 97.
15. Duwe AK, and JC Roder. 1984. In: Abstracts of the Second International Workshop on NK Cells, p. 44.
16. Ernst M, A Lange, H-D Flad, A Havel, J Ennen, and AJ Ulmer. 1984. Eur. J. Immunol. 7: 634-639.
17. Pohajdak B, JL Gomez, JA Wilkins, and AH Greenberg. 1984. J. Immunol. 135: 2430.
18. Ramstedt U, P Rossi, C Kullman, E Warren, J Palmblad, and M. Jondal. 1984. Scand. J. Immunol. 19: 457.
19. Koren HS, SJ Anderson, DG Fischer, CS Copeland, and PJ Jensen. 1981. J. Immunol. 127: 2007.
20. Ottenhof PC, A Morales, and MG Baines. 1981. J. Immunol. Methods 42: 305.
21. West WH, GB Cannon, HD Kay, GD Bonnard, and RB Herberman. 1977. J. Immunol. 118: 355.

22. Kay HD, GD Bonnard, WH West, and RB Herberman. 1977. J. Immunol. 118: 2058.
23. Tanaka Y, T Sakano, and T Usui. 1981. Hiroshima J. Med. Sci. 30: 135.
24. Abrams SI, RA Bray, and Z Brahmi. 1983. Cell. Immunol. 80: 230.
25. Seaman WE. 1983. J. Immunol. 131: 2953.
26. Kay HD, and DL Smith. 1983. J. Immunol. 130: 475.
27. Seaman WE, TD Gindhart, MA Blackman, B Dalal, N Talal, and Z Werb. 1981. J. Clin. Invest. 67: 1324.
28. Quie PG, JG White, B Holmes, and RA Good. 1967. J. Clin. Invest. 46: 668.
29. Curnutte JT, DM Whitten, and BM Babior. 1974. N. Engl. J. Med. 290: 593.
30. Donowitz GR, and GL Mandell. 1982. Blood 60: 1151.
31. Gallin JI, ES Buescher, BE Seligmann, J Nath, T Gaither, and P Katz. 1983. Annals Int. Med. 99: 657.
32. Baehner RL, and DG Nathan. 1968. N. Engl. J. Med. 278: 971.
33. Whitin JC, CE Chapman, ER Simons, ME Chovaniec, and HJ Cohen. 1980. J. Biol. Chem. 255: 1874.
34. Lopez C, D Kirkpatrick, PA Fitzgerald, CY Ching, RN Pahwa, RA Good, and EM Smithwick. 1982. J. Immunol. 129: 824.
35. Kay HD, DL Smith, G Sullivan, GL Mandell, and GR Donowitz. 1983. J. Immunol. 131: 1784.
36. El-Hag A, and RA Clark. 1984. J. Immunol. 132: 569.
37. Katz P, CB Simone, PA Henkart, and AS Fauci. 1980. J. Clin. Invest. 65: 55.
38. Goldfarb RH, T Timonen, E Pick, T Hoffman, JR Ortaldo, and RB Herberman. 1984. In: Abstracts of the Second International Workshop on Natural Killer Cells, p. 41.
39. Goldfarb RH, T Timonen, and RB Herberman. 1984. J. Exp. Med. 159: 935.
40. Pick E, and D Mizel. 1981. J. Immunol. Methods 46: 211.
41. Wayner EA, and CG Brooks. 1984. In: Mechanisms of Cell-Mediated Cytotoxicity. (P. Henkart, editor). Plenum Press, New York (in press).
42. Abrams SI, and Z Brahmi. 1984. J. Immunol. 132: 3192.
43. Nathan CF, JA Mercer-Smith, NM Desantis, and MA Pallodino. 1982. J. Immunol. 129: 2164.

44. Suthanthiran M, SD Solomon, PS Williams, AL Rubin, A Novogrodsky, and KH Stenzel. 1984. Nature 307: 276.
45. Anbar M, and P Neta. 1967. Int. J. Appl. Rad. Isotopes 18: 493.
46. Dorfman LM, and GE Adams. 1973. NSRDS, National Bureau of Standards, 46, Washington, DC.
47. Cederbaum AI, E Dicker, E Rubin, and G Cohen. 1977. Biochem. Biophys. Res. Commun. 78: 1254.
48. Kendall RA, and S Targan. 1980. J. Immunol. 125: 2770.
49. Brunda MJ, RB Herberman, and HT Holden. 1980. J. Immunol 124: 2682.
50. Saxena QB, E Mezey, and WH Adler. 1980. Int. J. Cancer 28: 413.
51. Saxena QB, RK Saxena, and WH Adler. 1982. In: NK Cells and Other Natural Effector Cells (RB Herberman, editor). Academic Press, New York, p. 651.
52. Mavier P, and TS Edgington. 1984. J. Immunol. 132: 1980.
53. Weinberg JB, and AF Haney. 1983. J. Nat'l Cancer Inst. 70: 1005.
54. Devlin RG, CF Lin, RJ Perper, and H Dougherty. 1981. Immunopharmacology 3: 147.
55. Salin ML, and JM McCord. 1975. J. Clin. Invest. 56: 1319.
56. Koppel P, E Peterhans, G Bertoni, R Keist, P Groscurth, R Wyler, and R Keller. 1984. J. Immunol. 132: 2021.
57. Chen TR. 1977. Exp. Cell. Res. 104: 255.
58. Marcus M, U Lavi, A Nattenberg, S Rottem, and O Markowitz. 1980. Nature 285: 659.

OXIDATIVE BURST DOES NOT MEDIATE HUMAN MONOCYTE TUMORICIDAL ACTIVITY[1]

Allen R. Chen
Hillel S. Koren

Division of Immunology
Duke University Medical Center
Durham, North Carolina

I. INTRODUCTION

Like NK cells, mononuclear phagocytes have been ascribed a role in surveillance against tumors (Hibbs et al., 1978; Adams & Snyderman, 1979). As is true of NK, their mechanism of lysis remains a subject of investigation and controversy. Our working hypothesis is that damage to target cells is mediated by soluble molecules produced by the effector cells.

The versatility of mononuclear phagocytes as secretory cells (reviewed in Nathan et al., 1980) provides many possible cytolytic mediators, listed in Table I. Of these, hydrogen peroxide and other reactive oxygen species, whose role in microbial defense is well established (reviewed by Klebanoff, 1982), have become a focus of attention.

Several observations suggest that the oxidative burst may also mediate monocyte tumoricidal activity. Mammalian cells are sensitive to lysis by reactive oxygen species (Clark & Klebanoff, 1975). Moreover, lysis by peroxide is rapid enough (Nathan et al., 1979) to account for the kinetics of antibody dependent cellular cytotoxicity (ADCC) and antibody independent cytotoxicity by fresh peripheral blood monocytes (Horwitz et al., 1979; Fischer et al., 1981). Monocytes and

[1]Supported by NIH grants CA 29589 and 5T326M07171.

ISBN 0-12-341370-2

macrophages efficiently generate reactive oxygen metabolites (Johnston et al., 1976; Nathan & Root, 1977). Moreover, macrophages activated for cytotoxicity by IFN-gamma, bacterial products such as lipopolysaccharide and muramyl dipeptide, and inflammatory proteases, have increased levels of oxidative burst activity (Nathan et al., 1983; Lepoivre et al., 1982; Pabst et al. 1982; Speer et al., 1984). Also, fresh blood monocytes have relatively high oxidative burst potential which declines gradually in culture (Nakagawara et al, 1981) roughly in parallel with their loss of spontaneous cytotoxic potential (Fischer et al., 1983).

TABLE I. Monocyte Macrophage Products with Reported Cytotoxic Effects on Neoplastic Cells

PROPOSED MEDIATOR	REFERENCE
Lysozyme	Osserman et. al., 1973
Lysosome transfer	Hibbs, 1974
Tumor necrosis factor	Carswell et. al., 1975
Interferon	Stewart et. al., 1976
C_3A	Schorlemmer & Allison, 1976
Thymidine	Stadecker et. al., 1977
Arginase	Currie, 1978
Hydrogen peroxide	Nathan et. al., 1979

Nathan and co-workers (1979) have shown that lysis of neoplastic targets by activated murine macrophages stimulated by phorbol myristate acetate (PMA) depends on hydrogen peroxide production. These observations have been extended to fresh human peripheral blood monocytes stimulated by PMA (Weiss & Slivka, 1982; Mavier & Edgington, 1984). Although these studies demonstrate that the oxidative burst of monocytes and macrophages <u>can</u> lyse neoplastic cells, they do not establish that the oxidative burst <u>does</u> mediate tumor lysis under physiologic conditions. Indeed, Weinberg and Haney (1983) found that no scavengers of reactive oxygen species could inhibit spontaneous cytolysis of HeLa target cells by unstimulated human monocytes or peritoneal macrophages. These results suggest that reactive oxygen species play no role in cytolysis unless the oxidative burst is stimulated extrinsically. However, it is conceivable that in these experiments, the scavengers of reactive oxygen intermediates inhibited monocyte/macrophage mediated lysis, and NK contaminants accounted for the observed lytic activity. Another possibility is that hydrogen peroxide mediates lysis both with and without PMA, but without extrinsic stimulation, the effects of hydrogen peroxide are confined to a contact zone from which catalase may be excluded (Wright and Silverstein, 1982).

We therefore designed experiments to clarify the role of reactive oxygen species in tumor cytolysis without extrinsic stimulation of the oxidative burst. In order to estimate the contribution of NK contaminants, we performed cytolytic assays with nylon wool passed cells alongside our monocyte preparations, using both K562 targets and NK insensitive Actinomycin D pre-treated WEHI-164 targets (Ziegler- Heitbrock & Reithmuller, 1984; Colotta et al., 1984). An alternative to inhibiting the effects of reactive oxygen metabolites is to prevent their generation. This approach does not require access to the interaction zone between effectors and targets. We have taken advantage of the impairment of oxidative burst in patients with chronic granulomatous disease (CGD) and have assessed the lytic potential of normal monocytes whose oxidative burst was prevented by anaerobic conditions.

II. Materials and Methods

A. Donors

1. Patients

The CGD patients used in this study were known to have diminished intracellular bacterial killing and defective oxidative burst activity. They were kindly referred to us by Dr. H. Gallis, Duke University, and Dr. M. Cohen, University of North Carolina at Chapel Hill.

2. Normal

For controls, we selected donors whose monocyte tumoricidal activity was in the upper quartile of our normal population of healthy university students and employees.

B. Monocyte isolation

Monocytes were purified by adherence to autologous serum-coated plates and were detached with EDTA, using a modification of a technique developed in our laboratory (Fischer et al., 1981). Both the RPMI-1640 medium (B&B/Scott Laboratories, Fiskeville, RI) and the fetal calf serum (Hyclone, Logan, UT) were free of endotoxin as detected by the Limulus amoebocyte lysate assay (Sigma, St. Louis, MO).

Purity of the adherent population was assessed by peroxidase staining (Kaplow, 1966) and routinely exceeded 97%. Viability as assessed by trypan blue exclusion was over 99%.

C. Target cells

The human erythromyeloid leukemia line K562 (Lozzio & Lozzio, 1975) was obtained mycoplasma free from the American Type Culture Collection (Rockville, MD) and maintained in log phase suspension cultures.

WEHI-164 (a generous gift of Dr. John Ortaldo, NCI), an adherent murine sarcoma line, was passaged by detachment with trypsin-EDTA, followed by washing. For cytotoxicity assays, sub-confluent cultures were pretreated with 1 ug/ml Actinomycin D in supplemented tissue culture media for 3 hours at 37°C, which was removed by washing three times, as previously described (Ziegler-Heitbrock & Riethmuller, 1984).

D. Cytotoxicity Assay

Tumoricidal activity was measured using short-term ^{51}Cr release assays as previously described (Fischer et al., 1981). Assays were harvested after 4 hours for K562 target cells and 6 hours for actinomycin D pre-treated WEHI-164 target cells.

Lytic units (LU) were determined as described by Pross et al. (1981) using an exponential fit computer program and a threshold of 20% specific ^{51}Cr release.

The contribution of NK cells was estimated as the product of percent peroxidase negative contaminants and the number of $LU/10^6$ nylon wool passed cells. The validity of this estimate is based on cytofluorometric observations that the proportion of NK cells (as defined by Leu 11b expression) is the same in nylon wool passed cells and in non-monocytic contaminants in the adherent population. NK cells recovered from the adherent preparations by fluorescent cell sorting did not express particularly high activity. Also, there was no synergistic interaction of small numbers of NK cells mixed with large numbers of monocytes.

E. Oxidative Burst

Hydrogen peroxide release was measured by phenol red oxidation in the presence of horseradish peroxidase (Pick & Keisari, 1980).

F. Anaerobiosis

Media, tubes, and tissue culture plates were pre-incubated overnight in a GasPak anaerobic container. Freshly isolated monocytes and tumor targets were transported in a GasPak container to an anaerobic bacteriologic glove chamber (kindly provided by Dr. Dolph Klein, Duke University), where they were allowed to pre-equilibrate for four hours before tumoricidal assay. The atmosphere in the chamber was 85% N_2, 10% H_2, and 5% CO_2, with platinum catalyst to prevent O_2 accumulation. Resazurin indicator remained colorless; moreover, oxygen tension in the chamber remained low enough to permit the growth of fastidious obligate anaerobic clostridia. To confirm the profound effect of oxygen deprivation on the monocytes in the chamber, we stimulated monocytes with 100 ng/ml PMA and measured hydrogen peroxide secretion as described above. There was no detectable oxidative burst within the anaerobic chamber.

G. Lysis of targets by peroxide generating enzyme system

Glucose oxidase (provided by Dr. M. Cohen) was titrated against 10^4 ^{51}Cr-labeled targets per microtiter well in the presence of 16 mM glucose in 200 μl complete RPMI-1640 medium. Plates were incubated at 37^oC for 4 hours in a 5% O_2 environment and the supernatant media was harvested for gamma counting.

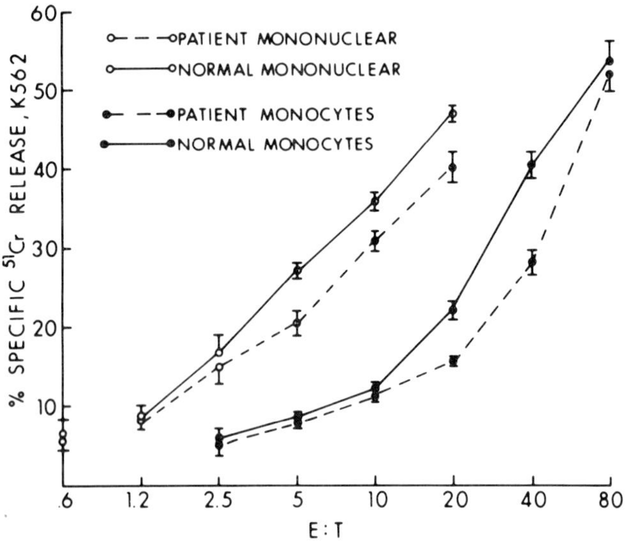

Figure 1. Lysis of K562 targets by CGD and normal monocytes and mononuclear cells, against 5×10^3 targets per well in a 4 hr. ^{51}Cr-release assay. Results are plotted as the mean $\pm$ 1 SEM of three experiments. Monocytes were obtained by adherence and mononuclear cells were obtained directly from the LSM interface (from Chen & Koren, in press).

III. Results

A. Lysis of K562 targets by CGD monocytes

Figure 1 compares the tumoricidal activity against K562 of a CGD patient with a normal control donor. This CGD patient had no detectable oxidative burst after stimulation by 100 ng/ml PMA in any of three experiments. The sensitivity of the assay was 5 nmoles of H_2O_2/mg protein/hr. and a normal oxidative burst was 518 nmoles H_2O_2/mg protein/hr. Thus, this patient had at least a 99% deficit in stimulated oxidative burst. Nevertheless, the CGD patient manifested undiminished K562 killing. Indeed, the lytic activity of this patient's monocytes consistently fell above the 95th percentile of all normal donors. In this particular experiment, the patient's monocyte tumoricidal activity was 4.3 LU/10^6 cells; in comparison, the mean activity of our normal population was 1.3 LU/10^6 cells with a standard deviation of 0.5 LU/10^6 cells (n=26).

Contamination by NK cells cannot account for the tumoricidal activity of this patient's monocyte fraction. Figure 1 shows that the NK cells' specific activity reflected by the mononuclear cell fraction (16.9 LU/10^6 cells) exceeded that of the monocyte fraction. However, since all contaminants comprise less than 4% of our adherent preparation, in this experiment, the extrapolated contribution of contaminants to activity in the monocyte fraction is 1.35 LU/10^6 cells. The net activity attributable to monocytes is 2.95 LU/10^6 cells.

In another experiment (Figure 2), the K562 lytic activity of the patient's adherent 97% pure monocyte fraction (1.96 LU/10^6 cells) approached the activity of his NK cells enriched by passage through a nylon wool column (2.46 LU/10^6 cells). Within the monocyte preparation, in this particular experiment, NK contaminants may only contribute up to 0.06 LU/10^6 cells and monocytes accounted for at least 1.90 LU/10^6 cells. On repeated testing in three other experiments, this patient displayed a typical degree of variation

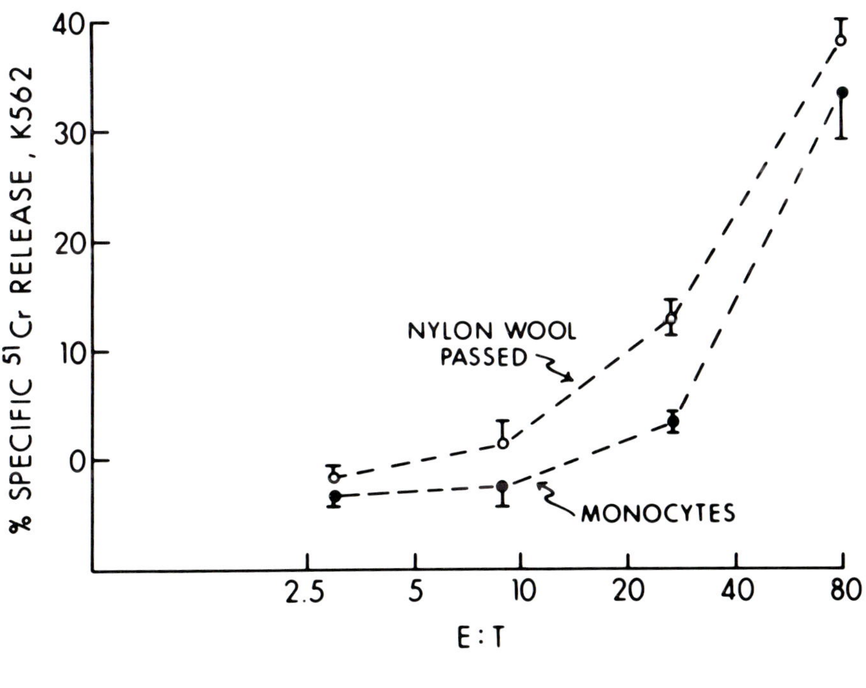

Figure 2. Lysis of K562 targets by a CGD patient's monocytes and NK cells enriched by passage through a nylon wool column (from Chen & Koren, in press).

in specific tumoricidal activity for a single donor on different occasions, but high levels of monocyte-mediated lysis of K562 were reproducibly demonstrated.

B. Lysis of NK-insensitive targets by monocytes of CGD patients

Actinomycin D pretreated WEHI-164 targets, completely insensitive to NK, provide a rapid, sensitive assay for lysis by monocytes (Ziegler-Heitbrock & Riethmuller, 1984; Colotta et al., 1984). As shown in Figure 3, the CGD patient's monocyte tumoricidal activity was superimposable on a normal donors activity. NK cells of the patient and the control donor manifested no activity against this target (data not shown).

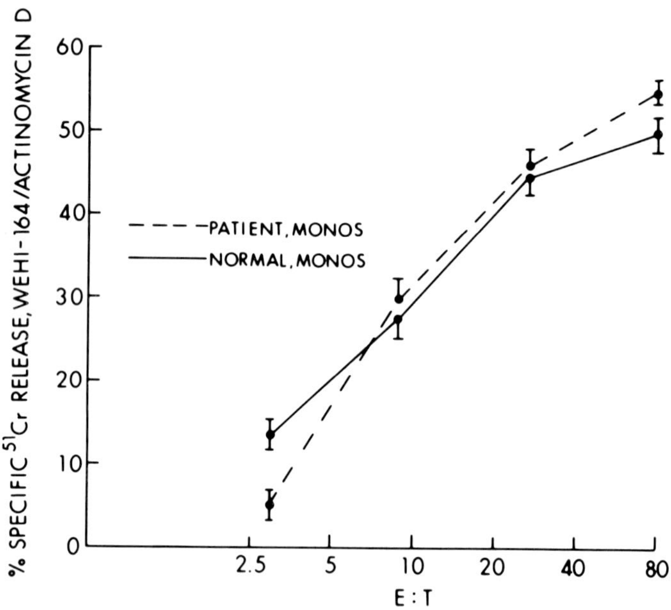

Figure 3. Lysis of WEHI-164 targets by monocytes of a CGD patient and a normal donor. Targets were pretreated with 1 ug/ml actinomycin D for 3 hours prior to ^{51}Cr-labeling. Cytotoxicity was determined in a 6 hour assay against 5000 targets per well (from Chen & Koren, in press).

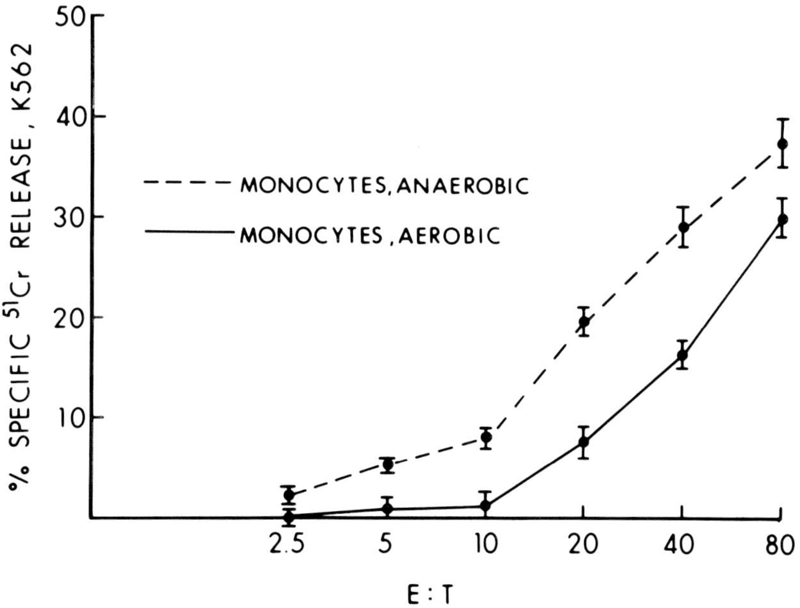

Figure 4. Lysis of K562 targets by monocytes from a normal donor. After isolation by adherence, monocytes were divided into two aliquots. One group was pre-equilibriated and assayed in an anaerobic glove chamger; the other was held and assayed in a normal oxygen containing atmosphere (from Chen & Koren, in press).

C. Effect of anaerobiosis on monocyte tumoricidal activity

Normal monocytes incubated under strictly anaerobic conditions failed to generate a detectable oxidative response to PMA (less than 15 nmoles H_2O_2/mg protein/hr). Nevertheless, to our surprise, monocytes pre-equilibrated and assayed in the anaerobic bacteriological chamber had significantly enhanced ($p < 0.05$) lytic activity against K562 targets (3.30 LU/10^6 cells) compared with control monocytes in an aerobic atmosphere (1.96 LU/10^6 cells) (Figure 4).

D. Susceptibility of targets to enzyme generated peroxide

In order to rule out extreme susceptibility even to undetectably low levels of hydrogen peroxide, we tested the cytolytic capacity of limiting concentrations of glucose oxidase. Lysis required levels of glucose oxidase which produced easily detectable quantities of hydrogen peroxide.

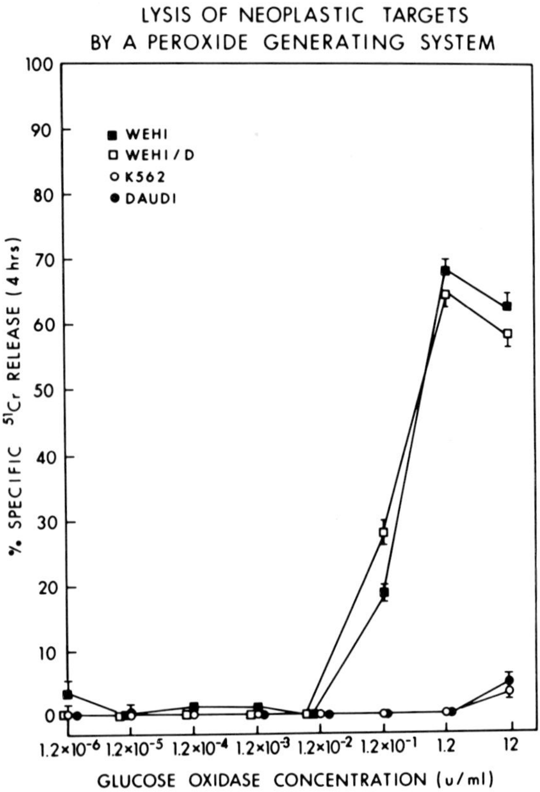

Figure 5. Lysis of neoplastic targets by a peroxide generating system. 10^4 ^{51}Cr-labeled targets per well were added to a mixture of 16 mM glucose with varying concentrations of glucose oxidase. Cytolysis was determined after a 4 hour incubation.

Furthermore, Actinomycin D pre-treatment of WEHI-164 cells did not render them more sensitive to the effects of hydrogen peroxide (Figure 5).

IV. DISCUSSION

We have studied tumoricidal activity of two different types of monocyte preparations which have severely impaired oxidative burst. Neither normal monocytes under anaerobic conditions nor monocytes from CGD patients produced a detectable oxidative burst in response to PMA. We cannot rule out the possibility that the CGD patients tested had an atypical form of the disease with normal NADPH oxidase activity but abnormal transduction (Harvath & Anderson, 1979). If so, they may have a normal oxidative burst in response to physiologic stimuli, including perhaps tumor cell binding. However, the more common, classic form of CGD is characterized by incapacity to generate any oxidative burst; also, tumor cell binding has not been shown to stimulate an oxidative burst; furthermore, under strictly anaerobic conditions, monocytes could not generate an oxidative burst.

It is conceivable that even undetectable levels of reactive oxygen metabolites suffice to produce maximal target lysis. However, our experiments with peroxide generated by glucose-glucose oxidase ruled out extreme sensitivity of the targets tested to hydrogen peroxide. We therefore believe that the levels of reactive oxygen metabolites produced by the CGD patients' monocytes and by the normal monocytes under anaerobic conditions were insufficient to lyse tumor cells.

If reactive oxygen metabolites did mediate tumor lysis, we would expect to see diminished killing by monocytes with impaired oxidative burst. Since both types of impaired effectors efficiently lysed tumor cells, we conclude that monocytes possess a non-oxidative tumoricidal mechanism which can account for all the activity observed. We found no evidence for involvement of reactive oxygen metabolites in direct cytolysis of tumor cells by monocytes.

Reactive oxygen species have been more extensively tested that the other proposed mediators. Previous experiments to confirm their physiologic role have yielded conflicting results. The experience in ADCC is summarized in Table II. The consistent inhibition by hypoxia of ADCC against erythrocytes, confirmed by most studies using CGD monocytes, suggests that the oxidative burst does mediate ADCC. The generally negative data using scavengers of reactive oxygen species may be explained by inaccessibility of the zone of interaction to large molecules such as catalase (220 KD) and superoxide diemutase (138 KD) (Adams et al., 1980; Wright & Silverstein, 1982).

However, there is no evidence to suggest a role for the oxidative burst in ADCC against nucleated targets. The work of Katz et al. (1980) provides a clue to the difference: in the absence of phagocytosis, non-oxidative mechanisms

TABLE II. Effects of Inhibiting Reactive Oxygen Species on ADCC

TARGET	SPECIFIC SCAVENGERS OR INHIBITORS					GENERAL INHIBITION		REFERENCE
	$O_2^{\cdot -}$	H_2O_2	$\cdot OH$	O_2^1	MPO	CGD	HYPOXIA/METABOLIC	
RBC	0	0	0	0	0	0		Fleer et al., 1979
RBC	0	0	↓35%		↑40%	↓50%	↓50% (hypoxia)	Borregaard and Krag-balle, 1980
intracellular RBC						↓85%		Katz et al, 1980
extracellular RBC						0		
autologous lymphocytes						0		
RBC	0	0	0	0	0	↓50-75%	↓80% (hypoxia)	Klassen and Sagone, 1980
CEM	0	0	0	0	0	0	0% (hypoxia)	
RBC	0	0	0	0	0	0	↓40% (vincristine, SH, EDTA, EGTA, 2-deoxy glucose)	Koller and LoBuglio, 1981
K562	0	0		0	0			Seim and Espevik, 1983

TABLE III. Effects of Inhibiting Reactive Oxygen Species on Antibody Independent Cytotoxicity by Mononuclear Phagocytes

EFFECTORS	ASSAY	SPECIFIC SCAVENGERS OR INHIBITORS					GENERAL INHIBITION	REFERENCE
		$O_2\cdot^-$	H_2O_2	$\cdot OH$	O_2^1	MPO		
C. Parvum infected murine macrophages	cytostasis		0				↓17% (hypoxia)	Sorrell et al., 1978
BCG infected murine macrophages	lysis w/PMA	0	↓100%	0	0	↓ 0-30%	↓100% (hypoxia)	Nathan et al., 1979b
Human monocytes	RBC lysis w/ PMA	↓20%	↓90%	0	0	±		Weiss et al. 1980
Human monocytes	lysis w/PMA	0	↓80%	0	0	↓80-95%		Weiss and Slivka, 1982
BCG infected murine macrophages	lysis w/eosinophil peroxidase		↓ 95%			↓100%		Nathan and Klebanoff, 1982
Human monocytes	lysis	0	0	0	0			Weinberg and Haney, 1983

(continued)

TABLE III. (continued)

EFFECTORS	ASSAY	SPECIFIC SCAVENGERS OR INHIBITORS					GENERAL INHIBITION	REFERENCE
		$O_2^{\cdot}$	H_2O_2	$\cdot OH$	O_2^1	MPO		
Nocardia cell wall Skeleton treated murine macrophages	lysis	0	0					Ito et al., 1983
Paraffin elicited murine macrophages	stasis w/ PMA	↓40%	↓45%				↓90% (tocopherol)	Flescher et al., 1984
Human monocytes	lysis w/ PMA	↓20%	↓89%	0	0	↑45%		Mavier and Edgington, 1984
BCG infected murine macrophages	CFU inhibition						0	Freedman et al., 1984

apparently predominate in ADCC.

In antibody independent systems, summarized in Table III, Nathan and co-workers (1979) demonstrated that in the presence of PMA, activated murine macrophage cytotoxicity is inhibited by catalase but not by scavengers of oxidative metabolites other than hydrogen peroxide. Weiss and Slivka (1982), and Mavier and Edgington (1984) performed similar experiments using PMA-stimulated human monocytes and also observed involvement of hydrogen peroxide.

However, in the absence of extrinsic stimulation of the oxidative burst by PMA, scavengers of reactive oxygen intermediates had significant effects only if targets were sensitized to hydrogen peroxide with adsorbed eosinophil peroxidase (Nathan and Klebanoff, 1984). Our results, using monocytes with impaired oxidative burst, complement those of Weinberg and Haney (1983) who used scavengers of reactive oxygen products, and Freedman et al. (1984) who studied activated murine macrophages under anaerobic conditions. It remains indisputable that mononuclear phagocytes have the _capacity_ to produce lytic quantities of hydrogen peroxide and, at different stages of maturation or activation, or with different targets, reactive oxygen metabolites may mediate the cytotoxic effects of mononuclear phagocytes. However, we support the conclusion that without PMA, an oxidative burst is not necessary for tumor lysis by monocytes.

What does mediate monocyte cytotoxicity? Early EM observations from two groups suggested lysis may involve exocytosis and transfer of lysosomes (Hibbs, 1974; Bucana et al., 1976), a possibility that has not been vigorously pursued. Thymidine and arginase have cytotoxic effects in some _in vitro_ systems (Stadecker et. al., 1977; Currie, 1978), but it is questionable whether _in vivo_, thymidine could accumulate to cytostatic levels or arginase could be depleted to cytolytic levels. The effects of C_3A at physiologic rates of production have been disputed (Goodman et al., 1980). Lysozyme, tumor necrosis factor (TNF), and interferon (IFN) have all been reported to have selective cytotoxic effects on neoplastic cells; their role in monocyte cytotoxicity merits further investigation.

An alternative approach to the problem of defining lytic mediators appears promising. Several investigators have detected cytotoxic activity in the supernatants of cultured monocytes and macrophages (Currie and Basham, 1975; Sharma et al., 1980; Adams et al., 1980; Matthews, 1981; Cameron,

1982; Hammerstrom, 1982; Reidarson et al., 1982 ab; Sone et al., 1984; McKinnon et al., submitted for publication). Varying degrees of progress have been made toward isolating and characterizing the active factors. Much work remains to identify these factors molecularly and assess their roles in cell-mediated cytotoxic systems.

ACKNOWLEDGMENTS

Dr. Dolph Adams and Susan McCord provided help with hydrogen peroxide release assays; Mark Dorton and Lida Sionit maintained our cultured cells; and Carol Peters provided efficient secretarial assistance. We thank Drs. Myron Cohen, Karen McKinnon, and Ralph Snyderman for their helpful suggestions.

REFERENCES

Adams, D.O., Kao, K.J., Farb, R., and Pizzo, S.V. 1980. J. Immunol. 124, 293.
Adams, D.O., and Snyderman, R. 1979. JNCI. 62, 1341.
Alexander, P., and Evans. 1971. Nature (New Biol). 237,76.
Arrick, B.A., Nathan, C.F., Griffith, O.W., and Cohn, Z.A. 1982. J. Biol. Chem. 257, 1231.
Borregaard, N., and Kragballe, K. 1980. J. Clin. Invest. 66, 676.
Bucana, C., Hoyer, L.C., Hobbs, B., Breesman, S., McDaniel, M., and Hanna, M.G., Jr. 1976. Cancer Res. 36, 4444.
Cameron, D.J. 1982 J. RES. 32, 247.
Carswell, E.A., Old, L.J., Kassel, L.J., et al. 1975. Proc. Natl. Acad. Sci. USA. 72, 3666.
Chen, A.R., and Koren, H.S. J. Immunol., in press (1985).
Clark, R.A., and Klebanoff, S.J. 1975. J. Exp. Med. 141, 1442.
Colotta, F., Peri, G., Villa, A., and Mantovani, A. 1984. J.I. 132, 936.
Currie, G.A. 1978. Nature. 272, 758.
Currie, G.A., and Basham, C. 1975. J. Exp. Med. 145, 1600.
Fischer, D.C., Golightly, M.G., and Koren, H.S. 1983. J. Immunol. 130, 1220.
Fischer, D.G., Hubbard, W.J., and Koren, H.S. 1981. Cell. Immunol. 58, 426.
Fleer, A., Roos, D., von dem Borne, A.E.G., and Engelfriet, C.P. 1979. Blood. 54, 407.

Flescher, E., Gonen, P., and Keisari, Y. 1984. JNCI. 72, 1341.
Freedman, V.H., Gorrell, T.E., Nathan, C.F., Copeland, C.S., and Silverstein, S.C. 1984. J. Exp. Med. 94, 107.
Golightly, M.G., Fischer, D.G., Ohlander, C., and Koren, H.S. 1983. Blood. 61, 390.
Goodman, M.G., Weigle, W.O., and Hugli, T.E. 1980. Nature. 283, 78.
Hammerstrom, J. 1982. Scand. J. Imunol. 15, 311.
Harvath, L, and Anderson, B.R. 1979. N. Engl. J. Med. 300, 1979.
Hibbs, J.B., Jr. 1974. Science. 184, 466.
Hibbs, J.B., Jr., Chapman, H.A., Jr., and Weinberg, J.B. 1978. J. Reticuloendothel. Soc. 24, 549.
Horwitz, D.A., Knight, N., Temple, A., and Allison, A.C. 1979. Immunology. 36, 221.
Ito, M., Suzuki, H., Nakano, N., Yamashita, N., Sugiyama, E., Maruyama, M., Hoshino, K., and Yano S. 1983. Gann. 74, 128-36.
Johnston, R.B., Jr., Lehmeyer, J.E., and Guthrie, L.A. 1976. J. Exp. Med. 143, 155.
Kaplow, L.S. 1966. Blood. 26, 215.
Katz, P., Simone, B., Henkart, P.A., and Fauci, A.S. 1980. J. Clin. Invest. 65, 55.
Klassen, D., and Sagone, A.L. 1980. Blood. 56, 985.
Klebanoff, S. J. 1982. Advances in Host Defense Mechanisms, eds. J. I. Gallin and A.S. Fauci. p. 111.
Koller, C.A., and LoBuglio, A.F. 1981. Blood. 58, 293.
Lepoivre, M., Tenu, J.P., Lemaire, G., and Petit, J.F. 1982. J. Immunol. 129, 860-6.
Lozzio, C.B., and Lozzio, B.B. 1975. Blood. 45, 321.
Matthews, N. 1981. Immunology. 44, 135.
Mavier, P., and Edgington, T.S. 1984. J.Immunol. 132, 1980.
Nakagawara, A., Nathan, C.F., and Cohn, Z.A. 1981. J. Clin. Invest. 68, 1243.
Nathan, C.F., Brukner, L.H., Silverstein, S.C., and Cohn, Z.A. 1979a. J. Exp. Med. 149, 84-99.
Nathan, C.F., and Klebanoff, S.J. 1982. J. Exp. Med. 155, 1291-1308.
Nathan, C.F., Murray, H.W., and Cohn, Z.A. 1980. 303, 622.
Nathan, C.F., Murray, H.W., and Wiebe, M.E. 1983. J. Exp. Med. 158, 670
Nathan, C.F., and Root, R.K. 1977. J. Exp. Med. 146, 1648.
Nathan, C.F., Silverstein, S.C., Brukner, L.H., Cohn, Z.A. 1979b. J. Exp. Med. 149, 100.
Osserman, E.F., Klockars, M., Halper, J., and Fishel, R.E. 1973. Nature. 243, 331.
Pabst, M.J., Hedegaard, H.B., and Johnston, R.B., Jr. 1982. J.Immunol. 128, 123.

Pick, E., and Keisari, Y. 1980. J. Immunol. Methods. 38, 161.
Pross, H.F., Baines, M.G., Rubin, P., Shragge, P. and Patterson, M.S. 1981. J. Clin. Immunol. 1, 51.
Reidarson, T., Granger, G.A., and Klostergaard, J. 1982b. J. Natl. Cancer Inst. 69, 889.
Reidarson, T., Levy, W., Klostergaard, J., and Granger, G.A. 1982a. J. Natl. Cancer Inst. 69, 879.
Schorlemmer, H.U., and Allison, A.C. 1976. Immunology. 31, 781.
Seim, S., and Espevik, T. 1983. J.RES. 33, 417.
Sharma, S.D., Piessens, W.F., and Middlebrook, G. 1980 Cell. Immunol. 49, 379.
Sone, S., Tachibana, K., Ishii, K., Ogawara, M., and Tsubura, E. 1984. Cancer Research 44, 646.
Sorrell, T.C., Lehrer, R.I., and Cline, M.J. 1978. J. Immunol. 120, 347.
Speer, C.P., Pabst, M.J., Hedegaard, H.B., Rest, R.F., and Johnston, R.B., Jr. 1984. J.Immunol. 133, 2151.
Stadecker, M.J., Calderon, J., Karnovsky, M.L., Unanue, E.R. 1977. J. Immunol. 119, 1738.
Stewart, W.E., Gresser, I., Tovey, M., et al. 1976. Nature. 262, 300.
Weinberg, J.B., and Haney, A.F. 1983. JNCI. 70, 1005.
Weiss, S.J., LoBuglio, A.F., and Kessler, H.B. 1980. Proc. Natl. Acad. Sci. USA. 77, 584.
Weiss, S.J., and Slivka, A. 1982. J. Clin. Invest. 69, 255.
Wright, S.D., and Silverstein, S. 1982. Cell. Biol. 95, 443a (abst.)
Zacharchuk, C.M., Drysdale, B-E., Mayer, M.M., and Shin, H.S. 1983. Proc. Natl. Acad. Sci. USA. 80, 6341.
Ziegler-Heitbrock, H.W.L., and Riethmuller, G. 1984. JNCI. 72, 23.

THE ROLE OF CYTOPLASMIC GRANULES IN NK CELL CYTOTOXICITY

Pierre Henkart
Maryanna Henkart
Paul Millard

Immunology Branch, NCI
NIH, Bethesda, Md. 20205

Robert Blumenthal

Laboratory of Theoretical Biology, NCI
NIH, Bethesda, Md. 20205

Craig W. Reynolds

Biological Therapeutics Branch, NCI
Frederick Cancer Research Facility, Frederick, Md. 21701

The mechanism of lymphocyte cytotoxicity has been the subject of investigation for well over ten years but remains controversial. Over the past few years, several groups studying NK cell cytotoxicity have shown that many of the features of this process are similar to those of cytotoxic T lymphocytes (CTL) (1,2). Thus, the initial step is a specific recognition of the target cell by unknown receptors resulting in a detectable adhesion. This is followed by a calcium- and energy-requiring step known as programming for lysis during which the target cell becomes irreversibly damaged. In the final step, killer-cell independent lysis (KCIL), the damaged target cell dies. Most of the measured properties of the last two steps are common to cytotoxicity by CTL and NK cells.

The calcium dependence the the programming step led to the proposal that an effector cell secretory process was required for this stage of lymphocyte cytotoxicity (3,4). This concept was reinforced in the NK system by studies showing inhibition of the lytic process by drugs such as monensin (5), chloroquine (6),

ISBN 0-12-341370-2

other lysosotropic agents (7), and a variety of other drugs (8).

The recognition that NK and ADCC effector cells belong to a morphologically distinct subset of normal lymphocytes characterized by cytoplasmic granules has been important in defining the NK phenemon (9). These granules range in size from 0.2-1.0 u, contain lysosomal enzymes (10), and in the human contain morphologically distinctive paracrystalline tubular arrays (11). The hypothesis that a secretory process involving these granules plays a role in NK cytotoxicity has been suggested by a number of lines of evidence. A role for some component of the lysosomal system was originally suggested by the defective NK activity of beige mice and patients with Chediak-Higashi syndrome, both of which are characterized by abnormally large lysosomes (12,13). More directly, Neighbour and Huberman (14,15) showed that NK cell pretreatment with strontium caused a time-dependent inactivation of their lytic capacity at a post-binding step. Cytoplasmic granule depletion was closely correlated with this inactivation, which appeared to parallel to the strontium induced degranulation of mast cells (16). EM studies of NK-target conjugates incubated at 37° demonstrated a granule rearrangement suggestive of secretion (17) and the concomitant formation of pore-like structures (see below). Other evidence for a role of these granules has come from morphological studies (18,19) and other correlations of granule quantity or quality and NK activity (20-23).

A granule exocytosis as part of the programming for lysis phase of NK cytotoxicity is compatible with the cytoplasmic rearrangement described in CTL after binding target cells (24). A comparable rearrangement in NK cells has been demonstrated by an assymmetric localization of the effector cell Golgi apparatus (5,25), cytoskeletal proteins (25,26), and granules (19,27) after binding relevant but not irrelevant target cells. LGL cytoplasmic rearrangement also occurs after contact with surfaces coated with antigen-antibody complexes simulating an ADCC reaction (28). It seems likely that the cytoplasmic rearrangement would allow the local exocytosis of secretory products in the vicinity of the bound target cell. This could occur via either small Golgi-derived secretory vesicles (5), or with the granules, which are larger and more electron dense than the vesicles.

The target cell damage arising from the effector cell secretory event has been proposed to be caused by lysosomal enzymes (6,19), NK cell cytotoxic factors (NKCF) (29), reactive oxygen intermediates (30), and effector cell granule derived membrane pore-forming substances (17). We will present additional evidence here in favor of the latter proposal.

Given all of the above lines of evidence for a role of LGL granules in the lethal event, we have begun a study of the biochemical and functional properties of these granules. As a first step in this project, we have developed a method for their purification from rat LGL tumor cells (Table I, 31). These tumor cells have been shown to be similar to normal LGL with respect to their morphology, surface markers, and NK and ADCC functions (32,33). They have the great advantage of being available in large amounts suitable for biochemical studies. The purification scheme is largely dependent upon a self-formed Percoll gradient, which results in pure granules banding near the bottom of the gradient. The presence of the granules was initially detected by assays for lysosomal enzymes and by EM morphology, and later by the cytolytic activity described below. Fig. 1 shows the fractionation of the Percoll gradient, with the cytolytic activity restricted to tubes 2-6 with a concomitant peak in lysosomal enzyme activity. A variety of marker enzymes were

Table I. Purification of Granules from Rat LGL Tumors

Spleen from Leukemic Rat

↓ Harvest cells and purify on Ficoll-Hypaque

Lymphoid Cell Suspension (~10^9 per rat, ~80% LGL)

↓ Wash and resuspend @ 10^8/ml in Disruption Buffer:
0.25M sucrose, 4mM EGTA, 1000U/ml heparin
0.01M Hepes, pH 7.4

↓ Disrupt cells by N_2 cavitation after equilibration
at 30 atm, 20 min. Incubate with DNAse I
(800u/ml, 22°)

Crude Homogenate

↓ Filter through 5μ and 3μ Nucleopore filters

Nuclei-Free Homogenate

↓ Layer on 20ml 48% Percoll.
Spin at 60,000XG, 30 min.
Pump out gradient, collect fractions.
Assay for B-glucuronidase, etc.

Pool Granule Fraction

↓ Percoll removal by ultracentrifugation
Collect fluffy pellet

Purified Granules

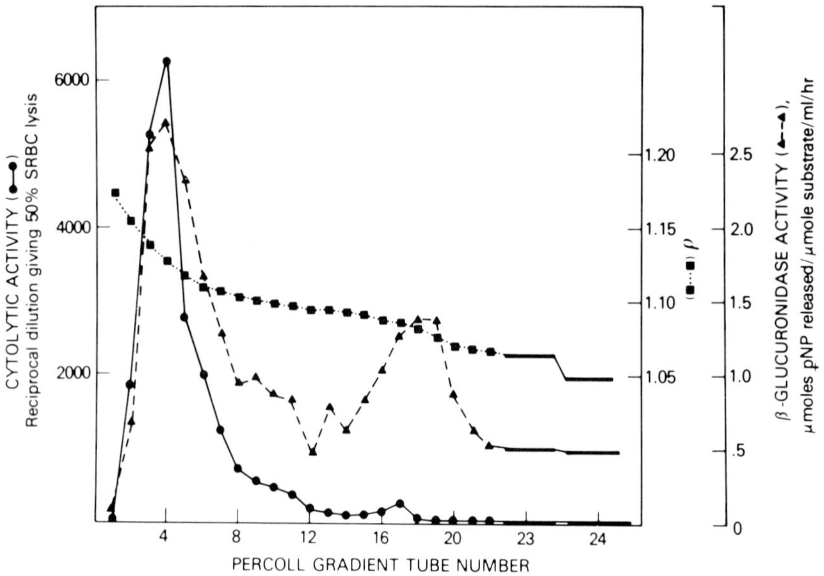

Fig. 1. Percoll gradient fractionation of LGL tumor homogenate. See Table 1 for conditions. Each fraction contained 0.8 ml, except 23 and 24, which contained 2.4 ml. From (31).

also measured on these gradients to localize other cellular components such as plasma membrane and mitochondria, which were found to occur higher up in the gradient (31). Fig. 2 shows that the pool of lytically active material in tubes 2-5 contains a pure population of granules which appear very similar morphologically to those in rat LGL tumors and blood LGL. The heterogeneity in granule morphology reflects that seen in the cells, and leaves open the possibility that more refined separation techniques could resolve the granules into subpopulations. No significant contamination by other cellular organelles was observed.

Analysis of the proteins in the Percoll gradient tubes by SDS gel electrophoresis gives further evidence of the purity of our granule preparation. As shown in Fig. 3, the granules contain 5 major protein bands of 60, 58, 30, 28 and 27 kd. The purity of the granule preparation is shown by the lack of proteins found maximally in other regions of the gradient, and even with the sensitive silver stain, minimal contamination is detectable. Analysis of these protein patterns and the enzymatic properties of the granule fractions leads to the conclusion that they are distinct from other known cytoplasmic granules such as those from mast cells and polymorphonuclear leukocytes (31).

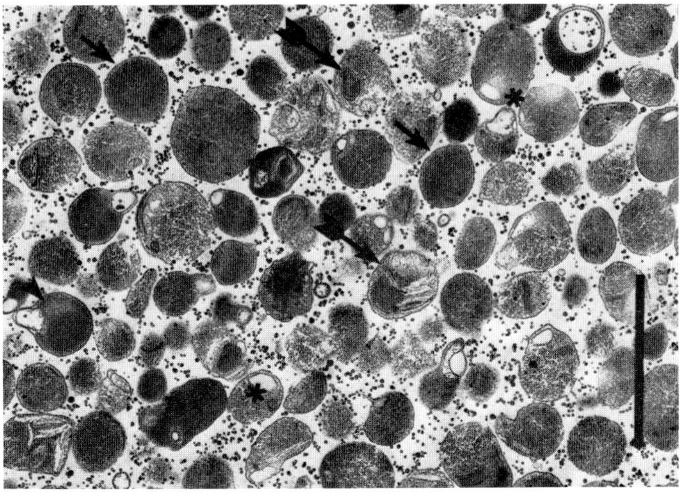

Fig. 2. Appearance of Percoll gradient purified granule fraction in the electron microscope. Arrows and star indicate different morphological forms of granules. Bar=1µm. From (31).

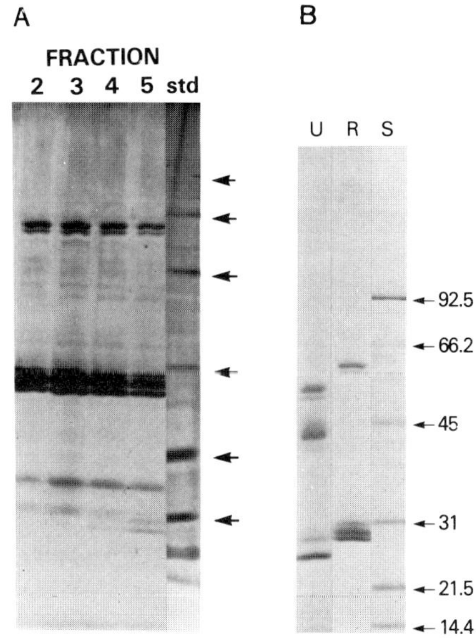

Fig. 3. SDS gel pattern of proteins from Percoll fractions in the granule region. A, reduced gels, silver stain, with]1 µg protein/lane. B. Coomassie stained gels showing reduced (R) vs. unreduced (U) pattern. Molecular weights of protein standards (S) are shown for both A and B in panel B. From (31).

From the start of our efforts to purify LGL tumor granules, we sought to detect a lytic activity which could fit into a secretory model for cytotoxicity. Our initial attempts did not succeed, and others in the past have been unable to detect plausible levels of lytic activities in cytotoxic lymphocyte homogenates. After we did succeed (34), it became clear that the interesting properties of the lytic agent, which we call a cytolysin, may well have prevented its discovery earlier. The lytic agent of LGL tumor granules is present in equal quantities in normal LGL and is shared only with CTL and a few exceptional cell lines which may be related to CTL. It is not in general found in the granules of non-cytotoxic lymphoid cells.

As shown in Fig. 4, the major controlling variable in cytolysin activity is calcium. In the absence of calcium, no lytic activity can be detected on any target cell. The activity is maximal in the physiological extracellular calcium range, and while magnesium and barium cannot replace calcium, strontium will substitute at ten-fold higher concentrations. We feel that this control makes physiological sense for a lytic agent stored in the effector cell interior, where the calcium concentration is very low and where its destructive power must be restrained. After exocytosis into the extracellular space where the calcium concentration is in the millimolar range, the lytic potential of the cytolysin is unleashed.

Another interesting finding with respect to divalent ions was that cytolysin activity is not stable in the presence of calcium. In order to assay cytolysin lytic activity, our routine procedure is to dilute the granules or extract into calcium-free PBS, followed by addition of an equal volume of target cells in

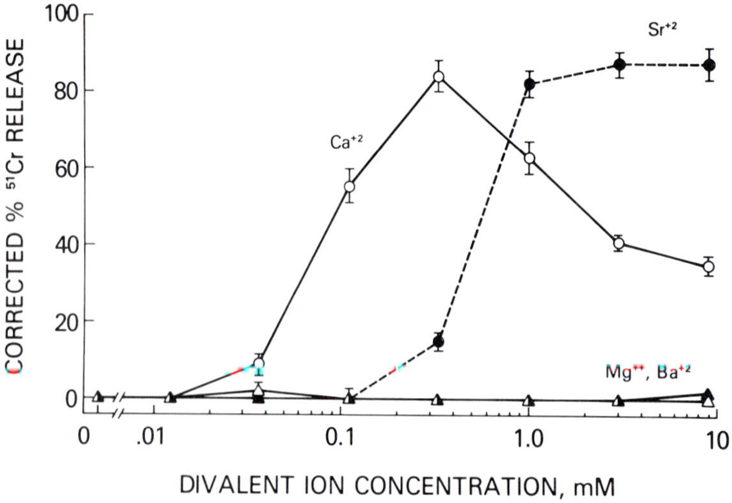

Fig. 4. Effect of divalent cations on LGL tumor granule cytolysin activity. Divalent cations were added with the target cells. From (34).

balanced salt solution containing calcium. If the granules are exposed to 2mM calcium prior to addition of the target cells, atime dependent loss of activity occurs, with a half life of about 10 minutes at room temperature. One interpretation of this behavior is that an unstable intermediate is generated when calcium. It again seems plausible that the physiologically relevant lytic agent should have such a property, since it would clearly be a problem for a highly lethal agent to be able to diffuse and so destroy "innocent bystander" cells, which does not seem to happen with NK cytotoxicity (6).

The cytolysin from LGL tumor granules does not show a specificity similar to NK cells. As shown in Fig. 5, the most sensitive targets we have found are red cells, which are not recognized by NK cells. In general, nucleated cells are about 10-100 fold less sensitive, with some nucleated cells still less sensitive. Among nucleated cells, normal spleen cells are relatively sensitive. However, if one considers only the lymphoid tumor cells, there is a general correlation of their lysability by LGL granules and by NK cells. If one considers the red cell lysis by ug/ml levels of granules, it can be said that the whole granules are comparable in potency to purified complement proteins such as C9, and more lytic than known hemolysins and toxins from bacteria.

One of the most striking features of the granule lytic activity is its rapid kinetics. As shown in Fig. 6, the lysis of red cells and nucleated cells is complete within a few minutes at

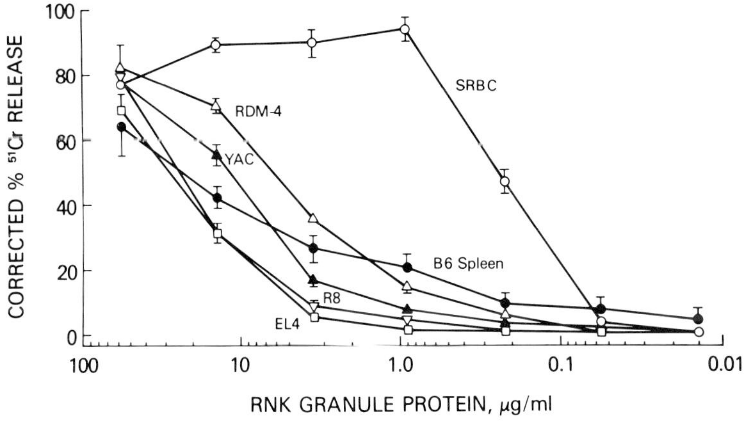

Fig. 5. Lytic activity of purified LGL tumor granules. Granules were diluted in PBS, followed by addition of ^{51}Cr labelled target cells in BSS. After 20 minutes, supernatants were harvested and counted. RDM-4, YAC, R8, and EL4 are lymphoid tumor cell lines. From (34).

room temperature, and is faster at 37°. This rate is in dramatic contrast to NKCF lysis, which takes days to achieve completion. This rapid rate is compatible with a detergent type of lytic mechanism, or with the creation of large membrane pores which allow the direct egress of cytoplasmic components. However, it is clearly faster than lysis of cells by complement, which occurs by a colloid osmotic mechanism requiring many minutes to achieve release of cytoplasmic proteins (35). The rate of target cell lysis during the NK cell KCIL has recently been determined, and is close to the speed of the cytolysin induced lysis (36).

The possibility that the NK lethal hit occurred via the creation of large pores in the target cell membrane was attractive to us because of a series of experiments on target cell damage induced by LGL in ADCC. In 1975, we showed that planar lipid bilayers modeling target cells were attacked so that their ionic permeability increased by several orders of magnitude (37). More recently, a study of resealed erythrocyte ghosts containing molecular size markers showed that ADCC induced large pores as evidenced by sieving of the released markers (38). When compared to complement, the LGL induced the release of larger markers, and a surprisingly large maximal pore size of 14 nm was estimated to be induced by the cytotoxic lymphocytes in these studies. Later, we showed that pore-like structures with this internal diameter were visible in the red cell ghost targets

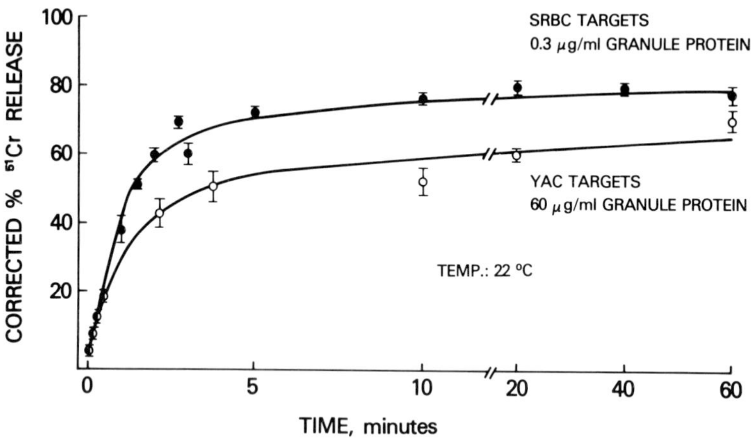

Fig. 6. Kinetics of lysis by LGL tumor granules. Cr labeled target cells in BSS were mixed with an equal volume of diluted granules in PBS at time zero, and the cells were briefly centrifuged at 12,000 rpm followed by harvesting the supernatant at the indicated time. From (34).

after negative staining and EM (39). These structures were generally similar to those associated with complement lysis, but were larger. These same structures were visualized in thin sections made of NK-target conjugates incubated at 37° (17), where they appeared in material secreted from the LGL granules into the space between the effector and target cells. Similar structures were shown to be associated with cytotoxicity by cloned NK cells (40).

Fig. 7 shows the appearance of red cell membranes exposed to LGL tumor granules in the presence of calcium. The rings on the membrane surface are filled with the dark negative stain, as would be expected from an aqueous pore. However, these images cannot be unequivocally interpreted as transmembrane pores because a crater-like depression would give approximately the same image. These structures are clearly identical to those we have previously seen (17,39). It should be emphasized that the appearance of these pore-like structures requires the presence of calcium, and the structures appear rapidly after its addition, correlating nicely with the lytic potential of the granules. If no target cells are present, the pore-like structures become associated with the granule membranes.

It was obviously of interest to see if the whole granules are required for the lytic effect, or if a soluble molecule(s) within the granules were responsible. The involvement of a protein was suggested by the temperature lability and Pronase sensitivity of the cytolysin activity (34). When we attempted

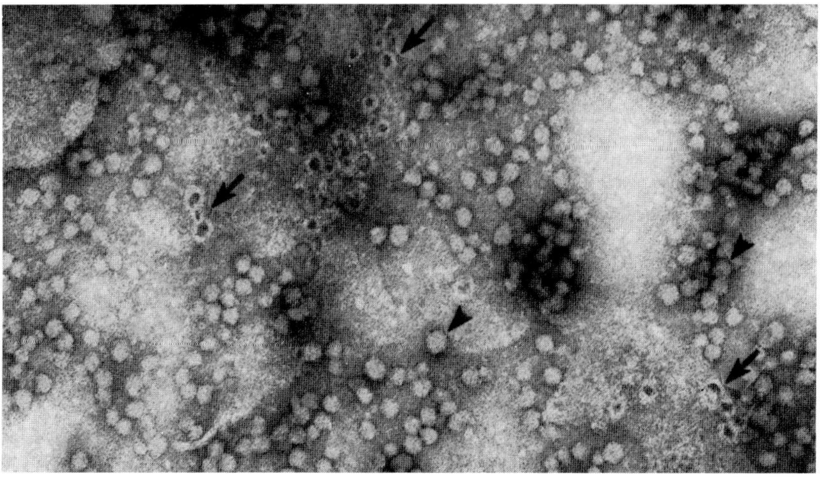

Fig. 7. Appearance of red cell ghosts after lysis by LGL tumor granules. Electron micrograph after negative staining. Arrowheads indicate Percoll particles. Arrows indicate ring structures of average diameter 15 nm. From (34).

to solubilize the lytic activity from the granules, simple membrane rupture by freezing or mild sonication failed to make the activity non-sedimentable. Non-denaturing detergents also did not have this effect, although their use was hampered by their own lytic effect. We found that 1-2 M NaCl or KCl would successfully solubilize the activity, and in fact we typically double our activity during this step. Solubilization of the cytolysin shows that the intact granule structure is not required for the lytic activity, and allows for the purification of the active molecule(s). As a first step in this purification, we have performed gel filtration chromatography on ACA 54, as shown in Fig. 8. This shows that the active cytolysin behaves like a protein of 50-60 kd. The SDS gels of the active peak of this column show that the components of 60, 58, and 28 kd are still present. The other major granule proteins are gone, along with most of the lysosomal enzyme activities we have measured. At this stage we obviously cannot say whether one or all of the remaining protein chains are required for activity, but further purification is currently underway. The biochemical results so far can be summarized by saying that the cytolysin behaves as if it is a single protein molecule of 60,000 molecular weight. It does not seem to be a complex set of interacting soluble proteins like complement.

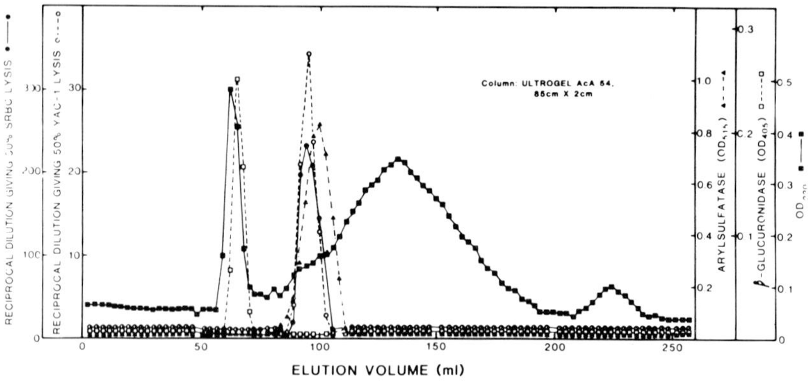

Fig. 8. Gel filtration of LGL granule cytolysin on ACA 54. LGL tumor granules were extracted with 2M NaCl and the extract chromatographed at 5° on Ultrogel ACA 54 (85 X 2 cm). (●), activity (U/ml) on SRBC; (O), reciprocal dilution giving 50% YAC lysis; (■), OD, 220nm; (▲), aryl sulfatase activity (OD, 515nm); (□), β-glucuronidase activity (OD, 405nm).

As we have shown, the appearance of the pore-structures in the EM is strongly reminiscent of complement, and we felt that the cytolysin was likely to act by a similar membrane-insertion mechanism. In order to test this hypothesis, we carried out a series of experiments with liposomes, using the water soluble dye carboxyfluorescein as a marker for liposome permeability. We found that liposomes of all types make excellent targets for LGL granule cytolysin (41). A typical experiment is shown in Fig. 9, in which one can seen the carboxyfluorescein release occurring over the course of a few minutes. This marker release was absolutely dependent on divalent cations in the medium, with magnesium and barium being inactive and strontium being ten-fold less active than calcium. These titration curves were virtually identical to the data shown in Fig. 4. Thus the liposome permeabilization parallels the kinetics and divalent ion requirements of target cell lysis by the cytolysin. The liposome fluidity as determined by the fatty acids of the phospholipids had only a minor effect on cytolysin activity (Fig. 9), and variation of the lipid head group had negligible effect (41). The carboxyfluorescein technique enabled us to show that marker release from individual liposomes was all-or-none, as predicted by a pore insertion model. The most striking evidence in favor of this hypothesis came from EM examination of the liposomes after treatment with partially purified cytolysin in the presence of calcium, and is shown in Fig. 10. Negative staining shows

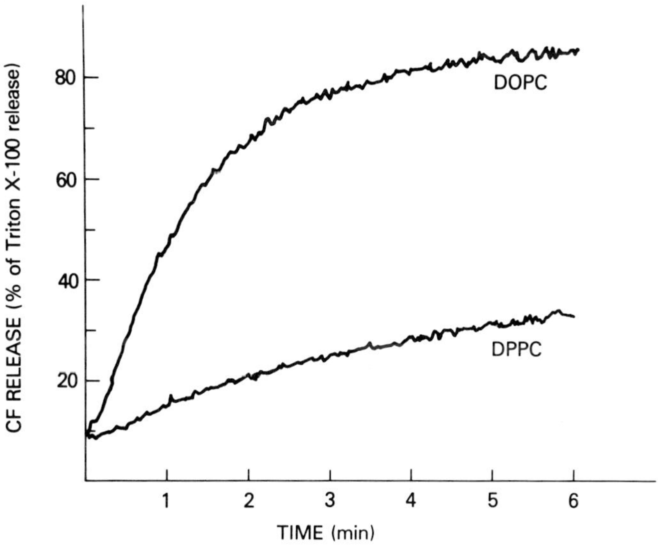

Fig. 9. Release of carboxyfluorescein from small unilamellar limposomes by partially purified LGL granule cytolysin. DOPC, dioleoyl phosphotidyl choline; DPPC, dipalmitoyl phosphatidyl choline. From ref. 41.

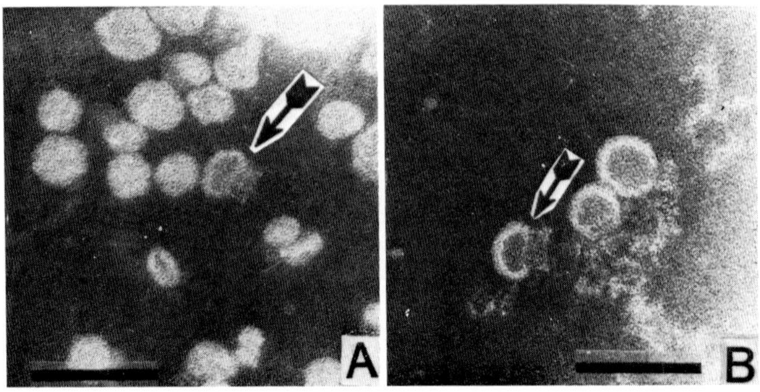

Fig. 10. Appearance of liposomes exposed to partially purified LGL granule cytolysin in the presence of calcium. Negative stain electron micrograph. Arrows show clear examples of inserted pore structures. Two other liposomes in B have inserted pore structures. Bar=60nm. From (41).

pore structures which appear deeply inserted into the lipid bilayer of these small unilamellar liposomes. The liposomes which contain inserted pore structures have become permeabilized to the negative stain so that their centers are dark, while the other liposomes have unstained, pale interiors. A similar correlation has been noted with complement (42).

The results which I have described so far strongly support a model of granule exocytosis for NK cytotoxicity. However, there is no direct evidence that the cytolysin is responsible for the lethal damage to the target cell. As one approach to provide such evidence, we have raised rabbit antibodies against the purified granules. These antibodies form two precipitin bands in agar gel diffusion against salt extracts of granules and do not react with comparable extracts of granules from the non-cytotoxic mouse tumor EL-4. On Western blots of SDS gels of granule proteins, the antibodies recognize four of the five major bands, and some minor ones as well. When immunofluorescence experiments were carried out, the antibodies do not detectably stain the surface membranes of any lymphoid cells including LGL tumors or LGL. However, if the cells are fixed and permeabilized, a bright and discrete cytoplasmic staining pattern reminiscent of granules is seen in rat LGL and LGL tumors. Other lymphoid cell granules do not stain under conditions when the LGL are brightly stained, but will stain if higher concentrations of antibody are used. Thus the antibodies appear to be specific for LGL granules.

Anti-LGL tumor granule antibodies will specifically block the lytic activity of the granules against sheep red cells or YAC tumor targets, as shown in Table 2. At present we cannot say which granule protein gave rise to this blocking antibody. The reason we have used $F(ab')_2$ antibodies is that we wanted to test their effects on LGL mediated cytotoxicity, and we wished to

Table 2

INHIBITION OF GRANULE CYTOLYSIN ACTIVITY BY ANTIBODY

$F(ab')_2$ Antibody	^{51}Cr Release from SRBC[1]	^{51}Cr Release from YAC-1[2]
none	77 ± 5	39 ± 1
anti-TNP	84 ± 5	47 ± 1
anti-EL-4 granules	78 ± 4	36 ± 1
anti-LGL granules	3 ± 1	0.3 ± .6

1. 1μg/ml LGL tumor granules; 5μg/ml $F(ab')_2$ (Final concentrations).
2. 40μg/ml LGL tumor granules; 62μg/ml $F(ab')_2$.

avoid complications due to the Fc receptor. Such an experiment with rat LGL in NK and ADCC assays is shown in Table 3, in which it is clearly seen that the anti-LGL granule antibodies will block the cell-mediated cytotoxicity as well as that mediated by granules. Again, we do not know which protein gave rise to the blocking antibody, or if is the same species responsible for blocking the cytolysin. However, the antibody must be present during the cytotoxicity reaction in order to block. If effector or target cells are pretreated and washed, no blocking occurs, as one would expect for an anti-granule antibody according to a granule exocytosis model for cytotoxicity. Furthermore, the blocking occurs at a post-binding step in the reaction, as shown by the inability of these antibodies to block the formation of killer-target conjugates in either the NK or ADCC system. This is again in accord with the granule exocytosis model. Nevertheless, only antibodies against a pure cytolysin molecule will answer the obvious question of whether cytolysin is responsible for the lethal damage to the target cell.

Table 3

EFFECT OF ANTI-GRANULE ANTIBODIES ON ADCC AND NK CYTOTOXICITY AND CONJUGATE FORMATION

Treatment	[51]Cr Release[1] YAC-1	[51]Cr Release[1] P815+Ab	%Conjugates[2] YAC-1	%Conjugates[2] P815+Ab
PBS control	34±3	37±3	20±2	33±2
Anti-TNP[3]	30±6	38±2	17±4	26±0
Anti-LGL granules[3]	3±1	14±1	18±4	35±2

1. E/T=25, 4 hour assay.
2. E/T=2, scored after 5 min at 37°.
3. $F(ab')_2$, 300 μg/ml in the assay.

In summary, the results presented here strongly support the hypothesis that lymphocyte cytotoxicity operates via a secretory process involving cytoplasmic granules of the effector cell. The results of various laboratories studying the NK cytotoxic process by different approaches are in accord with this hypothesis, and it appears to be applicable to cytotoxicity by CTL as well as LGL. At present, we have been concentrating on the cytolysin in these granules because its effects are so dramatic. However, it is very possible that other granule components may play a role in causing target cell damage, especially since the cytolysin may allow their access to the target cell interior. A further study of the other granule components may shed some light on this. It will be of interest to compare the biochemical properties of the granule cytolysin and NKCF. Although the behavior of their cytolytic effects is strikingly different and we thus tend to assume these are different molecules, only further biochemical analysis will determine if there is any connection. It will also be of interest to determine whether there are any relatioions between the receptor triggered granule exocytosis and the recently described secretion of interleukins by LGL (43,44).

1. Roder, J. C., R. Kiessling, P. Biberfeld, and B. Andersson. 1978. Target-effector interaction in the natural killer (NK) cell system. II. The isolation of NK cells and studies on the mechanism of killing. J. Immunol. 121:2509.
2. Hiserodt, J. C., L. C. Britvan, and S. R. Targan. 1982. Characterization of the cytolytic reaction mechanism of the human natural killer (NK) lymphocyte: Resolution into binding, programming and killer cell-independent steps. J. Immunol. 129:1782.
3. Henney, C. S. 1973. On the mechanism of T-cell mediated cytolysis. Transplant. Rev. 17:37.
4. Martz, E., W. L. Parker, M. K. Gately, and C. D. Tsoukas. 1982. The role of calcium in the lethal hit of T lymphocyte-mediated cytolysis. Adv. Exp. Med. Biol. 146:121.
5. Carpen, O., I. Virtanen, and E. Saksela. 1982. Ultrastructure of the human natural killer cells; nature of the cytolytic contacts in relation to cellular secretion. J. Immunol. 128:2691.
6. Roder, J. C., S. Argov, M. Klein, C. Peterson, R. Kiessling, K. Anderson, and M. Hanson. 1980. Target-effector cell interaction in the natural killer system. V. Energy requirements, membrane integrity, and the possible involvement of lysosomal enzymes. Immunol. 40:107.
7. Verhoef, J., and S. D. Sharma. 1983. Inhibition of human natural killer activity by lysosomotropic agents. J. Immunol. 131:125.
8. Quan, P. C., T. Ishizake, and B. Bloom. 1982. Studies on the mechanism of NK cell lysis. J. Immunol. 128:1786.
9. Timonen, T., J. R. Ortaldo, and R. B. Herberman. 1981. Characteristics of human large granular lymphocytes and relationship to natural killer and K cells. J. Exp. Med. 153:569.
10. Grossi, C. E., A. Cadoni, A. Zicca, A. Leprini, and M. Ferrarini. 1982. Large granular lymphocytes in human peripheral blood: Ultrastructural and cytochemical characterization of the granules. Blood. 59:277.
11. Payne, C. M., L. Glasser, R. Fiederlein, and R. Lindberg. 1983. New ultrastructural observations: Parallel tubular arrays in human T gamma lymphoid cells. J. Immunol. Methods. 65:307.
12. Roder, J. C., and A. K. Duwe. 1979. The beige mutation in the mouse selectively impairs natural killer cell function. Nature (Lond.). 278:451.
13. Klein, M., J. Roder, T. Haliotis, S. Korec, J. Jett, R. B. Herberman, P. Katz, and A. S. Fauci. 1980. Chediak-Higashi gene in humans II. The selectivity of the defect in natural killer and antibody-dependent cell-mediated cytotoxicity function. J. Exp. Med. 151:1049.

14. Neighbour, P. A., H. S. Huberman, and Y. Kress. 1982. Human large granular lymphocytes and natural killing: Ultrastructural studies of strontium-induced degranulation. Eur. J. Immunol. 12:588.
15. Neighbour, P. A., and H. S. Huberman. 1982. Sr+2-induced inhibition of human natural killer (NK) cell-mediated cytotoxicity. J. Immunol. 128:1236.
16. Foreman, J. C. 1977. Spontaneous histamine secretion from mast cells in the presence of strontium. J. Physiol. 271:215.
17. Henkart, M. P., and P. A. Henkart. 1982. Lymphocyte mediated cytolysis as a secretory phenomenon. Adv. Exp. Med. Biol. 146:227.
18. Frey, T., H. R. Petty, and H. M. McConnell. 1982. Electron microscopic study of natural killer cell-tumor cell conjugates. Proc. Natl. Acad. Sci. U. S. A. 79:5317.
19. Zucker-Franklin, D., G. Grusky, and J.-S. Yang. 1983. Arylsulfatase in natural killer cells: its possible role in cytotoxicity. Proc. Natl. Acad. Sci. U. S. A. 80:6977.
20. Nocera, A., E. Montesoro, P. Balbo, M. Ferrarini, A. Leprini, A. Zicca, and C. E. Grossi. 1983. Complement receptor distinguishes between two subsets of large granular lymphocytes with different natural killer activity and cytochemical and ultrastructural features. Scand. J. Immunol. 18:345.
21. Kay, N. E., and J. M. Zarling. 1984. Impaired natural killer activity in patients with chronic lymphocytic leukemia is associated with a deficiency of azurophilic cytoplasmic granules in putative NK cells. Blood. 63:305.
22. Itoh, K., R. Suzuki, Y. Umezu, K. Hanaumi, and K. Kumagai. 1982. Studies of murine large granular lymphocytes. II. Tissue, strain, and age distributions of LGL and LAL. J. Immunol. 129:395.
23. Abo, T., M. D. Cooper, and C. M. Balch. 1982. Characterization of HNK-1+ (Leu-7) human lymphocytes. I. Two distinct phenotypes of human NK cells with different cytotoxic capability. J. Immunol. 129:1752.
24. Zagury, D. 1982. Direct analysis of individual killer T cells: Susceptibility of target cells to lysis and secretion of hydrolytic enzymes by CTL. Adv. Exp. Med. Biol. 146:149.
25. Kupfer, A., G. Dennert, and S. J. Singer. 1983. Polarization of the Golgi apparatus and the microtubule-organizing center within cloned natural killer cells bound to their targets. Proc. Natl. Acad. Sci. U. S. A. 80:7224.
26. Carpen, O., I. Virtanen, V.-P. Letho, and E. Saksela. 1983. Polarization of NK cell cytoskeleton upon conjugation with sensitive targets. J. Immunol. 131:2695.

27. Acha-Orbea, H., P. Groscurth, R. Lang, L. Stitz, and H. Hengartner. 1983. Characterization of cloned cytotoxic lymphocytes with NK-like activity. J. Immunol. 130:1952.
28. Alexander, E., and P. Henkart. 1976. The adherence of human Fc receptoor-bearing lymphocytes to antigen-antibody complexes. II. Morphological alterations induced by the substrate. J. Exp. Med. 143:329.
29. Wright, S. C., and B. Bonavida. 1983. Studies on the mechanism of natural killer cell-mediated cytotoxicity. IV. Interferon-induced inhibition of NK target cell susceptibility to lysis is due to a defect in their ability to stimulate release of natural killer cytotoxic factors. J. Immunol. 130:2965.
30. Helfand, S. L., J. Werkmeister, and J. C. Roder. 1982. Chemiluminescence response of human natural killer cells. I. The relationship between target cell binding, chemiluminescence, and cytolysis. J. Exp. Med. 156:492.
31. Millard, P. J., M. P. Henkart, C. W. Reynolds, and P. A. Henkart. 1984. Purification and properties of cytoplasmic granules from cytotoxic rat LGL tumors. J. Immunol. 132:3197.
32. Ward, J. M., and C. W. Reynolds. 1983. Large granular lymphocyte leukemia. A heterogensous lymphocytic leukemia in F344 rats. Am. J. Path. 111:1.
33. Reynolds, C. W., E. W. Bere, and J. M. Ward. 1984. Natural killer activity in the rat. III. Characterization of transplantable large granular lymphocyte leukemias in the F344 rat. J. Immunol. 132:534.
34. Henkart, P. A., P. J. Millard, C. W. Reynolds, and M. P. Henkart. 1984. Cytolytic activity of purified cytoplasmic granules from cytotoxic rat LGL tumors. J. Exp. Med. 160:75.
35. Mayer, M. M. 1982. Membrane attack by complement (with comments on cell-mediated cytotoxicity). Adv. Exp. Med. Biol. 146:193.
36. Callewaert, D. M. 1984. Determination of rate constants for lytic programming and killer cell independent lysis. (This volume)
37. Henkart, P., and R. Blumenthal. 1975. The interaction of lymphocytes with lipid bilayer membranes: A model for the lymphocyte-mediated lysis of target cells. Proc. Natl. Acad. Sci. U. S. A. 72:2789.
38. Simone, C. B., and P. Henkart. 1980. Permeability changes induced in erythrocyte ghost targets by antibody-dependent cytotoxic effector cells: evidence for membrane pores. J. Immunol. 124:954.
39. Dourmashkin, R. R., P. Deteix, C. B. Simone, and P. A. Henkart. 1980. Electron microscopic demonstration of lesions on target cell membranes associated with antibody-dependent cytotoxicity. Clin. Exp. Immunol. 43:554.

40. Podack, E. R., and G. Dennert. 1983. Assembly of two types of tubules with putative cytolytic function by cloned natural killer cells. Nature (Lond.). 302:442.

41. Blumenthal, R., P. J. Millard, M. P. Henkart, C. W. Reynolds, and P. A. Henkart. 1984. Liposomes as targets for granule cytolysin from cytotoxic LGL tumors. Proc. Natl. Acad. Sci. U. S. A. 81:5551.

42. Bhakdi, S., and J. Tranum-Jensen. 1983. Membrane damage by complement. Biochim. Biophys. Acta. 737:343.

43. Kasahara, T., J. Y. Djeu, S. F. Dougherty, and J. J. Oppenheim. 1983. Capacity of human large granular lymphocytes (LGL) to rpoduce multiple lymphokines: Interleukin 2, interferon, and colony stimulating factor. J. Immunol. 131:2379.

44. Scala, G., P. Allavena, J. Y. Djeu, T. Kasahara, J. R. Ortaldo, R. B. Herberman, and J. J. Oppenheim. 1984. Human large granular lymphocytes are potent producers of interleukin-1. Nature (Lond.). 309:56.

FUNCTIONAL AND BIOCHEMICAL CHARACTERIZATION OF NATURAL KILLER CYTOTOXIC FACTORS AND THEIR ROLE IN THE NK LYTIC MECHANISM[1]

Susan C. Wright
Stanley M. Wilbur
Benjamin Bonavida

Department of Microbiology & Immunology
UCLA School of Medicine
Los Angeles, California

I. INTRODUCTION

Recently, NK cells have been identified and appear to play a major role in defense against viral infections and neoplasm. The mechanism by which NK cells lyse target cells has been examined using different strategies, and one model suggested that NK cells may secrete cytotoxic mediators which then lyse the target cell. This evidence was derived from experiments demonstrating that various agents known to inhibit a well-documented stimulus-secretion event, such as histamine release from mast cells, could also inhibit target cell lysis by NK cells (1,2). However, a soluble lytic mediator released from NK cells and specific for NK sensitive target cells had not yet been detected. Subsequent experiments carried out in our laboratory produced evidence that human, rat, or murine effector cells release soluble natural killer cytotoxic factors (NKCF) that lyse NK sensitive tumor cells (3). Analysis of the functional characteristics of these factors revealed may parallels to some of the known characteristics of the NK system. Based on

[1]Supported by PHS Grant CA35791 awarded by the National Cancer Institute (DHHS).

ISBN 0-12-341370-2

these findings, we have proposed a model for the NK lytic mechanism in which NKCF function as the lytic mediators (4,5). According to this model, the NK lytic pathway is composed of the following sequence of events. 1) The effector cells recognize and bind to the target. 2) The target cell stimulates the effector to release NKCF. 3) The effector cell releases NKCF. 4) NKCF bind to the target cell membrane and cause cell death. The present communication will review the evidence that led to this model as well as some of the more recent biochemical studies characterizing NKCF.

II. MATERIALS AND METHODS

Cell lines. All cell lines used in this study as either stimulator or target cells were cultured in the absence of antibiotics and screened at least twice a month for the presence of mycoplasma according to Chen's (6) technique using the Hoeschst stain and fibroblast indicator cells. All cell lines were found to be mycoplasma-free, and this result was confirmed using the adenosine phosphorylase-mediated nucleoside toxicity test as described by McGarrity and Carson (7).

Production of NKCF. Methods to produce NKCF have been described in detail previously (8). Plastic nonadherent rodent spleen cells were stimulated with YAC-1 cells or ConA (2.5 µg/ml) whereas human PBL were stimulated with U937 cells or ConA (2.5 µg/ml). After 20-24 h of culture in serum-free media, supernatants were harvested and tested for NKCF activity.

NKCF assay. The human and rodent NKCF assay has been described in detail previously (8). The rodent assay employed YAC-1 target cells that were labeled with ^{51}Cr (used prior to calibration date) by an overnight labeling procedure. The rodent assay was incubated 20 h whereas the human NKCF assay using ^{51}Cr-labeled U937 target cells was incubated 24 h.

III. RESULTS AND DISCUSSION

A. Characterization of NKCF Producing Cell and Conditions for Production

Experiments performed to characterize the NKCF producing cells are summarized in Table 1. It can be seen that all the characteristics listed for the murine NKCF producing cells (9) correlate with the known characteristics of the NK effector cell. We have also found that the human NKCF producing cell is enriched in low density fractions of Percoll density gradients (10), as is the NK effector cell population (11). Furthermore, we have found that the RNK tumor cell line that has been demonstrated to exhibit NK effector cell function (12) also produces NKCF (13). Taken altogether, the data strongly suggest that rodent and human NK effector cells produce NKCF.

TABLE I. Phenotypic characteristics of the NKCF Producing Cell

Murine:
Present in spleen but not thymus
Plastic non-adherent
Present in nylon wool non-adherent nude spleen cells
Thy1.2 negative
Asialo - GM1 positive
Human:
Enriched in low density LGL fractions

TABLE II. Conditions for Production of NKCF

Produced over 20-48 h culture period in absence of FCS
Release stimulated by lectin (PHA or ConA)
Release stimulated by many different types of NK-sensitive or resistant mycoplasma-free cultured cell lines
IFN pretreatment of effector cells augments production
Pretreatment of stimulator cells with IFN inhibits their ability to stimulate release
Pretreatment of effector cells with TPCK or ATEE inhibits production

The conditions under which NKCF are produced and some of the agents that affect levels of production are listed in Table 2. Although in some experiments high levels of NKCF are released spontaneously, usually the addition of a stimulatory agent results in optimal levels of production. Many types of cultured cell lines have been found to stimulate NKCF release regardless of NK sensitivity or species of origin (8). The use of mycoplasma infected stimulator cells should be avoided since one report indicated that mycoplasma organisms in the supernatants of such cultures may be toxic to YAC-1, thus mimicking NKCF activity (14). We have observed considerable variation from experiment to experiment in the ability of YAC-1 or U937 to stimulate release of NKCF in contrast to ConA which more consistently induces high levels of NKCF production.

Pre-treatment of murine spleen cells with semi-purified mouse fibroblast IFN (Calbiochem) or with human cloned recombinant IFNα A/D (donated by M. Brunda, Hoffman-La Roche) augments NKCF production stimulated by either YAC-1 or ConA (15 and unpublished observations). Augmentation of NKCF production requires protein synthesis during the IFN pre-treatment period, as is true for NK CMC. If YAC-1 stimulator cells are pre-treated with IFN, they can still form conjugates with effector cells although they are unable to stimulate release of NKCF (4). These findings account in part for the IFN induced augmentation of NK CMC as well as the observation that IFN pre-treated targets become relatively NK resistant (16).

The production of NKCF appears to involve one or more effector cell associated serine or sulfhydryl dependent protease since TPCK or ATEE inhibit NKCF production. This may be the basis for the inhibition of NK CMC reported by other investigators using these reagents (2,17).

TABLE III. NKCF Lytic Mechanism

Selectively toxic to NK sensitive tumor cells
Rapidly binds (within 1 h) to target cell
Binding and lysis inhibited by α-methyl-D-mannoside, fructose-6-phosphate and mannose-6-phosphate
Target cell zeosis occurs prior to cell death
Lysis proceeds at 37°C but not at 25°C
Lysis not affected by TPCK, TLCK, ATEE, α-1-anti-trypsin, soybean trypsin inhibitor, or lima bean trypsin inhibitor

B. Mechanisms of Target Cell Lysis by NKCF

Experiments designed to investigate the NKCF lytic mechanism are summarized in Table 3. Our previous work has demonstrated that NKCF selectively lyse only NK sensitive tumor cells (9,10). These factors bind rapidly to the membrane of NK sensitive and some resistant target cells. However, we have also observed some cases where lack of NKCF binding sites confers NK resistance to a cell. Thus, YAC-R variants selected for resistance to NKCF are also resistant to NK cells due to a deficiency of NKCF binding sites (5). In contrast, YAC-1 cells passaged *in vivo* in ascites form are resistant to lysis by NKCF or NK cells although they still bind NKCF. Thus, binding alone is not sufficient to lead to lysis.

Recently studies have indicated that adsorption of NKCF to YAC-1 cells is inhibited by α-methyl-D-mannoside, fructose-6-phosphate, and mannose-6-phosphate. This may be the mechanism by which these carbohydrates inhibit NK CMC as reported by other investigators (18,19). These findings suggest that NKCF bind to carbohydrate determinants on the target cell or that there are carbohydrate residues on NKCF molecules that are essential for lytic activity. The latter possibility is supported by the observation that NKCF is inactivated by oxidation with sodium periodate (see next section). This finding, as well as the observation that mannose-6-phosphate inhibits both mouse and human NKCF activity, is in accord with another study suggesting that the NK lytic molecules bind to mannose-6-phosphate receptors on the target cell membrane (19). Some of our observations parallel those in studies of lysosomal enzyme uptake. For example, it has been shown that periodate treatment of hexosaminidase inhibits its uptake by fibroblasts (20). Other studies demonstrated that mannose-6-phosphate, fructose-6-phosphate, and α-methyl-D-mannoside inhibit lysosomal enzyme uptake by cultured fibroblasts (21). These findings suggest that NKCF may bind to target cell receptors for mannose-6-phosphate and are internalized via the lysosomal enzyme pathway.

Alternatively, it is possible that NKCF possess lectin-like receptors and bind to carbohydrate determinants on the target cell membrane. This interpretation is supported by the work of Blanca et al. (22) who found pre-treatment of K562 with tunicamycin caused a decrease in their sensitivity to NKCF. Further studies on the interactin of NKCF with target cells will be necessary to distinguish between the two hypotheses.

The events following binding of NKCF to the target cell membrane are not clear, however, time-lapse cinematography studies have indicated that YAC-1 cells go through a period of zeosis prior to death (unpublished observations). Like NK CMC, target cell lysis of NKCF occurs at 37°C but not at 25°C. We did not obtain any evidence that target cell lysis mediated by NKCF involves proteolytic activity since a number of protease inhibitors found previously to inhibit NK CMC (2,23) had no effect on NKCF activity. This inhibition appears to be at the level of NKCF production as suggested by experiments in the preceding section.

C. Biochemical Characteristics of NKCF

Recent work in our laboratory has focused on the biochemical characterization and purification of NKCF. Table 4 lists several characteristics of murine or rat NKCF as determined by testing whole supernatants. We have also found that human NKCF exhibit an apparent MW of 20-30,000, as do the rodent factors. These data indicate that NKCF are probably glycoproteins in which disulfide bonding may be essential for lytic activity.

By all parameters tested thus far, we have not detected any differences in NKCF produced by the three different species, nor were there differences in factors generated by tumor cell or ConA stimulation. We have also reported that ConA-induced supernatants exhibit the same target specificity as do tumor cell induced factors (8). The biochemical and functional similarities exhibited by these supernatants suggest that NKCF produced by stimulation with tumor cells or ConA are identical. However, further purification and characterization will be necessary to determine if this is true.

TABLE IV. Biochemical Characteristics of Murine or Rat NKCF Produced by Stimulation with Tumor Cells or ConA

Inactivated by heating to 63°C 2 h
Stable for 1 month at -20°C
Trypsin sensitive
Not affected by neuraminidase
Inactivated at pH 2
Inactivated by reduction and alkylation
Inactivated by Na periodate (0.01 M)
Apparent MW of 20-30,000 by HPLC gel filtration

D. Evidence Supporting our Model for NK CMC

Our model for the NK lytic mechanism is based on the studies of NKCF summarized in the present report as well as some of the known characteristics of the NK system. Following the first step of the reaction involving the binding of the effector to the target cell, we postulate that a second signal must be delivered to the NK cell to activate the NKCF release mechanism. This is based on the observation that IFN-pretreated YAC-1 cells can still bind effector cells but are unable to stimulate release of NKCF (4). The delivery of this second signal may occur through the interaction of membrane determinants distinct from those involved in the initial binding reaction. The production of NKCF appears to require proteolytic activity which may be involved in the release mechanism or else may convert inactive precursors into lytically active NKCF. NKCF may be stored within cytoplasmic granules and the synthesis of these factors may be augmented by pretreatment with IFN (15). Released NKCF then bind sites on the target cell membrane, probably involving interaction of carbohydrate determinants. NKCF binding sites are distinct from those determinants involved in conjugate formation or in stimulation of NKCF release. This is based on the observation that the YAC-R clones can perform these functions, although they are deficient in NKCF binding sites (5). Binding alone is not sufficient to lead to cell death and possible subsequent events may involve pore formation, internalization, and/or enzymatic activity. This model accounts for many of the known phenomena of the NK system and provides a framework to determine which steps in the lytic pathway are affected by agents known to modulate NK CMC.

D. Relationship of NKCF to Other Cytotoxins

Recent progress in the biochemical characterization of NKCF now allows comparison, on a molecular level, to other cytotoxic lymphokines. Lymphotoxin has been proposed by a number of different investigators to be involved in various types of CMC reactions (24-26). LT is released from lectin-stimulated lymphocytes and is usually measured against L929 fibroblasts. Human LT consists of four MW classes: complex (>200,000), α (70 to 90,000), β (25-50,000), and γ (10-20,000) (27). According to our results, NKCF would correspond to γ or β LT, however, the comparison may not be valid since the MW of NKCF was determined by HPLC gel filtration as opposed to the conventional gel filtration techniques used for LT. Results from another laboratory using TSK-HPLC indicated that purified human LT has a MW of 60,000 (28). However, the same molecule exhibited a MW of 20,000 on SDS-PAGE, suggesting that a 60,000 MW oligomer may be dissociated under reducing conditions to a monomer.

Recent work has indicated that certain forms of LT of undetermined MW can lyse human NK sensitive targets (29). The same study reported that absorption experiments suggest that L929 targets are lysed by LT forms different from, or in addition to, those that lyse NK targets. Similar to the NKCF system, lysis of NK targets by whole supernatants containing LT is inhibited by α-methyl-D-mannoside (24), although lysis of L929 is weakly or not at all inhibited by the same sugar (30). These differences suggest that the use of L929 target cells to measure NKCF activity may not yield the same results as conventional NK target cells.

These studies suggest that NKCF may be a subclass of conventional LT or else a related molecule. However, it appears that the conventional LT assay will detect lytic molecules with differences in MW and functional characteristics from NKCF. Therefore, the use of purified material and a standardized assay system is required to clarify the possible relationship between NKCF and LT.

Another mechanism which has been proposed for NK CMC involves the release of pore-forming molecules from the effector cell which then polymerize and form pores in the target membrane resulting in cell lysis, analogous to the complement system (31-34). It is possible that this mechanism may be employed by one population of NK cells whereas another population may release NKCF. Alternatively, NKCF may be derived from the NK cytoplasmic granules and may be the subunits which form ring structures on the target cell membrane. We have demonstrated that RNK tumor cells produce NKCF (13), and both ring structures and soluble cytolysins have been recovered from purified cytoplasmic granules from the RNK tumor cell line (32,33). Both NKCF and ring structures can be induced under the same conditions, i.e., by stimulating with NK targets or with ConA (31). Tubules have been isolated and SDS gel analysis has revealed several proteins (34) within the MW range of NKCF, as is also true for several proteins found in RNK granules (32). Furthermore, antisera prepared against RNK granules has been shown to inhibit the lytic activity of NKCF (33). One major difference between RNK granules and NKCF is that the granule cytolysin lyses targets much more rapidly (within 1 h). It is possible that whole supernatants containing NKCF are too dilute and/or may contain inhibitors preventing rapid target lysis. Semi-purified preparations of NKCF exhibit more potent lytic activity and a faster rate of lysis than whole supernatants. The granule cytolysin and NKCF may also differ in that the cytolysin is not NK specific. However, it has been shown that, with the exception of red blood cells, the order of target sensitivity to the cytolysin is the same as that for NK sensitivity (33).

It has been reported that RNK granules in the presence of calcium rapidly lose cytotoxic activity (33). Since NKCF is produced in the presence of calcium, it could be that these factors are break-down products of the granule-derived cytolysins.

Another possibility is that NKCF are different from pore-forming molecules but interact with them to lyse the target. Thus, it is possible that perforins produce a pore in the target membrane to facilitate the rapid entry of NKCF into the cytoplasm where it inflicts lethal damage.

Further work is necessary to analyze the possible relationships between NKCF, lymphotoxin, granule cytolysins, and ring structures. The initial results of our biochemical characterization of NKCF provide the basis to examine this question and to further analyze the role of NKCF in the NK lytic mechanism.

ABBREVIATIONS

ATEE, acetyl-L-tyrosine ethyl ester; IFN, interferon; LT, lymphotoxin; NK, natural killer; NKCF, natural killer cytotoxic factors; TLCK, tosyllyslchloromethylketone; TPCK, toxylphenylalanylchloromethylketone.

ACKNOWLEDGEMENTS

The authors thank Michael Brunda (Hoffman-La Roche) for the recombinant IFN and Craig Reynolds for the RNK tumor cells.

REFERENCES

1. Roder, J. C., Kiessling, R., Biberfield, P., and Anderson, B. (1978). J. Immunol. 121, 2509.
2. Quan, D. C., Ishizaka, T., and Bloom, B. R. (1982). J. Immunol. 128, 1786.
3. Wright, S. C., and Bonavida, B. (1981). J. Immunol. 126, 1516.
4. Wright, S. C., and Bonavida, B. (1983). J. Immunol. 130, 2965.
5. Wright, S. C., and Bonavida, B. (1983). Proc. Natl. Acad. Sci. USA 80, 1688.
6. Chen, T. P. (1977). Exp. Cell. Res. 104, 255.
7. McGarrity, G. J., and Carson, D. A. (1982). Exp. Cell. Res. 139, 199.
8. Wright, S. C., and Bonavida, B. (1984). J. Immunol., in press.
9. Wright, S. C., and Bonavida, B. (1982). J. Immunol. 129, 433.
10. Wright, S. C., Weitzen, M. L., Kahle, R., Granger, G. A., and Bonavida, B. (1983). J. Immunol. 130, 2479.
11. Timonen, T., Ortaldo, J. R., and Herberman, R. B. (1981). J. Exp. Med. 153, 569.
12. Reynolds, C. W., Bere, E. W., and Ward, J. M. (1984). J. Immunol. 132, 534.
13. Wright, S. C., Reynolds, C. W., and Bonavida, B. (1984). In "Natural Killer Activity and its Regulation" (T. Hoshino et al., eds.). Excerpta Media, Amsterdam.
14. Wayner, E. A., and Brooks, C. G. (1984). J. Immunol. 132, 2135.
15. Wright, S. C., and Bonavida, B. (1983). J. Immunol. 130, 2960.
16. Trinchieri, G., and Santoli, D. (1978). J. Exp. Med. 147, 1314.
17. Hudig, D., Redelman, D., and Minning, L. (1982). In "NK Cells and Other Natural Effector Cells" (R. B. Herberman, ed.) p. 923. Academic Press, New York.
18. Stutman, O., Dien, P., Wisum, R. E., and Lattime, E. C. (1980). Proc. Natl. Acad. Sci. USA 77, 2895.

19. Forbes, J. T., Bretthaer, R. K., and Oeltmann, T. N. (1981). Proc. Natl. Acad. Sci. USA 78, 5797.
20. Hickman, S., Shapiro, L. S., and Neufeld, E. F. (1979). Biochem. Biophys. Res. Commun. 57, 55.
21. Kaplan, A., Achord, D. T., and Sly, W. S. (1977). Proc. Natl. Acad. Sci. USA 74, 2026.
22. Blanca, I. R., Ortaldo, J. R., and Herberman, R. B. (1984). In "Natural Killer Activity and its Regulation" (T. Hoshino et al., eds.). Excerpta Medica, Amsterdam.
23. Hudig, D., Haverty, T., Fulcher, C., Redelman, D., and Mendelsohn, J. (1981). J. Immunol. 126, 1569.
24. Weitzen, M. L., Innis, E., Yamamoto, R. S., and Granger, G. A. (1983). Cell. Immunol. 77, 42.
25. Walker, S. M., and Lucas, Z. J. (1973). Transpl. Proc. V, 137.
26. Kondo, L. L., Rosenau, W., and Wara, D. W. (1981). J. Immunol. 126, 1131.
27. Granger, G. A., Yamamoto, R. S., Fair, D. S., and Hiserodt, J. C. (1978). Cell. Immunol. 38, 338.
28. Aggarwal, B. B., Moffat, B., and Harkins, R. N. (1984). J. Biol. Chem. 259, 686.
29. Weitzen, M. L., Yamamoto, R. S., and Granger, G. A. (1983). Cell. Immunol. 77, 30.
30. Toth, M. K., and Granger, G. A. (1979). Molec. Immunol. 16, 671.
31. Podack, E. R., and Dennert, G. (1983). Nature 302, 442.
32. Millard, P. J., Henkart, M. P., Reynolds, C. W., and Henkart, P. (1984). J. Immunol. 132, 3197.
33. Henkart, P., Henkart, M., Millard, P., and Reynolds, C. W. (1984). In "Natural Killer Activity and its Regulation" (T. Hoshino et al., eds.) p. 156. Excerpta Medica, Amsterdam.
34. Podack, E. R. (1984). In "Natural Killer Activity and its Regulation" (T. Hoshino et al. eds.) p. 101. Excerpta Medica, Amsterdam.

CHARACTERISTICS OF HUMAN NATURAL KILLER CYTOTOXIC FACTOR (NKCF) LYSIS AND POSSIBLE RELATIONSHIP TO LYSIS BY NK CELLS

John R. Ortaldo, Isaac Blanca, and Ronald B. Herberman

Biological Therapeutics Branch
Biological Response Modifiers Program
National Cancer Institute
Frederick Cancer Research Facility
Frederick, Md.

Soluble natural killer cytotoxic factors (NKCF) have been detected in the supernatants of normal mouse, rat, and human lymphocytes stimulated in vitro for 1 to 3 days in serum-free medium. Stimulation of purified large granular lymphocytes (LGL) with NK-sensitive targets or mitogens has resulted in high levels of NKCF production. Previous studies have analyzed the cells responsible for production, specificity, and general characteristics of mouse and human NKCF. In the present study, using human NKCF as a model for cytolysis by LGL, we have analyzed a variety of agents previously demonstrated to inhibit NK activity. All inhibitors were tested for their effects on production of NKCF after target cell interaction, binding of NKCF to target cells, and target cell lysis (after 6-hr NKCF absorption and washing of targets). Phosphorylated sugars and antibodies to rat LGL granules were found to inhibit the binding of NKCF to targets, whereas the other agents tested had no detectable effect (ATP, cyclic AMP, protease inhibitors, prostaglandin E_2). Monensin inhibited post-NKCF binding, indicating a late target cell effect. In regard to the production of NKCF, the data indicated that (i) the absence of calcium and magnesium, (ii) prostaglandin E_2, and (iii) ATP inhibited production, whereas phosphorylated sugars did not. Studies with these types of agents will now enable us to dissect the sites at which these agents function within the lytic process. In addition to the above studies, purification studies were

ISBN 0-12-341370-2

performed to begin biochemical characterization of human NKCF. The results indicated that radiolabeled NKCF has an apparent molecular weight between 20,000 and 40,000. This material demonstrated a pattern of binding to target cells which was similar to the pattern of lysis by NKCF. Such approaches with radiolabeled NKCF should be useful for the further study of the biochemical characteristics of human NKCF and of its mechanism of action. In addition, experiments using enzyme-treated targets indicate that proteolytic (tryptic) enzymes may be involved in generation of target cell susceptibility to NKCF.

INTRODUCTION

During the past few years, natural killer (NK) cells have attracted considerable attention because of their potential role in antitumor defenses in humans and animals.[1] Recently, the cells responsible for NK activity in rodents and humans have been shown to be large granular lymphocytes (LGL).[2,3] However, the mechanism of their lytic activity is still under active investigation. Studies of the effects of metabolic inhibitors on NK activity support the stimulation-secretion model originally postulated[4,5] to explain the mechanism of killing by CTL.[6] In addition, Wright and Bonavida[7] have described a soluble factor, which is produced during the interaction of mouse spleen cells or human peripheral blood lymphocytes[7-10] with NK-susceptible target cells. They termed this NK cytotoxic factor (NKCF). Although several lines of correlative evidence suggest a role for NKCF in mechanism of NK cell-mediated cytolysis, a variety of important issues still remain to be settled. In the present study, NKCF was utilized in an attempt to dissect the various steps that have been postulated to be involved in NK cell-mediated lysis. Inhibitors of NK and NKCF were tested for their site of action, in the generation of factor and in its interaction with the targets. Studies were performed with a variety of agents which have been previously demonstrated to inhibit LGL-mediated cytolysis, as well as to begin initiated biochemical characterization of human NKCF.

RESULTS

Characteristics of NKCF Produced by Human LGL

If NKCF is involved in the mechanism of lysis by NK cells, one would predict that the factor would be produced from purified populations of NK cells. Therefore, LGL were tested for their ability to produce NKCF. The data in Table 1 represent a typical experiment in which Percoll density gradient fractions containing a high percentage of LGL, upon incubation for 18 hr with K562 (myeoplasm-free), produced high levels of NKCF. In contrast, with highly purified T-cell populations (or populations depleted of NK cells), did not release significant levels of NKCF activity. Stimulation of LGL with NK-susceptible targets, membrane fragments of such target cells, or with mitogens resulted in a substantial release of NKCF. In contrast, low or insignificant amounts of NKCF above those spontaneously released by LGL were found when human NK-resistant target cells such as RL♂1, YAC-1, and Raji were used as stimuli. It should be noted that although stimulation of NK cells usually was necessary for obtaining high levels of NKCF, some spontaneous NKCF was produced by the LGL in 10% of donors (data not shown). Table 2 summarizes some of these properties of NKCF. In addition to these general properties, some physicochemical properties have been examined. NKCF produced in serum-free conditions generally was rather stable when maintained at low temperatures (4°C or lower). In addition, it appears to be a protein since it was inactivated by trypsin treatment and was quite unstable when maintained at high temperatures (56°C or higher).

Table 1. Release of NKCF by Percoll Fractions of Human Lymphocytes

Cell Population	% T	% LGL	% Lysis K562 by supernatant*
Unseparated lymphocytes	85	7	15
Percoll fraction			
1	4	15	20
2	10	85	62
3	45	35	27
4	85	8	14
5	95	1	9
6	95	1	5

*Factor was generated by cells stimulated with K562 (50 lymphocytes/1 K562) for 18 hr. NKCF was assayed in a 24-hr ^{51}Cr release assay using factor concentrated 4-fold.

Table 2. Properties of NKCF

General Properties	
1. Source	LGL
2. Effective stimuli for release	NK-sensitive targets and mitogens
3. Lytic activity	Slow (>10 hr) and inhibited by serum proteins
4. Specificity	Similar to that of LGL (NK cells)
Biochemical Properties	
1. Stability at	
−20°C	weeks to months (range 3-7 weeks)
4°C	days to weeks (range 14-30 days)
56°C	hours (range 0.5-1 hr)
>65°C	<10 minutes
2. Inhibition by enzymes	Sensitive to trypsin Insensitive to neuraminidase

If NKCF plays an important role in NK cell-mediated lysis, then it would be predicted that the factor would bind to the target cells and kill them in a time comparable to that required by intact NK cells.[11] Although previous studies have demonstrated that NKCF could be adsorbed,[9] no direct evidence that binding was subsequent to target lysis. In order to test this prediction, target cells were incubated with NKCF for various intervals at 37°C and the incubation was either continued in the presence of NKCF or the targets were washed and incubated with culture medium or in the presence of fresh NKCF. At various times thereafter, the level of cytotoxicity was assessed. The data shown in Fig. 1 indicate that lysis of target cells by NKCF occurred slowly, with maximal lysis requiring a period of incubation of approximately 48 hr (with ^{111}In-labeled target cells) or 24 hr (with ^{51}Cr targets) (Fig. 2). These kinetics of lysis were similar for all of the supernatants tested against ^{111}indium-labeled targets. In contrast, studies on the kinetics of absorption showed that within 2 hr, and at a maximum of between 4 and 6 hr, after incubation with NK-susceptible targets, NKCF activity was removed from supernatants. After incubation of target cells with NKCF, the cells could be washed and lysis proceed with kinetics similar to that observed when the cells were maintained in the

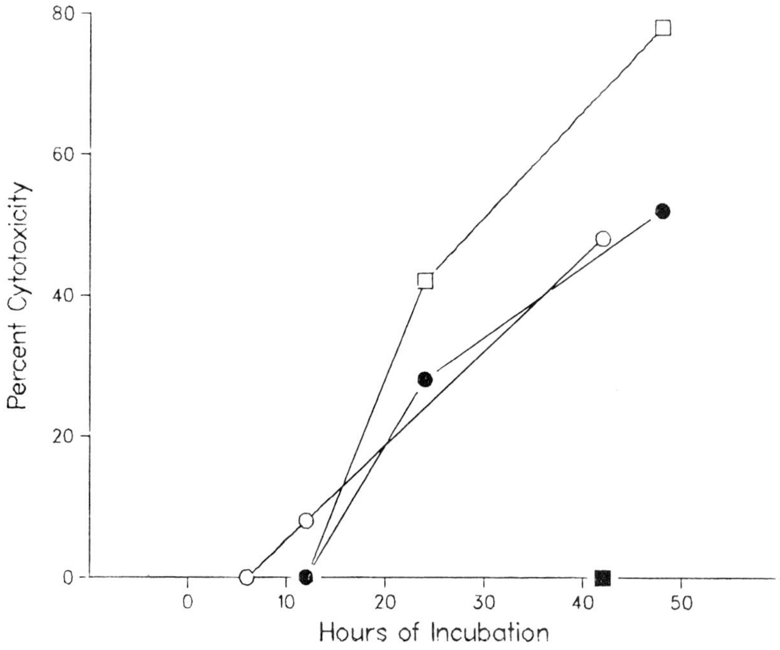

Fig. 1. Kinetics of K562 lysis by NKCF (● ^{111}In assay) when added to assay (o), when adsorbed into targets for 6 hr (●), or when adsorbed for 6 hr and additional NKCF added (□), or with adsorption-depleted factor (■).

presence of NKCF (Fig. 2). However, addition of fresh NKCF increased the percentage of lysis of target cells, indicating that maximal saturation of the target cells by NKCF was not maintained when the cells were incubated for several hours in the absence of factor. Overall, these results indicated that, an important step is the interaction of NKCF with targets, being the binding of NKCF to the targets, and then once bound, the targets go on to lyse without the need for continued exposure to NKCF. However, the possibilty of a rapid nonbinding interaction of NKCF with targets followed by inactivation rather than adsorption can only be elucidated by direct binding and elution studies.

A variety of agents were tested for their ability to block either (i) production of NKCF, when present during a 24-hr period of the effector-target cell interactions; (ii) binding of NKCF, when present during 6-hr period of incubation with target cells; or (iii) the lytic phase, by adding

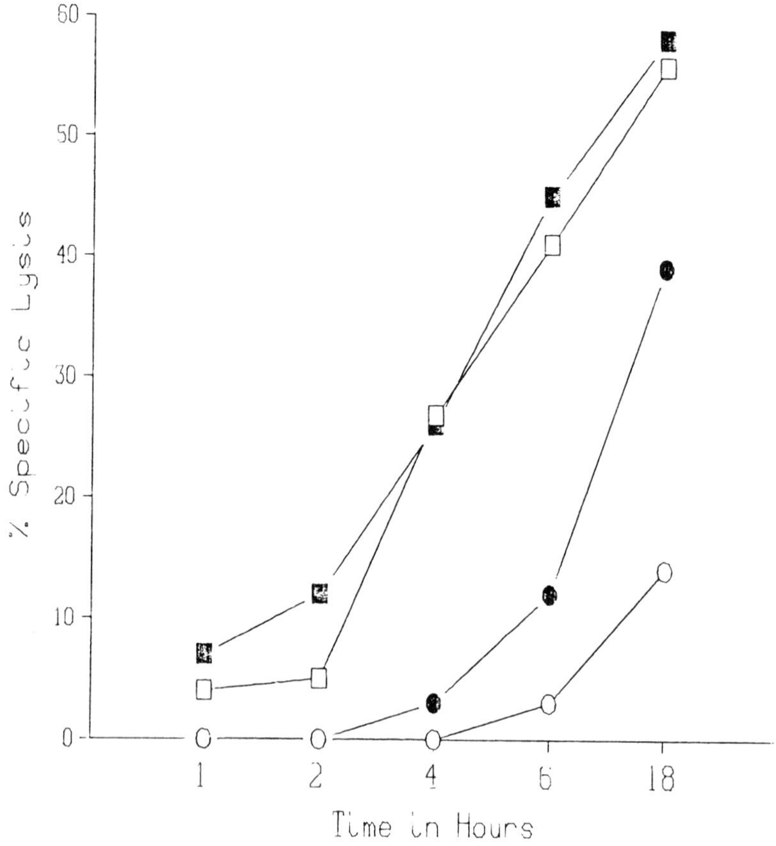

Fig. 2. Kinetics of lysis by PBL, with (■) or without (□) trypsin treatment and by NKCF, with (●) or without (○) trypsin treatment. Assay performed with ^{51}Cr-labeled K562 targets.

the agents after NKCF pretreatment of the targets (Table 3). All of the agents tested had previously been demonstrated to inhibit post-binding events involved in NK cell-mediated lysis by human LGL. When phosphorylated sugars (such as mannose-6-phosphate), were tested,[12-14] they were found to inhibit binding of NKCF to target cells. However, they had no effect on NKCF production or on lysis after absorption to targets. Other sugars previously shown not to inhibit NK lysis, such as fructose phosphate,[12-14] did not inhibit at any step involved with NKCF. Antisera to isolated granules from rat LGL tumors have previously been demonstrated to

Table 3. Effects of Agents (Which Inhibit NK Lysis) on NKCF

Agent	Concentration Active Range	Effect* on: Production	Binding	Lysis
1. Sugar-PO4				
a. Mannose-PO4	10-75 mM	–	↓	–
b. Fructose-PO4	10-75 mM	–	–	–
2. Antigranule sera	1-5 μg/ml IgG	?	↓	–↑
3. Protease inhibitors	1-100 μg/ml	–	–	–
4. CA^{++}/Mg^{++}	1-50 mM	↓	?	?
5. ATP---cAMP	1-100 mM	↓	–	–
6. PGE-2	10^{-3}-10^{-8} M	↓	–	–
7. Ammonium chloride	10-0.1 μg/ml	↓	–	–
8. Strontium chloride	10-0.1 μg/ml	↓	–	–
9. Monensin	10-0.1 μg/ml	↓	–	↓

*Summary of results of 5-10 experiments. "–" indicates no effect; "↓", consistent decrease in effect; "↑", consistent increase; "?", unable to determine due to technical reasons. Combinations indicate variable results, such as "–↑" indicates that the effect was either no change or increase.

block the lytic effect of granules (see chapter by Reynolds and Herkart). When these anti-granule antibodies were evaluated in the various steps with human NKCF, inhibition of binding was consistently seen. The effects of the antisera on production was not assessable because of the inability to remove the antibody at the end of the incubation period, as was achieved by dialysis with all the other agents tested. Interestingly, the addition of anti-rat LGL granule antibodies after absorption of NKCF resulted in variable results and often increased cell lysis. Contrary to previous reports[8] which claimed that protease inhibitors, TLCK and TPCK were found to decrease NKCF lysis, when tested at a variety of concentrations, they were not inhibitory in the present studies at any of the lytic phases examined.

Previous studies in a variety of cell-mediated cytotoxic systems have demonstrated the requirement for calcium and magnesium during the early events leading to lysis.[6] Similarly, the presence of calcium and magnesium appear to be necessary for the production of NKCF, as demonstrated by the results of experiments performed in calcium- and magnesium-free media. However, the effects of such media on the binding and subsequent lytic phases were technically not possible to assess because of the susceptibility of the target cells to the toxic effects resulting from the absence of calcium and magnesium for extended time periods.

A wide variety of other agents which have been demonstrated to inhibit postbinding events,[1] including cyclic nucleotides, ATP, prostaglandin E_2, ammonium chloride, strontium chloride, and monensin, were tested, and all were found to inhibit the production of NKCF but not the binding of NKCF to susceptible targets. None of these agents, with the exception of monensin, affected lysis of NK-susceptible targets after they were pretreated with NKCF. It is interesting that monensin did interfere with the subsequent lysis of NK-sensitive targets by NKCF. Little is known at present regarding what occurs in the target cell after binding of the soluble cytotoxic factors. However, the internal processes of the target cells following exposure to cytotoxic molecules appear to warrant a great deal of investigation. The consistent inhibition of NKCF action by monensin[19] suggests that receptor internalization or activation of lysosomes or some cytoskeletal changes may be involved in target cell lysis by NKCF.

Effect of Tunicamycin on Lysis by NKCF

Because phosphorylated sugars inhibit absorption of NKCF to susceptible targets, these sugars may compete with NKCF for binding to the receptors for glycosylated materials on target cells. Alternatively, NKCF might act in a lectin-like manner, reacting with phosphorylated sugars on the surface of target cells. To test this latter hypothesis, we studied the effects of NKCF on target cells pretreated with tunicamycin, which selectively inhibits the addition of asparagine-linked sugars to cell membranes.[15-18] NK-susceptible K562 target cells were pretreated for 18 hr with different doses of tunicamycin and then evaluated for tritiated glucosamine incorporation, for susceptibility to lysis by NK cells or NKCF, and for ability to absorb NKCF. The data demonstrated that treatment with tunicamycin, at levels which inhibited glycosylation but did not significantly inhibit protein synthesis, produced a significant decrease in susceptibility of K562 by lysis by NKCF and an impairment of the ability of these cells to absorb the factor (not shown). In addition, a significant decrease in the susceptibility of these cells to be lysed by intact LGL was also observed.

Combined Effects of NKCF and Proteolytic Enzyme

Whereas protease inhibitors had no detectable effect on any phase of NKCF production or activity, they have been previously reported and confirmed here in parallel experiments to inhibit lysis by intact LGL. In an attempt to reconcile these results, we examined the effect of pretreatment of target cells with various proteolytic enzymes. Fig. 2 demonstrates the typical kinetics of lysis by intact PBL or by concentrated crude NKCF, on target cells pretreated with 0.01% of crystallized trypsin. There were small but significant early increases in susceptibility to lysis by PBL when trypsin-treated targets were used. However, by 4 hr, no significant difference in lysis was seen. In contrast, with NKCF, which ordinarily induces significant lysis only after 12 hr, the trypsin-pretreated targets were lysed more rapidly and to a greater extent. Significantly greater levels of lysis were observed at both 6 and 18 hr. It should be noted that no significant toxicity was observed when low doses of trypsin were used (<0.05%). This enzyme-enhancing effect was restricted to trypsin, whereas other enzymes tested, including chymotrypsin, lipase, neuraminidase, pronase, papain, hyaluronidase, did not result in increased lysis and often resulted in a depression or abrogation of the ability of NKCF to result in target lysis (presumably due to modification of some surface structures required for interaction with NKCF or NK cells). Based on these results, we postulate that in addition to an NKCF, a tryptic protease released by the LGL is a critical element for rapid cell-mediated lysis.

Initial Biochemical Characterization of Purified NKCF

In an attempt to determine the molecular weight of NKCF and to begin purification efforts of NKCF, HPLC separation of serum-free factor was begun. These fractions were prepared using high-performance liquid chromatography at neutral pH on a molecular sizing column. Fig. 3 demonstrates a typical profile of protein, based on size separation, as well as relative cytotoxic activity, measured by dilution of cytotoxic activity, measured by dilution of purified fractions of NKCF. As consistently observed in several experiments, NKCF activity was observed in the molecular weight range between 18,000 and 40,000, with peaks of activity usually seen at approximately 20,000 and 40,000 daltons. These results suggest the possibility of polymerization of a 20,000 m.w. subunit. In addition, NKCF activity was consistently found in the void volume (i.e., far left of Fig. 3), indicating some high molecular weight (>300,000 daltons) aggregates of NKCF. This purification methodology provides an approach for further purification of NKCF and use of radiolabeled factors (data not shown) for subsequent binding studies and/or further analysis of the mode of action of NKCF. Preliminary studies with partially purified NKCF indicate that an accelerated level of lysis, detected as early as 6 hr (data not shown) was seen. These results support the hypothesis that one reason for the slow rate of lysis by NKCF may be the usual low concentration of the factor. Therefore, more highly purified and concentrated factors may minimize the differences that exist between the kinetics of lysis by intact NK cells versus NKCF.

DISCUSSION

Since the original descriptions of human and mouse NKCF,[7-11] several lines of evidence have suggested that this factor(s) might be involved in the mechanism of lysis by NK cells. Studies by Wright et al.[8] and Farram et al.[10] have indicated that NKCF is released during coculture of NK-sensitive target cells with NK cells. Although such evidence is consistent with the hypothesis that NKCF is involved in lysis by human NK cells, little or no information about the actual mechanism of lysis by human NKCF or its relationship to the mechanism of lysis by NK cells has been reported. In

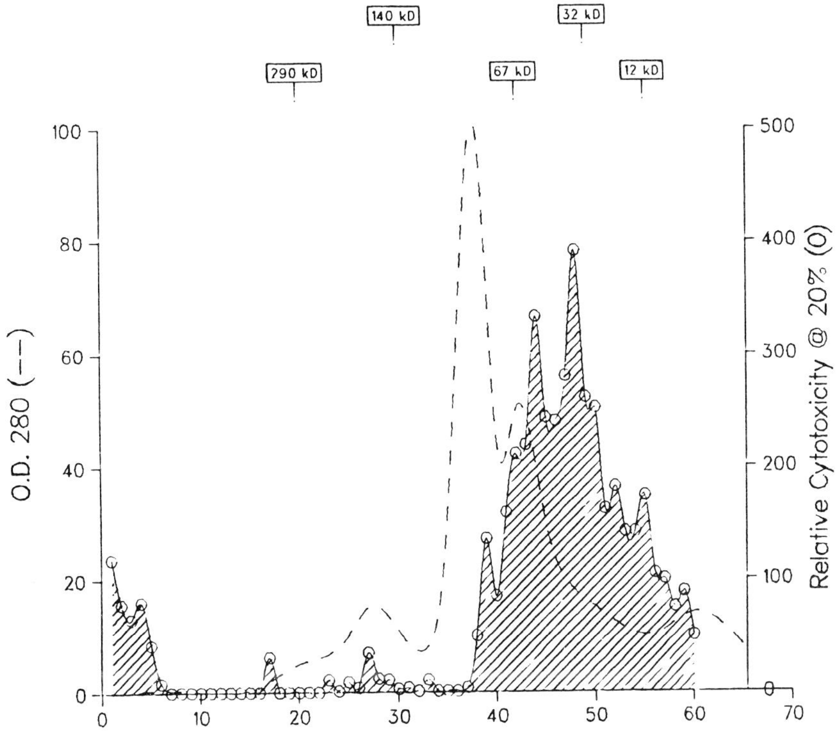

Fig. 3. Purification of NKCF from LGL was performed using HPLC molecular sizing. Protein profile (o) and NKCF activity (Δ) are shown. Typical experiment of 5 similar experiments. The LU 20% is the lytic unit equivalent, with the NKCF being diluted, and the lytic efficacy was calculated as the dilution resulting in 20% lysis.

the present study, using inhibitors which have been demonstrated previously to inhibit NK lysis, the possible relationships of NKCF and human NK cell-mediated cytotoxicity have been examined.

First, if NKCF is the mediator of lytic activity by human NK cells, it would be predicted that it would be produced and released by NK cells in the absence of other cell types. Our experiments certainly confirm previous reports that highly purified LGL are responsible for production of NKCF as well as for NK activity.[10]

Secondly, release of NKCF should be selectively triggered by NK-susceptible targets. Our results have shown that production of NKCF was elicited mainly by NK-susceptible target cells, with NK-insusceptible targets inducing little or no NKCF production. Thus, it is possible that the recognition receptors leading to release of NKCF and to binding of NK cells to susceptible target cells may be very similar or identical.

Studies have indicated that carbohydrate moieties are involved in some post-binding events of cell-mediated lysis.[12-14] The present results are consistent with the hypothesis that inhibition of NK activity by phosphorylated sugars occurs by blocking the uptake of the lytic factor (NKCF). In addition, the pattern of inhibition of NKCF binding by various sugars was similar to that seen in studies of inhibition of lysis by intact LGL, e.g., mannose-6-phosphate inhibited whereas nonphosphorylated mannose did not. In addition, phosphorylated fructose demonstrated no significant inhibition of either NK activity or of lysis by NKCF. These parallel patterns of inhibition by sugars for NK-mediated lysis and for NKCF are consistent with the hypothesis that NKCF is involved in the mechanism of lysis by NK cells. The susceptibility of target cells to lysis may depend, however, on several factors, including the expression of both a specific NKCF binding site and a sugar-related site. This hypothesis is based on the fact that NKCF demonstrates considerable specificity, with lysis of NK-susceptible targets and little or no activity on NK-insusceptible targets. The concept that a sugar-related site is involved in NKCF binding was also supported by the observation that tunicamycin-treated target cells, whose glycosylation of surface lipids or proteins was blocked, demonstrate deficient binding or subsequent lysis by NKCF. These results indicate that glycosylated materials on the surface of the target are somehow required for lysis by NKCF. Overall our data support the contention that although there are parallels between the specificity of NK and NKCF, there are target receptors for binding of NKCF which were distinct from the target structure recognized by NK cells.

Wright and Bonavida have previously demonstrated loss of NKCF activity upon incubation with NK-susceptible targets.[7-9] Our data support the contention that this is due to absorption rather than inactivation of the factor since this interaction leads to subsequent lysis. Our results also demonstrated that binding occurs gradually, with optimal binding of the preparations tested occurring by 6 hr. After this point, target cells were already programmed for subsequent lysis, without the need for continued presence of factor, which occurred with the same kinetics as those seen in the continued presence of NKCF. In contrast to the rapid killing seen by NK cells, which is detected in short-term chromium-release assays, high levels of NKCF lysis required 48 hr with ^{111}indium-labeled targets or 24 hr with ^{51}chromium-labeled targets. These results are consistent with previous results obtained in both the mouse and human[7-11]. One likely explanation for this slow cytotoxicity is that the rate of lysis of targets is largely controlled by the concentration of available factor and by the increase in target susceptibility by proteases released from, or present on the surface of, NK cells.

Fig. 4 demonstrates our current model for involvement of NKCF in mediation of lysis by natural killer cells. Because of the abrogation of cytolysis by both NK cells and NKCF by rabbit antibodies against LGL granules, our present hypothesis is that NKCF is derived from LGL granules and is released from the effector cell after initial contact with the NK target. This NKCF will bind directly to NK targets. This is supported by absorption data by Wright and Bonavida[11] as well as absorption data presented here. We postulate that this NKCF molecule may contain more than one binding site as indicated by (1) the specificity of NKCF for only lysing NK susceptible targets, and (2) by the ability of phosphorylated sugars to block the binding of NKCF to susceptible targets, which do not abrogate the binding of intact LGL to specific NK-susceptible targets. The possibility that a complex exists between the granular cytolysin and the NK as recognition structure, would be consistent with the results. Until recently one of the most disturbing differences between NKCF and lysis mediated by NK cells was related to the rates of lysis. However, our recent experiments with partially purified NKCF and with target cells pretreated with trypsin, tend to narrow the differences in kinetics. Our results were consistent with reports that tryptic proteases are involved in late KCIL (Targan et al. in this volume). We hypothesize that NKCF may act in synergy with a secreted tryptic protease, to elicit maximal cell lysis. In addition, concentrations of soluble material acheived at cellular junctions might be expected to exceed those which are capable of being attained in culture supernatants. Therefore, further purification and concentration of factor may be necessary in order to attain concentrations which begin to approach those which are isolated from LGL granules or present at cell-to-cell junctions. Consistent also with this hypothesis is the fact that serum rapidly inhibits the action of these soluble

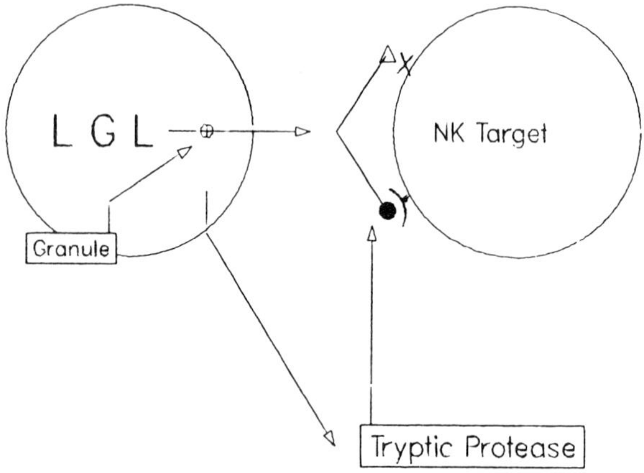

Fig. 4. A model for the role of NKCF in lysis by NK cells.

factors, therefore eliminating and minimizing side or bystander reactions which might occur due to release of soluble factors into the surrounding millieu.

The events which occur on and within the targets after initial interaction with NKCF still remain obscure. However, the experiments performed here may provide new insights to some of the interactions with NKCF and targets. The experiments with monensin support the active role of targets in their own destruction, and provide direction for studies of target cell requirements for lytic reactions to occur. However, the relationship between the mechanisms of lysis of targets by NKCF and by NK cells requires further investigation before it can be concluded that NKCF is definitely involved in the lysis by NK cells.

The possibility of producing NKCF under serum-free conditions and regularly obtaining large amounts of activity should facilitate its purification and biochemical characterization. The degree of purification achieved already by HPLC (approximately 10-fold) is consistent with the activity peak being separate from protein peak (Fig. 3). The results presented in the present studies provide information regarding the molecular size (20-40,000 m.w., with apparently a 20,000 m.w. subunit) and potentially useful methodologies for further purification and examination of binding, using radiolabeled NKCF. All of these collectively should provide information for further comparative dissection of the mechanism of lysis by NK cells and NKCF.

REFERENCES

1. R. B. Herberman (Ed.), "NK Cells and Other Natural Effector Cells," Academic Press, New York (1982).
2. T. Timonen, J. R. Ortaldo, and R. B. Herberman, Characteristics of human large granular lymphocytes and relationship to natural-killer and K cells, J. Exp. Med. 153:569 (1981).
3. C. W. Reynolds, T. Timonen, and R. B. Herberman, Natural killer (NK) cėll activity in the rat. I. Isolation and characterization of the effector cells, J. Immunol. 127:282 (1981).
4. J. C. Roder and T. Haliotis, A comparative analysis of the NK cytolytic mechanism and regulatory genes, in: "Natural Cell-Mediated Immunity Against Tumors," R. B. Herberman, ed., Academic Press, New York (1980).
5. P. C. Quan, T. Ishizaka, and B. R. Bloom, Studies on the mechanism of NK cell lysis, J. Immunol. 128:1786 (1982).
6. C. S. Henney, On the mechanism of T-cell mediated cytolysis, Transplant. Rev. 17:37 (1973).
7. S. C. Wright and B. Bonavida, Selective lysis of NK-sensitive target cells by a soluble mediator released from murine spleen cells and human peripheral blood lymphocytes, J. Immunol. 126:1516 (1981).
8. S. C. Wright and B. Bonavida, Studies on the mechanism of natural killer (NK) cell-mediated cytotoxicity (CMC). I. Release of cytotoxic factors specific for NK-sensitive target cells (NKCF) during coculture of NK effector cells with NK target cells, J. Immunol. 129:433 (1982).
9. S. C. Wright, M. L. Weitzen, R. Kahle, G. A. Granger, and B. Bonavida, Studies on the mechanism of natural killer cytotoxicity. II. Coculture of human PBL with NK sensitive or resistant cell lines stimulates release of natural killer cytotoxic factor (NKCF) selectively cytotoxic to NK-sensitive target cells, J. Immunol. 130:2479 (1983).
10. E. Farram and S. R. Targan, Identification of human natural killer soluble cytotoxic factor(s) NKCF derived from NK-enriched lymphocyte populations: specificity of generation and killing, J. Immunol. 130:1252 (1983).
11. S. Wright and B. Bonavida, YAC1 variant clones selected for resistance to NKCF are also resistant to natural killer cell-mediated cytotoxicity. Proc. Natl. Acad. Sci. USA 80(6):1688 (1983).
12. J. T. Forbes and T. N. Oeltmann, Carbohydrate receptors in natural cell-mediated cytotoxicity, in: "NK Cells and Other Natural Effector Cells," R. B. Herberman, ed., Academic Press, New York (1982).

13. J. R. Ortaldo, T. T. Timonen, and R. B. Herberman, Inhibition of activity of human NK and K cells by simple sugars: Discrimination between binding and post-binding events, Clin. Immunol. Immunopathol. 31(3):439 (1984).
14. O. Stutman, P. Dien, R. E. Wisum, and E. C. Lattime, Natural cytotoxic cells against solid tumors in mice: blocking of cytotoxicity by D-mannose, Proc. Natl. Acad. Sci. USA 77:2895 (1980).
15. J. S. Tkacz and J. O. Lampen, Tunicamycin inhibition of polyisoprenyl-N-acetylglucosamyl pyrophosphate formation in calf liver microsomes, Biochem. Biophys. Res. Commun. 65:248 (1975).
16. P. K. Keller, D. Y. Boon, and F. C. Crum, N-acetyl glucosamine-2-phosphate transferase from hen oviduct: Solubilization, characterization, and inhibition by tunicamycin, Biochemistry 18:3946 (1979).
17. G. W. Hart, The role of asparagine-linked oligosaccharides in cellular recognition by thymic lymphocytes. Effects of tunicamycin on the mixed lymphocyte reaction, J. Biol. Chem. 257:151 (1982).
18. J. A. Werkmeister, J. C. Roder, C. Curry, and H. F. Pross, The effect of unphosphorylated and phosphorylated sugar moieties on human and mouse natural killer cell activity: is there selective inhibition at the level of target recognition and lytic acceptor site?, Cell. Immunol. 80:172 (1983).
19. E. Saksela, O. Carpén, and I. Virtanen, Cellular secretion associated with human natural killer cell activity, in: "NK Cells and Other Natural Effector Cells," R. B. Herberman, ed., Academic Press, New York (1982).

Role of Soluble Lytic Factors in NK Activity

William H. Chambers
James T. Forbes
Thomas N. Oeltmann

Division of Oncology
Department of Medicine
Vanderbilt University
Nashville, Tennessee

Natural cell-mediated cytotoxicity (NCMC) is thought to be important as a T-cell independent mechanism for immune surveillance against certain tumors and virus-infected cells (1-4). Natural killer (NK) cells carry out this function without prior immunization and are therefore believed to perhaps represent a first line of defense in the immune system (5,6). NK cells also recognize and lyse certain normal cells including subpopulations of thymocytes, macrophages and bone marrow cells (1). The reactivity against bone marrow cells may account for part of the natural resistance to engraftment, either allogeneic or xenogeneic, in bone marrow.

NK cells develop in the fetal liver of man as early as nine weeks of gestation (7) and have recently been described morphologically as large, granular lymphocytes (LGL) (8). In addition, NK cells are nonadherent, nonphagocytic and bear Fc receptors for IgG (9). NK cell activity can not be separated from LGLs by several experimental techniques including differential adherence to target cells (8), and sedimentation in Percoll density gradients (10). It appears likely that most, if not all, NK cells belong to the population of LGLs (8).

At present, NK cell activity is arbitrarily divided into 4 or 5 discrete steps (Figure 1). These include: 1) recognition and binding to the target cell; 2) a triggering step which induces production or release of a lytic factor (NKCF); 3) NKCF release by NK Cell; 4) NKCF binding to the target cell and lysis or 5) bound NKCF is internalized by the target cell and causes lysis (11 and our observations). Our main research interest has been related to the post-recognition steps or, more specifically, the release and binding of a lytic factor from NK cells that leads to target cell lysis.

Although various types of NCMC have been extensively studied in recent years, the molecular mechanism(s) of target cell lysis is unknown. One possibility is that lysis occurs after the transfer of a cytotoxic molecule from the NK cell to the target cell. However, it has been extremely difficult

ISBN 0-12-341370-2

Figure 1.

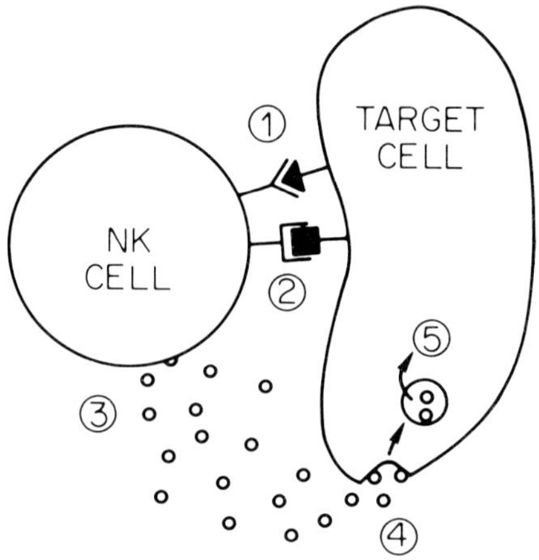

1. NK cell recognizes and binds Target Cell
2. Binding stimulates release of NKCF ($1 \stackrel{?}{=} 2$)
3. NKCF is released by NK Cell
4. NKCF binds to Target Cell and causes Lysis, or
5. Bound NKCF is taken up by Target Cell and causes Lysis

to obtain direct evidence for the involvement of such a molecule. In another system, it has been shown that target cell lysis by antigen-sensitized cytotoxic T-lymphocytes required cell-cell contact, and soluble mediators could not be demonstrated (12,13). These findings do not, however, exclude the involvement of a cytotoxic mediator in NCMC. In fact, recent studies by Wright and Bonavida (14) and Saksela et al (15) have demonstrated that a soluble mediator is likely involved with selective NK lysis of target cells.

A stimulus:secretion model for NK activity has been proposed by Roder et al (16,17) and is based on the finding that target cell binding and target cell lysis are separable phenomena. In this model, effector cell-target cell contact triggers secretion of the lethal substance(s) which then bind to the target cell, ending in its lysis (18,19). The recent studies of Saksela et al (15) have clearly demonstrated an

NK cell secretion associated with NCMC. Their studies demonstrated that in NK cell-target cell conjugates, an orientation of the NK cell Golgi apparatus occurred towards the contact surface and that inhibition of the fusion of the Golgi derived vacuoles with the NK cell membrane prevented lysis. The nature of the secretory substance(s) released has not been characterized.

Many investigations indicate an important role for secretion in the cytotoxic pathway. Treatment of human LGLs with $SrCl_2$ led to degranulation, with a concomitant loss of cytotoxic activity (20,21). Monensin, a carboxylic ionophore, which blocks secretion by LGL, also inhibited cytotoxicity (19). Compounds that block lysosomal secretions in other cells, also blocked cytotoxicity of NK cells (cAMP, PGE, theophylline, histamine) (22). Similarly, compounds that increase lysosomal discharge enhanced NK cell activity (cGMP, carbamylcholine) (22). It has also been shown that chloroquine and NH_4Cl, selective inhibitors of lysosomes, caused a decrease in NK cytotoxic activity (17,23). Additionally, secretion is energy dependent and DNP or CCCP, which uncouple oxidative phosphorylation, blocked NK cell activity and secretion (16,19).

Recently, much attention has been focused on a role for carbohydrate receptors in natural cell-mediated cytotoxicity. Many investigators have reported that certain hexose phosphates and some neutral sugars inhibit NK cell cytolysis (23, 24,25,26). There are conflicting reports concerning the specificity involved in this inhibition, but there is general agreement that mannose 6-phosphate is capable of mediating this effect. Further, it has been demonstrated that hexose phosphates do not interfere with target cell-effector cell conjugation (24,25, 26). This suggests that a cytolytic factor, produced by NK cells, may contain a hexose 6-phosphate residue and may be taken up by target cells through a receptor for hexose 6-phosphate.

The presence of a functional mannose 6-phosphate receptor on K-562 and F-265 target cells has been demonstrated by toxic effects of gelonin (a seed protein which inactivates the eukaryotic ribosome but is nontoxic to intact cells) covalently linked to monophosphopentamannose by reductive amination. This conjugate (gel-M6P) was toxic to intact target cells and this toxicity was inhibited by mannose 6-phosphate (23). In addition, we have directly demonstrated the presence of a mannose 6-phosphate receptor on target cells. F265 or K562 cells were cultured (24 hrs) in the presence of high specific activity [^{35}S] methionine and their plasma membranes subsequently isolated. These membranes were solubilized, applied to a mannose 6-phosphate Sepharose 4B

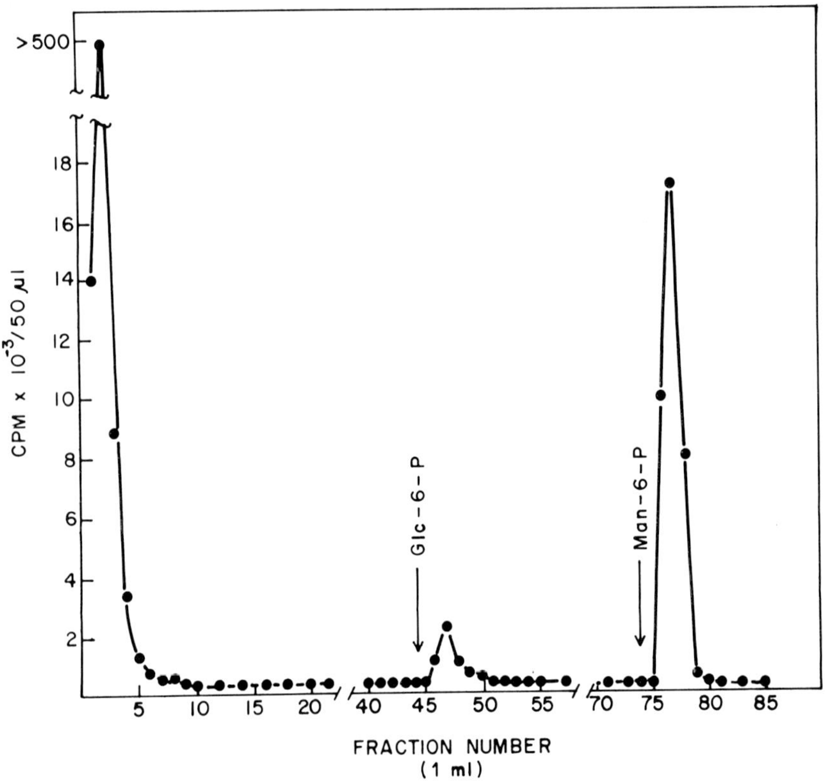

Figure 2. Elution profile of solubilized, [^{35}S] methonine labeled, K-562 plasma membranes applied to a Mannose 6-phosphate affinity column equilibrated with 10 mM phosphate , pH 7.2, containing 0.1% Triton X-100.

affinity column and washed extensively until only background radioactivity was observed. The column was then eluted first with 20 mM glucose 6-phosphate and then 20 mM mannose 6-phosphate. As seen in Figure 2, for K562 cells, only mannose 6-phosphate eluted a significant amount of radiolabled proteins. Subsequent analysis revealed the presence of three [^{35}S] methionine-labeled proteins with molecular weights of 25,000, 90,000 and 200,000 daltons. This data strengthens the proposal for the involvement of a mannose 6-phosphate receptor in NK cell-mediated cytolysis.

Clones of NK sensitive and NK resistant YAC lymphoma target cells (27) were tested for their resistance to the seed toxin gelonin-monophosphopentamannose conjugates as described above. These cells were provided by Dr. John Roder and their important properties are described in Table I (27, and our unpublished data).

TABLE I

Target Cell	NK Cell-Target Cell Conjugate Formation (% of total cells)	Susceptibility to Lysis by Poly I-C boosted NK Cells[a] without	with	Susceptibility to gelonin - M6P[b]
YAC 1 (parent)	20 ± 3	100 ± 12	1000 ± 97	53%
YAC 19	22 ± 1	119 ± 5	2050 ± 163	48%
YAC 28	24 ± 3	11 ± 2	230 ± 41	100%
P815 (non target control)	4 ± 2	ND	ND	ND

(a) Expressed by lytic units
(b) Expressed as % of control (non-treated cells) amino acid incorporation during a 1 hour pulse following a 24-hour incubation in the presence of 10^{-6} M gelonin-M6P. There was a dose dependence when tested from 10^{-4} to 10^{-10} M toxin.

As can be seen from these results, YAC 19 and YAC 28 form target cell-effector cell conjugates to the same extent. However, YAC 28 is not a lysable target cell, even when the NK cells have been boosted by an interferon inducer. Also, YAC 28 is resistant to the toxic effects of gelonin-M6P whereas YAC 19 is sensitive in a dose dependent manner. Since mannose 6-phosphate inhibits the toxicity of the gel-M6P (23), YAC 28 may be defective, or altered, in the mannnose 6-phosphate receptor. Modification or loss of this receptor may also decrease YAC 28 susceptibility to lysis by NK cells and this, in fact, is observed. YAC 28 is a target cell that binds, or is recognized, by NK cells, but is not lysed, and thus may be a powerful tool for the isolation of the soluble lytic factor.

In addition to inhibiton of NK cytolysis by mannose 6-phosphate, we have demonstrated (23) that the lysosomal inhibitors, NH_4^+ and chloroquine, block human NCMC as has been reported by Roder _et al_ (17) for the mouse system.

Because mannose 6-phosphate is involved in the targeting of certain molecules to the lysosome (28,29) and because chloroquine inhibits lysosomal function and receptor recycling, mannose 6-phosphate may prevent the direction of the lytic molecule into the target cell. The observations that a receptor on the surface of human fibroblast is specific for D-mannose 6-phosphate and is responsible for the pinocytosis of ß-glucuronidase (28) and α-L-iduronidase (28) add support for such a mechanism. It has been recently suggested (30) that mannose 6-phosphate represents a common recognition marker for intracellular transport of many lysosomal acid hydrolases. It is possible that such a receptor could represent a means of recognition between two cells such as an NK cell and a target cell, or could represent the mechanism of killing in NCMC. High uptake acid hydrolase (purified spleen ß-glucuronidase provided by Dr. William Sly) was also effective in blocking human NCMC against K-562 (24). Further, polyphosphomonoester fragments from the phosphomannans secreted by the yeast H. holstii also inhibit cytotoxicity of K-562 by human NK cells (24). The inhibiton of NCMC by mannose 6-phosphate may involve the local inhibition of uptake of a lytic molecule which contains hexose phosphate residues. It must be emphasized that while our initial experiments suggest that mannose 6-phosphate is present on a lytic molecule of NK cell origin which is released as a result of NK cell-target cell interaction, our data can not exclude the possibility that the receptor for the lytic molecule is different from the mannose 6-phosphate receptor reported for fibroblasts.

Using modifications of the methods of Wright and Bonavida (14), we have prepared [^{35}S]-labeled lytic factor(s) that both bind to, and cause lysis of target cells. This activity can be inhibited by several hexose phosphates, particularly mannose 6-phosphate. These data suggest that simple hexose phosphates inhibit at a receptor site which mediates the uptake of the NK lytic factor and which is different from the recognition and binding site(s).

Gel filtration of crude lytic supernatants indicated that the lytic activity co-fractionates with proteins of molecular weight 50-80,000 daltons (Figure 3). Fractions demonstrating lytic activity from several filtrations were pooled and concentrated. This pooled NKCF activity was applied to a second Sephadex G-200 column and fractions were assayed in 24 and 48 hour cytotoxicity assays. The results are shown in Figure 4. The lytic activity appeared to be stable and again to co-fractionate with proteins of molecular weight 50-80,000 daltons.

The working model, as illustrated in Figure 1, is consistent with most observations on NCMC. Thus, in Step 1, NK

Figure 3.

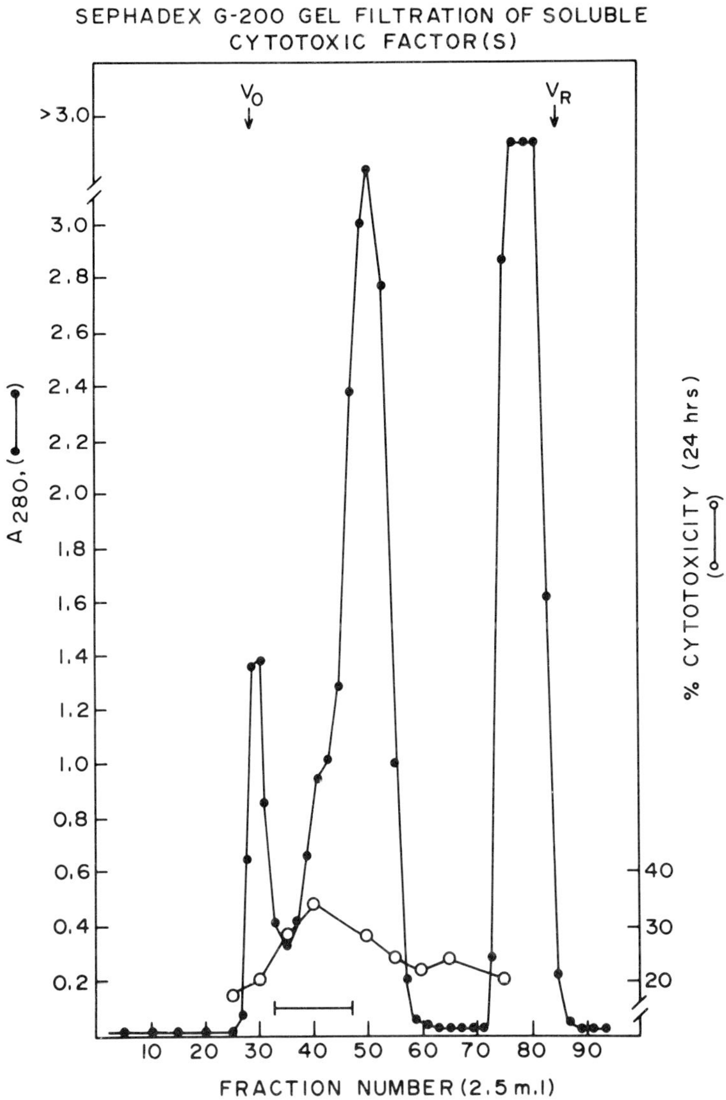
SEPHADEX G-200 GEL FILTRATION OF SOLUBLE CYTOTOXIC FACTOR(S)
V_O
V_R
>3.0
3.0
2.8
2.6
2.4
2.2
2.0
1.8
1.6
1.4
1.2
1.0
0.8
0.6
0.4
0.2
A 280, (●—●)
% CYTOTOXICITY (24 hrs) (○—○)
40
30
20
10 20 30 40 50 60 70 80 90
FRACTION NUMBER (2.5 m.l)

Figure 4.

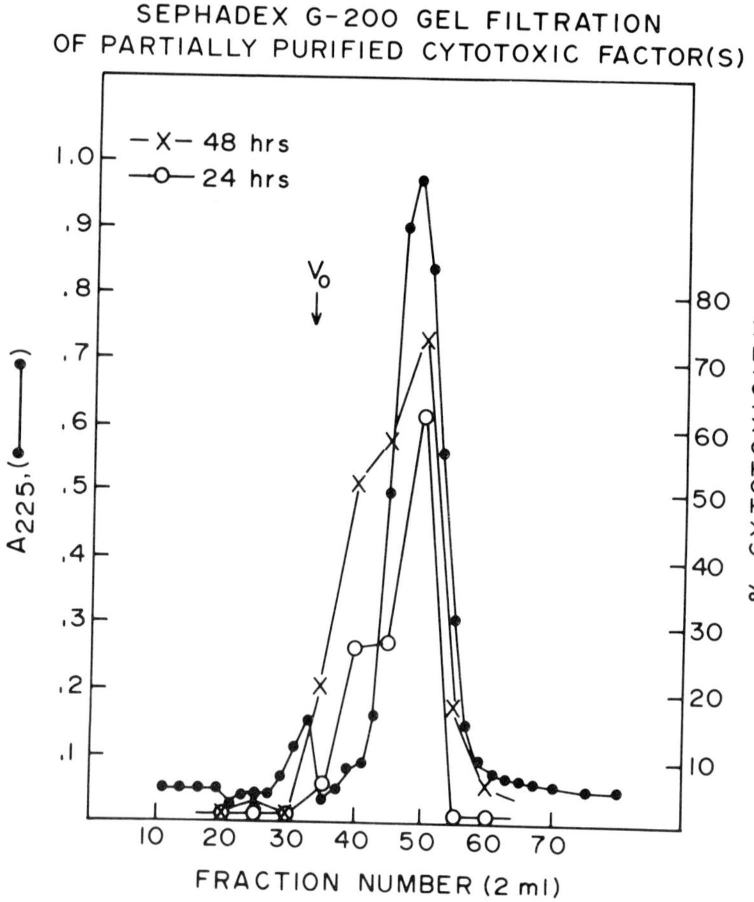

cell-target cell recognition and binding is dependent on cell surface proteins and/or glycoproteins and/or glycolipids. Some cells are not target cells because they are not recognized by NK cells. In addition, NK cell-target cell conjugate formation is necessary but not sufficient for lysis to occur. The binding of target cells by NK cells may, however, be sufficient for triggering of NKCF production/release (Step 2). In Step 3, NKCF is released by NK cells, possibly through degranulation. The binding of NKCF to the target cell (Step 4), can be inhibited by phosphorylated hexoses, phosphomannans, and high uptake phosphorylated lysosomal enzymes. Thus, this step may involve the transfer of a hexose-phosphate (mannose 6-phosphate ?) containing molecule from the NK cell to the target cell via the target cell hexose-phosphate receptor and lead to lysis. Another possibility (Step 5) is that the bound lytic molecule is transferred from the plasma membrane to the lysosome. This step may be inhibited by agents which raise the lysosomal pH or inhibit receptor recycling (NH_4Cl and chloroquine) or inactivate specific lysosomal proteases (chymostatin but not leupeptin) as well as inhibitors of energy production, both glycolysis and oxidative phosphorylation. In addition to the energy requirements of this step, results suggest that the lytic molecule could be presented to the target cell in an inactive zymogen form and require specific processing to become active (by Cathepsin D?). This also suggests the need of the low lysosomal pH for the escape of the lytic molecule to the cytoplasm similar to the recent results reported by Sandvig and Olsnes for diphtheria toxin (31). In this last step, the lytic molecule, now present in the cytoplasm, exerts its effects in some as yet undetermined mechanism which results in the ultimate lysis of the cell.

Clearly, our understanding of the molecular mechanism(s) in natural cell-mediated cytotoxicity is incomplete. However, a body of evidence implicating a role for a soluble lytic factor is accumulating.

ACKNOWLEDGEMENTS

This work was supported by NIH grants CA 33005 and CA 33589.

REFERENCES

1. Herberman, R.B. and Ortaldo, J.R. (1981) Science, 214, 24-30.
2. Herberman, R.B., Nunn, M.E., Holden, H.T., and Lavrin, D.H. (1975) Int J Cancer, 16, 230-239.
3. Pross, H.F. and Jondal, M. (1975) Clin Exp Immunol, 21, 226-235.
4. Kiessling, R., Klein, E., Wigzell, H. (1975) Eur J Immunol, 5, 112-117.
5. Anderson, J.J. (1978) Infect Immun, 20, 608-12.
6. Haller, O., Kiessling, R., Orn, A., and Wigzell, H. (1977) J Exp Med, 145, 1411-16.
7. Toivanen, P., Uksila, J., Leino, A., Lassilo, O., Hirvonen, T., and Ruuskanen, O. (1981) Immunol Rev, 57, 89-105.
8. Timonen, T., Ranki, A., Saksela, E., and Hayry, P., (1979) Cell Immunol, 48 121-132.
9. Herberman, R.B., Bartram, S., Kaskill, J.S., Nunn, M., Holden, H.T., and West, W.H. (1977) J Immunol, 119, 322-26.

10. Timonen, T. and Saksela, E. (1980) J Immunol Methods, 36, 285-292.
11. Herberman, R.B. (1980) in "Natural Cell-Mediated Immunity Against Tumors" (ed. R. Herberman) Academic Press, New York.
12. Wilson, D.B. (1965) J Exp Med, 122, 143.
13. Rosenau, W. (1968) Fed Proc, 27, 34.
14. Wright, S.C. and Bonavida, B. (1981) J Immunol, 126, 1516-1521.
15. Saksela, E., Carpen, O., and Virtanen, I. (1982) in "NK Cells and Other Natural Effector Cells" (ed. R. Herberman) Academic Press, New York, pp 983-987.
16. Roder, J.C., Kiessling, R., Bikenfeld, P. and Andersson, B. (1978) J Immunol, 121, 2509-2517.
17. Roder, J.C., Argov, S., Klein, M., et al (1980) Immunol, 40, 107-116.
18. Roder, J.C., Kane, R., and Kiessling, R. (1981) Prog Allergy, 28, 66-159.
19. Carpen, O., Virtanen, I. and Saksela, E. (1981) Cell Immunol, 58, 97-106.
20. Neighbour, P.A., and Huberman, H.S. (1982) J Immunol, 128, 1236-1240.
21. Neighbour, P.A., Huberman, H.S., and Kress, Y. (1982) Eur J Immunol, 12, 588-595.
22. Roder, J.C., and Klein, M. (1979) J Immunol, 123, 2785-2790.
23. Forbes, J.T., Bretthauer, R.K., and Oeltmann, T.N. (1981) Proc Natl Acad Sci USA, 78, 5797-5801.
24. Forbes, J.T. and Oeltmann, T.N. (1982) in "NK Cells and Other Natural Effector Cells" (ed. R. Herberman) Academic Press, New York, pp 977-982.
25. Werkmeister, J., Roder, J.C., Curry, C., and Pross, H.F. (1983) Cell Immunol, 80, 172-186.
26. Ortaldo, J.R., Timonen, T.T. Herberman, R.B. (1984) Clin Immunol Immunopath, 31, 439-43.
27. Roder, J.C., Beaumont, T.J., Kerbel, R.S., Haliotis, T., and Kozbor, D. (1981) Proc Natl Acad Sci USA, 78, 6396-6400.
28. Kaplan, A., Achord, D.T., and Sly, W.S. (1977) Proc Natl Acad Sci USA, 74, 2026-2030.
29. Sando, G.N. and Neufeld, E.F. (1977), Cell, 12, 619-627.
30. Fischer, H.D., Gonzalez-Noriega, A., Sly, W.S., and Morre, D.J. (1980) J Biol Chem, 255, 9608-9615.
31. Sandvig, K. and Olsnes, S. (1981) J Biol Chem, 256, 9068-9076.

CYTOTOXIC FACTORS FROM MARMOSET MONKEY NK-LIKE CELL LINES[1]

Donald R. Johnson

Department of Pathology & Laboratory Medicine
Department of Microbiology and
Eppley Institute for Research on Cancer & Allied Diseases
University of Nebraska Medical Center
Omaha, Nebraska 68105

I. INTRODUCTION

Marmoset monkey lymphocytes transformed by Herpesvirus ateles (HVA) or Herpesvirus saimiri (HVS) leads to continuously growing lymphoid cell lines. These lymphoblastoid cells have membrane markers of T cells and their proliferation is not dependent on lectins or exogenous growth factors (1-4). Many of these virus-transformed lymphoid lines have the capacity to spontaneously kill certain target cells (5,6). This cytotoxicity expressed by HVA and HVS cell lines resembles NK cell mediated cytotoxicity by: 1) not requiring lectins or antibodies attached to target cells for cytolysis, 2) sharing a spectrum of susceptible target cells similar to that of human and marmoset NK cells, and 3) having similar regulatory mechanisms as that seen with human and marmoset NK cells (7).

Natural killer cell cytotoxic factors (NKCF) released by stimulated NK cells have recently been described by Wright and Bonavida (8,9,10) and Farram and Targan (11). NKCF may be responsible for the terminal lytic event in NK cellular cytolysis. It is released by human and mouse NK cells after

[1] Supported by Grant #CA-32969 from the National Institutes of Health and the Lymphoproliferative Research Fund of the University of Nebraska.

ISBN 0-12-341370-2

incubation with suitable target cells, or with the lectins phytohemagglutin and Concanavalin A. This soluble factor is active against NK-sensitive but not NK-resistant target cells and can be augmented by pretreatment of the effector cells with interferon (12). Purification and characterization of NKCF would allow studying directly the mechanism of NK mediated cytotoxicity. Such experiments would be facilitated if large amounts of the factor(s) could be generated by lymphoid cell lines similar to the production of interleukin-2 by thymoma cells (13).

Since marmoset HVA and HVS transformed lymphoid cell lines appear to have NK-like properties, we have investigated whether they could produce soluble cytotoxic factors resembling those produced by mouse and human NK cells obtained from fresh lymphocytes. Co-culture of HVA and HVS cell lines with NK-sensitive target cells for 48 hr produced soluble factors causing cytolysis of NK-sensitive target cells in an 18 hr ^{51}Cr release assay. The cell lines did not spontaneously produce cytotoxic factors not could Epstein-Barr virus (EBV) transformed marmoset B cell lines (B95-8) be stimulated to produce NK cytotoxic factors. These studies suggest that the HVA and HVS cell lines can serve as a model for studying NK activity and as a resource for producing NKCF for biochemical and immunological analysis.

II. MATERIALS AND METHODS

Permanent marmoset lymphoid cell lines were established by transformation of marmoset lymphocytes with cell-free HVA obtained from supernatants of HVA-infected marmoset kidney fibroblasts or by cultivation with HVA-producing tumor or lymphoid cells (4). HVS-transformed marmoset cell lines were obtained from HVS-induced lymphomas of marmosets (1). The origins of the HVA and HVS lymphoid cell lines and marmoset EBV lymphoblastoid cell lines are shown in Table 1. All cell lines were propagated at 37°C and 5% CO_2 in RPMI-1640 medium (Gibco, Grand Island, NY), supplemented with 10% fetal bovine serum (FBS, Irvine Scientific, Santa Anna, CA), 100 units penicillin/ml, and 100 μg streptomycin/ml. Effector and target cells were tested 3 times for mycoplasma contamination and were negative by the method of Kaplan et al.(14).

Peripheral blood lymphocytes from normal human volunteers were separated by centrifugation on Ficoll-Hypaque and passaged through nylon wool as described (15,16).

Lymphoid cell lines or human NK cells were used to produce cytotoxic factors similar to those described (10).

TABLE 1. ORIGINS OF CELL LINES

Cell Line	Origin
70-KI-1	HVA-transformed *in vitro*, marmoset
22CM-37	HVA-transformed *in vitro*, marmoset
1670	HVS-derived tumor, marmoset
70N-2	HVS-derived tumor, marmoset
B95-8	EBV-transformed *in vitro*, marmoset
Molt-4	Acute lymphocytic leukemia (ALL)
1301	ALL
CCRF-CEM	ALL
HSB-2	ALL
RAMOS	Burkitt's lymphoma
Daudi	Burkitt's lymphoma
Raji	Burkitt's lymphoma
K562	Chronic myelocytic leukemia (CML)
YAC-1	Mouse T cell line
EL-4	Mouse T cell line

Briefly, 2.5×10^6 effector cells/ml were co-cultured with stimulator cells at a ratio of 25:1 in tissue culture medium with 2% FBS at 37°C in a 5% CO_2 incubator. The cell-free supernatant was harvested at 48 hr of culture by centrifugation and filtered through a 0.45 μM membrane. Supernatants from the stimulated effector cells were concentrated 5x on Minicon B-15 concentrators or a Diaflow UM-10 membrane (Amicon, Danders, MA) prior to assay.

The cytotoxicity assay was done in standard V-shaped microplates (Limbro/Titertec, McLean, VA) using 10^4 target cells/well in a total volume of 150 μl of medium with 10% FBS. Target cells were labelled with sodium ^{51}Cr (100 μCi/1 $\times 10^6$ cells) for 1 hr at 37°C. One hundred μl of effector cells or supernatant from stimulated effector cells were added and incubated with target cells for 6 or 18 hr before harvesting 50 μl of supernatant for determination of released radioactivity. The percent of cytotoxicity of triplicate wells was calculated by the formula:

$$\% \text{ cytotoxicity} = \frac{\text{test release - spontaneous release}}{\text{80\% of total label - spontaneous release}} \times 100$$

Each experiment was repeated at least three times.

III. RESULTS

A. Production of cytotoxic factors by virus-transformed marmoset cell lines

Human and marmoset peripheral blood NK cells, as well as HVA- and HVS-transformed cell lines, are highly cytotoxic towards the human tumor cell line, Molt-4 (5-7). Initial experiments were concerned with establishing the culture time and the effector to target cell ratio necessary for maximum cytotoxic activity in the culture medium. Maximum cytotoxicity was obtained when the effector and target cells were co-cultured for 48 hr, with a decrease in activity at longer culture periods. A plateau of cytotoxic activity was seen at effector cell to target cell ratios of 25:1 and higher. As such, subsequent experiments reported in this study used a 48 hr culture period at an effector to target cell ratio of 25:1. Supernatant-mediated lysis as determined by the ^{51}Cr release assay required an 18 hr incubation culture period for optimal detectable activity compared to the usual 6 hr used for direct cell cytotoxicity.

As reported by others (9-12), nylon wool-treated human lymphocytes were similarly able to produce active cytotoxic factors when stimulated by NK-sensitive stimulator cells (Table 2). The NK-insensitive target cells Daudi and YAC were not lysed by the cytotoxic factors from any of the transformed marmoset cell lines or human NK cells (Table 2), and little cytotoxic activity was found in the culture medium from the transformed cells not stimulated with target cells (data not shown). No activity was detected in the culture medium from non-stimulated human PBL, indicating that release of cytotoxic factors required a period of contact with stimulator cells and were not spontaneously released (data not shown). Stimulation of the transformed cell lines with NK cell-insensitive cells did not yield supernatants with cytotoxic activity to any other target cells tested. Similarly, EBV-derived marmoset B cells (B95-8) stimulated with Molt-4 cells produced no cytolytic material (Table 2).

When a non-stimulated marmoset HVA-transformed cell line was frozen and thawed three times, significant cytotoxic activity was found in the supernatant, indicating that the cytotoxic factors released by these effector cells may have already been formed in the effector cells and incubation with stimulator cells was required for release of these factors (Table 2).

B. Specificity of cytotoxic supernatants released by marmoset transformed cell lines

To determine the target cell specificity of the soluble factor mediated killing, the supernatants of co-culture from one of the HVA-transformed marmoset cell lines (70-KI-1) stimulated with Molt-4 was compared to direct cellular cytotoxicity by these cells against a variety of tumor cell targets. The spectrum of target cells killed by the cytotoxic supernatants was similar to that obtained by assay direct cellular cytotoxicity (Table 3). The NK-sensitive target T cell lines such as Molt-4, 1301, CCRF-CEM, and HSB-2 were susceptible in varying degrees to the cytotoxic supernatants. NK-insensitive target cells, including an EBV-carrying Burkitt lymphoma B cell line (B95-8), human T cell blasts, and two mouse T cell-lymphoma cell lines (YAC-1 and EL-4) were not killed by the cytotoxic factors. Although there was a discrepancy in the target cell specificity toward the EBV-negative B cell tumor, RAMOS, target cell susceptibility to cellular cytotoxicity implied sensitivity to cytotoxic factors.

TABLE 2. CYTOTOXICITY OF FACTORS RELEASED FROM TRANSFORMED MARMOSET AND HUMAN NK CELLS

Supernatant Production		Cytotoxicity*			
Effector Cells	Stimulator Cells	NK-Sensitive Target Cells		NK Non-Sensitive Cells	
Marmoset cell line	NK sensitive	Molt-4	K562	Daudi	YAC
70-KI-1	Molt-4	31*	5	2	0
70-KI-1	K562	7	3	0	0
22CM-37	Molt-4	13	9	0	0
1670	Molt-4	30	5	0	0
70N-2	Molt-4	44	15	0	0
	NK insensitive				
70-KI-1	YAC	0	0	0	0
70-KI-1	Daudi	0	0	0	0
70-KI-1	Freeze-thaw	23	0	0	0
EBV-transformed B cell (B95-8)	Molt-4	0	0	0	0
Human PBL	Molt-4	36	18	0	0
Human PBL	K562	20	10	0	0

* % ^{51}Cr release after 18 hr incubation; mean of 3 experiments

TABLE 3. COMPARISON OF CYTOTOXICITY EXPRESSED BY HVA-TRANSFORMED CELLS AND RELEASED FACTORS

Target Cell		Cellular Cytotoxicity*	Cytotoxic Factors**
Molt-4	ALL	69	31
1301	ALL	49	13
CCRF-CEM	ALL	32	36
HSB-2	ALL	55	30
K562	CML	17	5
Daudi	Burkitt's lymphoma	0	2
Raji	Burkitt's lymphoma	0	0
RAMOS	Burkitt's lymphoma	16	2
P3HR-1	Burkitt's lymphoma	0	1
B95-8	EBV-B cell line	6	1
Human T cell	Blast cells	0	4
YAC-1	Mouse lymphoma	0	0
EL-4	Mouse lymphoma	0	1

*Percentage ^{51}Cr release in 6 hr assay effector; target cell, 25:1. Mean of 3 experiments.
**Percentage ^{51}Cr release in 18 hr assay from culture medium of HVA-transformed cell incubated with Molt-4 cells for 48 hr at 25:1 effector:target cell ratio. Mean of 3 experiments.

IV. DISCUSSION

We have proposed that Herpesvirus ateles and Herpesvirus saimiri-transformed marmoset lymphocytes are similar to native NK cell populations, and that these virus-transformed lymphoid cell lines have functional properties similar to marmoset and human NK cells (5-7). Alternatively, HVA and HVS cell lines may represent transformed T lymphocytes that express NK activity and specificity because of their activated state. Such NK activity has been reported for activated T lymphocytes in man and cloned murine T lymphocytes (17,18). Killing by the HVA- and HVS-transformed lines is a stable characteristic unrelated to the age of the culture or virus-producer status of the cells. In addition, there seems to be little variation in target cell specificity among different cell lines. If HVA- and HVS-transformed cell lines use a similar kind of lytic mechanism that NK cells use, and if, as seems likely, soluble cytotoxic factors released by mouse and human NK cells are responsible for the lytic event in NK cell mediated cytotoxicity, then marmoset NK cell-like lines should produce and release similar factors.

The present investigation demonstrates that the transformed marmoset cell lines produce cytolytic mediators similar to those observed from murine and human NK cells. The method of production of these soluble mediators was similar to that of the NKCF described by other investigators (8-11). The supernatants from the stimulated HVA or HVS cell lines were cytotoxic to NK-sensitive target cells. Non-stimulated effector cells produced little or no cytotoxic activity in an 18 hr ^{51}Cr-release assay. The target cells susceptible to lysis by the soluble factors were those likewise killed by direct cellular cytotoxicity. EBV-transformed B cells were not cytotoxic and did not produce cytolytic factors when incubated with target cells, a finding that reduces the possibility that the production of the cytotoxic factors by the transformed marmoset cell lines was due only to Herpesvirus transformation.

Transformed cell lines producing large amounts of cytotoxic factors were obtained. The observation that the same target cells are susceptible both to direct cellular cytotoxicity and to NKCF from stimulated effector cells suggest that there are specific receptors for binding of the effector cells and the NK cytotoxic factors (9,11). Cytotoxic supernatants were also found in supernatants of cell lines that were frozen and thawed repeatedly, indicating that the cytolytic material was preformed in the effector cells and only required stimulation to release these factors into the culture medium.

Cytotoxic factors produced by the HVA and HVS cell lines and the NKCF reported by Wright and Bonavida (8-10) may in fact be similar. Alternatively, the cytotoxic factors produced by the HVA and HVS lines could be non-NK cell-derived lymphotoxin-like factors (19,20). The mechanism of cellular cytotoxicity mediated by NK cells is not yet clear but evidence is mounting that factors released from NK cells may mediate the final lytic event. If the cytotoxic factors produced by the virus-transformed cell lines with NK-like activity are similar to the cytotoxic factors responsible for NK-cell mediated lysis, then large amounts of cytotoxic supernatants can be prepared and analyzed for their biochemical properties and mechanisms of cytotoxicity.

ACKNOWLEDGMENTS

I thank Drs. Kent Thomas, David T. Purtilo, Lynell W. Klassen, and David Kay for helpful comments; Dorothy Oleson and Julie Whitted for technical assistance; and Jan Leemkuil for preparation of this manuscript.

REFERENCES

1. Falk, L., Wolfe, L., Marczynska, B., and Deinhardt, F. (1972). Bacteriol. Proc. 34, 191.
2. Wallen, W., Newbauer, R., Rabin, H., and Cicmanec, J. (1973). J. Natl. Cancer Inst. 51, 967.
3. Johnson, D., Falk, L., Yang, R., and Deinhardt, F. (1975). Fed. Proc. 994.
4. Falk, L., Johnson, D., and Deinhardt, F. (1978). Intl. J. Cancer 21, 652.
5. Johnson, D., and Jondal, M. (1981). Nature 291, 81.
6. Johnson, D., and Jondal, M. (1981). Proc. Natl. Acad. Sci. USA 78, 6391.
7. Johnson, D., Jondal, M., and Biberfeldt, P. (1982). In "Natural Cell-Mediated Immunity" (R. B. Herberman, ed.), Vol. 2, p. 835. Academic Press, New York.
8. Wright, S. C. and Bonavida, B. (1981). J. Immunol. 126, 1516.
9. Wright, S., Hiserodt, J., Bonavida, B. (1981). Transplantation Proc. 13, 770.
10. Wright, S., and Bonavida, B. (1982). J. Immunol. 129, 433.

11. Farram, E., and Targan, S. (1983). J. Immunol. 130, 1252.
12. Wright, S. C., and Bonavida, B. (1983). J. Immunol. 130, 2960.
13. Farrar, J. J., Fuller-Farrar, J., Simon, P., Hilfaker, M., Stadler, B., and Farrar, W. (1980). J. Immunol. 125, 2555.
14. Kaplan, D., Henkel, T., Braciale, U., and Brachiale, T. (1984). J. Immunol. 132, 9.
15. Boyum, A. (1968). J. Clin. Lab. Invest. 21: 77.
16. Julius, M.H., Simpson, E., and Herzenberg, L.A. (1973). Eur. J. Immunol. 3, 645.
17. Klein, E. (1980). Immunol. Today 1, iv.
18. Brooks, C. A. (1983). Nature 305, 155.
19. Berke, G. (1979). Prog. Allergy 27, 69.
20. Hiserodt, J. C., Tiango, G. J., and Granger, G. A. (1979). J. Immunol. 123, 332.

Part IV.

Regulation of NK Activity and Regulation by NK Cells

REGULATION OF NK CELLS[1]

Hillel S. Koren

Department of Microbiology and Immunology
Duke University Medical Center
Durham, North Carolina

This session concentrated on both the positive and negative modulation of NK activity by biological response modifiers (BRMs) and other endogenous and exogenous agents which may affect NK function.

Interferon (IFN) remains a focus of attention in the regulation of NK activity. Current issues include functional differences between types of interferon, mechanisms whereby IFN may augment activity, and the roles of IFNs in inducing cell proliferation and differentiation. Gamma or immune interferon (IFN-gamma) is produced by lymphoid cells in response to mitogenic or antigenic stimulation. This interferon has been shown to be different from alpha (leukocytes) or beta (fibroblast) interferons on the basis of different antigenic properties, target cell specificity and low pH treatment susceptibility. The availability of purified preparations of alpha and beta interferons permitted the study of their biological properties. These interferons exhibited profound effects on the cells of the immune system, including significant and rapid augmentation of natural killer cytotoxicity _in vitro_ and _in vivo_. In contrast, studies on the effects of IFN-gamma on the immune system were significantly hindered by the presence of numerous cytokines with immunomodulatory activities in IFN-gamma preparations. This

[1]Names followed by city are used when referring to statements made at the meeting. Names of authors not followed by city refer to quotes from the literature not necessarily presented at the meeting.

ISBN 0-12-341370-2

This problem was resolved with the introduction of recombinant DNA methods. Cloning of the IFN-gamma gene and expression in *E. coli* provided large amounts of recombinant human IFN-gamma for studies of its biological properties.

Human alpha interferons (IFNs) are a family of proteins that have a variety of biological activities. Recently, the genes for these proteins have been cloned, expressed in bacteria and the proteins purified in large quantitites. In addition to the subtypes of IFN-alpha, molecular hybrids between subtypes have been constructed which retain biological activity. Although the alpha IFN subtypes are species specific, the constructed hybrid proteins have gained the ability to act on cells from other species.

Working with Minnesota miniature swine, Kim (Chicago) noted that all the human IFN preparations he tested (IFN-alpha, recombinant IFN-alpha-A, recombinant IFN-alpha-D, and recombinant IFN-gamma) shared the property of boosting NK but not ADCC activity.

Ortaldo (Frederick) reported that the potential to boost human NK cytotoxicity corresponded to antiviral titer for all types of IFN-alpha except type J which required prolonged exposure and higher antiviral titers to activate NK. This work suggests that different types of IFN may be selected for different therapeutic effects. In this regard, it is interesting to note that IFN-gamma and alpha-A/D had been shown by Brunda to down regulate NK activity as measured by both an *in vivo* and an *in vitro* assay against B16 and FIO melanoma cells when the mice were treated with 10 daily injections compared to 4 day treated mice. Based on the data presented in the workshop and published results, it would be fair to conclude that recombinant IFNs used under appropriate conditions can regulate the development of metastases. However, in order to maximize the anti-tumor effects of IFN against metastases either the appropriate administration of IFN or its use in combination with other drugs may be necessary. Ortaldo (Frederick) reported that mice treated with various BRMs whose effects at the cellular level may be mediated by IL-2 and IFN have augmented hepatic NK cytotoxicity. This increase in NK activity was associated with an increase in the number of hepatic LGLs (over 50% of nylon wool passed cells had LGL morphology). Interestingly, treated mice resisted experimental metastases to a degree which correleated with the elevation of their hepatic NK activity.

Issues of proliferation and differentiation of NK cells recurred during discussion of the role of IL-2 in NK regulation. Recently, techniques have been developed to continuously culture human and murine lymphoid cells with T-cell growth factor (interleukin-2 "IL-2"). The cultured lymphoid cells have been shown to exert some T-cell functions, including helper and cytotoxic activities. The retention of these functional capacities during proliferation has made the cultured cells potentially useful tools in the dissection of T-cell responses and in adoptive immunotherapy. Although several groups have demonstrated NK-like cytotoxic reactivity among cultured cells, it has remained unclear whether this results from differentiation during culture of T cells or whether NK cells active in fresh blood can proliferate and retain function in the presence of IL-2. This question has been addressed by culturing highly purified human NK cells (large granular lymphocytes "LGL"), and T cells, obtained by discontinuous density gradient centrifugation on Percoll, in the presence of IL-2 containing conditioned medium (CM) derived from PHA-stimulated cultures of human peripheral blood lymphocytes. The results indicate that both LGL and T cells proliferate in the presence of CM. The cultured LGL retained the typical morphology and function of NK cells, distinguishable from those of cultured T cells. However, the surface antigenic phenotype of LGL changed toward a T cell pattern during the culturing, suggesting an association of the proliferating LGL with T cells. Clones of LGL have also been shown to produce mutiple cytokines.

The ability of endogenous NK cells to proliferate remains controversial. NK cells proliferate in virus-infected mice. Here, Santoni (Rome) has observed that mice treated with IFN-alpha, poly I:C, or C. parvum have an increased number of proliferating LGLs in the spleen, bone marrow, and peritoneum. In the development of NK-like cytotoxicity in culture with fetal calf serum, Lattime (New York) found that anti-Ia treatment blocked the production of IL-2 at day 1 and the subsequent appearance of NK-like cells at day 3. Lattime was unable to detect IFN in his culture system. Kumagai (Sendai) has shown that human cell populations enriched for NK cells proliferate in response to recombinant IL-2 and produce IFN-gamma, and acquire capacity to kill NK-resistant cell lines. Antibody to IFN-gamma blocked proliferation but IFN-gamma alone did not induce proliferation of killer cells. This work suggests that both IL-2 and IFN are necessary. Brunda (Nutley) presented data that recombinant IL-2 alone augmented murine and human NK cytotoxicity. Recombinant IFN-alpha-A had an additive or synergistic effect. However, experiments to assess the contributions of endogenous IFN production have not been done.

These findings have several important implications: 1) Production of these various cytokines by perhipheral blood mononuclear cells appears attributable, at least in part, to LGL. Therefore, these cytokines can no longer be considered to be associated only with T cells or monocytes. 2) The ability of NK cells and perhaps other LGL to produce cytokines indicates that this population may play an important immunoregulatory role as well as mediate natural cytotoxic activity. The potential importance of such functions is supported by the finding of accessory cell activity associated with a subset of LGL. 3) The ability of LGL to produce cytokines suggests that this population is capable of self-regulation of cytotoxic reactivity. IFN-gamma and IL-2 have been shown to strongly augment NK activity, with the latter presumably acting by stimulating LGL to produce IFN-gamma. IL-1 has also been reported to contribute to augmentation of NK activity, when present along with IFN and/or IL-2.

There has been general interest in whether IFN activation involves more efficient lysis or an increased number of active NK cells. Mahle (Rochester, MI) discussed a kinetic model which favored the former mechanism for increased cytotoxicity. At the molecular level, Djeu (Bethesda) has shown that IFN-beta induces (or enhances) the expression of at least ten new proteins in purified human NK cells using protein synthetic labeling. The total synthetic capacity of the cells was, however, decreased by IFN-beta treatment.

Mononuclear phagocytes and other cells can interact with NK cells and affect the expression of natural cytotoxicity. In infant and senescent normal mice as well as NK deficient SJL/J mice, Riccardi (Perugia) has identified suppressor cells. The number of NK-precursor cells was diminished in infant mice and in SJL/J mice, and infant mouse NK was not activated by IFN or IL-2 This type of suppression has been associated with T cells. In man, other reports have suggested that cells that bear the HNK-1 differentiation antigen, whether found within the peripheral blood or as cells activated in late MLC, can regulate the cytotoxic activity of NK cells. These suppressor cells, or similar ones, can also affect the kinetics of development of activated killer (AK) cells generated in MLC. The findings of D'Amore and Golub indicate that NK cells or cells in the NK lineage can also function as immunoregulators. These results also suggest a close functional relationship between cells mediating NK and AK cytotoxicity as they both are subject to similar control mechanisms.

Macrophages play a central role in the regulation of immune responses and cell differentiation through the interaction with lymphoid cells. It has been shown that macrophages are required for boosting of murine NK cells induced by a variety of stimulating agents. The direct interation of macrophages with NK cells received relatively little attention. Seaman (San Francisco) explored possible mechanisms of interactions between phagocytes and NK. Specifically, PMNs exposed to opsonized particles activate monocyte-mediated suppression of NK cells, probably by releasing H_2O_2 release. Other studies have shown an augmentation of NK activity after being in contact with activated macrophages (Suzuki et al.). In those studies, activated macrophages were co-cultured with NK-enriched murine spleen cells. These macrophages rapidly and greatly augmented NK activity against target cells including those insusceptible to endogenous NK activity. This augmented cytotoxicity against the targets insusceptible to endogenous NK cells could not be induced by IFN. This rapid and strong augmentation of NK cell activity by activated macrophages appears to involve contact of viable macrophages with NK cells.

The remainder of the session sampled a broad spectrum of issues in NK regulation. One of the issues discussed was the possibility of regulation of NK activity be sensory- autonomic nervous system. Recent work has caused one to rethink our traditonal views on immunomodulation in which nervous system influences on immune function are mediated by neuroendocrine mechanisms, i.e., the hypothalmic-pituitary-adrenal axis. A decade of research has led to the discovery and characterization of receptors on both lymphocytes and granulocytes for neurotransmitters. In a general sense, substances which increase intracellular cyclic AMP levels tend to be immuno-suppressive, and in contrast, substances which increase intracellular cyclic GMP tend to be immunoenhancing. Tanaka and his group have recently reported that the antibody formation and graft versus host reactions (GVHR) were augmented by stimulation of sensory-autonomic nervous system, especially by sympathetic nerves. In this session, Santoni (Rome) evaluated the effects of lesions induced in the brains of mice. Hypothalmic lesions in young mice resulted in prolonged, profound depression of activity and number of NK cells, which could be restored by IFN injections. Clearly, this is an important and relatively underworked area of research that warrants further investigation.

Research that was reported on the metabolism of NK cells may shed light on modulatory mechanisms. Several groups discussed the role of lipoxygenase metabolites and phospholipids in NK function. Arachidonic acid (AA) metabolites play a central role in modulating a variety of immune cell functions. There are two major pathways of AA metabolism: The cyclooxygenase pathway and the lipoxygenase pathway. Of the cyclo-oxygenase products, PGE_2 has been shown to inhibit a great number of immune reactions, including NK activity. The role of lipoxygenase products in NK regulation is poorly understood. Lipoxygenase converts AA into hydroperoxyeicosatetraenoic acid (HPETE); HPETE are unstable and are reduced to hydroxy fatty acids (HETE). 5-HETE, 12-HETE, 15-HETE have been identified in various cell types. 5-lipoxygenase activity also lead to formation of a variety of biologically active HETE, called leukotrienes (LTA_4, LTB_4, LTD_4, LTE_4, LTF_4). The transformation of LTA_4 to LTC_4, a slow-reacting substance of anaphylaxis, is carried out by glutathione-S-transferase. Various lipoxygenase products have been isolated from macrophages, polynonphonunclear cells and other inflammatory cells.

Seaman (San Francisco) reported earlier in the meeting that lipoxygenation is essential for NK activity. Koren (Durham) confirmed that lipoxygenase inhibitors reduced endogenous and culture- induced NK activity; however, addition of leukotrienes and HETEs did not reverse the effect. Moreover, NK cells that were activated by IFN were less responsive to lipoxygenase inhibitors. Wigzell (Stockholm) has reported some suppression of NK by HETE. In addition, Hudig (Reno) showed that inhibition of phospholipase A_2 also decreased NK levels.

The cellular effects of exposure to exogenous modulators of NK and ADCC activities <u>in vitro</u> such as UV-irradiation and tumor-promoting agents such as phorbols were also discussed. Bonavida (Los Angeles) described the suppressive effect of UV-irradiation on NK cells. Single cell conjugate assays in agarose have demonstrated that the inhibition was at the lytic stage.

Recently, several investigators have reported that the addition of phorbol esters to cytotoxic assays inhibits NK cell activity. Seaman et al. previously reported that tumor promoting phorbol diesters suppress natural killing indirectly via monocytes. NK was inhibited by pretreatment of either the leukocytes or the target cells with TPA, though this was less effective than direct addition of TPA to the assay. Moreover, PMA inhibited the cytotoxic activity of purified human LGL directly, as was also shown by Abrams et al. Goldfarb and Herberman have demonstrated that PMA but not non-tumor promoting structural analogs inhibits both murine and human NK activity. The inhibitory effects of phorbols may be explained by the aggregation of either the effectors or the target cells, which inhibit the NK cells from acting as killer cells based on reent wojrk by Pataroyo et al. showing aggregation of lymphocytes following a 20 minute pretreatment of the cells with phorbol dibutyrate (PDBU). Elevated NK activity mediated by Balb/c nude splenocytes preincubated with PMA has also been reported.

In this meeting, the laboratories of Klein (Stockholm), Trinchieri (Philadelphia), and Koren (Durham) discussed their findings on the efects of phorbols during _in vitro_ culture. Trinchieri proposed that augmented NK levels following exposure to TPA could be explained by the the result of modulation of the F_c receptor as detected by B73.1 monoclonal antibody. Klein reported that the water soluble phorbol PDBU can augment or suppress NK cytotoxicity depending on the duration of the exposure. Short incubations decreased NK, possibly as the result of reduced recycling. Longer exposure to PDBU could have increased NK activity due to increased binding and "activation of the lytic machinery". Argov (Durham) reported that PDBU induces selective adherence of NK cells to plastic. The selectively adherent cells exhibited enhanced NK cytotoxicity due to specific enrichment rather than activation.

Experiments discussed in this session generated information about the mechanism of regulation of natural killing at the levels of various species and various organs, and further described regulation of NK cells at cellular and molecular levels. In addition, the investigators posed questions which initiated new approaches to understanding the processes involved in the normal, activated, and suppressed natural killing.

KINETIC MODELS FOR NATURAL CYTOTOXICITY AND THEIR USE FOR STUDYING ACTIVATED NK CELLS[1]

Denis M. Callewaert[2]
Norma H. Mahle

Department of Chemistry
Oakland University
Rochester, Michigan 48063

I. INTRODUCTION

Natural cytotoxicity reactions, when performed under suitable conditions, exhibit kinetic behavior that appears analogous to that observed for simple enzyme-catalyzed reactions (1-3). Specifically, if one determines the cytotoxicity of increasing numbers of target cells that is effected by a constant number of lymphocytes and plots the rate of target cell lysis, v, versus the initial target cell concentrations tested, a hyperbolic curve is obtained for which v approaches a maximum value, V_{max}, as the concentration of target cells becomes very large. Because of the gross similarities between the kinetic behavior of NK reactions and simple enzymes, initial studies of NK kinetics borrowed heavily on the model and equations developed for the study of simple enzymes (4). However, NK reactions have several distinguishing features, some of which were recognized immediately while others have become apparent more recently. Since the interpretation of experimentally-determined kinetic

[1]Supported in part by grants AI-12766 and CA-34112 from the National Institutes of Health.

[2]Research Career Development Awardee (CA-00744) of the National Cancer Institute.

ISBN 0-12-341370-2

parameters is critically dependent upon the model used to describe a given phenomenon, our understanding of the significance and utility of kinetic parameters for NK reactions has evolved along with our understanding of the details of the mechanism involved.

II. SIMPLE MODELS FOR NK KINETICS

Because NK reactions exhibit hyperbolic kinetics that are similar to that of simple enzymatic reactions, the simple model and equations originally developed by Michaelis and Menten (4) were tested a number of years ago to determine if they were applicable for NK reactions (3). For cytotoxicity reactions, this simple kinetic model is

$$E + T \underset{k_{-1}}{\overset{k_1}{\rightleftharpoons}} ET \xrightarrow{k_2} E + T_L \qquad (1)$$

wherein E, T, ET and T_L represent effector, target, effector-target conjugate, and lysed target cells respectively and k_1, k_{-1}, and k_2 are rate constants for the steps indicated. Based on this model, a fairly simple rate equation can be derived (see 5):

$$v = \frac{V_{max}[T]}{K_M^{app} + [T]} \qquad (2)$$

in which the rate of target cell lysis, v, is related to the concentration of target cells as well as to V_{max} and an apparent Michaelis constant, K_M^{app}. For simple enzyme-catalyzed reactions in which the enzyme concentration is negligible compared to that of the substrate, $K_M^{app} = K_M$, and

$$K_M = (k_2 + k_{-1})/k_1. \qquad (3)$$

Values for both V_{max} and K_M can be easily obtained by application of linear transformations of equation 2 (6). Furthermore, for most (but not all) simple enzyme-catalyzed reactions involving a simple substrate, k_2 is much smaller than k_{-1} so that K_M is approximately equal to the dissociation constant for the enzyme-substrate complex. By analogy, values for K_M^{app} obtained for NK reactions were initially interpreted in terms of the affinity of NK receptors for target cell antigens (1-3). However, for NK reactions involving human peripheral blood lymphocytes (PBL) we have demonstrated that the experimentally-determined apparent Michaelis "constant" K_M^{app}, is actually not constant but varies in direct proportion to the concentration of lymphocytes added to the assay mixture (5,7). This also appears to be the case for antibody-dependent cellular cytotoxicity (ADCC) reactions mediated by human PBL (1,5).

These observations led us to reexamine the model for NK kinetics and the assumptions involved in deriving kinetic equations based on this model. In particular the assumption that the enzyme concentration is negligible compared to the concentration of the substrate is suitable for most simple enzymatic reactions conducted *in vitro* , but the analogous assumption may not be valid for NK reactions. We therefore derived more general equations for the kinetics of cytotoxicity reactions that do not require such assumptions (5,8). One consequence of this approach is

$$K_M^{app} = K_M + [E] \tag{4}$$

wherein [E] is the concentration of active effector cells in the assay mixture.

We subsequently performed a series of experiments designed to determine the relative magnitude of K_M and [E], and found that extrapolation of K_M^{app} versus [E] to [E] = 0 resulted in values for $K_M = 0$ within experimental error. Therefore, based on this simple model, we concluded that $K_M^{app} \simeq [E]$, and attempted to use this relationship to simultaneously determine the concentration and lytic activity of NK cells in human peripheral blood (5,7). Unfortunately, values for the frequency of NK cells in peripheral blood determined by this method were generally considerably larger than those reported using other methods including the single cell assay (9-11). These results, along with recent results obtained in other laboratories, suggest that the model presented as equation 1 may not adequately represent NK reactions.

III. MUTLTISTEP MODELS FOR NK KINETICS

In the past few years considerable progress has been made in elucidating the details of the mechanism whereby NK cells recognize and lyse target cells. In particular, Targan and coworkers (12) have demonstrated that NK reactions involve several discrete steps, including conjugate formation, lytic programming and killer cell independent lysis (KCIL) of programmed target cells. Additional details concerning the component steps in this process will undoubtedly be discovered. It was therefore necessary to determine if the simple kinetic model presented as equation 1 was an adequate representation of natural killing, or if a more complex model and associated equations were required.

A more general model for the kinetics of cellular cytotoxicity reactions has recently been published by Merrill (13). This model includes all of the known steps in the overall reaction:

$$E + T \underset{k_{-1}}{\overset{k_1}{\rightleftharpoons}} ET \xrightarrow{k_2} ET_P \xrightarrow{k_3} E + T_P \xrightarrow{k_4} T_L \quad (5)$$

In this model, E, T, ET, T_L, k_1, k_{-1}, and k_2 are defined as for equation 1, T_P represents a target cell that has been programmed for lysis, and k_3 and k_4 are rate constants for the dissociation of ET_P complexes and killer cell independent lysis respectively. Merrill's general model also includes consideration of inhibition of cytotoxicity by nonlytic "bystander" lymphocytes, that can bind to but cannot kill target cells:

$$T + B \underset{k'_{-1}}{\overset{k'_1}{\rightleftharpoons}} TB \quad (6)$$

wherein B is a noncytotoxic bystander lymphocyte, and k'_1 and k'_{-1} are rate constants for the steps depicted. This latter consideration may be quite significant since upwards of 30% of peripheral lymphocytes can form conjugates with commonly used NK target cells, while only a small percentage of the lymphocytes that bind to target cells appear to be active NK cells (9-11).

Based on this model, Merrill derived a general equation for the kinetics of cytotoxicity reactions that gives rise to the following expressions (13):

$$V_{max} = k_2 \alpha f[E_T] \quad (7)$$

and

$$K_M^{app} = \frac{K_M + [E_T]\left[(1-\alpha)\frac{K_M}{K_I} + \alpha\left(1 + \frac{k_2}{k_4}\right)\right]}{n} \quad (8)$$

wherein n is the number of cytotoxic lymphocytes that simultaneously bind to a single target cell, K_I (the dissociation constant for bystander-target complexes) is equal to k'_{-1}/k'_1, $[E_T]$ is the total concentration of lymphocytes in the assay mixture that can interact with (bind to) target cells, and α is the fraction of these lymphocytes that lyse target cells.

In order to test the applicability of this general multistep model to actual NK reactions conducted *in vitro*, it was necessary for us to consider some additional facts that are known about NK reactions. First, the majority of peripheral lymphocytes do not appear capable of binding or killing commonly used target cells under conditions tested to date (9-11). In other words,

$$[L] = [E_T] + [N] \tag{9}$$

wherein [N] is the concentration of lymphocytes that do not interact with target cells. Therefore, if we let f equal the fraction of the total lymphocyte population that can interact with the target cells:

$$[E_T] = f[L] \tag{10}$$

Second, several laboratories have now reported that conjugates between a single target cell and multiple effector cells are extremely rare under assay conditions generally employed (10,11). It is therefore reasonable to let n = 1. Finally, we have previously demonstrated that extrapolation of plots of K_M^{app} versus [L] to [L] = 0 result in values for K_M equal to zero within experimental error (5). With these additional considerations, equations 7 and 8 can be written as

$$V_{max} = k_2 \alpha f[L] \tag{11}$$

and

$$K_M^{app} = f[L]\left[(1 - \alpha)\frac{K_M}{K_I} + \alpha\left(1 + \frac{k_2}{k_4}\right) \right] \tag{12}$$

Note that, even for this complex multistep model, the expression for V_{max} (equation 11) is essentially identical to that derived based on a simple kinetic model ($V_{max} = k_2[E]$). However, the expression for K_M^{app} (equation 12) is considerably more complex.

IV. MEASUREMENT OF RATE CONSTANTS FOR LYTIC PROGRAMMING AND KCIL

To determine whether the expression for K_M^{app} might justifiably be simplified to a point where experimental values could be assigned some physical significance, we sought to determine the relative magnitude of the component terms. Standard methods were used to measure V_{max}, K_M^{app} (6) and the fraction of lymphocytes that form conjugates with or lyse target cells (9). In addition, since calcium is required for cytotoxic lymphocytes to initiate lytic programming (14) and this requirement has been used to discriminate between steps in the natural cytotoxicity process

(13), we adapted a calcium pulse technique for use in the quantitative determination of the values of the rate constants for lytic programming (k_2) and killer cell independent lysis (k_4). The details of these procedures have recently been reported (15), and are briefly summarized below.

A. Experimental Procedures

Prior to initiation of an assay for measurement of k_2 or k_4, lymphocytes and target cells were washed and resuspended in RPMI medium containing 5 mM EGTA, 2.5 mM $MgCl_2$, and 10% newborn calf serum. Aliquots of both lymphocytes and target cells were then mixed and preincubated for one hour at 37°C to allow conjugate formation. Subsequetly, an equal amount of a calcium-dextran solution (5mM $CaCl_2$ and 10% dextran T-500 in RPMI-1640) that had been prewarmed to 37°C was added to each tube to initiate lytic programming but prevent recycling. For the determination of k_2, 1.5 ml of prewarmed RPMI-NCS containing EDTA (final concentration 10 mM) were added to each of four replicate tubes at several precise time intervals subsequent to the addition of the calcium-dextran mixture, the tubes mixed by inversion and incubated for an additional 120 minutes to allow KCIL of programmed targets to reach completion. The tubes were then mixed again, centrifuged, and ^{51}Cr-release determined on a 1.0 ml aliquot of the supernate.

For the determination of the rate of KCIL, k_4, lytic programming was first allowed to proceed for 90 minutes in the presence of calcium-dextran. EDTA was then added to each tube and the contents mixed by vortexing. At several time intervals, KCIL was terminated by addition of 2.0 ml of ice-cold medium, with rapid mixing by inversion and immediate centrifugation. An aliquot was then removed from each tube for gamma counting.

A second set of tubes containing target cells but not lymphocytes were included for each time interval in either assay for determination of spontanous release of the isotope. Maximal release of ^{51}Cr was determined by sonicating an aliquot of the target cells. The percentage of cytotoxicity (%C) was then determined using the standard formula

$$\%C = \frac{E - S}{M - S} \times 100 \qquad (13)$$

Now, according to the model presented as equation 5, both lytic programming and KCIL are predicted to follow first order kinetics. Therefore, values for k_2 and k_4 were determined using the equation 14:

$$Y = 1 - \left[\frac{C_t - C_o}{C_\infty - C_o} \right] = e^{-kt} \tag{14}$$

wherein the cytotoxicity that occurs during a given time interval, C_t, between initiation and termination of the reaction being studied is compared to the cytotoxicity, C_o, observed at t = 0 (achieved by simultaneous addition of initiation and termination reagents), and an estimate of the cytotoxicity at infinite time, C_∞, obtained by extrapolation. The best fit of the data to equation 14 was determined by nonlinear least squares analysis of plots of mean Y values versus t.

B. Values Obtained for the Rate Constant for Lytic Programming

In the presence of calcium, target cells complexed with active NK effector cells are predicted to undergo lytic programming at random time intervals typical of a first order process. Several experiments were performed in order to determine if the reaction actually does follow first order kinetics under the conditions employed. The generally excellent fit of the data from such experiments to equation 14 indicates that this is indeed a first order process (15). Values obtained for k_2 in a number of typical experiments are presented in Table I. While additional experiments are clearly required prior to making broad generalizations, the data obtained to date suggest that the mean time required for lytic programming of K-562 cells (after specific recognition) is on the order of 15 minutes at 37°C. Preliminary data also suggest that a major difference between the kinetics of lysis for highly susceptible versus less susceptible cell lines is in the rate of lytic programming, but this needs to be verified.

TABLE I. Rate Constants for Lytic Programming

Target	$k_2 (h^{-1})$	r^2 *
K-562	4.31	0.88
K-562	7.13	0.99
K-562	4.14	0.96
K-562	4.47	0.76
K-562	4.29	0.88
Jurkat	1.55	0.96

*Correlation coefficient for the fit of the data to equation 14.

C. Values Obtained for the Rate Constant for KCIL

Based on the multistep model presented as Equation 5, dissociation of the effector-(programmed)target complex, with rate constant k_3, occurs prior to ^{51}Cr-release from the programmed target cells, i.e. KCIL. However, we are not aware of any experimental evidence supporting ET_p complex dissociation as a prerequisite to ^{51}Cr release. On the contrary, substantial evidence supports the ability of a programmed target cell to undergo lysis while attached to an effector cell (10-12). As the term killer cell independent lysis implies, attachment to the effector cell is apparently no longer necessary after programming has occurred (13). Assuming that the actual release of ^{51}Cr is independent of association between the programmed target cell and the effector cell, KCIL should be a simple first order process dependent on $[T_P]_{total}$. To test this prediction, effector-target conjugates were formed in the presence of EGTA + Magnesium, followed by addition of calcium and dextran to initiate lytic programming but prevent recycling. After a fixed time interval, EDTA was added to stop lytic programming. KCIL was then stopped at different times as described above, and values obtained for cytotoxicity at different time intervals were plotted and used to compute a value for k_4. When sufficiently short time intervals (2.5 minutes) were tested, the data obtained fit the curve described by equation 14 very well (15) - again indicative that this process does follow first order kinetics. Values for k_4 obtained in several typical experiments are listed in Table II.

TABLE II. Rate Constants for Killer-Cell-Independent Lysis

Experiment	Donor	Target	k_4 (h^{-1})
1	A	K-562	15.6
2	B	K-562	34.2
3	C	K-562	27.0
	D	K-562	29.4
4	E	K-562	21.6
	E	CEM	27.9

Values obtained for k_4 using the K-562 cell line were typically 4 to 8 times greater than values obtained for k_2. In several experiments we consistently observed that the KCIL rate was essentially independent of the donor lymphocytes. In addition, preliminary data suggests that the CEM cell line, for which the overall lytic rate has been reported to be considerably slower than for K-562 cells (16), exhibits a comprable KCIL rate, which suggests that the primary difference in lytic rate may be due to differences in the rates of recognition and/or lytic programming.

V. EXPERIMENTAL EVALUATION OF A MULTISTEP KINETIC MODEL

The simple model for NK kinetics (equation 1) gave rise to a simple and apparently useful (7) expression for K_M^{app} (equation 4). However, values obtained for the NK cell frequency using this expression were generally considerably larger than those obtained using other methods including the single cell cytotoxicity assay. On the other hand, the multistep kinetic model for natural cytotoxicity (equation 5) and consideration of lymphocytes that bind but do not kill labeled target cells (equation 6) result in complex expressions for K_M^{app} (equations 8 and 12). In order to experimentally evaluate the multistep model, simultaneous determinations of k_2 and k_4 were made for single donor-target combinations. In parallel experiments, the kinetic parameters K_M^{app} and V_{max} were also determined, as were the frequency of target binding cells, and the frequency of lytic conjugates. The values obtained for all of these parameters were used to determine the relative magnitude of the component terms in the expression for K_M^{app}, and thereby glean an indication of the physical significance of this expression and its component terms (15). Results obtained in a typical experiment of this type are presented in Table III.

TABLE III. Evaluation of K_M^{app} for A Multistep Kinetic Model

Parameter	Value
[L]	5×10^5 cells/well
k_2 (h^{-1})	1.26
k_4 (h^{-1})	15.66
V_{max} (T lysed/hr)	6.67×10^4
K_m^{app} (cells)	4.72×10^4
Target Binding Cells (f)	0.31

In addition to an estimate of the frequency of target binding cells, the single-cell cytotoxicity assay yields an estimate for the fraction of the total lymphocytes that form lytic conjugates, which should be equal to αf, the fraction of the lymphocytes that are active NK cells (9). In the experiment presented in Table III, the value obtained for αf was 0.032.

Now, given values for V_{max}, k_2, and [L], one can also calculate a value for αf using equation 7. For the experiment presented here, the value thus obtained was 0.034. Note that for this experiment there is excellent agreement between the values obtained for αf obtained using this approach and that obtained using the single cell cytotoxicity assay.

Based on the values obtained in such an experiment, one can also compute the relative magnitude of the component terms in the expression for K_M^{app} (equation 12). First, comparison of the values for k_2 and k_4 presented in tables I and II, as well as those obtained in parallel experiments such as presented in Table III indicate that k_2 is consistently considerably smaller than k_4. Therefore, the term $\alpha(1 + k_2/k_4)$ is approximately equal to α. Second, the term $(1 - \alpha)(K_M/K_I)$ is considerably larger than the term $\alpha(1 + k_2/k_4)$, with the ratio of these two terms equal to 7.3 for the experiment presented in Table III. Third, The sum of the terms $(1 - \alpha)(K_M/K_I)$ and $\alpha(1 + k_2/k_4)$ is equal to 0.31 for this experiment, and was between 0.3 and 0.9 for a number of similar experiments. Finally, from the experimental data and equation 12, the value for K_M/K_I calculated for this experiment is 0.28 and for similar experiments always was less than 1.

While values for αf obtained from single-cell cytotoxicity assays and from application of equation 7 do not always agree as well as in this experiment, they are in the same order of magnitude, suggesting that both approaches are reasonably sound. The finding that k_4 is consistently larger than k_2 supports lytic programming as the rate-determining step in NK cytolysis and, in the absence of other information, indicates that application of a simple model for NK kinetics (equation 1) would be a valid approximation. However, we consistently find that the term involving K_M/K_I is significantly larger than the term involving k_2/k_4. This is the opposite from what would be expected for a simple model devoid of interference from lymphocytes that bind but cannot lyse the target cells (for which K_M/K_I should equal zero). In fact, the presence of non-lytic target binding lymphocytes has a very significant effect on the value for K_M^{app}. On the other hand, the fact that K_M/K_I values ranged from 0.2 to 0.9 indicates that the avidity of the NK effector cells for the target cells is greater than that of lymphocytes that bind but do not lyse the target cells. Finally, given the information obtaine to date, it is not possible to assign a simple physical or biological significance to values obtained for K_M^{app}.

VI. EFFECTS OF INTERFERON ON NK KINETIC PARAMETERS

It is now well known that natural cytotoxicity is enhanced by short term exposure to interferon or interferon inducers (16-20), and the mechanism of this enhancement has been investigated in several laboratories. The results of a number of experiments suggest a dual role for interferon: acceleration of the rate of target cell lysis by competent NK cells and recruitment of pre-NK cells (21-24). However, results of other studies suggest that interferon acts solely by recruitment of pre-NK cells (25,26). Comparison of the kinetic parameters for natural cytotoxicity mediated by interferon treated versus control lymphocytes should therefore prove useful for distinguishing between these proposed mechanisms. Ortaldo, et. al. have previously studied the effect of interferon on NK kinetic parameters, and reported that interferon treatment results in an increase in V_{max} but no change in K_M^{app} (27). They interpreted these results as supporting the activation of the lytic machinery of NK cells by interferon with no change in the affinity of NK cells for antigens on the target cells. We have also reported that interferon increases the V_{max} for natural cytotoxicity, but found that K_M^{app} also increases (28). With the development of methods that allow for the direct measurement of the rate constants for the individual steps in the NK reaction (i.e. k_2 and k_4), it was of interest to reinvestigate the effect of interferon on the kinetics of natural cytotoxicity. In several experiments that have been performed to date, interferon pretreatment of human PBL caused essentially no change in the rate constant for KCIL (Table IV), but did produce an approximately parallel increase in V_{max}, K_M^{app}, and k_2 (Table V).

TABLE IV. Effect of Interferon Pretreatment on KCIL

Experiment	Interferon Pretreatment	k_4 (h^{-1})
1	none	19.5
	2 h at 37°C	19.6
2	none	22.2
	2 h at 37°C	27.2

Now, since $V_{max} = k_2\,\alpha f[L]$, the percentage of the total lymphocyte population that express NK activity in these assays can be estimated by the formula:

$$\% \text{ NK} = \alpha f \times 100 = \frac{V_{max}}{k_2[L]} \times 100 \tag{15}$$

For the experiments presented in Table V, $[L] = 5 \times 10^5$. Thus, for experiment 1, the calculated value for NK cell frequency (4.0%) is unaltered by interferon pretreatment. For experiment 2, 0.16% of the control lymphocytes and 0.13% of the interferon pretreated lymphocytes are estimated to be NK cells (these values are not significantly different).

TABLE V. Kinetic Parameters Altered by Interferon

Experiment	Interferon Pretreatment	V_{max} (T lysed/h)	k_2 (h^{-1})
1	-	3.42×10^4	1.55
	+	5.51×10^4	2.82
2	-	0.56×10^4	7.14
	+	1.31×10^4	19.78

These results strongly suggest that interferon enancement of cyotoxicity is due primarily to accelerated lytic programming (k_2) rather than an increase in the fraction of the lymphocytes that are NK cells (αf). This conclusion is in contrast to previous reports that suggest recruitment of pre-NK cells as a primary mechanism for interferon enhancement of NK activity of PBL. In addition to major differences in experimental techniques, the present studies involved only relatively short pretreatment with interferon (2 h). It is possible that more prolonged treatment with interferon may have additional effects on PBL, including the differentiation of precursor cells into functional NK cells. Given that interferon treatment of PBL results in a rapid increase in the rate of protein synthesis (29-31), it also remains to be determined if the observed increase in lytic programming is due to the synthesis of molecules directly involved in the programming step or a different molecular mechanism.

VII. SUMMARY AND PROGNOSIS

When cytotoxicity assays are performed under appropriate conditions it is possible to obtain reasonably precise estimates for the kinetic parameters V_{max} and K_M^{app}. The interpretation of a V_{max} value for NK reactions is straightforward in that, whether a simple or mutlistep model is applied, it is equal to the concentration of NK cells tested times the mean lytic activity per NK cell. With available methods for statistical analysis (6), one can readily obtain very precise estimates for V_{max}, so that this is a very useful measure of the overall cytotoxic activity of a given preparation of lymphocytes for a particular target cell line. On the other hand, interpretation of the physical and biological significance of experimentally-determined K_M^{app} values is highly dependent upon the kinetic model that is applied. Earlier interpretations that related K_M^{app} values to dissociation constants for effector-target complexes are now known to be incorrect (7). Based on the experiments reported herein, it is also clear that K_M^{app} cannot be simply related to the concentration of NK effector cells in the cytotoxicity assay system. Given the available multistep model for NK cytotoxicity (equation 12) and our experimental values for k_2 and k_4, the following general observations are appropriate: (a) the steps associated with the rate constants k_2 and k_4 are (or can be approximated as) first order processes, (b) comparison of k_2 and k_4 values indicates that lytic programming is the rate-determining step in NK reactions, (c) nonlytic target binding lymphocytes appear to have a significant effect on the value of K_M^{app}, (d) the ratio K_M/K_I is consistently less than unity suggesting that the affinity of NK effector cells for target cells is greater than the affinity of nonlytic target binding cells, and (e) while directly proportional to the frequency of target binding lymphocytes, K_M^{app} values for NK reactions are also dependent on other parameters and therefore cannot be assigned a simple biological interpretation.

Investigation of the effect of interferon pretreatment of human PBL on the kinetic parameters for natural cytotoxicity suggests that the primary mechanism for enhancement of NK by interferon involves acceleration of the rate of lytic programming. This result is in contrast to several previous studies that suggest a dual role for interferon (21-24). This discrepancy may be due to the different experimental approaches used. It is also possible that the enhancement produced by longer exposure to interferon may involve additional mechanisms. Clearly, further experimentation along these lines is needed.

The methods that have been developed for measuring the rate constants k_2 and k_4 for natural cytotoxicity reactions should prove useful in a variety of studies of cytotoxicity reactions.

For example, values obtained thus far for k_2 and k_4 for natural cytotoxicity indicate that the lytic step is extremely rapid, which favors certain mechanisms for the lytic event over others that have been proposed. In addition, these methods should prove valuable for investigation of the mechanism of action of various agents that either enhance or suppress NK reactions. Comparisons of k_2 and k_4 values for different effector - target cell combinations may also help to explain the relative susceptibility of various cell lines to natural killing. Determination of k_2 values, in conjunction with determinations of V_{max} and single cell cytotoxicity assays should also help to clarify the nature of the defects that reduce NK activity in certain disease states.

REFERENCES

1. Callewaert, D. M., Johnson, D. F., and Kearney, J. (1978). J. Immunol. 121, 710.
2. Pollack, S. B., and Emmons, S. L. (1979). J. Immunol. 123, 160.
3. Forbes, J. T., and Oldham, R. K. (1980). In "Natural Cell-Mediated Immunity Against Tumors" (R. B. Herberman, ed.), pp. 819-834. Academic Press, New York.
4. Michaelis, L., and Menten, M. L. (1913). Biochem. Z. 49, 333.
5. Callewaert, D. M., Genyea, J., Mahle, N. H., Dayner, S., Korzeniewski, C., and Schult, S. (1983). Scand. J. Immunol. 17, 479.
6. Callewaert, D. M., Smeekens, S. P., and Mahle, N. H. (1982). J. Immunol. Methods 49, 25.
7. Pross, H. F., Callewaert, D. M., and Rubin, P. (1984). In "Immunobiology of Natural Killer Cells" (E. Lotzova and R. B. Herberman, eds.) CRC Press, (in press).
8. Callewaert, D.M., Genyea, J., Mahle, N.H., and Korzeniewski, C. (1983). J. Immunol. Methods 65, 351.
9. Bonavida, B., Bradley, T. P., and Grimm, E. A. (1983). Immunol. Today 4, 196.
10. Rubin, P., Pross, H. F., and Roder, J. C. (1982). J. Immunol. 128, 2553.
11. Targan, S. (1980). J. Clin. Lab. Immunol. 4, 165.
12. Hiserodt, J. C., Britvan, L. J., and Targan, S. R. (1982) J. Immunol. 129, 1782.
13. Merrill, S. J. (1982). Math. Biosciences. 62, 219.
14. Martz, E. (1977). In "Contemporary Topics in Immunobiology" (O. Stutman, ed.), Vol. 7. pp. 301-361. Plenum Press, New York.
15. Callewaert, D. M., Mahle, N. H., Wilusz, J. R., Chores,

J. B., Baker, S., Thomas, R., and Ruedisueli, E. (1984) Natural Immunity (in press).

16. Callewaert, D. M., Kaplan, J., Peterson, W. D., and Lightbody, J. J. (1977). Cell. Immunol. 33, 11.
17. Djeu, J. Y., Heinbaugh, J. A., Holden, H. T., and Herberman, R. B. (1979). J. Immunol. 122, 175.
18. Herberman, R. B., Ortaldo, J. R., and Bonnard, G. D. (1979). Nature 277, 221.
19. Heron, I., Hokland, M., Moller-Larsen, A., and Berg, K. (1979). Cell. Immunol. 42, 183.
20. Trinchieri, G., Santoli, D., and Koprowski, H. (1978). J. Immunol. 120, 1849.
21. Gustafsson, A., and Lundgren, E. (1981). Cell. Immunol. 62, 367.
22. Trinchieri, G., Santoli, D., Granato, D., and Perussia, B. (1981). Fed. Proc. 40, 2705.
23. Silva, A., Bonavida, B., and Targan, S. (1980). J. Immunol. 125, 479.
24. Targan, S., and Dorey, F. (1980). J. Immunol., 124, 2157.
25. Saksela, E., Timonen, T., and Cantell, K. (1979). Scand. J. Immunol. 10, 257.
26. Berthold, W., Masucci, M. G., Klein, E., and Strander, H. (1981). Human Lymph. Differ. 1, 1.
27. Ortaldo, J. R., Lang, N. P., Timonen, T., and Herberman, R. B. (1981). J. Interferon Res. 1, 253.
28. Callewaert, D. M., and Mahle, N. H. (1983) Fifth Int. Cong. Immunol. Abstract 501-11.
29. Senik, A. Kolb, J. P., Orn, A., and Golub, M. (1980). Scand. J. Immunol. 12, 51.
30. Ortaldo, J. R., Phillips, W., Wasserman, K., and Herberman, R. B. (1980). J. Immunol. 125, 1839.
31. Gustafsson, A., Sundstrom, S., Ny, T., and Lundgren, E. (1982). J. Immunol. 129, 1952.

Augmentation of Natural Killer Cell Activity by Recombinant Interleukin-2 and Recombinant Interferons

Michael J. Brunda
Viveca Davatelis

Department of Experimental and Applied Biology
Roche Research Center
Hoffmann - La Roche Inc.
Nutley, NJ 07110

I. INTRODUCTION

Interleukin-2 (IL-2) is a lymphokine which was first described because of its ability to act as a second signal and induce the proliferation of activated T cells (1). Subsequently, in addition to its effects on cytotoxic T cells, IL-2 has been shown to initate the generation of, or to modulate the functional activity of, several other cytotoxic lymphocyte populations. It has been shown that various preparations of natural IL-2 augment cytotoxic activity *in vitro* of both murine and human natural killer (NK) cells (2-8) and murine natural cytotoxic cells (9). Furthermore, IL-2 can induce the proliferation of NK cells (4,10,11). Following culture for several days with IL-2, lymphokine-activated killer (LAK) cells, which are cytotoxic to NK-resistant target cells, also have been generated from human or murine cells (12,13). Therefore, IL-2 appears to have a multiplicity of effects on cytotoxic cell populations.

Recently the sequence of the DNA encoding the gene for human IL-2 has been determined (14). Using recombinant DNA technology, it has been possible to express and produce human IL-2 protein in *Escherichia coli* and purify it in large quantities. The recombinant IL-2 (rIL-2) appears to have a similar amino acid sequence to natural IL-2 (14,15,16) although, in contrast to natural IL-2, rIL-2 is not glycosylated. rIL-2 joins a growing list of immunomodulators produced by recombinant DNA technology, such as recombinant interferon alpha (rIFN-α) and gamma (rIFN-γ), that are available in pure form in large amounts.

ISBN 0-12-341370-2

In the following study, experiments were undertaken to evaluate the ability of rIL-2 to modulate NK activity of both murine and human lymphocytes. In addition, the interaction of rIL-2 with rIFN-α and rIFN-γ on the *in vitro* modulation of NK activity was also examined. The results presented in this study indicate that rIL-2 can cause a marked increase in NK activity which is, in terms of the level of cytotoxicity, higher than that seen with rIFN-α. Furthermore, rIL-2 can interact with either rIFN-α or IFN-γ to result in a synergistic increase in NK activity. The data are consistent with the hypothesis that rIL-2 can augment NK activity under some conditions independently from its ability to stimulate IFN-γ production.

II. RESULTS

The ability of purified human rIL-2, produced in *E. coli*, to augment murine or human NK activity *in vitro* was evaluated and the data are presented in Table I. When murine spleen cells were incubated for 18 hrs. with rIL-2, a dose dependent augmentation of NK activity was observed. Similarly, rIL-2 was able to increase the cytotoxic activity of human peripheral blood lymphocytes (PBL). To determine if this augmentation was a direct effect on NK cells, large granular lymphocytes (LGL) from PBL were enriched for by discontinuous Percoll gradients (17) and exposed to rIL-2. As with whole PBL, marked augmentation of NK activity was found, suggesting that the rIL-2 acted directly on LGL. Human rIL-2 prepared in COS cells, and thus presumably glycosylated, gave an equivalent augmentation of NK activity dependent upon the number of units of activity added (data not shown) . These results confirm the results of prior studies indicating that natural or rIL-2 can augment NK activity (2-8).

Since human rIL-2 was active *in vitro* on murine cells and was available in sufficient quantities, the ability of rIL-2 to modulate NK activity *in vivo* was tested using CBA/J mice. Treatment of mice with 3 daily intraperitoneal (ip) injections of 100,000 U of rIL-2 resulted in increased cytotoxic activity both in the spleen and peritoneal cavity (Table II). Treatment of mice with lower doses or fewer injections resulted in little or no augmentation in the spleen. These latter treatments resulted in increases in the peritoneal cavity but these were less marked than that observed with 3 daily injections of rIL-2 (data not shown). These results indicate that human rIL-2 can augment NK activity *in vivo* as well as *in vitro.*

TABLE I

Augmentation of Murine and Human NK Activity by rIL-2 *In Vitro*[a]

Cell Source	rIL-2 (Units/ml) 0	10	100	1,000
Mouse Spleen	0	2	29	68
Human PBL	10	19	23	48
Human LGL	19	31	51	60

[a] CBA/J spleen cells, human PBL or human PBL enriched for LGL were incubated for 18 hrs. with medium or rIL-2, washed and assayed for cytotoxicity. Spleen cells were assayed against YAC-1 at an E/T of 200/1. Human PBL and LGL were assayed against K562 at an E/T of 50/1 and 12.5/1 respectively.

TABLE II

Augmentation of NK Activity *In Vivo* by rIL-2[a]

Treatment	% Cytotoxicity Spleen	Peritoneal Cavity
PBS	22	3
100,000 U rIL-2	47	75

[a] CBA/J injected ip with PBS or 100,000 U rIL-2 daily for 3 days. Cytotoxicity assayed against YAC-1 at E/T of 100/1.

It has been previously demonstrated that IFN can also augment NK activity of both human and murine cells. Using recombinant DNA technology, human alpha IFNs have been previously expressed and produced in *E. coli* and purified in large quantities. In addition, molecular hybrids between subtypes of IFN-α have been constructed (18); one of these hybrids, designated rIFN-αA/D, has gained the ability to augment NK activity of murine cells *in vitro* and in vivo (19,20). We have compared the ability of rIL-2 and rIFN-α to augment NK activity using both murine and human lymphocytes (Table III). With murine spleen cells, both human rIL-2 and rIFN-αA/D were active but rIL-2 consistently induced higher levels of cytotoxicity than rIFN-αA/D. It should be emphasized that units of IL-2 and IFN-α are not equivalent. However, the maximum level of NK cytotoxicity is obtained following incubation of spleen cells in 1,000 U of IFN-α and higher levels of cytotoxicity were routinely observed when cells were incubated with rIL-2. Likewise, rIL-2 induced higher levels of cytotoxicity than rIFN-αA when incubated with human PBL or LGL.

TABLE III

Augmentation of Murine and Human NK Activity by Human rIL-2 or rIFN-α In Vitro[a]

		Units/ml				
Cell Source	Treatment	0	10	100	1,000	10,000
Mouse Spleen	rIL-2	0[b]	-[c]	16	44	69
	rIFN-αA/D	0	0	26	37	39
Human PBL	rIL-2	6	15	16	41	82
	rIFN-αA	6	27	37	39	46
Human LGL	rIL-2	27	27	51	67	71
	rIFN-αA	27	33	48	54	-

[a] CBA/J spleen cells, human PBL or human PBL enriched for LGL were incubated for 18 hrs. with medium, rIL-2 or rIFN-α as listed. Murine spleen cells were assayed against YAC-1 at an E/T ratio of 100/1. Human PBL and LGL were assayed against K562 at an E/T of 25/1 and 12.5/1 respectively.

[b] % Cytotoxicity

[c] Not tested

It has previously been reported that partially purified murine IL-2 and partially purified murine IFN-β could synergistically augment NK activity (3). Since both rIL-2 and rIFN-αA/D could augment murine NK in vitro, the interaction of these two proteins on the modulation of NK activity was evaluated (Table IV). To maximize the detection of synergistic or additive effects a dose of rIFN-αA/D, 10 U/ml, was chosen that was inactive by itself. As shown in the two experiments, combining rIFN-αA/D and rIL-2 resulted in a synergistic increase in NK activity. The results presented in Tables I, III and IV clearly demonstrate that rIFN-α and rIL-2 can interact independently to increase NK activity and that combining these two proteins leads to a synergistic interaction.

Synergistic induction of NK (21) or anti-viral (22) activity has been previously reported when combinations of IFN-α and IFN-γ have been tested. Since IL-2 is known to help induce IFN-γ production by lymphocytes (23,24), it is possible that the enhanced cytotoxicity observed in these experiments might be the result of an induction of IFN-γ by rIL-2 and the subsequent synergistic interaction of IFN-γ with IFN-αA/D. To determine if the induction of IFN-γ was the pathway through which augmentation of NK activity by rIL-2 occurred, murine spleen cells were incubated with either rIL-2 alone or rIL-2 in the presence of 2,000 neutralizing units of anti-IFN-γ antibody (Table V). As expected, rIL-2 enhanced the cytotoxic activity of murine

TABLE IV

Synergistic Augmentation of Murine NK Activity *In Vitro* by rIL-2 and rIFN-αA/D[a]

Expt.	Treatment	% Cytotoxicity[b]	$LU_{20}/10^7$ Cells[c] Observed	Expected
1	Medium	1	0	0
	1,000 U/ml rIL-2	21	45	45
	10 U/ml rIFN-αA/D	0	0	0
	1,000 U/ml rIL-2+ 10 U/ml rIFN-αA/D	33	112	45
2	Medium	2	0	0
	1,000 U/ml rIL-2	38	99	99
	10 U/ml rIFN-αA/D	2	0	0
	1,000 U/ml rIL-2+ 10 U/ml rIFN-αA/D	53	333	99

[a] CBA/J spleen cells were incubated for 18 hrs. with medium, rIL-2 or rIFN-αA/D as listed.

[b] E/T ratio is 200/1 against YAC-1.

[c] Lytic units for 20% cytotoxicity per 10^7 cells.

spleen cells. In the presence of antibody to murine IFN-γ, there was little inhibition of NK augmenting activity of rIL-2. Therefore,the data suggest that, under these conditions, IFN-γ induction plays little or no role in the enhancement of NK activity by rIL-2.

TABLE V

Failure to Inhibit rIL-2 Induced Augmentation of Murine NK Activity by anti-IFN-γ[a]

Treatment	% Cytotoxicity[b]	$LU_{20}/10^7$	% Inhibition
Medium	4	0	-
1,000 U/ml rIL-2	68	50	-
1,000 U/ml rIL-2+ anti-IFN-γ	54	43	14

[a] CBA/J spleen cells were incubated for 18 hrs. with medium, rIL-2 or rIL-2 plus anti-IFN-γ

[b] E/T ratio is 200/1 against YAC-1.

Since the preceding data suggested that human rIL-2 could interact with rIFN-α to cause a synergistic increase of murine NK activity, the ability of these two recombinant products to interact with human LGL and modulate cytotoxicity was examined. In addition, the ability of human rIL-2 and rIFN-γ to interact was also tested. Doses of rIFN-αA (1 U/ml), rIFN-γ (10 U/ml) and rIL-2 (100 U/ml) were selected which alone gave only modest increases in cytotoxicity (Table VI). When rIL-2 was combined with rIFN-αA, a greater than expected increase in cytotoxicity was observed, similar to the data with murine spleen cells. Combining rIL-2 with rIFN-γ also resulted in a comparable synergistic increase of cytotoxicity. Those results indicate that rIL-2 can interact with either rIFN-αA or rIFN-γ to induce a synergistic increase in NK activity of human LGL.

The data presented earlier (Table V) suggested to us that at least with murine cells rIL-2 could have a direct augmenting effect on natural killer cells. To determine if this was also true with human LGL, we examined the time requirements necessary for augmentation of NK activity. As seen in Table VII, when LGL were incubated for 2 hrs. with rIFN-αA, cytotoxicity was enhanced. In contrast, this short treatment of LGL with rIFN-γ was not sufficient for augmentation to occur. rIL-2, like rIFN-αA, also induced marked augmentation of NK activity of LGL following a 2hr. incubation. These results clearly indicate that rIL-2 can activate NK cells under conditions where rIFN-γ is inactive.

TABLE VI

Synergistic Augmentation of Human NK Activity by rIL-2 and rIFN-αA or rIFN-γ[a]

Treatment	% Cytotoxicity[b]	$LU_{50}/10^7$ cells Observed	Expected
Medium	19	13	13
100 U/ml rIL-2	52	173	173
1 U/ml rIFN-αA	37	78	78
10 U/ml rIFN-γ	41	81	81
100 U/ml rIL-2+ 1 U/ml rIFN-αA	85	423	251
100 U/ml rIL-2+ 10 U/ml rIFN-γ	77	380	254

[a] LGL were incubated for 18 hrs. with medium, rIL-2, rIFN-αA or rIFN-γ as listed, washed, and assayed for cytotoxicity.

[b] E/T ratio is 12.5/1 against K562.

III. DISCUSSION

The experiments summarized in this report describe our preliminary studies on the ability of human rIL-2 to augment human and murine activity. It is clear from the data presented that pure rIL-2 can augment NK activity of human PBL or LGL *in vitro*. In addition, human rIL-2 can augment NK activity of murine spleen cells *in vitro* or following injection *in vivo*. Other workers have also recently demonstrated the ability of rIL-2 to augment human NK cells (7,8) or LAK cells (14) *in vitro*. Thus, rIL-2 appears to be, along with the IFNs, a major regulatory molecule for NK activity as well as other cytotoxic cells.

TABLE VII

Augmentation of Human NK Activity Following 2 hr. Incubation with rIL-2, rIFN-αA or rIFN-γ[a]

Treatment	% Cytotoxicity[b]	$LU_{50}/10^7$[c]
Medium	19	13
1,000 U/ml rIL-2	60	239
1,000 U/ml rIFN-αA	49	146
1,000 U/ml rIFN-γ	23	20

[a] LGL were incubated for 2 hrs with medium, rIL-2, rIFN-αA or rIFN-γ, washed, and assayed for cytotoxicity.

[b] E/T ratio is 12.5/1 against K562.

[c] Lytic units giving 50% cytotoxicity per 10^7 cells.

The mechanism of NK cell activation is not known but it has been suggested that IL-2 might cause an indirect augumentation of NK activity through the induction of IFN-γ. It has been established that either T cells or LGL can produce IFN-γ following stimulation with IL-2, (23, 24). A second stimulus, such as a mitogen, is required to obtain maximal amounts of IFN-γ but low levels of IFN-γ can be produced by LGL in the absence of another stimulant. To examine whether the augmentation of NK activity by rIL-2 was dependent on the induction of IFN-γ, we performed 2 types of experiments. In the first anti-murine IFN-γ was added to cultures of murine cells and rIL-2. This treatment resulted in little inhibition of the augmented NK activity (Table V) suggesting that the production of IFN-γ was not the major mechanism by which rIL-2 enhanced NK activity. Using human (7,8) or murine (5) cells other workers have done similar experiments; some obtained the opposite results (5,8) while others have obtained results similar to those presented here (7). The reason for this discrepancy is not known but may reflect slight differences in techniques used, the purity of the IL-2 or the antibodies tested.

A second approach used to examine the potential requirement for IFN-γ induction for augmentation of NK activity by human LGL involved testing the effect of short term exposure (2 hrs.) of LGL to rIL-2, rIFN-αA, and rIFN-γ. Under the conditions used, rIL-2 and rIFN-αA augmented NK activity while rIFN-γ was inactive. Similar results demonstrating short term augmentation of NK activity by rIL-2 but not rIFN-γ were reported recently (7). Since rIL-2 requires a minimum of a 3 hr. incubation with lymphocytes before any detectable levels of rIFN-γ are produced (7), the results here clearly indicate that induction IFN-γ is not absolutely necessary for augmentation of NK activity by rIL-2.

It is likely that there are at least two ways that rIL-2 can augment NK activity. The first is a direct stimulation of NK cells, which appear to have IL-2 receptors since they are able to grow _in vitro_ in response to IL-2 (4, 10, 11). A second pathway is the induction of IFN-γ from T cells or LGL which in turn stimulates NK activity. The primary pathway of stimulation could depend on the culture conditions used and may vary in different laboratories. No data is available to evaluate which pathway is of the most importance _in vivo._

Recombinant DNA technology is providing a growing list of proteins which are available in pure form and have potent biological activities. We tested 3 of these proteins, rIL-2, rIFN-α and rIFN-γ for their abilities to modulate NK activity. All 3 were active in augmenting cytotoxicity when tested alone. Comparing rIL-2 and rIFN-α, we observed that rIL-2 induced a higher maximal level of cytotoxicity than rIFN-α. In the mouse, some work has demonstrated that different subpopulations of NK cells respond to IFN-α or IL-2 (25). The maximal levels of cytotoxicity observed in our experiments may reflect differences in cytotoxic potential of such subpopulations after augmentation by rIFN-α or rIL-2. In addition, combining IFN-α with rIL-2 resulted in a synergistic increase in NK activity. This confirms prior work demonstrating a synergistic augmentation of murine NK activity by cruder preparations of IL-2 and IFN-β (3). In addition, using human LGL, rIL-2 synergized with rIFN-γ as well as rIFN-α resulting in a greater than expected rise in cytotoxicity. This has also been recently observed by other workers (7). If these _in vitro_ results can be extended to the _in vivo_ situation, it may be possible to treat animals or humans with combinations of recombinant proteins to maximize augmentation of NK activity while potentially minimizing other deleterious side effects. If so, combination treatments with rIL-2, rIFN-α and/or rIFN-γ could result in increased efficacy compared to treatment with any single recombinant protein.

ACKNOWLEDGEMENT

We thank Deborah Rosenbaum, Kim Kaffka and Deborah Tarnowski for their help in performing some of these experiments and Debra Guerriero for typing this manuscript. We also wish to thank all those involved at Hoffmann-La Roche Inc. in the cloning, production and purification of the recombinant IFNs and IL-2.

REFERENCES

1. Farrar, J. J., et al., Immunol. Rev. 63, 129, 1982.
2. Henney, C. S., et al., Nature 291, 335, 1981.
3. Kurabayashi, K., et al., J. Immunol. 126, 2321, 1981.
4. Timonen, T., et al., Cell. Immunol. 72, 178, 1982.
5. Weigert, D. A., Stanton, G. J. and Johnson, H. M., Infect. Immun. 41, 992, 1983.
6. Rook, A. H., et al., J. Clin. Invest. 72, 398, 1983.
7. Svedersky, L. P., et al., J. Immunol. 33, 714, 1984.
8. Ortaldo, J. R., et al., J. Immunol. 133, 779, 1984.
9. Lattime, E. C., Pecoraro, G. A. and Stutman, O., J. Exp. Med. 157, 1070, 1983.
10. Suzuki, R., et al., J. Immunol. 130, 981, 1983.
11. Handa, K., et al., J. Immunol. 130, 988, 1983.
12. Grimm, E. A., et al., J. Exp. Med. 155, 1823, 1982.
13. Rosenberg, S.A., et al., Science 223, 1412, 1984.
14. Taniguchi, T., et al., Nature 302, 305, 1983.
15. Robb, R. J., Kutny, R. M. and Chowdhry, V., Proc. Natl. Acad. Sci. USA 80, 5990, 1983.
16. Stern, A. S., et al., Proc. Natl. Acad. Sci. USA 81, 871, 1984.
17. Timonen, T. and Saksela, E., J. Immunol. Methods 36, 285, 1980.
18. Rehberg, E., et al., J. Biol. Chem 257, 11497, 1982.
19. Brunda, M. J., and Rosenbaum, D., Cancer Res. 44, 597, 1984.
20. Brunda, M. J., Rosenbaum, D. and Stern, L., Int. J. Cancer, In Press, 1984.
21. Weigert, D. A., et al., Infect. Immun. 40, 35, 1983.

22. Fleischmann, W. R., Jr., *et al.*, Infect. Immun. 26. 248, 1979.
23. Farrar, W. L., Johnson, H. M. and Farrar, J. J., J. Immunol. 126, 1120, 1981.
24. Herberman, R.B., *et al*, *In* Natural Killer Activity and Its Regulation. T. Hoshimo, H. S. Koren and A. Uchida (Eds.). Excerpta Medica, Amsterdam. p. 409, 1984.
25. Minato, N., Reid, L. and Bloom, B.R., J. Exp. Med. 154, 750, 1981.

LYMPHOKINES INVOLVED IN THE GENERATION AND REGULATION OF NK AND NC CELLS[1]

Edmund C. Lattime
Stuart Macphail
Osias Stutman

Immunobiology Program
Memorial Sloan-Kettering Cancer Center
New York, New York

I. INTRODUCTION

Initial studies from our laboratory (1,2), since confirmed and extended by others (3), demonstrated that natural cell mediated cytotoxicity (NCMC) in mice is the function of at least two distinct effector cell populations, natural killer (NK) (4) and natural cytotoxic (NC) (1,2) cells. While these two cells share the characteristics which have come to define the NCMC system, most notably, existence at high levels without the need for specific priming (1,4) and lack of major histocompatibility restriction for lysis (1,4), we have shown them to be distinct based on kinetics of lysis (1,4), cell surface antigen phenotype (5) and susceptibility to regulation by interferon (IFN) (6,7). Whereas lymphoid tumor targets are principally susceptible to NK cell lysis, we have shown that the majority of non-lymphoid tumor cell lines are sensitive to lysis by a combination of NK and NC cell activities (8). The term NCMC has also been used to encompass a number of effector cell populations induced in culture ie. activated killer (AK) cells, which, although they share a number of the characteristics of the traditional NK cell, require in vitro culture for their activation (reviewed in ref. 9).

1. Supported by National Institute of Health Grants CA-08748, CA-15988, CA-17818, CA-29803 and American Cancer Society Grant IM-188. E.C.L. is a scholar of the Leukemia Society of America.

ISBN 0-12-341370-2

In this report, we summarize our published studies into the regulation of NK and NC cell activities by IFN and the lymphokines interleukin-2 (IL-2) and interleukin-3 (IL-3). We then describe recent studies into AK cell activity with primary emphasis on (1) the relationship of AK cells to NK, NC and cytotoxic T lymphocytes (CTL); and (2) the mechanism involved in AK cell activation in culture.

II. REGULATION OF NK AND NC ACTIVITIES BY IFN, IL-2 AND IL-3

In order to further characterize the effector populations mediating NCMC, and in particular elucidate the mechanism(s) involved in their regulation, the effect of the lymphokines IFN, IL-2 and IL-3 on NK and NC cell activities were examined. NCMC effector cells can be distinguished by their expression of a number of cell surface antigens such as Qa-5, Qa-2,3, and asialo-GM-1 which are expressed on NK, but not NC cells (5). Using separation of NK and NC effector functions by anti-Qa-5 + C and the prototype NK (YAC-1) and NC (WEHI-164) target cell lines, it is possible to examine the regulation of both NK and NC activities in a single experiment using the same starting populations (7,8).

In vivo administration of the IFN inducing compound poly-IC augmented the lysis of the YAC-1 target by NK cells ($Qa-5^+$), while having no effect on the NC cell ($Qa-5^-$) mediated lysis of the WEHI-164 target (Table I Expt. 1). As we have previously reported (8), whereas WEHI-164 is the prototype NC target, it is, partially sensitive to highly activated NK cells. Poly-IC treatment resulted in an augmentation of WEHI-164 lysis (Expt. 1), however it was the function of a $Qa-5^+$ NK population, and not the $Qa-5^-$ NC cell (see the WEHI(Qa-5-) column in Table I). In vitro studies (Expt. 2, Table I) demonstrated that both NK ($Qa-5^+$) and NC ($Qa-5^-$) cell activities were augmented by IL-2, while only NC cell activity was increased by IL-3. As was the case with Poly-IC, a small percentage of augmented lysis of WEHI-164 was shown to be mediated by activated NK ($Qa-5^+$) cells in the IL-2 studies, while there was no augmented lysis of either target by NK ($Qa-5^+$) cells following culture with IL-3. Consistent with the separation of NK and NC cell activities by Qa-5 phenotype at the effector level, pretreatment of spleen cells with anti-Qa-5 + C (Expt. 3, Table I) eliminated both control and augmented NK cell activities, but failed to eliminate the activity of either control or augmented NC cells.

The findings described in Table I, give additional

Table I. NK and NC Cell Regulation by Lymphokines[a]

Exp.	Effector Cell[b]	Treatment[c]	Percent Specific Lysis YAC-1 (NK) 100:1	YAC-1 (NK) 50:1	WEHI-164 (NC) 100:1	WEHI-164 (NC) 50:1
1	C57BL/6J	C	20	6	35	26
		Qa-5 + C	0	0	30	28
	C57BL/6J (P-IC)	C	51*	45*	48	34
		Qa-5 + C	0	0	37	24
2	C57BL/6J (fresh)	C	27	24	47	37
		Qa-5 + C	8	3	47	33
	24 hrs.	C	9	9	46	29
		Qa-5 + C	4	4	48	33
	IL-2	C	47*	38*	79*	63*
		Qa-5 + C	9	5	70*	53*
	IL-3	C	11	8	62*	56*
		Qa-5 + C	4	5	66*	55*
3	C57BL/6J	(24 hr.)	9	7	37	30
	(C)	(IL-2)	42*	38*	74*	53*
	"	(IL-3)	nd	nd	62*	56*
	C57BL/6J	(24 hr.)	6	4	43	30
	(Qa-5-)	(IL-2)	6	4	61*	47*
	"	(IL-3)	nd	nd	56*	44*

[a] 18 hr. ^{51}Cr assays using 5 X 10^3 targets were used.
[b] Spleen cells from C57BL/6J mice untreated or injected with 100ug poly-IC ip 24 hrs. prior to the assay (Exp. 1); C57BL/6J spleen cells, fresh or cultured for 24 hrs. alone or with IL-2 (20 U/ml) or IL-3 (20 ED_{50}/ml) (Exp. 2); C57BL/6J spleen cells pretreated with C or anti-Qa-5+C prior to culture alone or with IL-2 or IL-3 as above (Exp. 3) (7). nd = not done.
[c] Prior to the ^{51}Cr assay, effector cells were treated with the indicated Ab+C (7). Pretreatment cell numbers were not restored prior to the assay.
* p<0.05 significant augmentation by Students t test

evidence that NCMC is mediated by at least two distinct effector cells, NK and NC. Furthermore, it is shown that the two activities are regulated by an overlapping group of lymphokines with IFN and IL-2 but not IL-3 augmenting NK and IL-2 and IL-3, but not IFN augmenting NC cell activity (7).

III. AK CELL GENERATION IN CULTURE AND ITS REGULATION

We have previously shown that NC cell activity is maintained in culture for at least six days (2). Although NK cell activity is lost within 24 hrs. of culture (2), cytotoxic cells with characteristics of the NK cell are generated in a number of culture systems (reviewed in ref. 9). These cells have been called by various names such as anomalous killer, and promiscuous killer, and, depending on the method of generation, express the cell surface antigen and target specificity characteristics of the NK cell (reviewed in refs. 9-11). In order to further characterize the regulation and differentiation of both NK and NC cells, we have examined the lytic populations generated in two culture systems, syngeneic culture in the presence of fetal calf serum (FCS) and allo MLR.

A. Heterogeneity of AK Cells Generated in Culture with FCS

Spleen cells from C57BL/6J mice were placed in culture, and their ability to lyse the YAC-1 (NK) and WEHI-164 (NC) targets was measured at daily intervals (Table II). As previously shown (7), lysis of the YAC-1 target declined at 24 hrs. and was regenerated by 2-3 days in culture, while the lysis of the WEHI-164 target was maintained, and in some experiments increased, during the culture period. YAC-1 lysis reached its peak at 3 days and then gradually declined (data not shown), therefore the day 3 time point was used for all further studies. It should be pointed out that whereas the AK anti-YAC-1 activity declined and was then regenerated or activated from a non lytic precursor, the AK anti-WEHI activity was maintained throughout the culture period. Therefore, although we may use the term activated to describe the AK activities, it is not clear that we are referring to the same mechanism for the two effectors.

When the AK cells were examined for their cell surface antigen expression, it was found that two types of effector cells were present in the cultures (Table III). The first,

Table II. Time Course of AK Generation and IL-2 Production[a]

Effector Cell		Percent Specific Lysis YAC-1 (NK) 100:1	50:1	WEHI-164 (NC) 100:1	50:1	IL-2 U/ml
C57BL/6J	fresh	66	48	46	41	--
	24 hrs.	15	12	50	47	0.9
	48 hrs.	20	16	65	61	4.2*
	72 hrs.	43*	32*	82*	70*	9.8*

[a]C57BL/6J spleen cells were assayed (fresh) or following culture for the indicated number of hours in RPMI-1640 with 10% FCS. For assay conditions see Table I. IL-2 levels were measured in the supernatants as described (13).
* $p < 0.05$ significant augmentation from 24 hr. control by Students t test (cytotoxicity) or probit analysis (IL-2).

which lysed the YAC-1 target expressed an NK-like phenotype (Qa-5+, Thy-1+, Lyt-2-) and the second which lysed the WEHI target, an NC-like phenotype (Qa-5-, Thy-1-, Lyt-2-) (5). Treatment with anti-Thy-1.2 + C only eliminated approximately 40% of fresh non-cultured NK cell activity while it consistently eliminated all YAC-1 lysis by the AK cells. Whether this was due to two different populations present in the fresh preparation (Thy-1.2+ and Thy-1.2-) or due to variability in the susceptibility of the two cells to antibody + C treatment, is currently under study. The results (Table III) do show, however that the two activities present in the AK cultures are distinct from one another, and not the function of CTLs (both are Lyt-2-).

B. Mechanism of Activation of AK Cells

Since the lymphokines IFN and IL-2 have been shown to regulate NK and NC cell activities (7, and this report) and we have shown that IL-2 and IFN are produced when splenic T cells are cultured with syngeneic macrophages (12), we examined the role of these lymphokines in the generation of AK lytic activity. Supernatants from the AK cultures were found to have significant levels of IL-2 (Table II), but no IFN activity (data not shown). We further found that the anti-YAC-1 lytic activity was directly proportional to the amount of IL-2 present during the three days of culture (Table II), suggesting that it was the generation of IL-2 that resulted in the activation of the NK-like cytotoxic activity.

Table III. AK Effector Cell Heterogeneity[a]

Exp.	Effector Cell	Treatment	Percent Specific Lysis YAC-1 (NK) 100:1	50:1	WEHI-164 (NC) 100:1	50:1
1	C57BL/6J (fresh)	C	26	19	41	34
		Qa-5 + C	0*	0*	35	28
		C	26	19	23	10
		Thy-1.2 + C	14*	13*	29	10
		C	16	11	23	10
		Lyt-2 + C	23	12	21	13
2	C57BL/6J (3 day FCS)	C	36	23	47	37
		Qa-5 + C	0*	0*	50	38
		C	36	23	29	23
		Thy-1.2 + C	0*	0*	42	39
		C	31	22	35	27
		Lyt-2 + C	28	24	43	30

[a]C57BL/6J spleen cells either fresh or cultured for 3 days as (Table II) were used in 18 hr. assays (Table I.) Ab + C treatment was done prior to the assay as described (5). Cell numbers were not restored prior to the assay.
* $p < 0.05$ significant reduction from C control.

Although we have shown that IL-2 augments NC cell activity (7, and this report), the finding that lysis of WEHI-164 was not lost in the absence of IL-2 does not preclude the involvement of IL-2 in the regulation of the NC cell in culture, but is consistent with NC being a long lived population not requiring IL-2 for its short term maintenance.

Our previous studies of IL-2 production in syngeneic cell interactions demonstrated that IL-2 production was dependent on self-Ia recognition and the consequence of class II MHC antigen recognition by Lyt-1+, L3T4+ T cells (13, and our unpublished results).

Table IV. AK(YAC) But Not AK(WEHI) Generation is Inhibited by the Addition of Anti-Ia and Anti-L3T4 Antibodies[a]

Effector Cell	Added Antisera	Percent Specific Lysis: YAC-1 (NK) 100:1	YAC-1 (NK) 50:1	WEHI-164 (NC) 100:1	WEHI-164 (NC) 50:1	IL-2 U/ml
C57BL/6J (3 day FCS)	--	25	12	61	50	4.8
	NMS 1:100	25	19	71	62	5.8
	1:200	21	19	68	56	6.0
	Ia[b] 1:100	0*	0*	57*	50*	0.4*
	1:200	0*	0*	58*	53	0.6*
C57BL/6J (3 day FCS)	--	28	13	28	21	2.8
	L3T4 1:100	7*	4*	27	22	<.1
	1:300	8*	2*	28	24	<.1

[a]C57BL/6J spleen cells were cultured for 3 days (Table II). The indicated NMS or antisera was added at the initiation of the culture period as described (12). Effector cells were assayed for cytotoxicity and IL-2 levels were measured as described in Tables I and II.
* $p < 0.05$ significant reduction from control by Students t test (cytotoxicity) and Probit analysis (IL-2)

In order to further assess the role of IL-2 in the regulation of AK cell activity, we examined the effects of inhibiting IL-2 production from the cultures on the two lytic activities. As was the case in our previous studies (13), the addition of either haplotype specific anti-Ia antisera or anti-L3T4 (14) antisera to the culture, resulted in the inhibition of IL-2 production (Table IV). This blockade of IL-2 production significantly reduced, and in most cases eliminated, the generation of the AK anti-YAC-1 cytotoxic activity. Similarly, the removal of either Lyt-1+ or L3T4+ T cells prior to culture (Table V), eliminated IL-2 production and reduced or eliminated AK anti-YAC-1 lysis (control experiments showed that AK anti-YAC-1 killer cells did not express these two markers, data not shown). Table V also shows that the elimination of Qa-5+ cells (putative pre-AK YAC-1) resulted in the elimination of YAC-1 lysis without affecting the production of IL-2, thereby showing that both IL-2 and a pre-AK cytotoxic population are required for the generation of the AK YAC-1 cytotoxic cell.

Table V. AK(YAC) But Not AK(WEHI) Require IL-2 Producing Cells For Activation[a]

Effector Cell	Pretreatment	Percent Specific Lysis YAC-1 (NK) 100:1	50:1	WEHI-164 (NC) 100:1	50:1	IL-2 U/ml
C57BL/6J (3 day)	C	28	13	43	25	2.8
	L3T4 + C	5*	4*	32	25	<.1*
	C	30	24	64	55	4.0
	Lyt-1 + C	0*	0*	40*	35*	<.1*
	C	31	25	nd	nd	7.8
	Qa-5 + C	6*	1*	nd	nd	7.7

[a]Prior to the initiation of 3 day culture C57BL/6J spleen cells were pretreated with the indicated Ab + C (5). Cells were assayed for cytotoxicity and supernatants were measured for IL-2 as described in Tables I and II.
* $p < 0.05$ reduction from C control by Students t test (lysis) and probit analysis (IL-2)

The studies described demonstrate that the activation of AK anti-YAC-1 cytotoxic cells in syngeneic culture is dependent on the production of IL-2. The activation, or perhaps maintenance of anti-WEHI-164 activity, however did not share this dependence, at least through 3 days of culture. While anti-WEHI activity did not require IL-2 production for maintenance, the magnitude of the lysis of WEHI did appear in most studies to be affected by the level of IL-2 in the cultures. For example, elimination of IL-2 production was associated with reductions in anti-WEHI cytotoxicity (see Table V, anti-Lyt-1 pretreatment and Table IV, addition of anti-Ia antisera). While this was not always statistically significant, it was frequently observed. This finding would be consistent with our demonstration that IL-2 augments NC cell activity (7, and Table I) and would suggest that although AK anti-WEHI activity is not dependent on IL-2 production, in the presence of IL-2, lytic activity is increased.

C. AK Cells Generated in Allogeneic MLC Express Increased Heterogeneity

We and others have previously shown that cytotoxic cells which lyse syngeneic tumors (AK), as well as specific CTL, are activated in allogeneic mixed lymphocyte reactions (MLR) (10,11,15). When allogeneic MLRs are done in normal mouse serum (NMS), thus eliminating the contribution of FCS induced activation, a heterogeneous group of effector cells, capable of lysing a single target can be demonstrated (reviewed in ref. 11).

AK anti-YAC activity generated from whole spleen cell responders in MLR was manifested by at least two distinct effector cells (Thy-1+, Lyt-2+, Qa-5- and Thy-1+, Lyt-2-, Qa-5-) (Table VI). When the MLR responder cell population was pretreated with anti-Lyt-2 + C, thus eliminating the pre CTL compartment, most of the cytotoxicity was mediated by Thy-1+, Lyt-2-, Qa-5+ cells which would appear to be the same cells as found in the culture generated effectors presented above (Table III). The data also suggests the presence of the Thy-1+, Lyt-2-, Qa-5- effector generated from whole spleen. The existence of this latter population is dependent on data in which only partial elimination of cytotoxicity is achieved using Ab + C treatment, and will require positive selection using FACS analysis for its verification. A similar positive selection procedure would also be required to confirm the presence of a Thy-1-, Lyt-2- killer cell suggested by the data in Exp. 2, Table VI. The results, however, clearly show an increased heterogeneity in the MLR generated AK cell pool. The reason for the lack of generation of the Qa-5+ AK cell in MLR without anti Lyt-2 + C treatment is not clear. It is possible that the expansion of Lyt-2+ CTL in the 5 day culture may overgrow the Qa-5+ component, or compete with it for available lymphokines. The presence of a Lyt-2+ suppressor cell of the Qa-5+ cell activity can not be ruled out.

As was the case in the FCS system described in section III A, the addition of anti-L3T4 to the MLR culture significantly inhibited both the production of IL-2 and the generation of the Qa-5+ AK cell produced from Lyt-2- responder cells (Table VII). Even though allospecific CTL activity was significantly reduced by anti-L3T4 in the NMS pretreated cultures, the generation of AK cell activity, which was shown to be mediated by a Qa-5- cell (see Table VI, Expt. 2, NMS cultures), was not significantly effected.

At least two possible explanations may exist for the increased heterogeneity found in the MLR activated cells. First, the Lyt-2+ Qa-5- population, not observed in the AK generated in syngeneic culture, may require recognition of the allogeneic histocompatibility antigens for its activation, and may recognize determinants on the YAC-1 target which cross react with those on the allo stimulator cell.

Table VI. Heterogeneity of Syngeneic Tumor Cytotoxic Cells Generated in Allogeneic MLR [a]

Expt.	Pre-treatment	Post-treatment	Percent Specific Lysis: YAC-1 100:1	YAC-1 50:1	DBA/1 blasts 100:1	DBA/1 blasts 50:1
1	NMS	NMS	17	13	71	58
	"	Lyt2.2	11*	6*	5*	1*
	Lyt2.2	NMS	46	34	1	2
	"	Lyt2.2	49	38	1	0
2	NMS	NMS	11	5	71	62
	"	Thy1.2	0*	1*	18*	8*
	"	Qa5	12	7	71	65
	Lyt2.2	NMS	39	24	13	3
	"	Thy1.2	12*	10*	3*	2*
	"	Qa5	8*	5*	5*	4*

[a] (C57BL/6 x A/J)F_1 (CAF1) (H-$2^{b/a}$) responder cells, pretreated with either normal mouse serum (NMS) or monoclonal anti-Lyt2.2 and C, were stimulated with irradiated allogeneic DBA/1 (H-2^q) spleen cells for 5 days. Effector cells were harvested, treated with the various reagents shown and tested for cytotoxicity in a conventional ^{51}Cr release assay against DBA/1 Con A blast and YAC-1 (H-2^a) target cells at the E:T ratios shown, as previously described (11). Cells harvested from control unstimulated (CAF1 anti-CAF1) MLR cultures were not cytotoxic for DBA/1 blast cells and in Expt.1 lysed YAC-1 targets at 5.1% (100:1) and 2.7% (50:1) and in Expt. 2 at 3.4% (100:1) and 1.8% (50:1).

*$p<0.05$ significant reduction from C control by Students T test.

Second, the generation of higher levels of IL-2, and perhaps other lymphokines may also play a role in the generation of the lytic cells not observed in the culture generated AK. This latter possibility is supported by studies, which show that cloned Lyt-2+ CTL lines, when cultured in increasing levels of IL-2 containing conditioned medium, lyse YAC-1 and other NK sensitive cell lines (16). While the FCS cultured cells above generated between 4 and 10 U/ml of IL-2, MLRs generated 15-40 U/ml. The IL-2 levels shown in Table VII were exceptionally low for the MLR studies, but were significantly higher than in the FCS cultures. Therefore, the Lyt-2+ AK cell which we describe, may be similarly derived from CTL under the influence of the increased IL-2 generated in the cultures.

Table VII. Anti-L3T4 Antibody Inhibits Both IL-2 Production and Activation of Qa-5+ Tumor Killer Cells in Allogeneic MLR[a]

Effector Pretreatment	Added Antisera		Percent Specific Lysis YAC-1 (NK) 50:1	10:1	DBA/1 blasts 50:1	10:1	IL-2 U/ml
NMS + C	Asc.	1:100	17	5	41	22	12.1
		1:300	19	6	55	26	13.6
	L3T4	1:100	17	3	17*	5*	1.8*
		1:300	16	3	23*	7*	3.7*
Anti-Lyt-2 + C	Asc.	1:100	32	14	2	1	15.7
		1:300	39	12	1	0	15.0
	L3T4	1:100	7*	4*	2	0	0.4
		1:300	8*	5*	0	0	0.6

[a]Effector cells were generated and tested as in Table VI. Indicated dilutions of either anti-L3T4 or myeloma ascites fluid were added to the generating cultures. Supernatant IL-2 levels were measured on day 3.
*p<0.05 significant reduction by Students t test

IV. CONCLUSIONS

Two main conclusions can be reached from the present studies: 1) that the effectors of NCMC as well of in vitro activated AK cells are heterogeneous and 2) that lymphokines, especially IL-2, appear to be important in the overall function of all of these effectors. It could be added that in the case of the two AK models studied, IL-2 was produced by Lyt 1/L3T4+ T cells in response to syngeneic class II (Ia) surface antigens (13 ,and Table IV) and that such production was mandatory for the demonstration of the AK activity.

REFERENCES

1. Stutman, O., Paige, C.J., and Figarella, E.F. (1978) J. Immunol. 121, 1819.
2. Paige, C.J., Figarella, E.F., Cuttito, M.J., Cahan, A. and Stutman, O. (1978). J. Immunol. 121, 1827.
3. Burton R.C. (1980). In "Natural Cell-Mediated Immunity Against Tumors" (R.B. Herberman ed.), p. 45. Academic Press, New York.
4. Herberman, R.B., and Holden, H.T. (1978). Advanc. Cancer Res. 27, 305.
5. Lattime, E.C., Pecoraro, G.A., and Stutman, O. (1981). J. Immunol. 126, 2011.
6. Senik, A., Gresser, I., Maury, C., Gidlund, M., Orn, A., and Wigzell, H. (1979). Cell Immunol. 44, 186.
7. Lattime, E.C., Pecoraro, G.A., and Stutman, O. (1983). J. Expt. Med. 157, 1070.
8. Lattime, E.C., Pecoraro, G.A., Cuttito, M.J., and Stutman, O. (1983). Int. J. Cancer. 32, 523.
9. Stutman O., and Lattime, E.C. In "Lymphokines" (M. Landy, ed.), vol. 12, (in press). Academic Press, New York.
10. Paciucci, P.A., Macphail, S., Zarling, J.M., and Bach, F.H. (1980). J. Immunol. 124, 370.
11. Macphail, S., Paciucci, P.A., and Stutman, O. (1984). J. Immunol. 132, 3205.
12. Lattime, E.C., Gillis, S., David, C., and Stutman, O. 1981. Eur. J. Immunol. 11, 67.
13. Lattime, E.C., Gillis, S., Pecoraro, G.A., and Stutman, O. (1982). J. Immunol. 128, 480.
14. Dialynas, D.P., Quan, Z.S., Wall, K.A., Pierres, A., Quintans, J., Loken, M.R., Pierres, M., and Fitch, F.W. (1983). J. Immunol. 131, 2445.
15. Karre, K., and Seeley, J.K. (1979). J. Immunol. 123, 1511.
16. Brooks, C.G., Urdal, D.L., and Henney, C.S. (1983). Immunological Rev. 72, 43.

REGULATION OF MOUSE NK ACTIVITY[1]

Carlo Riccardi
Graziella Migliorati
Antonio Giampietri
Emira Ayroldi

Institute of Pharmacology
University of Perugia
Perugia, Italy

Ronald B. Herberman

Biological Response Modifiers Program
Division of Cancer Treatment
National Cancer Institute
Frederick, Maryland

I. INTRODUCTION

An important aspect of the understanding of NK cell function has been the study of the possible regulatory factors involved in determining the levels of reactivity (1). In particular, suppressor cells, with the ability to inhibit NK activity, have been demonstrated in several situations in which NK activity was depressed by drug treatment but also in untreated animals with spontaneous low NK activity such as infant and old mice or SJL/J mice (2-4). This phenomenon, which has been extensively studied _in vitro_, has also been shown to be relevant _in vivo_ (5). Conversely, interferon (IFN) and IFN-inducers have been shown to be able to augment the spontaneous levels of NK activity both _in vivo_ and _in vitro_ in different species, including mouse and man (e.g., 6,7).

---

[1]These studies were supported by "Progretto Finalizzato Oncologia," contract no. 84.00762.44 (U.O.: Riccardi) C.N.R. Rome, Italy.

ISBN 0-12-341370-2

A. *In Vitro* Regulation at the Level of Progenitors of NK Cells

The above-mentioned observations dealt with studies on the regulation of spontaneously active NK cells or IFN-boostable cells, often called pre-NK cells (2-7). In addition, we have recently performed experiments designed to analyze the regulatory role of cells and factors on the growth and differentiation of progenitors of NK cells *in vitro* by using a limiting dilution assay (8-10). These studies, with IL-2-supplemented cultures of spleen cells, indicated that the limiting dilution assay correlates well with previously described features of NK activity. High-NK-reactive CBA/J mice were found to have a considerably higher frequency of IL-2-dependent NK cell progenitors than low-NK-reactive mice when assessed against NK-susceptible YAC-1 targets. The frequency of progenitors of cells cytotoxic against YAC-1 was higher in spleens of high-NK-reactive mice than that of cells reactive against the NK-insensitive targets P-815, P-388, and EL-4 (10-12). Furthermore, the phenotype of the progenitor cells and the cultured effector cells was consistent with that of NK cells rather than cytotoxic T cells, in that the cells expressed asialo-GM_1, some Thy 1, but no detectable Lyt-1 or Lyt-2 antigens. Using the limiting dilution assay, we have shown that IL-2 and β-IFN are able to modulate *in vitro* the growth and differentiation of NK progenitor cells into active NK cells and also that this phenomenon is under the influence of thymocytes as well as peripheral T cells (9,10).

B. *In Vivo* Regulation at the Level of Progenitor NK Cells

Based on the above observations, it appears clear that cells and factors are able to regulate NK activity by acting both at the level of mature effector cells as well as at the level of less-differentiated progenitor NK cells. Since the majority of data relative to the regulation of the growth of NK cell progenitors has been obtained by *in vitro* experiments, an important issue is to determine whether cells and factors, previously shown to be involved in NK cell regulation, are involved in the *in vivo* maturation process of NK activity.

To examine this problem, we performed *in vivo* experiments by using two experimental models in which the influence of different factors, possibly acting on NK maturation, could be studied: 1) Infant mice, since NK activity is virtually absent after birth but first appears spontaneously between the third and the fourth weeks of age; 2) lethally irradiated and bone marrow-reconstituted mice, in which the NK activity, completely abrogated by total body irradiation (950-1000 rads), gradually recovers spontaneously within two weeks.

C. Modulation of NK Cell Maturation of Infant Mice by IL-2 and β-IFN

To analyze the possible regulatory role of IL-2 and β-IFN on the maturation of NK cells of infant mice, experiments were performed in which mice were treated with IL-2 or β-IFN and splenic NK activity was evaluated at different times. In Fig. 1 are shown the results of a representative experiment in which C3H/HeN mice were treated intraperitoneally with IL-2 (10 U/day) or β-IFN (500 U/day) from day 0 through day 7 and NK cell activity was measured at several times after birth. The medium-injected controls did not show any NK activity until day 15. In contrast, mice treated with IL-2 had detectable cytotoxic activity against YAC-1 by day 10 after birth, high activity by day 15, and their NK activity reached the levels of young-adult mice by day 18, clearly indicating that IL-2 is able to induce an accelerated maturation of the activity of infant mice. This effect was not associated with a decline of the suppressor cell activity, previously described to be present in the spleen of infant mice (2,4), but rather correlated with an increase in the number of splenic LGL (data not shown). In contrast to the effects of IL-2, treatment with β-IFN caused an inhibition of NK cell maturation (Fig. 1). These data are reminiscent of our previous experiments showing that IL-2 is able to induce the maturation of NK cell progenitors _in vitro_ and that the β-IFN is able to inhibit this process (8,9). These results clearly indicate that IL-2 and β-IFN are able to modulate the _in vivo_ maturation of NK cells. However, further studies are necessary to better define these modulatory effects. In particular, the use of different doses and schedules of treatment should give us more information on the possible complexity of this phenomenon. Studies along this line are in progress in our laboratory.

D. Effects of IL-2 and β-IFN on the Growth and Maturation of Bone Marrow Progenitors of NK cells

To further analyze the regulation at the level of NK progenitor cells _in vivo_, we also examined the reconstitution of NK activity of lethally irradiated (950 R) mice by reconstitution with syngeneic bone marrow (10^7-2 x 10^7 cells/mouse).

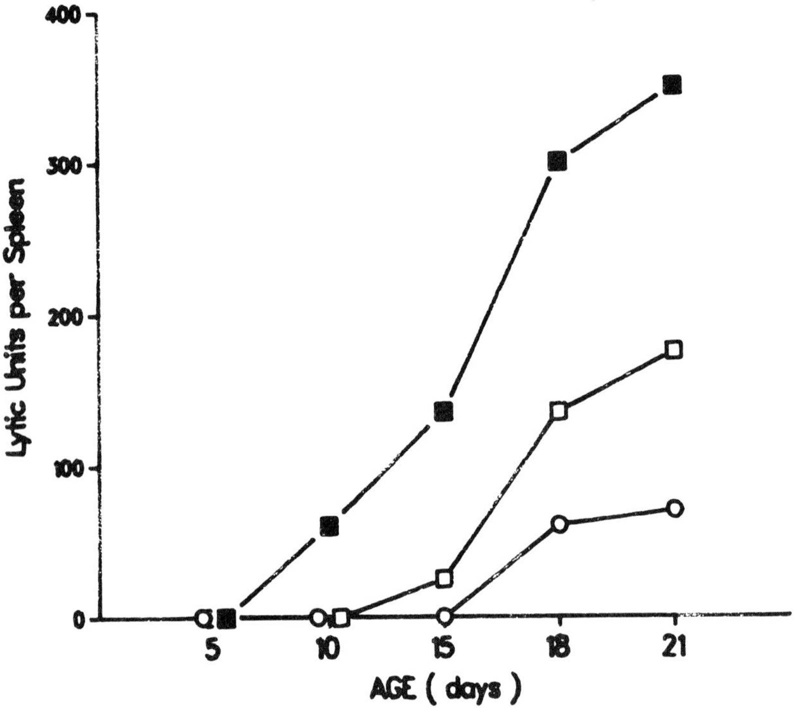

Fig. 1. Maturation of NK cell activity in infant C3H/HeN mice. □——□ medium-injected controls. ■——■ IL-2-treated mice. ○——○ β-IFN-treated mice.

Reconstitution to the levels of the untreated controls, or to higher levels, was obtained within 12-14 days after irradiation and bone marrow grafting. This phenomenon was shown to be due to reconstitution by donor bone marrow cells. In Table I are reported the results of a representative experiment, showing that injection of C57BL/6 bone marrow cells into C57BL/6 x DBA/2 F_1 mice gave rise to NK cells which were insensitive to treatment with anti-H-2^d antibodies plus complement. Parallel treatment of $BD2F_1$ spleen cells resulted in complete elimination of NK activity (data not shown). These data, along with those previously reported (13,14), exclude a contribution of radioresistant recipient-derived NK cells in the assay. This phenomenon of NK reconstitution is dependent on cell proliferation in that treatment of chimeras with antiproliferative drugs such as hydroxyurea inhibited the NK reconstitution (data not shown and 13).

TABLE I. Effect of Treatment with Anti-H-2^d Antiserum Plus Complement on Splenic NK Activity of $BD2F_1$ Mice Lethally Irradiated and Reconstituted with B6 Bone Marrow Cells

Treatment of spleen cells	% Cytotoxicity[a]
-	35.7
Anti-H-2^d + C'	38.0

[a]Effector to target ratio 100:1.

Using this system of lethal irradiation and bone marrow reconstitution, we performed experiments to evaluate the possible modulating effects of recombinant human IL-2 or β-IFN on the *in vivo* development of bone marrow progenitors responsible for the NK regeneration. Fig. 2 shows data of a representative experiment in which different groups of $BD2F_1$ mice were treated intraperitoneally from day 0 through day 3, after lethally irradiation (950 rads) and bone marrow grafting (10^7 cells/mouse iv), with IL-2 (10 U/mouse/day) or β-IFN (10^3 U/mouse/day) or medium alone. The NK activity of all the groups of mice was strongly depressed after lethal irradiation (days 4 and 7) and then progressively regenerated, reaching the level of untreated control mice, in the case of medium-injected animals, by day 14. In contrast, in IL-2- or β-IFN-treated mice, detectable levels of activity appeared on day 7 and reached peak levels on day 9 after irradiation and bone marrow graft, indicating that these agents induced an earlier maturation of bone marrow progenitor cells to NK effector cells.

The positive effects of β-IFN in these studies are in contrast to the inhibition of development of NK activity in infant mice (Fig. 1). The reasons for these divergent results are not clear. They may be attributed to some important differences in the NK progenitors in each situation, in regard to their susceptibility the regulatory effects of IFN or, more likely, may be due to the differences in the protocol of treatment with IFN (8 days in infant mice versus 4 days in chimeras).

Also of note in these experiments was a significant depression in NK activity at day 14 in the groups of treated animals. In an attempt to explain this late inhibitory effect of β-IFN and IL-2, we evaluated the number of LGL

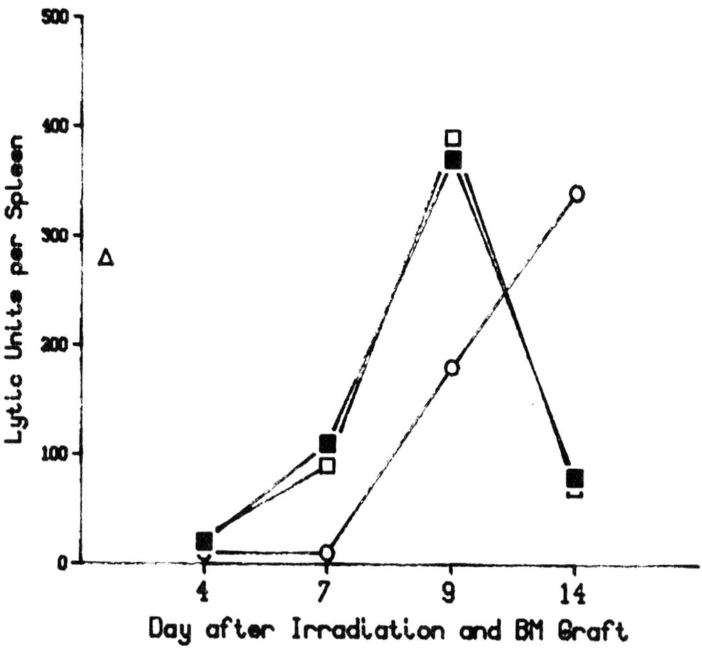

Fig. 2. Reconstitution of NK activity in lethally irradiated and BM-reconstituted BD2F$_1$ mice. ○———○ medium-injected control. ■———■ IL-2-treated mice. □———□ β-IFN-treated mice. △ Unirradiated control mice.

present in the spleens of IL-2 or β-IFN-treated animals and of medium-injected controls at different times after irradiation and bone marrow grafting. As shown in Table II, there was a significantly higher number of LGL in the spleens of treated animals on day 9, which appears to account for the higher NK activity as the result of accelerated maturation of bone marrow progenitors of NK cells to mature NK effectors. However, on day 14, when depressed activity was seen in the groups of IL-2- and β-IFN-treated animals, the number of LGL was comparable to that of the medium-injected controls. In an attempt to explain the discrepancy between the number of LGL and the levels of NK activity, we performed experiments to evaluate if suppressor cells were present in the spleen of treated animals. Nonadherent suppressor cells, which distributed in the high-density fraction 5 of discontinuous Percoll gradients were detected in the spleens of

TABLE II. Number of LGL/Spleen of Untreated or Treated Bone Marrow Chimeras

Group	Number ($x 10^{-6}$) of LGL/Spleen		
	Day 4	Day 9	Day 14
Control	0.1	0.4	3.1
IL-2-treated	0.2	3.1	4.2
IFN-treated[a]	0.1	3.8	3.6

[a]For details of treatment, see text.

IL-2- or β-IFN-treated mice (data not shown). The development of these suppressor cells may therefore account for the low NK activity on day 14.

These data demonstrate that IL-2 of β-IFN are able to stimulate the growth and differentiation of bone marrow progenitors of NK cells but also lead to the development of suppressor cells which may be responsible for subsequent inhibition of NK cell activity.

E. Role of the Thymus in Maturation of Bone Marrow Progenitors of NK Cells

In our *in vitro* studies, we have shown that the regulation of IL-2-dependent growth and differentiation of progenitors of NK cells is under the regulatory control of the thymus (9-10). To analyze the possible *in vivo* relevance of this phenomenon, we performed experiments to study the possible influence of the thymus on reconstitution of NK activity in lethally irradiated and bone marrow grafted animals. We measured NK activity of thymectomized chimeras and also of thymectomized and thymus-implanted chimeras. In Fig. 3 is shown a representative experiment in which NK levels of normal and thymectomized (15 days before irradiation and bone marrow graft) mice were evaluated at different days after irradiation and bone marrow reconstitution. The NK reconstitution was delayed in thymectomized mice, suggesting that the thymus plays some role in NK cell maturation (see day 7 and day 9). However, at day 12 the mice lacking a thymus developed higher NK activity and higher numbers of

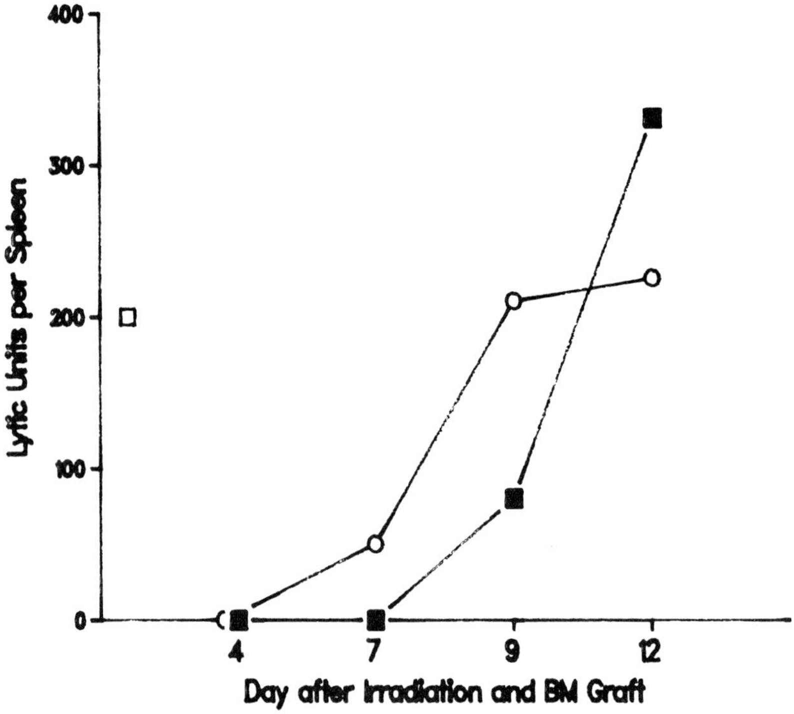

Fig. 3. Effect of thymectomy on NK regeneration in lethally irradiated and BM-reconstituted $BD2F_1$ mice. ○———○ sham-operated control. ■———■ thymectomized mice. □ normal, unirradiated control mice.

LGL (data not shown) relative to the levels seen in the control mice, which may be analogous to the well-known development of high NK activity in nude athymic mice (15). The *in vivo* influence of the thymus on reconstitution of NK activity was further indicated by the experiment in Fig. 4, in which we evaluated the NK activity of thymectomized mice, and thymectomized and thymus-implanted (subcutaneous) mice (thymectomy and thymus implant were performed 15 days before irradiation and bone marrow graft). The implantation of the thymus led to more rapid development of NK activity. However, again by day 11, when the NK activity in the thymus-implanted mice was back to the levels of untreated controls, the NK activity of mice lacking a thymus was higher than that of thymus-bearing mice.

These data confirm the previous concept that thymus can

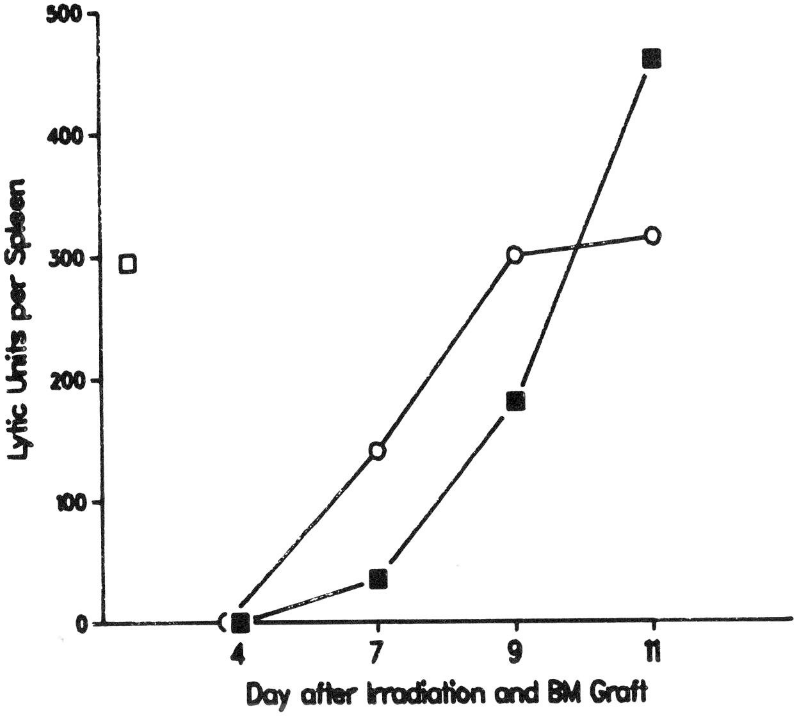

Fig. 4. Effect of thymus implantation on NK regeneration in thymectomized, lethally irradiated, and BM-reconstituted $BD2F_1$ mice. ■——■ thymectomized. ○——○ thymectomized and thymus-implanted. □ Normal, untreated mice.

base a negative regulatory role on NK activity but also indicate that there also may be a positive regulatory role of the thymus on the initial development of bone marrow progenitor cells into active NK cells.

II. CONCLUSIONS

It now appears that mouse NK activity is regulated at least at two different levels: 1) at the level of differentiated effector cells; 2) at the level of less differentiated progenitors of NK cells.

Soluble factors, such as IL-2 and β-IFN, as well as cells, such as thymus and suppressor cells, may be involved in the regulation at either or both levels, sometimes with contrasting effects. It is also of interest to note that the *in vivo* use of some factors such as IL-2 may not only directly affect NK cells but also affect other, NK-regulatory cells, such as nonadherent suppressor cells. All these parameters will need to be taken into account for a complete understanding of the regulatory mechanisms involved in determining the levels of reactivity, as well as for appropriate therapeutic approaches involving the modulation of NK activity.

REFERENCES

1. Herberman, R. B. (1982). "NK Cells and Other Natural Effector Cells." Academic Press, New York.
2. Cudkowicz, G., and Hochmann, P. (1979). Immunol. Rev. *44*, 13.
3. Riccardi, C., Santoni, A., Barlozzari, T., Cesarini, C., and Herberman, R. B. (1981). Int. J. Cancer *28*, 811.
4. Riccardi, C., Giampietri, A., Migliorati, G., Guadagni, F., and Herberman, R. B. (1983). Proc. 1st International Congress of the Italian Society of Immunopharmacology. Perugia, Italy, p. 66.
5. Riccardi, C., Barlozzari, T., Santoni, A., Cesarini, C., and Herberman, R. B. (1982). *In* "NK Cells and Other Natural Effector Cells" (R. B. Herberman, ed.), p. 549. Academic Press, New York.
6. Trinchieri, G., Santoli, D., Dee, R. R., and Knowles, B. B. (1978). J. Exp. Med. *147*, 1299.
7. Herberman, R. B., Ortaldo, J. R., and Bonnard, G. D. (1979). Nature *277*, 221.
8. Riccardi, C., Vose, B. M., and Herberman, R. B. (1983). J. Immunol. *130*, 228.
9. Riccardi, C., Vose, B. M., and Herberman, R. B. (1983). *In* "NK Cells and Other Natural Effector Cells" (R. B. Herberman, ed.), p. 909. Academic Press, New York.

10. Riccardi, C., Migliorati, G., and Herberman, R. B. (1984). Natural Immunity Cell Growth Regul. 3, 7. and Herberman, R. B. (1984). Int. J. Tissue React., in press.
12. Riccardi, C., Migliorati, G., and Frati, L. (1984). In "Hormonally Defined Media. A Tool in Cell Biology" (G. FIscher and R. J. Wieser, eds.), p. 317. Springer Verlag, Berlin.
13. Hurme, M. (1984). Scand. J. Immunol. 17, 379.
14. Hurme, M., and Sihvola, M. (1984). Cell. Immunol. 84, 276.
15. Herberman, R. B., Ortaldo, J. R., Holden, H. T., Djeu, J. Y., Mattes, J., Brunda, M., Riccardi, C., and Santoni, A. (1982). In "Thymus, Thymic Hormones and T Lymphocytes" (F. Aiuti and H. Wigzell, eds.), p. 165. Academic Press, New York.

ALTERATIONS IN NUMBERS OF LGLs AS IMPORTANT COMPONENT OF IN VIVO REGULATION OF NK ACTIVITY

Angela Santoni[1]
Mario Piccoli[1]
Guido Forni[2]
Ronald B. Herberman[3]
Luigi Frati[1]

[1] Istituto di Patologia Generale Università
delgi Studi "La Sapienza"
Rome, Italy

[2] Istituto di Microbiologia Università
degli Studi di Torino
Turin, Italy

[3] Biological Therapeutics Branch
BRMP, DCT, NCI-FCRF,
Frederick, Maryland USA

I. INTRODUCTION

Natural killer (NK) cells have cytotoxic activity against a variety of tumor cells and also against some micro-organisms or cells infected by microbial agents (1). They may thereby serve as important effector cells in host resistance against disease. NK cells are present in virtually all individuals and can be activated rapidly by a wide variety of stimuli. Since NK cells have been detected by their cytotoxic reactivity, the definition of these cells has been mainly a functional one. However, extensive efforts have been made to determine the cell surface markers and other morphologic and functional properties of NK cells.

Of particular interest has been the quest for unique features of these effector cells that would allow them to be

ISBN 0-12-341370-2

readily and definitively distinguished from other cytotoxic effector cells. In addition to their importance in understanding the nature of various cell-mediated cytotoxic effector cells, NK-specific markers would allow identification and enumeration of this subpopulation by direct means, separate from functional activities.

One of the most significant contributions to the study of NK cells has been their identification as a relatively homogeneous cell type on the basis of their morphology.

This important development stemmed from the initial observation by Timonen et al. (2) that the majority of human lymphocytes binding NK-sensitive targets and thereby forming conjugates, were large lymphocytes with a high cytoplasm to nucleus ratio and discrete azurophilic granules in the cytoplasm (and therefore termed large granular lymphocytes, LGL).

It has been possible to separate and enrich for LGL on discontinuous Percoll density gradients (3-5) and this has proven very useful to analyze the degree of association between LGL and NK cells.

The central role of LGL as NK cells has been supported by the finding that in each of the species studied, most of the NK activity against a variety of NK-susceptible target cells and against antibody-coated tumor target cells was recovered in the low density Percoll fractions enriched for LGL, whereas the higher density fractions containing most of the small-medium lymphocytes have been shown to be devoid of NK activity. In studies with human peripheral blood LGL, at least 70% of these cells have been shown to have NK activity (6) and in the rat, about one-quarter of LGL have been shown to be lytic (4). Since their initial discovery, 1-10% LGLs have been found in the peripheral blood and/or spleen of most mammalian species and chickens (7).

Much evidence suggests that NK activity is a highly regulated system. Regulatory signals can affect the NK system at various levels including differentiation and proliferation of NK precursors, proliferation of active NK cells, and modulation of lytic function.

The ability to identify the NK cells as LGL and separate them by centrifugation on discontinuous Percoll density gradients has allowed the assessment of alterations in the size of the NK-related cell population which occur in pathological conditions or can be induced by treatment with several biological response modifiers (BRMs).

In this review we will focus primarily on the alterations in the LGL number observed following treatment with several augmenting or inhibitory agents.

2. AUGMENTING AGENTS

A wide variety of viruses, adjuvants, and immunomodulators have been found to augment NK activity (8-11). Most of these agents are interferon (IFN) inducers, and IFN itself has been found to enhance the cytotoxic activity of NK cells (12,13). Recently some evidence indicates that IFN might act *in vivo* not only by directly increasing the activity of mature NK cells or pre-NK cells, but also by affecting the growth and differentiation of NK cells. Biron and colleagues (14-16) have shown that stimulation of NK activity by *in vivo* administration of various viruses or poly I:C or IFN itself is associated with some proliferation of NK cells and the appearance of blast cells with NK activity. Similarly, we have found (17,18) that after administration of C. parvum, MVE-2, poly I:C or IFN α, β, there was a significant increase in LGL in the spleens of nude as well as of euthymic mice. A substantial proportion of the augmented NK activity was associated with LGL of lower density, larger size and more vacuolation than are characteristic of the LGL of unstimulated mice. This increase (Table 1) and change in characteristics (Fig. 1) seemed not to be attributable to a preferential migration of LGL into the spleen since it could be observed in other sites such as peripheral blood, bone marrow and peritoneal cavity. Rather it appeared that a portion of activated killer cells, in contrast to endogeneous NK cells, were in the process of proliferating. Indeed, *in vivo* treatment with hydroxyurea, a cell cycle-specific toxic drug, caused a marked decrease of NK activity and of LGL in the spleens of boosted mice (Fig. 2), but had no detectable effect on the activity of untreated mice (19).

TABLE I. *In Vivo* boosting of NK activity by poly I:C[a] at various anatomical sites

Treatment	Effector Cells	% cytotoxicity[b] 50:1[c]	25:1	12:1	%LGL
none	spleen	11.6	5.7	2.9	<0.5
poly I:C		21.4	15.1	10.9	2.5
none	PBL	9.1	5.1	4.2	<0.5
poly I:C		N.T.	48.5	33.5	29.0
none	PBL	11.5	6.8	1.3	1.0
poly I:C		77.4	75.7	63.8	28.0

[a] 7-wk-old BABL/c mice were injected i.p. on day -2 with 4 mg/kg poly I:C
[b] Cytotoxicity was assessed in a 4 hr ^{51}Cr-release assay against YAC-1 targets
[c] E:T ratio

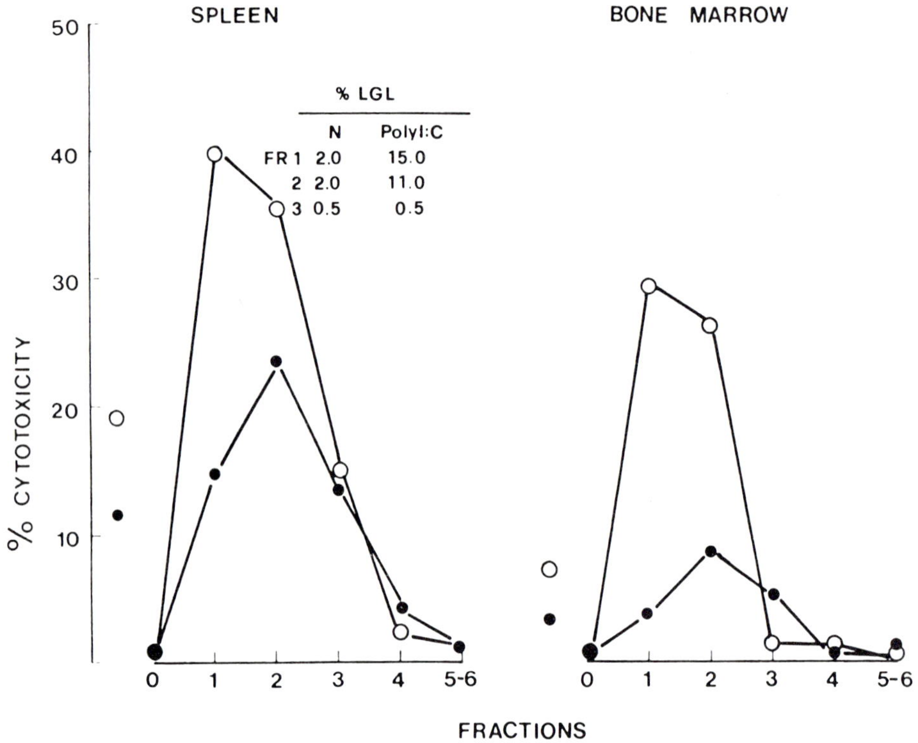

Fig. 1. Distribution of NK activity on Percoll density gradient after augmentation with poly I:C. Spleen (left panel) and bone marrow (right panel) cells from normal BALB/c mice (●) or BALB/c mice treated 2 days earlier with 4mg/kg poly I:C, i.p. (O) were separated and tested for NK activity. On the left are shown the levels of NK activity of the unseparated cells.

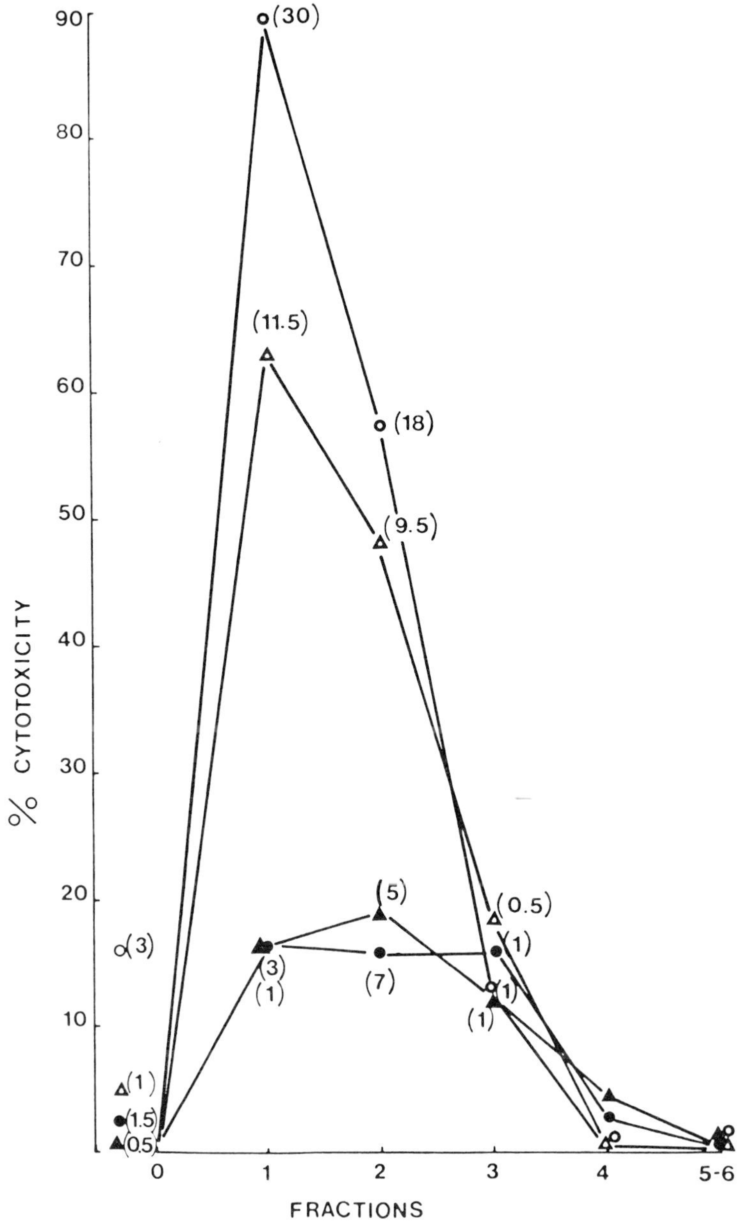

Fig. 2. Effects of in vivo treatment with hydroxyurea (HU, 1g/kg, i.p., -9 hr, -2 hr) on boosting of NK activity by poly I:C (4 mg/kg, day -2). Spleen cells from normal BALB/c (●) and BALB/c treated with HU (▲), poly I:C (O), or poly I:C plus HU (△) were separated on Percoll density gradients and tested for NK activity. In parentheses are reported percentages of LGLs. On the left are the values for the unseparated cell populations.

All these observations indicate that an increase in NK cell number, as well as activation of preexisting NK cells, may contribute to the augmentation of NK activity *in vivo* by IFN or IFN-inducers. However, it is quite uncelar how IFN could expand the pool of NK cells. IFN is not known to be a mitogen but rather at high concentrations it tends to inhibit cell division (20).

Our hypothesis is that IFN affects growth and differentiation of NK cells indirectly by interaction with other mediators. One likely candidate is interleukin-2 (IL-2). IFN treatment *in vitro* has been shown to increase the number of receptors for IL-2 on T and NK cells (21,22) and, since IL-2 stimulates NK cell proliferation *in vitro* (e.g. 23,24), it is possible that IFN treatment *in vivo* induces or augments the expression of receptors for IL-2 or other growth factors, and the cells then proliferate more actively in response to endogenous growth factors. Alternatively, IFN may induce the *in vivo* release and/or production of growth factors for NK cells or NK cell precursors.

3. INHIBITING FACTORS

A variety of negative regulatory mechanisms have been described which can result in reduced NK cell activity: a) direct inhibition of development or function of NK cells; b) generation of suppressor cells; and c) inhibition of the function of accessory cells.

In some situations, reduced NK activity has been found to be accompanied by a decrease in LGL number, perhaps due to interference with proliferation and/or differentiation of NK cells.

The substantial decrease in NK activity observed 10-21 days after treatment with C. parvum (Cp) has been found to be associated with the presence of splenic suppressor cells, capable of inhibiting the lytic activity of NK cells (25-27). Moreover, we have recently observed that such treatment, in addition to induction of suppressor cells, results in a decrease in the number of splenic LGLs (27). Experiments with adoptively transferred radiolabeled LGL-enriched cell populations tend to rule out that this reduction is attributable to a reduced localization of LGL in the spleen following treatment with C. parvum. It is therefore likely that Cp or the induced suppressor cells could exert inhibitory effects on the growth and/or differentiation of NK cells.

Another immunomodulator, MVE-2, which induced substantial augmentation of mouse splenic NK activity after one injection, when repeatedly injected caused little or no augmentation of cytotoxic activity or of the number of LGL (28,29). A quite similar pattern of results has been obtained in studies of the effects of recombinant human IFN α on the NK activity of cancer patients. In contrast to early increases in NK activity, frequent repeated treatments with IFN α resulted in no augmentation of reactivity or even a decrease below pretreatment levels (30). In a study of the possible mechanisms for this hyporesponsiveness to IFN, the treated patients were found to have reduced numbers of lytically active LGL (31).

How might these BRMs regulate NK activity at the level of proliferation and/or differentiation? It seems possible that suppressor cells, in addition to inhibiting effector cell activity, may also interfere with the growth and differentiation of NK cells. Alternatively, prolonged exposure of NK cells to IFN *in vivo*, which might be associated with the later time point with *C.* parvum or multiple doses of MVE-2, might have a suppressive rather than a stimulating effect on NK cells. IFN *in vitro* is known to block division of some tumor cells (32) and lymphocytes (20) and to cause depression of bone marrow function (32). Moreover, recent studies of the IL-2-dependent *in vitro* growth of mouse NK cells (34) have shown that the presence of IFN during the assay resulted in a considerable decrease in the detected frequency of NK cell progenitors, with the inhibitory effect dependent on the presence of T cells. It is possible that initial exposure to IFN *in vivo* results in blastogenesis of NK cells. More prolonged exposure of IFN *in vivo* may induce feed-back regulation, resulting in inhibition of LGL proliferation and a consequent decrease in the size of the population. In this regard, Biron and Welsh reported that infection with high doses of emphocytic choriomeningitis virus, which induces high levels of IFN (>10^4 units/spleen), resulted in a reduced blastogenesis of NK cells (35). Also, Riccardi et al. (36) has observed that treatment of mice for four days with IFN resulted in a more rapid development of NK activity, whereas in a different protocol involving treatment with IFN for eight days, a delayed development of NK activity was observed. Consistent with the *in vitro* indication for a role of the T cells for negative regulation by IFN (34), it has been our observation that the inhibition of NK activity and the decrease in LGL numbers in the spleens of nude mice were less profound than in the spleens of euthymic mice.

A reduction of numbers of LGL has also been observed in mice with surgically induced lesions of the tuberoinfundibular nuclei of the hypothalamus, which had a profound and persistent inhibition of NK activity (37). These results suggest that this region of the hypothalamus in some way controls or at least affects the complete activation and maturation of NK cells, either directly or through some complex hormonal interactions. Of interest has been the observation that treatment of mice with the hypothalamic lesion, with poly I:C _in vivo_ or with IFN or IL-2 _in vitro_, resulted in a rapid restoration of NK activity and LGL numbers, indicating that the hypothalamus may exert its influence at a relatively late stage of differentiation of NK cells. This critical step has been described to be mediated by IFN (38) and has been suggested to depend on the transition of large granular lymphocytes into functionally active LBL (39).

4. CONCLUSIONS

The recent ability to morphologically identify and highly enrich for LGL has permitted an analysis of changes in the size of this potential effector population which might be associated with altered _in vivo_ regulation of NK activity. In several situations described here, augmented NK activity _in vivo_ was found to be closely associated with expanded numbers of LGL in various organs, appraently due to proliferation of this cell population. Conversely, a lack of augmentation or a depression of NK activity, after BRM treatment _in vivo_, was associated with a lack of expansion in the numbers of LGL. Thus, regulation of the size of the LGL population appears to be a major determinant of the _in vivo_ levels of NK activity. However, it should be noted that not all LGL have NK activity and that IFN can convert lytically inactive LGL into NK cells (6). Therefore, it seems very likely that an important additional component of _in vivo_ regulation of NK activity will be at the level of the proportion of lytically active LGL. Along these lines, one would predict that there will be situations in which the number of LGL will not correlate well with NK activity and substantial attractions in cytotoxic reactivity will occur without parallel changes in the size of the LGL population.

REFERENCES

1. Herberman, R.B. (ed.) (1982). "NK Cells and Other Natural Effector Cells." Academic Press, New York.
2. Timonen, T., Saksela, E., Ranki, A., and Hayry, P. (1979). Cell. Immunol., 48, 133.
3. Timonen, T., Ortaldo, J.R., and Herberman, R.B. (1981). J. Exp. Med. 153, 569.
4. Reynolds, C.W., Timonen, T., and Herberman R.B. (1981). J. Immunol. 127, 282.
5. Luini, W., Boraschi, D., Alberti, S., Aleotti, A., and Tagliabue, A. (1981). Immunology 43, 663.
6. Timonen, T., Ortaldo, J.R., and Herberman, R.B. (1982). J. Immunol. 128, 2541.
7. Herberman, R.B., Brunda, M.J., Domzig, W., Fagnani, R., Goldfarb, R.H., Holden, H.T., Ortaldo, J.R., Reynolds, C.W., Riccardi, C., Santoni, A., Stadler, B.M., Taramelli, D., Timonen, T., and Varesio, L. (1982). In "The Biological Significance of Immune Regulation" (Gershwin, E. and Ruben, L. eds.). Dekker, New York.
8. Herberman, R.B., Nunn, M.E., Holden, H.T., Staal, S. and Djeu, J.Y. (1977). Int. J. Cancer 19, 555.
9. Wolfe, S.A., Tracey, D.E. and Henney, C.S. (1977). Nature 262, 584.
10. Santoni, A., Puccetti, P., Riccardi, C., Herberman, R.B. and Bonmasser, E. (1979). Int. J. Cancer 24, 656.
11. Welsh, R.M. and Zinkernagel, R.M. (1977). Nature 268, 646.
12. Djeu, J.Y., Heinbaugh, J.A., Holden, H.T. and Herberman, R.B. (1979). J. Immunol. 122, 175.
13. Gidlund, M., Orn, A., Wigzell, H., Senik, A., and Gresser, I. (1979). Nature 223, 175.
14. Biron, C.A. and Welsh, R.M. (1982). J. Immunol. 129, 2788.
15. Biron, C.A., Turgiss, L.A. and Welsh, R.M. (1983). J. Immunol. 131, 1539.
16. Biron, C.A., Sonnenfeld, G., and Welsh, R.M. (1984). J. Leuk. Biol. 35, 31.
17. Santoni, A., Herberman, R.B., Piccoli, M., and Ortaldo, J.R. (1983). Fed. Proc. 42, 1380.
18. Santoni, A., Piccoli, M., Ortaldo, J.R. and Herberman, R.B. (1985). J. Immunol., in press.
19. Miller, S.C. (1982). J. Immunol. 129, 2282.
20. Heron, I., Berg, K. and Cantell, K. (1976). J. Immunol. 117, 1370.
21. Johnson, H.M. and Farrar, W.L. (1983). Cell. Immunol. 75, 154.
22. Kuribayashi, K., Gillis, S., Kern, D.E. and Henney, C.S. (1981). J. Immunol. 126, 2321.

23. Dennert, G. (1980). Nature 287, 47.
24. Riccardi, C., Allavena, P., Ortaldo, J.R. and Herberman, R.B. (1982). In "NK Cells and Other Effector Cells." (R.B. Herberman, ed.). Academic Press, New York, p. 873.
25. Savary, C.A. and Lotzova, E. (1978). J. Immunol. 120, 239.
26. Santoni, A., Riccardi, C., Barlozzari, T. and Herberman, R.B. (1982). In "NK Cells and Natural Effector Cells." (R.B. Herberman, ed.). Academic Press, New York, p. 519.
27. Santoni, A., Wiltrout, R.H. and Herberman, R.B. Submitted for publication.
28. Piccoli, M., Santoni, A., Velotti, F., Frati, L. and Chirigos, M.A. (1983). Fed. Proc. 43, 1755.
29. Saito, T., Welker, R.D., Fukui, H., Herberman, R.B. and Chirigos, M.A. Cell. Immunol, in press.
30. Maluish, A.E., Ortaldo, J.R., Conlon, J.C., Sherwin, S.A., Leavitt, R., Strong, D.M., Weirnik, P., Oldham, R.K. and Herberman, R.B. (1983). J. Immunol. 131, 503.
31. Hizuta, A., Maluish, A.E., Ortaldo, J.R. and Herberman, R.B. (1984). In "Natural Killer Activity and Its Regulation." (T. Hoshino, H.S. Koren and A. Uchida, eds.). Excerpta Medica Ltd., Tokyo, p 453.
32. Gresser, I., Brouty-Boye, D., Thomas, M. and Marcieira-Coelho, A. (1970). Proc. Natl. Acad. Sci. 66, 1052.
33. Nissen, C., Speck, B., Emodi, G. and Iscove, N.N. (1977). Lancet 1, 203.
34. Riccardi, C., Vose, B.M. and Herberman R.B. (1983). J. Immunol. 130, 228.
35. Biron, C.A. and Welsh, R.M. (1982). In "NK Cells and Other Natural Effector Cells." (R.B. Herberman, ed.). Academic Press, New York, p. 493.
36. Riccardi, C., Migliorati, G., Giampietri, A., Ayroldi, E. and Herberman, R.B. (1985). In "Mechanisms of Cytotoxicity by NK Cells." (R.B. Herberman, ed.). Academic Press, New York.
37. Forni, G., Bindoni, M., Santoni, A., Belluardo, N., Marchese, A.E. and Giovarelli, M. (1983). Nature 306, 181.
38. Minato, N., Reid, L., Cantor, H. and Bloom, B.R. (1980). J. Exp. Med. 152, 124.
39. Itoh, K., Suzuki, R., Umezu, Y., Hanaumi, K. and Kumagai, K. (1982). J. Immunol. 129, 395.

EFFECTS OF PHORBOL ESTERS ON NK ACTIVITY [1]

Eva Klein
Maria-Grazia Masucci
Oscar Ramos

Department of Tumor Biology
Karolinska Institutet
Box 60 400
S-104 01 Stockholm
Sweden

Ever since the discovery of the natural killer effect attempts were made to distinguish it from the effect of lytic cells which appear after immunization i.e. cytotoxic T cells (CTL).

The definition "natural" killing is operational. It refers to experiments where effectors are derived from unmanipulated donors, and are used directly after separation without being exposed to culture conditions or to immunomodulators. In vitro manipulated cells can be more appropriately referred to as "activated" killer cells, AK (1). For IFN treated lymphocytes we used the designation IAK, interferon activated killers (2). Corresponding definitions were adopted by others e.g. LAK for lymphokine (IL-2) activated killers (3). The activated killer cells can act on a broader target panel than NK cells. We have found that all non-B subsets of human blood lymphocytes separated on the basis of surface markers such as E and Fcγ receptors can be triggered for cytotoxicity (4). As an example IFN and PHA induced activation can be mentioned. IFN enhanced effect of the NK subset while PHA induced cytotoxicity in the Fcγ receptor negative, E receptor positive lymphocyte population devoid of NK activity.

---

[1] This investigation was supported by PHS grant number IMB 5 R01 CA25250-06 and by the Swedish Cancer Society.

ISBN 0-12-341370-2

Natural and activated killer effects are independent on clonal expansion. This contrasts with antigen triggered cytotoxic reactions, which involve the proliferation of those T cells that carry the relevant receptors. This difference can be used for the definition of NK and AK, on the one hand, and CTL, on the other hand (5).

Phorbol esters are polyclonal activators for human lymphocytes. They induce DNA synthesis and enhance lectin induced blastogenesis (6,7,8). The plasma membrane of phorbol ester treated lymphocytes undergo conspicuous morphological and biochemical changes (9). Already after short time exposure the lymphocyte aggregate and can attach to other cells also, provided these are derived from the same or closely related species (10). The species specificity of the binding probably reflects a phenomenon that is important in immunological recognition, and represents its first step, followed by the interaction between antigens expressed on the target and the corresponding receptor on the lymphocytes. Species specific binding of lymphocytes to various types of cells occurs also with unmanipulated T cells (11). The phorbol ester treatment potentiates the recognition between complementary cell surface proteins. Relevant for the phorbol ester treatment induced changes of lytic potential is that it enhances the production of cytokines (12,13,14,15).

The mitogenic property, the enhancement of target cell binding and the secretory activitites by phorbol esters can be expected to cause an augmentation of the cytolytic function of lymphocytes. The reports about the modification of the natural killer assay however vary, both enhancement and suppression were observed (16,17).

In experiments aimed to reveal the mechanism of NK suppression in response to phorbol ester treatment the inhibition was abolished or reduced when adherent cells were removed from the population (18). Phorbol esters act on the majority of cell types and influence their maturation state (19,20,21). Consequently in a cell mixture that contains interacting components, the interactions can be modified by the impact of the phorbol ester effect on each (22). Thus the composition of the effector cell population may determine the modification of the lytic function (17).

We have exposed subsets of human blood lymphocytes separated on the basis of cell density to P(Bu)2, phorbol 12,13-dibutyrate, one of the phorbol esters with tumor promoting activity, and investigated the expression of certain surface markers and cytolytic function (23,24). This compound was selected because it is a less lipophylic phorbol ester derivative and binds therefore only to a specific receptor on the plasma membrane. The compound often used,phorbol 12-myristate 13-acetate has substantial lipophilicity and binds therefore also non-specifically to plasma membranes (25).

Lymphocytes cultured in the presence of P(Bu)2 showed altered expression of certain surface markers, some of which can be observed after short time incubation, others require longer exposure (26). Decrease of Fc-γ receptor and OKT4 expression occurred already after 24 h culture in the presence of 30 nM P(Bu)2 (27). The latter effect is reversible, OKT4 reactivity reappears after incubation in fresh medium (27).

A corollary of the activation of lymphocytes was the induction of strong reactivity with the monoclonal antibody directed to IL-2. This could be detected after 12 hours. The number of OKT3 positive cells decreased and the OKT8 positive cells increased. These changes required longer time, 48 hs exposure to P(Bu)2 was not sufficient to induce them. It is possible that the change in OKT4 reactivity was due to a direct effect on the cell membrane. In addition to the morpholocigal changes, P(Bu)2 treatment enhances the lateral mobility of certain cell surface molecules (28). On the other hand alterations on OKT3 and OKT8 expression probably reflects the effect on cell maturation.

The modifying effect of P(Bu)2 on the lysis of K562 depended on the time period of lymphocyte pretreatment. Short term treatment caused a suppression but after 3 hours this subsided and the direction of the influence changed to potentiation (23).

After further incubation the maximal effect was reached after 24 hs, and 2 days later the lytic potential started to decline (Table I).

TABLE I. Cytotoxic effect of lymphocytes preexposed to 30nM P(Bu)2 for various length of time

Target	Lymphocytes	Days of preincubation					
		0	1	2	3	5	7
K562	Control	35	35	30	20	10	8
	P(Bu)2		57	56	45	35	22
Daudi	Control	15	15	10	5	0	0
	P(Bu)2		30	34	34	22	0

Effector target ratio 25:1, 4 hours cytotoxic assay, the results are expressed in % specific ^{51}Cr release.
The incubation medium contained 10% human serum.

Throughout the whole period the efficiency of the low density cells was higher than that of the high density cells. Thus exposure to phorbol esters promoted the lytic mechanism of the cell subset that have the innate capacity to express this function.

Other tumor promoters belonging to chemically unrelated compounds, such as indole alkaloids and polyacetates derived from Streptomyces and the blue green alga Lyngbya majuscula that bind to common cell surface receptors with phorbol esters and induce similar biological effects, activated also the lymphocytes for cytotoxicity (24,29,30,31). On the other hand chemically related compounds that do not bind with high avidity to the cell surface receptors and are devoid of tumor promoting effect were inactive in this assay system also. The potentiation for cytotoxicity did not take place when monocytes were present in the effector population (23). Monocytes are known to suppress lymphocyte mediated cytotoxicity (32).

Studies on the conditions for the elevation of cytotoxicity suggested that phorbol esters do not induce an alternative cytotoxic mechanism (24). Similar to the effect of untreated lymphocytes, contact with the target was required and the secretary function had to be intact. The cytolysis was inhibited by addition of cytochalasin B, colchicine, cheleating agents and monensin. P(Bu)2 induced lytic function only in metabolically active cells, but did not require DNA synthesis.

Concerning the mechanism of lymphocyte activation for cytotoxicity it had to be decided whether it is a direct or an indirect event. It can be attributed to the induction of IFN or IL-2 production in a subpopulation of cells. They are known to be induced by phorbol esters and they are both potent activators of the cytolytic function (13,14,15). Their contribution was excluded however on the following basis: Presence of antibodies directed either against IFN or IL-2 did not abrogate the P(Bu)2 induced increase of cytotoxicity (24,33).

The presence of antiserum against IFN in the incubation medium reduced however the cytotoxic potential of control cells, suggesting that endogenous IFN production contributes to the maintenance of lytic function in cultured cells (24). If this mechanism is counteracted by addition of anti-IFN serum, the phorbol esters can provide an alternative activation signal. When P(Bu)2-activated lymphocytes were subsequently treated with IFN-α or IFN-γ, their lytic capacity was further increased (33). These results indicate that P(Bu)2 and IFN activate the cytotoxic potential of lymphocytes through different pathways. This conclusion was reached previously on the basis of experiments performed with mouse cells (17).

Comparison of results obtained in the immobilised agarose assay and the 4 hours ^{51}Cr release assay with various effector target ratios and the kinetics of target lysis allowed to dissect some of the events which are influenced by treatment of lymphocytes with P(Bu)2 and/or IFN (33). We confirmed that IFN treated lymphocytes attach to target in proportion similar to untreated ones but the proportion of lytic interactions occurring during 4 hours is increased (Table II)(34).

TABLE II. Lysed K562 cells in the conjugates with lymphocytes previously exposed for 60 minutes to IFN or P(Bu)2[a]

Effectors	TBC[b]	DTC[c]	LC[d]	LU/10^6 lymphocytes[e]
Control	14	16	2.2	120
P(Bu)2	31	8	2.5	60
IFN	15	30	4.5	250

[a] 10^6 lymphocytes were preincubated with or without 30nM P(Bu)2 or 10^3 U/ml α-IFN for 60 minutes. Mean values of the results obtained in 4 experiments performed at 1:1 effector target ratio in the immobilised agarose assay. They were evaluated after 4 hours of incubation. For comparison the results obtained in parallell suspensions assays are also shown.

[b] % target binding cells

[c] % conjugates containing dead targets

[d] % total lytic cells, calculated from b and c

[e] Cytotoxic activity in the 4 hours ^{51}Cr release assay at 25:1 effector target ratio. One LU corresponds to the number effector cells required for 20% target cell lysis.

The decrease of the lytic effect after short time P(Bu)2 treatment was considerable in the ^{51}Cr release assay but it did not appear in the immobilised conjugate assay (Table II). The number of lytic cells in the conjugates that performed the function during the 4 hours observation was similar to that of the control samples, 2% of all lymphocytes. Because the number of functioning cells was not decreased, P(Bu)2 treatment seems to inhibit the capacity of the effectors to lyse more than 1 target. Lysis of several targets by one effector can occur in suspension but not in the agarose where the established contacts are stabilised and recycling cannot occur.

The activation of lytic function by prolonged P(Bu)2 treatment was manifested by the elevated ^{51}Cr release values and by the decrease in time after which damaged cells could be detected. This was more obvious with Daudi than with K562.Lysis of Daudi by unmanipulated lymphocytes occurs after 2 hours of interaction while lysis of K562 starts promptly. Treatment with P(Bu)2 for 15 hours increased also the number of lymphocytes that attached to K562 (Table III).

TABLE III. Lysed 562 cells in the conjugates with lymphocytes previously exposed for 15 hours to P(Bu)2 and/or IFN

Effectors	TBC[b]	DTC[c]	LC[d]	LU/10^6 lymphocytes[e]
Control	15	16	2.3	110
P(Bu)2	24	27	6.6	350
IFN	16	28	4.2	250
P(Bu)2+IFN	24	28	6.6	500

10^6 effector cells were preincubated without or with 30 nM P(Bu)2 during 15 hours. 10^3 U/ml-IFN-α were added for additional 3 hours. Mean of 10 experiments performed at 1:1 effector target ratio in the immobilised agarose assay, evaluated after 4 hours. For comparison, results in suspension assays are also shown.

[b], [c], [d], [e] See Table II.

In addition, higher number of lymphocytes were cytotoxic, their proportion increased from 2 to 7% in the agarose assay.

The cytotoxic potential of P(Bu)2 treated cells was further enhanced by subsequent IFN treatment. After 60 minutes interaction with K562 the ^{51}Cr release values with the control, P(Bu)2, IFN and combined P(Bu)2-IFN treated lymphocytes at 25:1 effector target ratios were 10, 30, 30 and 50% respectively. This was measured under conditions which inhibited the development of further interactions but allowed the release of isotope from the damaged cells. The elevated lysis performed by lymphocytes treated with both agents was not seen in the agarose assay (Table III).The percentage of lytic lymphocytes in the P(Bu)2 and the P(Bu)2-IFN treated samples estimated after 4 hours were identical, 6.6%. The difference between the results of the immobilised and the suspension assays suggest that the mechanism for the IFN induced additional target lysis is enhanced recycling. The capacity to lyse higher number of targets was also shown by the maximal lytic capacity in the tests performed with 25:1 effector target ratios. The plateau was reached after 150 minutes with the treated cells. It reached 50% lysis with P(Bu)2 or IFN treated cells and 60% after the combined treatment. At that time the control cells lysed 20% of the targets and their effect continued to increase during the experimental period, 270 minutes, when the lysis was 30%.

The results show thus that P(Bu)2 treatment activates the lymphocytes for lytic function. The time of exposure is critical, the optimum period is between 24-48 hours. The cell membrane changes that enhance intercellular contact may hamper the lytic efficiency in the suspension assay. Subsequent exposure to IFN elevates the recycling capacity of the lymphocytes with the result that additional targets are lysed.

Phorbol esters inhibit the generation of antigen specific cytotoxicity in mixed lymphocyte cultures. Both allospecific and tumor specific (syngeneic) systems were shown to be inhibited (35, 36). TPA enhanced the Epstein-Barr virus induced transformation of B cells when the total blood lymphocyte population was exposed to the virus (37). In this condition the T cells have an inhibitory effect on the growth of B cells (38). Phorbol ester treatment probably abrogated the T cell response directed against EBV determined antigens expressed on the B cell surface. We have confirmed the inhibitory effect of P(Bu)2 on the generation of allospecific killer cells in MLC and in addition showed that the anomalous killer activity that accompanies the generation of allo-killer cells was also impaired in these cultures (26). The anomalous killer activity was defined earlier as the lytic effect exerted on cells that are unrelated to the stimulus, most importantly autologous PHA blasts or EBV transformed lymphoblastoid lines (39). Activity against K562 or Daudi is also a good indicator for this effect because these lines do not express Class I MHC antigens, thus do not cross react with the specific target and they are lysed by activated T cells. The P(Bu)2 induced inhibition of lytic activity in MLC was not the result of cell damage or lack of activation because cell recovery and blastogenesis were not altered (26). Addition of IFN-γ or IL-2 alone and in combination did not overcome the P(Bu)2 mediated inhibition, thus disturbance in the production of these lymphokines was not involved in the suppression mechanism. The cell surface markers of the MLC lymphocytes cultured in the presence of P(Bu)2 differed from the control MLC showing the alterations described above for the control cultures i.e. low OKT3 and OKT4 and elevated OKT8 reactivity. An additional phenotypic change inflicted by P(Bu)2 on the allo-stimulated mixed cultures was that their lymphocytes did not react with the OKI1 monoclonal reagent that detects expression of Class II MHC antigens. These antigens are not present on resting T cells but appear after activation. The generation of suppressor cells was not impaired in the P(Bu)2 containing mixed cultures. In summary, presence of P(Bu)2 in the MLC inhibited the generation of cytotoxic cells - both antigen specific and anomalous,and the induction of Dr antigens on the cells but did not impair proliferation and the generation of suppressor cells. Its presence altered the expression of maturation markers detected by the monoclonal antibodies OKT3, OKT4 and OKT8.

The 7 day old MLC cells carried in the presence of P(Bu)2 could not be activated for allocytotoxicity by exposure to IFN or by renewed exposure to P(Bu)2 after an interposed washing. Thus the effect of P(Bu)2 on the generation of allospecific killer cells has to be searched in the blockade of the maturation of antigen specific cytotoxic precursors. This was a specific event and not a general effect on the lytic machinery because such cells could be induced to perform lysis of the AK sensitive targets such as K562 and Daudi.

We have also studied the impact of P(Bu)2 preexposure of lymphocytes on the generation of killer cells in MLC. ter 72 hours treatment the lymphocytes were washed and incubated for further 48 hours in fresh medium without the compound. At this time the lymphocytes reexpressed regular amounts of the OKT4 marker but the expression of OKT3 and OKT8 markers still differed from the control cells. Exposure to alloantigens triggered regular allospecific cytotoxicity with elevated anomalous effect on which the allospecificity was superimposed.

The activation and manifestation of lytic potential against a given target can be dissected into afferent and efferent events. The afferent arm involves recognition of the antigen and the receipt of the triggering signals. The activation turns on the lytic machinery in the precursors. The efferent events are also initiated with the recognition of the target followed by the delivery of the lethal hit. Non specific effects that do not involve the antigen receptors can occur at both the afferent and efferent levels. The maturation and activation state of the lymphocytes and the interaction between various subsets in the course of activation determines the final outcome of a cytolytic system. Phorbol esters have specific receptors on cell surfaces and they act on cells in low concentrations. It is assumed that biological compounds with similar action exist. Their effect are highly variable depending on the cell type and the particular maturation stage. These effects mimick or potentiate cellular changes that occur normally, i.e. those which are in the program of the cell. By studying the impact of phorbol ester treatment on the cytolytic potential of lymphocytes, the complexity of events that leads to this easily recorded function is accentuated due to the action of the compound on the different steps.

REFERENCES

1. Klein, E. (1980). Immunol. Today 1, IV.
2. Vánky, F., Argov, S., Einhorn, S., and Klein, G. (1980). J. Exp. Med. 151, 1151.
3. Grimm, E.A., Mazumder, A., Zhang, H.Z., and Rosenberg, S.A. (1982). J. Exp. Med. 155, 1823.
4. Masucci, G., Masucci, M.G., and Klein, E. (1982). Cellular Immunol. 69, 21.
5. Klein, E. (1982). In "NK cells and other natural effector cells" (R.B.Herberman,ed.)p. 239.Academic Press,New York.
6. Touraine, J.L., Hadden, J., Touraine, F., Hadden, E.M., Estensen, R., and Good, R.A. (1977). J. Exp. Med. 145, 1960.
7. Abb, J., Bayliss, G.J., and Deinhart, F. (1979). J. Immunol. 122, 1639.
8. Kabelitz, D., Totterman, T.H., Gidlund, M., Nilsson, K., and Wigzell, H. (1982). Cell. Immunol. 70, 277.
9. Patarroyo, M., Yogeeswaran, G., Biberfeld, P., Klein, E., and Klein, G. (1982). Int. J. Cancer 30, 707.
10. Patarroyo, M., Biberfeld, P., Klein, E., and Klein, G. (1983). Cell. Imunol. 75, 144.
11. Galili, V., Galili, N., Vánky, F., and Klein, E. (1978). Proc. Natl. Acad. Sci. 75, 2396.
12. Mizel, Z.B., Rosenstreich, D.L., and Oppenheim, J.J. (1978). Cell. Immunol. 40, 230.
13. Yip, Y.K., Pang, R.H.L., Oppenheim, J.D., Nachbar, M.S., Henriksen, D., Zerobeckyi-Eckhardt, I., and Vilcek, J. (1981). Infect. Immun. 34, 131.
14. Johnson, H.M., and Torres, B. (1982). Infect. Immun. 36, 911.
15. Fuller-Farrar, J., Hilfiker, M.L., Farrar, W.L., and Farrar, J.S. (1981). Cell. Immunol. 127, 1983.
16. Goldfarb, R.H., and Herberman, R.B. (1981). J. Immunol. 126, 2129.
17. Kolb, J.D., Senik, A., and Castagna, M. (1981). Cell. Immunol. 65, 258.
18. Seaman, W.E., Gindhart, T.D., Blackman, M.A., Dalal, B., Talal, N., and Werb, Z. (1981). J. Clin. Invest. 67, 1324.
19. Mastro, A.M. (1982). Lymphokines 6, 263.
20. Weinstein, I.B., Lee, L.S., Fischer, P.B., Mufson, A., and Yamasaki, K. (1979). J. Supramol. Struct. 12, 19.
21. Keller, R. (1982). In: "NK Cells and other Natural Effector Cells" (R.B. Herberman, ed.), p. 601. Academic Press, New York.
22. Allison, A.C. (1978). Immunol. Rev. 40, 3.
23. Ramos, O.F., Masucci, M.G., Bejarano, M.T., and Klein, E. (1983). Immunobiol. 165, 403.

24. Ramos, O.F., Masucci, M.G., and Klein, E. (1984). Cancer Res. 44, 1857.
25. Blumberg, P.M., Delclos, K.B., Dunphy, W.G., and Jaken, S. (1982). In "Carcinogenesis" (E. Hecker et al., ed), Vol. 7, p. 519. Raven Press, New York.
26. Masucci, M.G., Torsteinsdottir, S., Pear, W., Carstensen, A., and Klein, E. (1984). J. Immunol., in press.
27. Solbach, W. (1982). J. Exp. Med. 156, 1250.
28. Patarroyo, M., and Gahmberg, C.G. (1984). Eur.J. Immunol. 14, 781.
29. Fujiki, H., Mori, M., Nakayasu, M., Terada, M., Sugimura, T., and Moore, R. (1981). Proc. Natl. Acad. Sci. 78, 3872.
30. Umezawa, K., Weinstein, B., Horowitz, A., Fujiki, H., Matsushima, T., and Sugimura, T. (1981). Nature 290, 411.
31. Yip, Y.K., Kelker, H.C., Stone-Wolff, D.S., Pearlstein, K., Urban, K., and Vilcek, J. (1983). Cell. Immunol. 79, 389.
32. Seaman, W.E., Gindhart, T.D., Blackman, M.A., Dalal, B., Talal, N., and Werb, Z. (1982). J. Clin. Invest. 69, 876.
33. Ramos, O., Masucci, M.G., and Klein, E. Cell. Immunol. in press.
34. Ullberg, M., and Jondal, M. (1981). J. Exp. Med. 153, 615.
35. Andreotti, F.P. (1982). J. Immunol. 129, 91.
36. Fredrikson, G.G., and Bennet, M. (1982). Cancer Res. 42, 3601.
37. Yamamoto, H., Katsuki, T., Harada, S., and Hinuma, Y. (1981). Int. J. Cancer 27, 167.
38. Klein, E., and Masucci, M.G. (1982). In: "Springer Seminars in Immunopathology" (Miescher and Müller-Eberhard, ed.), Vol. 5, p. 67. Springer International.
39. Seeley, J.K., and Golub, S.H. (1978). J. Immunol. 120, 1415.

CHARACTERIZATION OF SOLUBLE SUPPRESSOR FACTORS PRODUCED BY UNSTIMULATED NORMAL HUMAN LYMPHOCYTES[1]

Stanley A. Schwartz[2]
Madhavan P. N. Nair

Departments of Pediatrics and Epidemiology
The University of Michigan
Ann Arbor, Michigan

I. INTRODUCTION

Down regulation of the immune response by suppressor cell mechanisms has been a continuing subject of interest for well over a decade. Previous investigations have demonstrated that suppressor cell activity can be either antigen specific or nonspecific and may or may not show genetic restriction. Various methods were previously employed to measure suppressor cell function including inhibition of: DNA synthesis, polyclonal B cell differentiation, and cytotoxic T lymphocyte activation. Natural killer (NK) cells have come under intense recent investigation because of their likely role in host defense against cancer and viral infections (for a review, 1). Moreover, Cudkowicz and Hochman (2) have proposed that NK cells may also

[1]Aided by National Institutes of Health Grants CA35922 and AI19890; and by the Children's Leukemia Foundation of Michigan and the Michigan Diabetes Research and Training Center, Ann Arbor.
[2]Recipient of NIH Research Career Development Award CA00896.

ISBN 0-12-341370-2

exert immunoregulatory functions. It is therefore likely that NK cells may themselves be subject to immunologic control mechanisms.

Initial assays of suppressor cell functions often utilized mixing or reconstitution experiments wherein putative suppressor cells were added directly to effector cells whose functions were monitored. For a considerable time it was speculated that suppressor cell activity required close juxtaposition of the regulatory cell with the effector cells being modulated. Early experiments in the murine system described a soluble immune response suppressor which could be induced by treatment of lymphocytes with Con A (3). These observations were subsequently extended to humans when Shou et al. (1) demonstrated that Con A could also activate suppressor cells among human peripheral blood lymphocytes (PBL). Only recently have we developed a good molecular definition of suppressor cell activity in the human system with the isolation and characterization of soluble suppressor factors either directly from serum or from the supernates of cultured PBL.

II. SUPPRESSOR CELLS

The original studies examining inhibitory immunoregulatory mechanisms utilized cellular mixing experiments. With the development of a human Con A-activated suppressor cell system we could routinely induce suppressor activity using PBL. This method could suppress blast transformation responses of PBL to mitogens, antigens and allogeneic cells (3) as well as polyclonal B cell differentiation (5). Subsequently, Mawas et al. (6) demonstrated suppression of cytotoxic T lymphocyte activity using phytohemagglutinin stimulated cells. More recently, we demonstrated that a relatively faster sedimenting subpopulation of thymocytes separated by velocity sedimentation suppressed the NK activity of both autologous and allogeneic spleen cells in mice, suggesting a role for the thymus in regulating NK activity *in vivo* (7). Further, we demonstrated that precultured PBL from healthy human donors could also suppress the NK and antibody-dependent cellular cytotoxic (ADCC) activities of both autologous and allogeneic PBL (8). This activity could be substantially increased when cells were precultured in the presence of supramitogenic doses of Con A. However, when submitogenic doses were utilized, enhancement of cytotoxicity was observed. A similar dose-dependent, biphasic suppressor activity against other lymphocyte functions has been previously reported (9). It is interesting to contrast Con A-induced suppressor cells capable of inhibiting NK and ADCC activities with Con A activated suppressors of other lymphocyte functions. Whereas the former are resistant to ionizing radiation up to 3000 rads (8), the latter are exquisitely sensitive to much lower doses of radiation starting at approximately 100 rads (10).

III. SOLUBLE SUPPRESSOR ACTIVITIES

A. Serum Factors

Serum factors from patients and animals with malignancies have long been known to inhibit the proliferative responses of normal allogeneic lymphocytes in vitro. However Nair and his colleagues (11) previously showed that sera from mice with transplanted and spontaneous tumors could also suppress the NK and ADCC activities of syngeneic spleen cells from healthy young animals. They further observed that although splenocytes from normal young mice positive for mammary tumor virus demonstrated high NK and ADCC activities, these functions declined with age concomitant with an increase in serum inhibitory activity. This serum effect was shown to be independent of the appearance of circulating immune complexes. We have fractionated this murine serum suppressor activity by Sepharose 6B gel filtration and obtained the elution profile shown in Fig 1. Thus the inhibitory activity elutes in the region of globular proteins with a molecular weight of approximately 150,000 daltons.

As demonstrated in Fig. 2, sera from patients with various cancers also showed selective inhibition of NK and ADCC activities of allogeneic PBL from healthy donors (12).

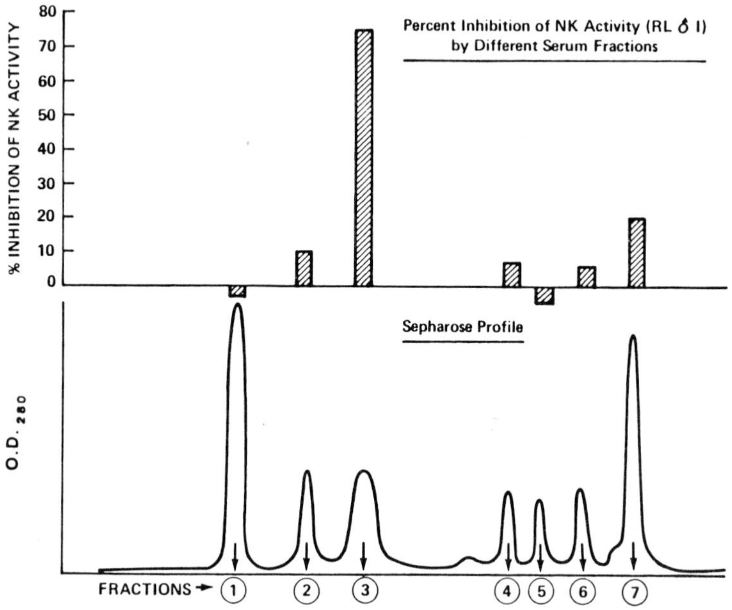

Fig. 1. Molecular sieve chromatography of murine serum suppressor factors. Heat inactivated serum samples were passed over a column of Sepharose 6B gel and eluates were monitored by absorbance at 280 nm. Each peak was individually concentrated and tested for an inhibitory effect on the NK activity of syngeneic spleen cells against RL♂1 target cells.

Although normal donor serum inhibited both NK and ADCC activities of allogeneic PBL when substituted for fetal calf serum, significantly greater suppression was observed with cancer patients' sera. These data suggest that soluble factors capable of modulating NK and ADCC activities of normal PBL are present in sera from healthy donors, but the significantly increased inhibition of cytotoxicity manifested by cancer patients' sera may represent a mechanism of decreased host resistance to tumor growth.

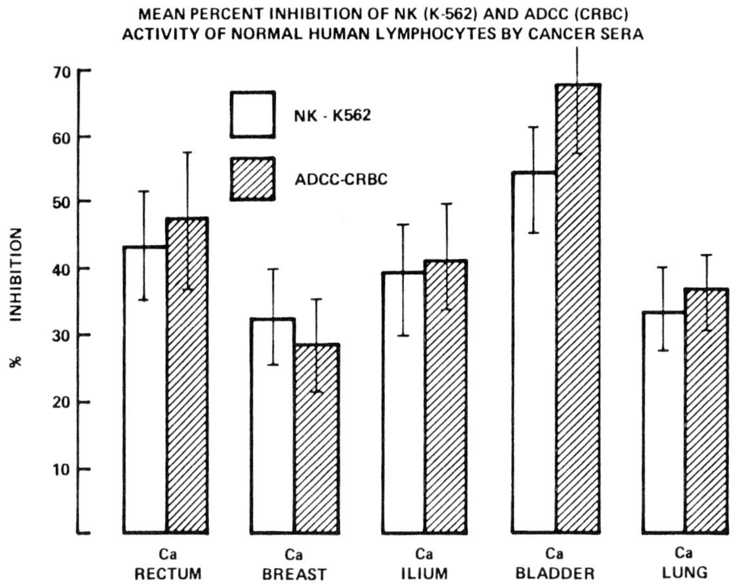

Fig. 2. Inhibition of NK and ADCC activity of allogeneic normal human lymphocytes against K562 and Ab-coated chicken RBC targets respectively by cancer sera. All sera were heat inactivated at 56°C for 30 min, centrifuged at 105,000 x g for 1 hr and used at a final concentration of 20% (V/V). The percentage suppression of cytotoxicity was calculated on the basis of the cytotoxicity obtained when normal human sera were used as controls.

B. Suppressor Factors from Cultured Cells

Having reported that human Con A-activated suppressor cells could inhibit lymphocyte activities (4, 8), initial attempts to demonstrate soluble suppressor activity (SSF) in their supernates were unsuccessful. It was subsequently observed that the presence of fetal calf serum could prevent any supernatant suppressor activity. Thus when lymphocytes were cultured in serum-free medium a potent soluble suppressor factor (SSF) capable of inhibiting the proliferative responses of T cells and polyclonal B cell activation, could be consistently found in the supernatant fluid (13). SSF appears to be the product of an autologous mixed lymphocyte reaction as neither T nor non-T cells alone were capable of its synthesis. Further, we were able to show that SSF could suppress the NK and ADCC activities of allogeneic lymphocytes (14). Preliminary characterization demonstrated two active moieties eluting from gel filtration on Sephacryl S-200 (Figs 3 and 4). The first active fraction eluted in the region bounded by authentic markers of 13,700 and 25,000 daltons; the second eluted after the 13,700 dalton marker. Most significantly, SSF inhibition of NK activity could be blocked by three monosaccharides, α-methyl-D-mannoside, L-fucose and L-mannose, in a dose-dependent manner. These results suggest that SSF may act through binding to unique glycosylated residues on the surface of effector cells.

The immunoregulatory activity of SSF could itself be modulated by the biological modifier substance, α-interferon (IFN). Thus suppression of NK cytotoxicity could be almost completely reversed by incubation of SSF pretreated cells with IFN for an additional 24 hr. These findings contrast with the earlier studies of Kadish et al. (15) who proposed that the active fraction of Con A induced suppressor cells could be inhibited with anti-IFN antibodies. Recently we observed that interleukin 2 (IL2) is also a powerful inhibitor of SSF activity (Table I). However, IL2 is not only capable of reversing suppression mediated by SSF but as shown in Figure 5, IL2 can also inhibit SSF production when added to precultured PBL.

Other properties of SSF not previously reported include its sensitivity to proteolytic enzymes suggesting that, at least in part, it is a polypeptide (Table II). Although acidic conditions slightly reduce SSF activity, it is almost completely inactivated when subjected to an alkaline pH.

As noted above, lymphocytes preincubated for 72 hr with or without Con A can effectively suppress cytotoxicity even when irradiated up to 3000 rads (8). In like fashion, SSF production by lymphocytes cultured in serum-free medium was unaffected by preirradiation (Table III). These observations are consistent with the hypothesis that suppressor

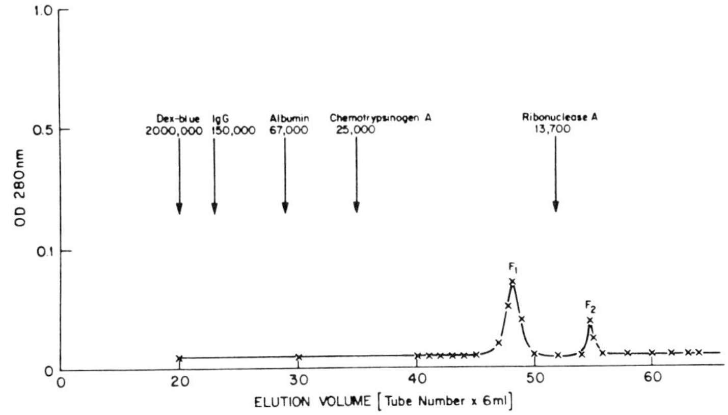

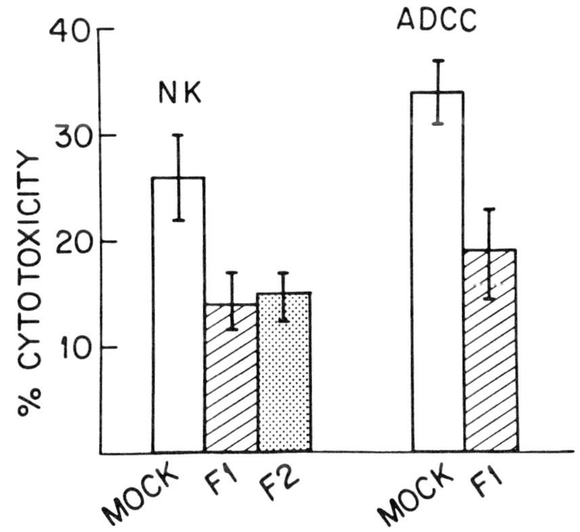

Figs. 3 and 4. Molecular sieve chromatography of human SSF using Sephacryl S-200. The eluates were collected and monitored for protein by absorbance at 280 nm. The fractions were tested for their inhibitory effects on NK and ADCC activity of allogeneic normal human lymphocytes. The column was previously calibrated with known molecular weight markers.

cells capable of inhibiting NK and ADCC functions may be a distinct subpopulation different from suppressor lymphocytes regulating T cell proliferation and polyclonal B cell activation.

Table I. Effect of IL-2 on Inhibition of NK Activity by SSF[a]

SSF Sample	% Cytotoxicity Incubation with IL-2 (% v/v)[b]					
	0		5		10	
Mock SSF	37.5	----	49.4	----	55.2	----
SSF 14	22.8	(39.2)	49.3	(0.2)	51.0	(7.6)
SSF 15	20.8	(44.5)	47.1	(4.6)	49.0	(11.2)
SSF 16	22.1	(41.0)	46.5	(5.8)	51.1	(7.4)
SSF 17	15.8	(57.8)	47.2	(4.4)	51.5	(6.7)
SSF 18	17.0	(54.6)	48.5	(1.8)	49.3	(10.6)
SSF 19	18.6	(50.0)	48.6	(1.6)	52.4	(5.0)
SSF 20	17.9	(52.2)	47.2	(4.4)	50.5	(8.5)
SSF 21	20.2	(46.1)	47.7	(3.4)	51.7	(6.3)

[a]Human lymphocytes were precultured separately with either mock SSF or SSF alone, or SSF + IL 2, washed and tested for their cytotoxicity.
[b]Numbers in parentheses are % suppression.

IV. IMMUNOPHARMACOLOGIC MODULATION OF NK CELLS

IFN has clearly been shown to significantly enhance the cytotoxicity of NK cells (16-18). When cells infected with Herpes simplex virus type I were used as targets of NK activity, significant levels of IFN could be assayed in the culture supernates; however, cytotoxicity was observed to be independent of the amounts of IFN produced (18, 19).

Addition of amphotericin B to mixtures of effector and target cells was noted to cause not only direct inhibition of NK and ADCC activities but also increased suppressor cell activities against cytotoxic functions (20). Other drugs observed to directly inhibit NK activities include H1 and H2 histamine antagonists (21). This inhibitory effect of histamine antagonists could be decreased by the addition of histamine to the mixture. Although direct addition of histamine to effector-target mixtures had no effect on cytotoxicity, preculture of PBL with histamine caused both an increase in NK-directed suppressor cell activity concomitant with increased numbers of $OKT8^+$ lymphocytes.

Corticosteroids are potent inhibitors of many lymphocyte functions including NK and ADCC activities (22). Our results suggest that corticosteroid suppression of cytotoxicity may be due, in part, to decreased target binding activity of effector cells. Corticosteroid induced suppression of NK and ADCC activities could be reversed by subsequent treatment of PBL with either IFN or IL2.

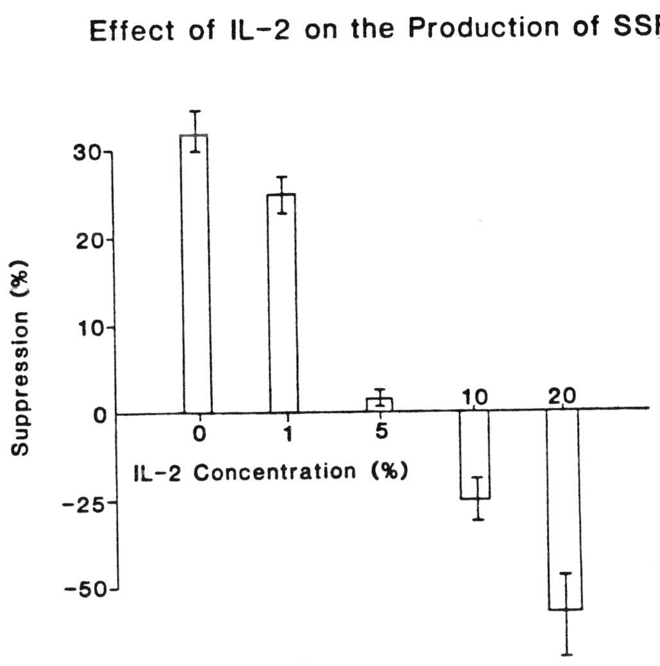

Fig. 5. Human PBL were precultured with different concentrations of IL2 and the culture supernates were tested for their inhibitory effect on the NK activity of allogeneic lymphocytes against K562 target cells.

Table II. Enzyme Sensitivity of SSF

Enzyme Treatment of SSF	Treated Sample	% Cytotoxicity	% Suppression
None	Mock SSF	36.2	----
	SSF	23.4	35.3
Trypsin	Mock SSF	35.4	----
	SSF	33.3	5.9
Chymotrypsin	Mock SSF	32.1	----
	SSF	29.8	7.1

Mock or real human SSF were treated with enzymes coupled to polyacrylamide beads (P-100) at 37°C for 1 hr on a rocking platform and the supernates were tested for suppressive activities. Allogeneic PBL were precultured with supernates for 72 hr, washed and tested for NK activity at 50:1 E:T cell ratios against K562 targets.

Table III. Effect of Irradiation of PBL on the Production of SSF

SSF Source	Radiation Dose (R)	NK Cytotoxicity (%)	Suppression (%)
Mock SSF	0	35.1	----
PBL	250	21.1	39.8
PBL	500	23.5	33.0
PBL	1000	19.5	44.4
PBL	2000	23.6	32.7

Human PBL (5 x 10^6/ml) were suspended in medium and

irradiated at ambient temperature by a ^{60}Co source at a rate of 76 rads/min. PBL were incubated for 5 days and culture supernates (SSF) were tested for their inhibitory effect on NK activity of allogeneic PBL at a 50:1 E:T cell ratio against K562 targets.

V. CONCLUDING REMARKS

NK cells appear to play a significant role in host surveillance against malignant and virus infected cells. Although NK cells may also have immunoregulatory functions, it is likely that they themselves may be subject to modulation. Our studies have demonstrated an endogenous SSF which may be the product of an autologous mixed lymphocyte reaction. Similar soluble suppressor activity can be observed in serum from healthy donors and may represent a normal immunologic control mechanism. In cancer patients and tumor-bearing animals, an abnormal increase in serum SSF may result in decreased immune surveillance and exacerbation of disease. In vitro experiments indicate that biological modifier substances such as IFN and IL2 may have therapeutic potential in reversing abnormal suppressor activity accompanying several disease states. Other drugs may also have a role in the clinical management of immunoregulatory dysfunctions.

VI. REFERENCES

1. Herberman, R. B. (ed) (1980). "NK and Other Natural Effector Cells", Acad. Press, New York.
2. Cudkowicz, G. and Hochman, P. S. (1979). Immunol. Rev. 44:13.
3. Rich, R. R. and Pierce, C. W. (1973). J. Exp. Med. 137:649.
4. Shou, L., Schwartz, S. A. and Good, R. A. (1976). J. Exp. Med. 143:1100.
5. Schwartz, S. A., Choi, Y. S., Shou, L. and Good, R. A. (1977). J. Clin. Invest. 59:1176.

6. Mawas, C., Charmot, D., Comoy, A. and Sasportes, M. (1977). Eur. J. Immunol. 7:11.
7. Nair, M.P.N., Schwartz, S. A., Fernandes, G., Pahwa, R., Ikehara, S. and Good, R. A. (1981). Cell. Immunol. 58:9.
8. Nair, M.P.N. and Schwartz, S. A. (1981). J. Immunol. 126:2221.
9. Gupta, S., Schwartz, S. A. and Good, R. A. (1979). Cell. Immunol. 44:242.
10. Siegal, F. P. and Siegal, M. (1977). J. Immunol. 118:642.
11. Nair, M.P.N., Fernandes, G., Onoe, K., Day, N. K. and Good, R. A. (1980). Int. J. Cancer 25:667.
12. Nair, M. P. N. and Schwartz, S. A. (1981). Fed. Proc. 40:1045.
13. Shou, L., Schwartz, S. A., Good, R. A., Peng, R. and Chen, C. L. (1980) Proc. Natl. Acad. Sci. U.S.A. 77:6096.
14. Nair, M. P. N., and Schwartz, S. A. (1982). J. Immunol. 129:2511.
15. Kadish, A.S., Tansey, F. A., Yu, G. S. M., Doyle, A. T., and Bloom, B. R. (1980). J. Exp. Med. 151:637.
16. Djeu, J. Y., Heinbaugh, J. A., Holden, H. T. and Herberman, R. B. (1979) J. Immunol. 122:175.
17. Herberman, R. B., Ortaldo, J. R. and Bonnard, G. D. (1979). Nature 277:221.
18. Bishop, G. A. and Schwartz, S. A. (1982). Clin. Immunol. Immunopathol. 25:374.
19. Bishop, G. A., Glorioso, J. C. and Schwartz, S. A. (1983). J. Immunol. 131:1849.
20. Nair, M. P. N. and Schwartz, S. A. (1982) Cell. Immunol. 70:287.
21. Nair, M. P. N. and Schwartz, S. A. (1983). Cell. Immunol. 81:45.
22. Nair, M. P. N. and Schwartz, S. A. (1984) J. Immunol. 132:2876.

ROLE OF TRANSFERRIN AND TRANSFERRIN-RELATED PROTEINS IN REGULATION OF NK ACTIVITY IN VITRO

H. David Kay
Howard T. Petrie
Philip J. Bierman
Lynell W. Klassen

Experimental Immunology Laboratory
Department of Internal Medicine
Veterans Administration Medical Center, and the
University of Nebraska Medical Center
Omaha, Nebraska

Part of the excitement, as well as the frustration, in working with natural killer (NK) cells derives from attempting to unravel the mystery with which the NK target structure(s) have been shrouded. The difficulty in resolving this enigma relates to a) the sheer diversity of cell types which are apparently vulnerable to NK cytolysis, which include hemopoietic stem cells and fetal thymocytes, as well as a fairly large variety of cultured tumor cell lines; and b) the well-documented heterogeneity of effector cell types (1) which react with NK-like activity against these various target cells.

As convenient as it would be, it has been difficult for investigators to accept that all NK-susceptible

*Supported by research grant RO1 CA36033 from the National Cancer Institute, National Institutes of Health, Bethesda, Maryland.

ISBN 0-12-341370-2

target cells have in common a unique membrane antigen recognizable by all NK cells (the so-called "minimalist" view of diversity) -- that is, until now. Now it seems that a formidable candidate for this target structure may have been found in the membrane-associated receptor for the iron-carrying protein transferrin (Tf) (2,3). The ubiquitous transferrin receptor (TfR) seems to have what it takes to be an NK target structure, since it is a cell-surface glycoprotein which is expressed on virtually all proliferating cells at levels proportional to the rate of proliferation; and rapidly proliferating cells are frequently targets of NK cell attack -- in vitro at least. Because virtually all cells and tissues require iron for growth and metabolism (2,4), the physiologic apparatus for iron delivery, to meet this need, utilizes Tf as the iron-binding protein in the serum, and a plasma membrane receptor for this iron-Tf complex to capture and carry the iron into the cell (5).

I. TRANSFERRIN (Tf)

The delivery of iron to human tissues is mediated primarily, if not solely, by transferrin (Tf), a monomeric glycoprotein found in serum at average normal concentrations ranging from 1.6 mg/ml to 4.0 mg/ml (6). With a molecular weight of 79,550 daltons (678 amino acid residues) (7), it contains two homologous, independent iron-binding sites which can bind two ferric (Fe^{3+}) ions with a K_A of approximately $10^{22} M^{-1}$ at neutral pH (7). Iron is then delivered to the cell when the $Tf\text{-}(Fe^{3+})_2$ complexes contact, are bound by, and move into the cellular cytoplasm with specific plasma-membrane receptors for Tf (5).

II. TRANSFERRIN RECEPTORS

There are considerable circumstantial data to support the notion that the transferrin receptor is (or is closely related to) the target antigen/structure being recognized by NK effector cells. First, NK cells have been described as being primarily involved in regulating the growth of autologous, rapidly-dividing normal cells (8). Second, in line with this, evidence suggests that most of these activated, rapidly-proliferating vertebrate cells express large numbers of membrane receptors for transferrin (9). Thirdly, there is evidence that the cultured cell lines which are most susceptible to NK cell-mediated lysis are those which express the highest numbers of TfR (3). In view of these data, it did not seem unreasonable to assume that the natural ligand for this receptor -- Tf itself -- would act as a natural, serum-borne soluble regulator of NK cell activity. Baines *et al* did the same (2), and found that Tf, when saturated with iron (FeTf), inhibited human NK cytolytic activity against K562 cells, if included in the NK assay mixtures at physiologically relevant levels. We, too, repeatedly observed that exogenous Tf significantly inhibited human NK cell activity in a dose-dependent manner when included in a ^{51}chromium-release assay (CRA) at physiologic (approximately 1.5 to 5.0 x 10^{-5} M) and suprophysiologic (>5 x 10^{-5} M) concentrations (Figure 1). Bovine serum albumin, ovalbumin with high mannose content and ovalbumin with low mannose content were used as protein controls, and were always found to exhibit little or no inhibition of NK cell activity, even at final molar concentrations as great as 24 x 10^{-5} M.

The question of the effects of iron-saturated versus non-saturated Tf was impossible to evaluate unambiguously in these studies, because we also learned that RPMI 1640 balanced salt solution used throughout these studies contained sufficient free iron (as trace metal in commercial reagents) to saturate most, if not all, of the exogenous apotransferrin (apoTf) during the 4-18 hr ^{51}chromium release assay (10). Therefore, what was called "apoTf" in our experimental setup may in fact have been rapidly saturated with iron during the assay, especially since the association constant for the binding of iron to Tf is so high (7). To avoid this problem, we repeated a large number of the experiments in serum-free, transferrin-free synthetic HB103 medium (Hana Biologics). By doing this, we could strictly quantitate the amount of Tf in the cytotoxicity assay. As shown in Figure 2, control NK cytotoxicity against K562 was excellent in this specially-prepared Tf-free synthetic medium. When exogenous Tf was added to the assay, NK cytotoxocity was significantly inhibited in a dose dependent manner, consistent with the concept that soluble Tf might exert a regulatory role in NK cytolysis. To verify that K562 cells were not uniquely influenced in this type of system, we also tested the effect of Tf on NK activity against the cultured MOLT-4 T-lymphoblastoid cell line. Again, Tf significantly inhibited NK activity against this cell line, in a dose-dependent fashion.

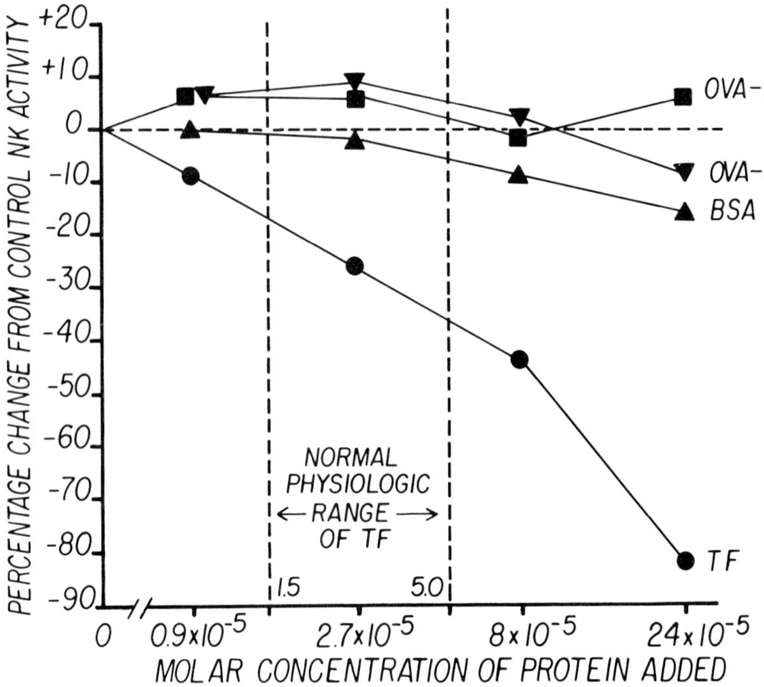

Figure 1. Soluble transferrin (Tf, Sigma lot #53-) inhibited NK cytotoxicity in a dose-dependent fashion when included in the NK assay at the concentrations shown. Control proteins ovalbumin with high or low mannose content (OVA-HM and OVA-LM) and bovine serum albumin (BSA) produced little effect on NK activity when tested at similar concentrations. (Effector-to-target ratio of 100:1.)

We discovered, much to our chagrin, that we could not reproduce our own experiments when we bought fresh stocks of Tf from our original supplier. Upon close examination, we determined that these differences reflected the fact that we were now using a different lot of Tf (Sigma lot #73F-0517, instead of Sigma lot #53F-9345 which had been used in the original experiments). On close examination, these lots differed considerably in color and consistency; i.e., Tf73 (our abbreviation) was whiter and more crystalline, and was listed as 98% Tf. Tf53, on the other hand, was slightly "tannish" in color, and more amorphous in consistency. Reproducibly, lot Tf53 exhibited a greater capacity to inhibit NK cytolysis than did lot Tf73 (Figure 3). In response to our enquiries, we were informed by the chemists at Sigma that the major difference between these lots was a protein contaminant in lot Tf53 called hemopexin.

III. HEMOPEXIN (HPx)

Hemopexin (HPx) is a single-chain β-globulin molecule with a molecular weight of approximately 60,000 daltons (11). It is synthesized in the parenchymal cells of the liver, and is present in normal human serum at a concentration of approximately 0.5 - 1.0 mg/ml (11). It binds the iron-carrying protein heme in equimolar ratios, with a $K_A = 10^{14}\ M^{-1}$. Its major function in the blood is to transport heme to the liver for degradation.

In collaboration with Dr. Ursula Muller-Eberhard and Valentina Goldfarb at the Cornell Medical Center in New York City, we have begun examining the effects of highly purified HPx on the in vitro NK response. As shown in Figure 4, purified HPx (a very generous gift of Dr. U. Muller-Eberhard) exhibited a reproducible, dose-dependent, potent inhibition of NK activity when included in the assay at the physiologic concentrations shown. This dose-dependent inhibition was seen both with K562 as well as with MOLT-4 target cells.

To determine if HPx was inhibiting NK cell function because it was toxic to cells, we decided to preincubate effector cell suspensions with HPx (1.0 mg/ml) for 18 hr prior to testing the cells for cytotoxic function. Such pretreated effectors were then washed twice in tissue culture medium and tested for the dual functions of NK capacity (against K562 cells) and antibody-dependent cellular cytotoxicity (ADCC; against rabbit IgG-coated RAJI cells). As shown in Figure 5, preincubation with HPx completely abrogated both NK and ADCC functions of these effector preparations; in contrast, preincubation for the same period of time in tissue culture media, or with Tf53 in TC media (4 mg/ml) did not alter NK or ADCC function. Treatment with Tf73 led to an enhancement of NK, as well as of ADCC, cytotoxicity activities. Viability of the HPx cultures was 80 ± 5% of control

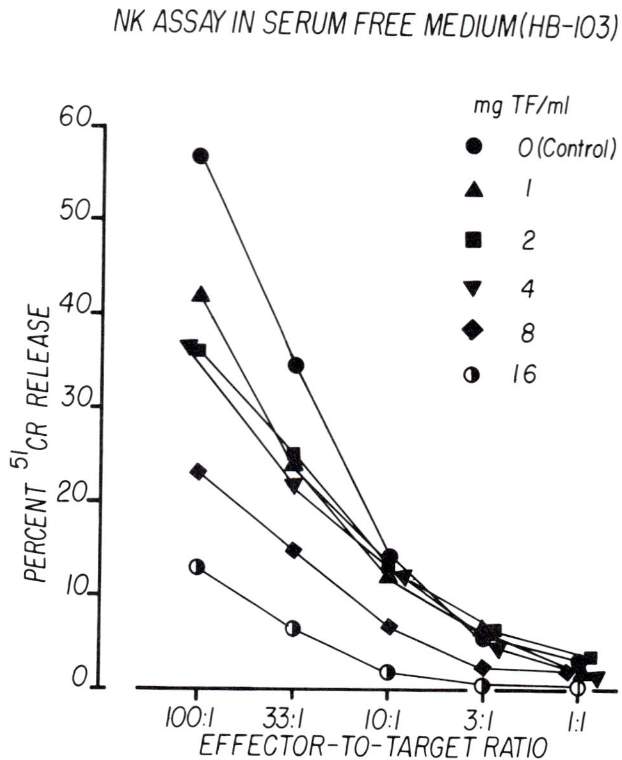

Figure 2

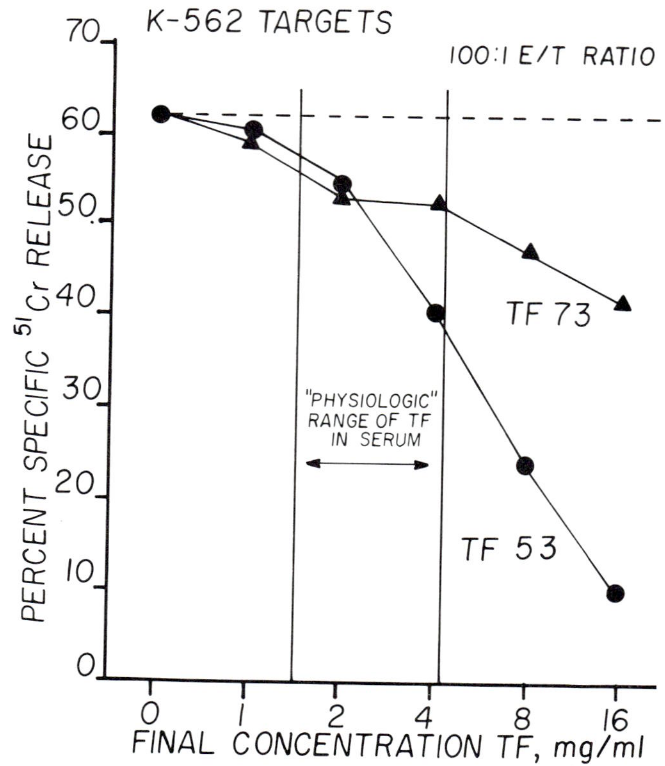

Figure 3

cultures; Tf53- and Tf73-treated cultures exhibited 95 ± 4% viability, compared to controls. This total inhibition of cytotoxicity by HPx, accompanied as it was by a loss of only 20-25% viability of effectors, suggests that the effect is either a) specific for a marker or receptor on NK/K cells, and leads exclusively to their destruction; or b) the NK/K cell subsets are inately more fragile than other PBLs, and are thus more easily inactivated by HPx. Since we are not aware of evidence to support the latter position, it seems not unreasonable to consider that HPx may have an affinity for NK/K cells, and thereby may selectively inhibit the functions of that subset. Very preliminary data (done once at the time of this writing) showed that overnight preincubation of PBL with HPx induced only a 25-30% loss of capacity to respond to PHA (as measured by ^{3}H-thymidine uptake at 72 hr), even though the 20-25% loss in viability of the treated PBL suspension completely abrogated NK activity. This preliminary data would tend to confirm the concept that the HPx effect might be restricted to non-T cells, since T cell and monocyte functions (in the blasto genesis assay) were only partially inhibited.

These data lead one to speculate that HPx itself might exhibit dual functions physiologically, acting as a heme carrier on one hand, and as a potent regulator of NK/K cell function on the other. It is always possible that this dual effect occurs because another protein is contained in the HPx preparations, and it is this second protein which is responsible for the toxic/inhibitory effects on NK cells which we are seeing. These concepts are being actively studied in our laboratory with the collaboration of Dr. Muller-Eberhard.

Figure 2. NK cells readily lysed K562 targets in an assay set up in serum-free, Tf-free synthetic media (HB103, Hana Biologics). Tf inhibited NK cell function in a dose-dependent manner, especially at concentrations > 4 mg/ml.

Figure 3. Capacity to inhibit NK activity varied with different lot preparations of Tf purchased from commercial source. Sigma Tf, lot 53F-9345 (=TF53), was always more inhibition in NK assays than was Sigma Tf, lots #73F-0517 (=TF73) (and #93F-0195, identical in response to TF73, not shown)

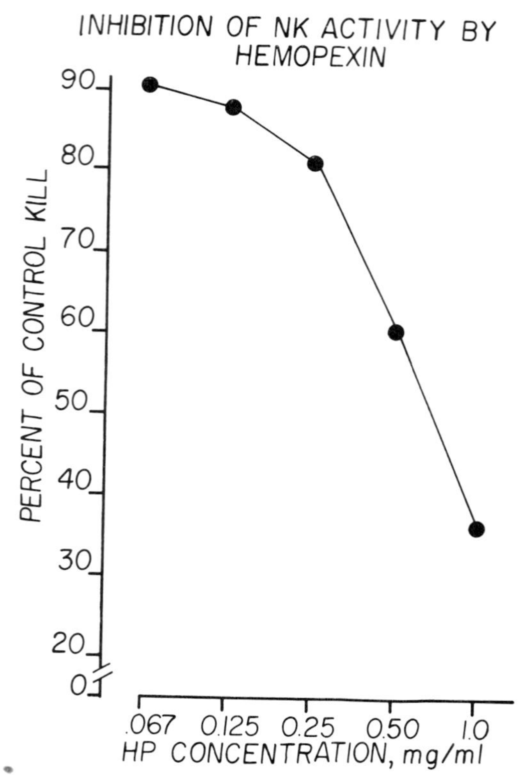

Figure 4

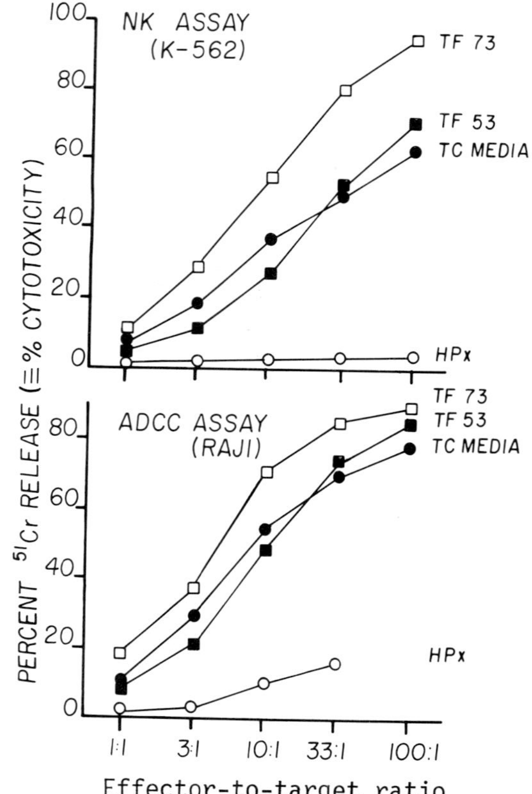

Figure 5

IV. Tf RECEPTORS (TfR) AND NK FUNCTION

There has been reported considerable circumstantial data implicating TfR as a potential target structure for NK cells (2,3,12). We have found that highly purified Tf (free of HPx) does not seem to interfere in NK function as we might expect it to, were the TfR to be the major target structure for NK cells (13).

The role of TfR as target antigen is also being questioned by data being reported from several laboratories. Monoclonal antibodies to human TfR have been raised by a number of investigators, and the 2 most commonly used are to monoclonals B3/25, and 42/6, raised by Dr. Ian S. Trowbridge (14). Contrary to early reports which implied that these antibodies, when included in an NK assay, inhibited lysis of targets, recent results from our laboratory showed that NK activity was not inhibited by these monoclonals, whether the antibodies were added to the assay, or used to pretreat the K562 target cells before the assay. Dokhelar _et al_ (15) reported similar findings.

Dokhelar _et al_ (15) further showed that low TfR expression and high NK sensitivity can occur concommitantly. They reported no change in NK sensitivity of K562 cells between cells in exponential growth phase (expressing high numbers of TfR) and cells incubated for several days after reaching the plateau phase (where TfR expression completely disappeared). Further, when TfR expression at the K562 cell surface was abolished by use of protein synthesis inhibitors, susceptibility to NK lysis was not effected. This may, however, also reflect an inhibition of membrane repair

---

Figure 4. NK cytolysis of K562 cells was inhibited by purified hemopexin (a generous gift from Dr. Ursula Muller-Eberhard) in a dose-dependent manner, at the concentrations shown. Control kill was 48% at an effector-to-target ratio of 33:1. Spontaneous release (with and without HPx) was 6% in the 4-hr assay.

Figure 5. Preincubation of effector cells with HPx (1 mg/ml) for 18 hr abrogated the NK responsiveness of the treated cells. Control cultures were preincubated with tissue culture media only, TF53, or TF73 in culture media. Viability of cultures were 80 ± 5% after HPx treatment, and >95 ± 3% after the other treatments.

mechanism by the protein synthesis inhibitors, so this data needs cautious interpretation. Lastly, hemin induction of K562 differentiation did not change NK susceptibility, even though it considerably reduced TfR expression.

Newman et al (16) support these conclusions, reporting no correlation between the amount of TfR expressed on murine targets and the susceptibility of these targets to murine NK lysis or NK binding in cold target competition assays. In contrast to this model system, however, they did find that TfR-positive mouse L-cell transformants (transfected with human DNA and selected for the expression of human TfR) could competitively inhibit (as cold targets) the lysis of K562 cells by human NK cells; the capacity to inhibit was proportional to the amount of TfR expressed on the transformants. Because inhibition was not complete, however, other cell surface antigens probably played a role in these human NK-target interactions.

The report of Hagner (17) is in complete agreement with these results. Again no correlation between expression of TfR on targets (as determined by reactivity with OKT9 monoclonal antibody) and the capacity to be lysed by NK cells was observed. They reported, though, that TfR may be involved in target cell recognition by effector cells which exhibit the $OKT8^+$ phenotype.

V. CONCLUSIONS

It might be concluded that, rather than being a specific target structure for NK mediated lysis, the TfR may serve primarily as a convenient (but nonessential) ligand for interaction between effector and target cells. Such a view is consistent with the recent report of MacDougall and Sullivan (18), as well as that of Baines et al (2) and Vodinelich et al (3).

Still to be resolved are problems such as why some cell lines which express abundant TfR are resistant to NK killing, while other cells and cell lines with very few TfR are highly sensitive to kill. Presumably, other factors related to intrinsic resistance to lysis, efficiency of membrane repair, or expression of alternative NK-associated target antigens, are also important in determining the outcome of an assay.

Considerably more data will be needed before a direct role of TfR as a unique NK target structure can be established. The work of Newman, Warner, and Dennert (16) is indeed heading us in the right direction. Perhaps highly purified TfR can be inserted in various numbers into the membrane structure of liposomes, which can then be examined for binding and lysability by appropriate effectors. This type of direct experimentation would go far toward elucidating the potential role of TfR as a target structure for NK cells.

REFERENCES

1. Ortaldo JR, and RB Herberman. 1984. Annual Rev. Immunol. 2: 359.
2. Baines MG, FL LaFleur, and BE Holbein. 1983. Immunol. Letters 7: 51.
3. Vodinelich L, R Sutherland, C Schneider, R Newman, and M Greaves. 1983. Proc. Natl. Acad. Sci. (USA) 80: 835.
4. Gatter KC, G Brown, IS Trowbridge, R-E Woolston, and DY Mason. 1983. J. Clin. Pathol. 36: 539.
5. Klausner RD, J Van Renswoude, G Ashwell, C Kempf, AN Schechter, A Dean, and KR Bridges. 1983. J. Biol. Chem. 258: 4715.
6. Markowitz H, and VF Fairbanks. 1983. Mayo Clinic Proc. 58: 827.
7. MacGillivray RTA, E Mendez, SK Sinha, MR Sutton, J Lineback-Zins, and K Brew. 1982. Proc. Natl. Acad. Sci. (USA) 79: 2504.
8. Cudkowicz G, and PS Hochman. 1979. Immunol. Rev. 44: 13.
9. Sutherland R, D Delia, C Schneider, R Newman, J Kemshead, and M Greaves. 1981. Proc. Natl. Acad. Sci. (USA) 78: 4515.
10. Brock JH, and T Mainou-Fowler. 1983. Immunol. Today 4: 347.
11. Muller-Eberhard U. 1970. N.E.J. Med. 283: 1090.
12. Editorial. 1983. Immunol. Today 4: 98.
13. Kay HD, HT Petrie, PJ Bierman, CL Scribner, KR Thomas, and LW Klassen. 1984. Immunobiology 167: 187.
14. Trowbridge IS, and F Lopez. 1982. Proc. Natl. Acad. Sci. (USA) 79: 1175.
15. Dokhelar M-C, D Garson, U Testa, and T Tursz. 1984. Eur. J. Immunol. 14: 340.

16. Newman RA, JF Warner, and G Dennert. 1984. J. Immunol. 133: 1841.
17. Hagner G. 1984. Immunology 52: 555.
18. MacDougall SL, and AK Sullivan. 1984. Immunobiology 167: 189.

EFFECT OF UVR AND MEMBRANE FLUIDITY ON NK CYTOTOXIC ACTIVITY[1]

Benjamin Bonavida
Monica Weitzen

Department of Microbiology & Immunology
UCLA School of Medicine
Los Angeles, California

Robert C. Roozemond

Laboratory of Histology and Cell Biology
University of Amsterdam
Amsterdam, The Netherlands

I. INTRODUCTION

Natural killer (NK) cells are present in peripheral blood and lymphoid organs and exhibit cytotoxic activity against susceptible tumor cell lines and virally infected cells (1,2). There has been a lot of interest in this form of cytotoxicity since NK do not require an active sensitization phase to express their cytotoxic function. For this reason, it has been proposed that NK cells might provide a first line of defense against cancer or other diseases (3).

The NK activity has been shown to be modulated under various conditions such as those involving cancer and immunodeficiency diseases. Furthermore, the NK activity is influenced by various regulatory factors *in vivo* or *in vitro* (1-3).

[1]Supported by Grant CA35791 from the National Cancer Institute (BB) and by Grant R93-112 from the Netherlands Organization for the Advancement of Pure Research (RCR).

ISBN 0-12-341370-2

The present study examines the effect of UVR and membrane fluidity on NK cytotoxic function. Previous studies have demonstrated that *in vivo* exposure to UVR alters the immune response in both experimental animals and humans (4,5). Such *in vivo* effects of UVR may be mediated by biological molecules which are chemically altered by UV light, for example DNA, protein or lipid (6-8). Other effects have also been noted, such as alteration of levels of prostaglandins, kinins, histamine and cholesterol, or release of lysosomal enzymes (9,10). Finally, these effects may be secondary to the direct effects of UVR on lymphoid cells as clarified by *in vitro* studies. Such studies have shown a decrease in lymphocyte viability (11), inhibition of lymphocyte response to allogeneic cells and mitogen (12,13), and inhibition of NK activity (13).

The mechanism of NK CMC has been resolved into 3 distinct stages, namely recognition and/or binding, programming for lysis of the lethal hit, and the killer cell independent target lysis (14). Since close contact between the effector and target occurs during the various stages of lysis, it is conceivable that any alteration of the physical state of the plasma membrane of both effector and target cells may alter the cytotoxic function. The physical state of the plasma membrane is determined by the fluidity of the lipid bilayer which, in turn, is dependent on the lipid composition. Therefore, modulation of lipid composition results in alteration of membrane fluidity. It has been reported that alteration of membrane fluidity affects T-cell mediated cytotoxicity (15-17). Only in a few reports have the effects of fatty acid treatment on NK cell-mediated cytotoxicity been described (18,19).

The present study investigates the stage at which UVR mediated inhibition of NK occurs, the effect of recombinant A/D IFN on protection against UV induced inhibition, and the effect of UVR on the release of cytotoxic factors by human lymphocytes. The present study also investigates the effect of lipid modulation and membrane fluidity of both NK effector cells and NK targets on cytotoxic function.

II. MATERIALS AND METHODS

Effector cells. Human peripheral blood lymphocytes and the LGL fractions were prepared as previously described (20,21).

Cell lines. The cell lines used in the study were maintained in suspension culture in RPMI 1640 supplemented with 10% fetal calf serum in a humidified atmosphere of 5% CO2 at 37°C.

Cytotoxic assays. The chromium ^{51}Cr release assay and the single cell assay were performed as previously described (22).

Irradiation. The cells used for irradiation were washed twice in PBS and suspended at $5X10^6$ cells per mililiter in PBS. Cells were placed in 35mm Petri dishes and exposed to a Westinghouse FS20 sunlamp with output of 0.5 MJ/cm^2/second, measured by an IL SEE 240 detector with a W225 infuser and a 313 filter. Dosimetry was achieved with a radiometer (IL model 700) and a photodosimeter (IL model 72). Doses were varied according to the time of exposure. Viability of the cells was greater than 95% for at least 4 hours after irradiation as assessed by trypan blue exclusion.

Micro-supernatant assay. This was performed as described previously (23).

Lipid modulation of target and effector cells. PVP-BSA lipid mixtures were prepared by dissolving the appropriate amount of lipid from a stock solution of ethanol in 3.5% PVP (dialyzed) + 1% BSA in PBS, pH 7.4 as described (24,25). The final lipid concentration was 0.5 mg PC or 0.25 mg CHS per mililiter. A PVP control solution contained only ethanol without lipids. K562 cells, PBL, or LGL (10^6/ml) were incubated in PVP-BSA lipid media at room temperature for 2 hours. After incubation, the cells were centrifuged, washed twice with RPMI, and resuspended in RPMI + 10% FCS to a concentration of $8X10^6$/ml (chromium release assay) or in RPMI alone to a concentration of $2X10^6$/ml (single cell assay). Fluorescence polarization measurements were performed as described (26).

The lethally hit programmed target cells were isolated according to the method described by Hiserodt et al. with minor modifications (27).

III. RESULTS AND DISCUSSION

A. Effect of UVR on NK Activity

Experiments were designed to investigate the mechanism by which UVR inhibits NK activity of human PBL. The following questions were examined: the stage at which inhibition takes place, the effect of αA/D interferon in reversing inhibition, and the effect of UVR on the release of cytotoxic factors.

1. Stage of the lytic pathway at which UVR inhibition takes place

Treatment of PBL or LGL with various doses of UVR resulted in inhibition of NK activity in the ^{51}Cr release assay. The inhibition of NK CMC by UVR may be the result of a number of factors including an inability of the effector cell to bind to the target cell (a step required for lysis to be achieved), or an alteration of the post-binding of the lethal hit stage of lysis. The single cell assay was used to distinguish between these two possibilities. The frequency of conjugates formed between NK effector and target cells was not altered following UVR, although killing was abrogated. These results indicate that UVR inhibits the lethal hit stage of lysis without inhibiting the initial recognition event (Figure 1).

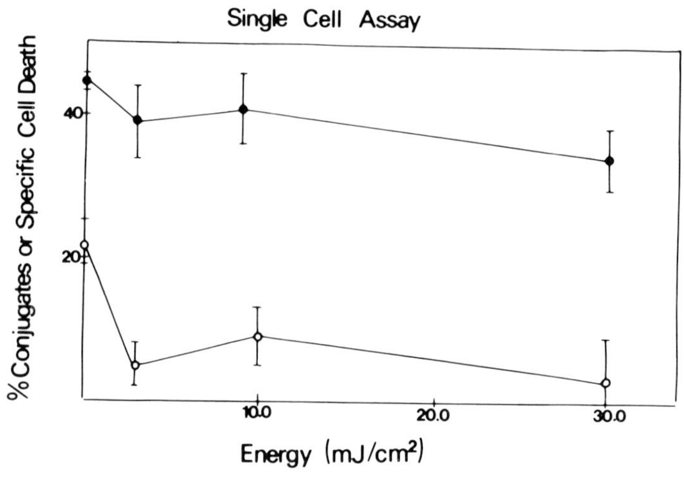

Fig. 1. The effect of various doses of UVR on the ability of NK cells to form conjugates and kill bound targets. Percoll enriched LGL were mixed with K562 target cells for conjugate (●——●) and for killed targets (○——○).

2. Effect of IFN treatment on UVR-induced NK inhibition

Previous studies have demonstrated that treatment of PBL with interferon results in augmentation of cytotoxic activity. This enhancement has been found to result from recruitment of pre-NK cells to NK, enhancement of the kinetics of lysis, and enhanced recycling (28,29).

Experiments were done to delineate whether cells pretreated with rαIFN are protected from the effect of UVR. Treated cells were inhibited similarly to control cells (Table 1). These results demonstrate that pre-NK cells are UVR sensitive. However, although the percentage of inhibition following interferon treatment was the same for treated as for untreated cells, the residual cytotoxic activity was higher. These results suggest that interferon may allow residual NK activity following UV treatment under conditions in which the activity of NK cells is minimal or abrogated. Therefore, it is possible that modulating agents which enhance NK activity may be effective following treatment of cells with UVR.

TABLE I. Interferon treated cells are susceptible to UVR induced inhibition of NK function[a]

IFN pretreatment	UVR dose MJ/cm^2	% Cytotoxicity (% Inhibition) E:T 50:1	25:1
None	None	27	20
None	3.0	18(32)	14(30)
None	30	10(62)	7(65)
50 units	None	46	31
50 units	3.0	30(34)	18(42)
50 units	30	13(71)	9(71)

[a]PBL plate nonadherent cells ($5X10^6$ml) were treated with recombinant αA/D interferon for 1 hr at 37° and then washed. The cells were then treated with UVR and tested for cytotoxicity against ^{51}Cr U937 target cells.

3. Effect of UVR on the release of cytotoxic factors

We have proposed that NK-CMC may be mediated by secreted cytotoxic factors, both in experimental animals and in man (23). Accordingly, it was possible that inhibition of NK by UVR was due to inability of the effector cells to release NKCF. Our experiments have demonstrated that UVR treatment does not affect release of NKCF but, in fact, induces the release of cytotoxic factors from unstimulated effector cells. Kinetics of the release have revealed that shortly after UVR treatment, the cells release cytotoxic factors detected in a 48-72 h viability assay (Figure 2). In an effort to determine the relationship of these UV induced cytotoxic factors (UVCF) to the NK reaction and also to the previously described NKCF, a spectrum of target cell sensitivities was examined. Our studies have shown that target cells, both NK sensitive and NK insensitive, are lysed by UVCF with the exception of K562 (Table 2). The nature and specificity of UVCF released by blood lymphocytes is not known. Further studies are needed to biochemically define the nature of these materials as well as the population of cells involved in the release of these materials.

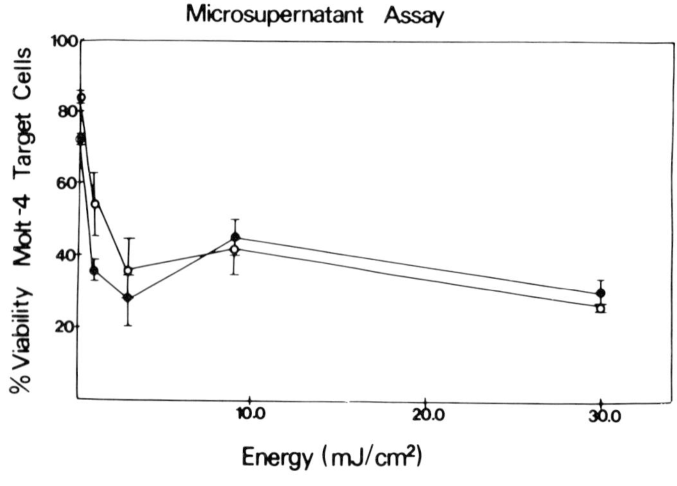

Fig. 2. Spontaneous release of cytotoxic materials from PBL following UVR. Supernatants were collected from PBL within 15 min. of exposure to graded doses of UVR (Time 0 supernatant). The PBL were then resuspended in RPMI 1640, incubated for 2 h at 37°, and supernatants were again collected (Time 2 h supernatant). All supernatants were tested on Molt-4 target cells in a 48 h microsupernatant assay. Control viabilities were 90% in PBS and 88% in RPMI 1640. (●——●), time 0 supernatant and (○——○) time 2 h supernatant.

TABLE II. Target cell spectrum of UVCF[a]

Target[c]	NK sensitivity	% Cytotoxicity ± SD[b]
Molt-4	+	47.1 ± 13.1
K562	+	0
U937	+	32.5 ± 18.3
Daudi	-	62.3 ± 3.7
YAC-1	-	46.2 ± 10.2
Raji	-	68.7 ± 8.8

[a]UVCF were derived from 2 h supernatant of PBL treated with 3.0 MJ/cm^2 UVR.

[b]Cytotoxicity was measured in a 68 h microsupernatant assay.

[c]Target viability for the cell line was as follows: Molt-4 (96%); K562 (38%); U937 (98%); Daudi (80%); YAC-1 (72%); Raji (95%).

B. Effect of Membrane Fluidity of NK Effector and Target on NK Function

The membrane fluidity of both NK effector and target cells were altered in an effort to delineate its effect on cytotoxic activity. The effects seen were analyzed in an effort to delineate the stage of the lytic pathway at which the inhibitory effect takes place.

1. Effect of membrane fluidity of target cell on its susceptibility to lysis

Treatment of K562 target with PVP/PC resulted in an increase of lipid fluidity. This treatment did not lead to any significant change in NK CMC. However, treatment in PVP/CHS resulted in decreased membrane lipid fluidity and inhibition of cytolysis (Figure 3). Using the single cell assay, we observed no significant change in the frequency of conjugates, but a significant decrease in the frequency in killer cells was noted (Table 3). Further analysis of the inhibitory effect was examined by using treatment of programmed target cells with PVP/CHS. This treatment resulted in significant inhibition of lysis. These results demonstrate that inhibition is at the killer cell independent lysis phase, and programming for lysis can be reversible.

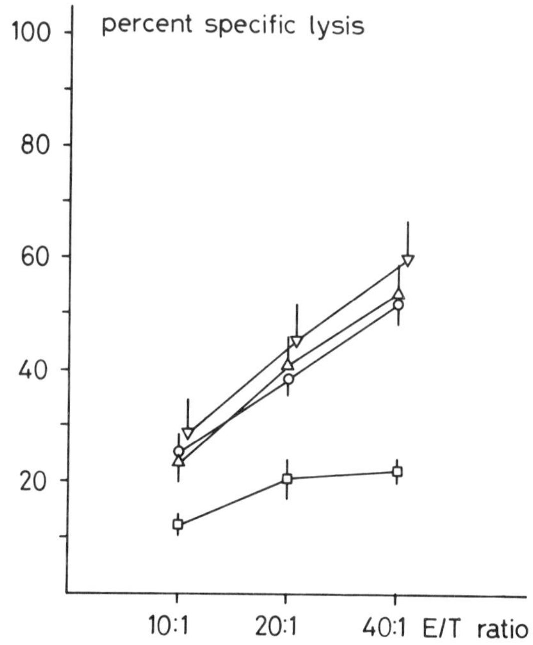

Fig. 3. K562 target cells were treated with PVP-BSA-lipid mixture and tested in the ^{51}Cr release assay. (○——○), medium; (△——△), PVP-medium without lipid; (▽——▽), PVP-PC medium; (□——□), PVP-CHS medium.

TABLE III. Effect of PVP-BSA lipid treatment of K562 target cell on conjugate formation and cytolysis by human LGL

Treatment[a]	% Conjugates	% Bound targets killed	
PVP-PC	25.5 ± 3.2	32.7 ± 5.9	
PVP-CHS	24.0 ± 1.8	12.7 ± 2.8	$p<0.01$
PVP-Control	27.0 ± 3.8	23.2 ± 2.8	
RPMI medium	28.7 ± 2.9	19.0 ± 4.3	

[a]Target cells were treated with PVP-BSA-lipid mixture and assayed in the sample cell assay as described in Methods.

2. Effect of membrane fluidity of NK effector cells on cytotoxic function

Treatment of effector PBL or LGL with PVP-BSA lipid mixtures resulted in either increase or decrease of membrane fluidity. In both instances, the cytotoxic activity was inhibited (Figure 4). In the single cell assay, the frequency of binders decreased when fluidity increased, whereas the frequency of binders increased when fluidity decreased. The frequency of killers was decreased in both cases. These results show considerable inhibition of the NK cell mediated cytotoxicity irrespective of an increase or a decrease in the lipid fluidity of the cell membrane.

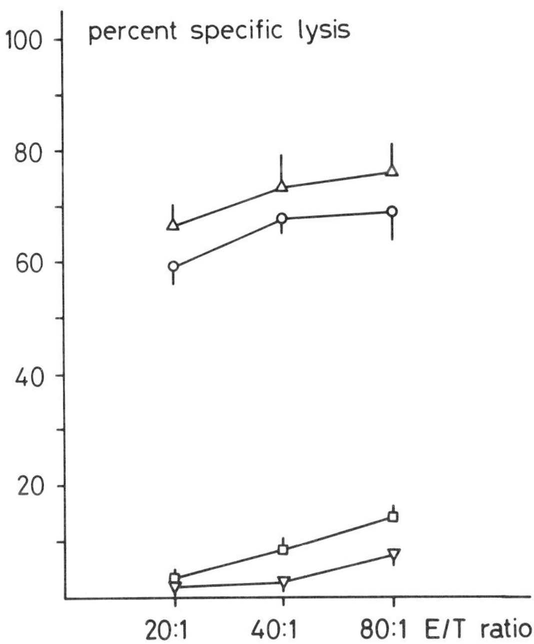

Fig. 4. Human PBL were treated with PVP-BSA-lipid mixture and tested for cytotoxic activity against ^{51}Cr labeled K562 cells. (○———○), medium; (△———△), PVP-medium without lipid; (▽———▽), PVP-PC medium; (□———□), PVP-CHS medium.

Changes in plasma membrane lipid composition have been shown to affect the expression of a variety of plasma membrane proteins due to vertical displacement of these proteins as a consequence of an altered fluidity of the lipid bilayer (30). These changes in antigenic expression may have an influence on the effectiveness of cell-cell interaction during the cytotoxic reaction. Therefore, similar changes involving NK target antigens, receptors, or other interaction molecules on the effector cells may affect the lytic pathway.

A more detailed analysis of the role of membrane fluidity on NK cytotoxic function is described elsewhere (31,32).

C. Concluding Remarks

The studies with UV-induced inhibition of NK activity have demonstrated a dichotomy between inhibition of NK activity in the short term assay, and release of cytotoxin after UVR detected in the long-term assay. Several possible mechanisms are proposed to explain this dichotomy. UVR may induce the release of inactive cytotoxic precursors which are activated either by the effector cell or prolonged incubation of the target cells. Alternatively, unstable inhibitors of lysis may be released along with the cytotoxic factors. Another possibility is that lysosomal enzymes are released which subsequently damage the effector cell and prevent lysis at some stage beyond binding to the target cells. Finally, UVR may cause non-directional release of materials which cause rapid lysis when delivered directly to the target cell membrane at high concentrations, and more protracted lysis at lower concentrations. Additional studies are being pursued to determine which model most accurately describes what is occurring.

The observed selective inhibition of specific stages of lysis of NK sensitive targets following changes in membrane fluidity is in agreement with our current understanding of the mechanism of NK cell mediated lysis. We have recently reported that target cell lysis of NK cells is a complex phenomenon which requires several interactions, namely, (a) binding of target to effector; (b) stimulation of effector to release NKCF; (c) binding of NKCF to target cells; and (d) susceptibility of the target cell to lysis by NKCF. It is possible that the defects described in the present reports are apparent in steps "b" to "d." The finding that rigidification of target cells affects programmed cells for lysis suggests that the membrane fluidity also plays a role in the post binding stages of lysis. Studies in these directions should provide information on the mechanism of the inhibitory activity seen by alteration of the plasma membrane fluidity.

ABBREVIATIONS

NK, natural killer; UVR, ultraviolet radiation; rαA/D, recombinant alpha A/D interferon; PVP-BSA, polyvinylpyrrolidone-bovine serum albumin; CHS, cholesteryl hemisuccinate; PC, phosphatidylcholine; FCS, fetal calf serum; PBS, phosphate buffered saline; LGL, large granular lymphocytes; CMC, cell-mediated cytotoxicity; UVCF, ultraviolet cytotoxic factors.

ACKNOWLEDGEMENTS

We wish to acknowledge Doctor Michael Brunda from Hoffman La Roche for supplying the recombinant interferon used in these studies.

REFERENCES

1. Roder, J. C., Karre, K., and Kiessling, R. (1981). Progress Allergy 28, 66.
2. Herberman, R. B. (ed.) (1982). "NK Cells and Other Natural Effector Cells." Academic Press, New York.
3. Herberman, R. B. (ed.) (1980). "Natural Cell-mediated Immunity Against Tumors." Academic Press, New York.
4. Morrison, W. L. (1981). Photoimmunology, J. Inv. Dermat. 77, 71.
5. Daynes, R. A., Bernhard, E. J., Gurish, M. F., and Lynch, D.H. (1981). J. Invest. Dermat. 77, 77.
6. Smith, K. C. (1974). In "Sunlight and Man" (T. B. Fitzpatrick et al., eds.). University of Tokyo Press, Tokyo.
7. Giese, A. C. (1950). Physiol. Rev. 30, 43.
8. Kochevar, I. E. (1980). J. Invest. Dermat. 74, 256.
9. Ogura, R. M., and Know, J. M. (1974). In "Sunlight and Man" (T. B. Fitzpatrick et al., eds.). University of Tokyo Press, Tokyo.

10. Johnson, B. E., and Daniels, F. (1969). J. Invest. Dermat. 53, 85.
11. Horwitz, S., Cripps, D., and Hong, R. (1974). Cellular Immunol. 14, 80.
12. Lindahl-Kressling, K., and Safwenberg, J. (1978). J. Int. Arch. Allergy 41, 670.
13. Schachter, B., Lederman, M. M., Levine, M. J., and Ellner, J. J. (1983). J. Immunol. 130, 2484.
14. Martz, E. (1977). Contemp. Topics Immunobiol. 7, 301.
15. Berke, G., Tzur, R., and Inbar, M. (1978). J. Immunol. 120, 1378.
16. Dabrowski, M. P., Peel, W. E., and Thomson, A. E. R. (1980). Eur. J. Immunol. 10, 821.
17. Heiniger, H. J., and Marshall, J. D. (1982). Proc. Natl. Acad. Sci. USA 79, 3823.
18. Rice, C., Hudig, D., Newton, R. S., and Mendelson, J. (1981). J. Clinical Immunol. Immunopath. 20, 389.
19. Yoo, T. J., Kuo, C. Y., and Spector, A. A. (1982). Cancer Res. 42, 3596.
20. Boyum, A. (1968). J. Clin. Lab Invest. 21 (Suppl. 97), 77.
21. Timonen, T., Reynolds, C. W., Ortaldo, J. R., and Herberman, R. B. (1982). J. Imm. Methods 51, 269.
22. Bradley, T. P., and Bonavida, B. (1981). J. Immunol. 26, 208.
23. Wright, S. C., Weitzen, M. L., Kahle, R., Granger, G. A., and Bonavida, B. (1983). J. Immunol. 130, 2479.
24. Shinitzky, M., Skornick, Y., and Haran-Ghera, N. (1979). Proc. Natl. Acad. Sci. USA 76, 5313.
25. Muller, C. P., Stephany, D. A., Shinitzky, M., and Wunderlich, J. R. (1983). J. Immunol. 131, 1356.
26. Roozemond, R. C., and Urli, D. C. (1979). Biochem. Biophys. Acta 556, 17.
27. Hiserodt, J., Britvan, L., and Targan, S. (1984). Cell. Immunol. 83, 43.
28. Silva, A., Bonavida, B., and Targan, S. (1980). J. Immunol. 125, 479.
29. Timonen, T., Ortaldo, J. R., and Herberman, R. B. (1982). J. Immunol. 128, 2514.
30. Borochov, H., and Shinitzky, M. (1976). Proc. Natl. Acad. Sci. USA 73, 4826.
31. Roozemond, R. C., and Bonavida, B. (1984). Immunobiology 167, 174.
32. Roozemond, R. C., and Bonavida, B. (submitted).

IMMUNOREGULATORY EFFECTS OF NK CELLS

Katsuo Kumagai
Ryuji Suzuki, Satsuki Suzuki

Department of Microbiology
Tohoku University
School of Dentistry
Sendai, Japan

Sumio Arai

Department of Laboratory Medicine
Kumamoto University Medical School
Kumamoto, Japan

I. INTRODUCTION

Certain species of experimental animals are more effective in induction of immune responses than other, and this can often be traced to either the genetic or the physiologic status of the species. A variety of humoral and cellular factors may be contributing to this genetic and physiologic status involved in immune responses. Of the various cellular factors contributing to this genetic and physiologic status, i.e., macrophages, polymorphonuclear leukocytes and natural killer (NK) cells, the NK cells appear to be most sensitive to genotype, age, physiologic, and hormonal variations (1-2). Accumulative evidence suggests that NK cells may play an important physiologic or pathologic role in the regulation of hematopoiesis (3-6). Therefore, it is possible that NK cells contribute to genetic and physiologic status in regulation of specific

ISBN 0-12-341370-2

humoral and cell-mediated immunity, although less information has been available concerning the direct participation of NK cells to immune responses.

As to the possible contribution of NK cells to humoral immunity, ealier study (7) showed that non-T, non-B human lymphocytes (L cells) might play a role in feed-back regulation of Ig synthesis by virtue of having abundant Fc IgG receptors. Recently, we have demonstrated a suppressor property of human NK cells isolated from the peripheral blood on pokeweed mitogen (PWM)-driven polyclonal Ig production of B cells (8). We have also shown that murine NK cells exhibit a suppressor property on B cells, which are undergoing differentiation spontaneously in vivo or stimulated by PWM in vitro (9,10). Abruzzo and Rowley (11) showed that induction of NK activity in mice immunized 2 or 3 days previously, or the addition of NK cells to cultures, promoted early termination of the ongoing primary immunoglobulin M antibody response.

On the other hand, direct evidence linking NK cells with effective induction of T cell-mediated immune responses has been lacking although a number of indirect approaches have support the concept. NK cells become activating during the early stages of tumor or allogeneic cell inoculation or virus infection (12,13). Addition of IFN into the mixed lymphocyte cultures, which can induce augmented NK activity, enhances generation of alloimmune cytotoxaic T lymphocytes (CTL) (14). IFN administration prior to viral infection in vivo is also associated by effective induction of virus-specific CTL (15). NK cell-depletion of mice enhances infection with mouse hepatitis virus, murine cytomegalovirus and other viruses in which virus specific CTL may be required for the protection of virus infection (16,17). These evidence led us to the speculation that NK cells may display, together with macrohages, an initial interaction with certain cellular antigens, providing a triggerring of CTL generation against specific cellular antigens. This hypothesis was reinforced by work that showed NK cells when incubated with tumor cells (18), viruses (19) or a lymphokine, interleukin 2 (IL-2) (20) produced IFN. IFN produced by NK cells may be involved in differentiation of CTL for which collaboration of IFN and IL-2 is required (21).

In this communication, we summarized the recent data undertaken in the authors' laboratories concerned with (1) the suppressor effects of NK cells on the antibody production by human and murine lymphocytes, and (2) the accessory activity of NK cells on generation of cytotoxic T lymphocytes.

II. EFFECT ON ANTIBODY RESPONSE

We first examined whether the NK cells purified from human peripheral lymphocytes by Percoll-density gradient centrifugation suppress pokeweed mitogen (PWM)-induced Ig production by purified B cells with help of T cells (9,10). When the fractionated mononuclear cells were co-cultured with purified autologous B cells in the presence of PWM, and were examined for their helper and suppressor activities on differentiation of B cells to immunoglobulin-(IgM and IgG) producing cells by a reversed hemolytic plaque assay. The NK-depleted high density fractions provided help for B cell differentiation to levels higher than that of unfractionated mononuclear cells. On the other hand, the NK-enriched low density fractions did not show helper activity, and when added to the culture of B cells plus helper T cells, they markedly suppressed B cell differentiation. This suppressive activity, as well as the NK cytotoxicity of NK-enriched fractions, was abrogated by treatment of the cells with monoclonal antibody against human NK cells (Leu7), but not against T cells (OKT3) in the presence of complement. Removal of NK cells by complement and anti-Leu11 monoclonal antibody showed a more marked reduction of suppressive activity on PWM-induced Ig production.

NK cells showed a suppression of PWM-driven B cell differentiation in the presence of helper T ($T4^+$) cells, but no inhibition of soluble factor-dependent B cell differentiation in the absence of helper T cells. It is thus concluded that human peripheral blood NK cells exhibit an ability to suppress PWM-driven B cell differentiation, possibly by acting through the effect on helper T cells but not directly on B cells. Tilden et al. (22) also reported that human NK cells purified by a fluorescence-activated cell sorter (FACS) using Leu7 monoclonal antibody efficiently suppressed PWM-induced IgG production by B cells. Manifestations of this suppressor cell activity required immune complex activation and was partially sensitive to 2000 rad irradiation. In contrast to $Leu7^+$ cells, the suppressive activity of $Leu2a^+$ T cells was not augmented by immune complex activation and was absolutely radiosensitive in PWM assays. Further studies will be needed on the difference and similarity that exist between suppressor functions of NK cells and suppressor T cells.

We have extended the work in human system into the experiments of murine system, using an antiserum to gangliotetraosylceramide (asialoGM_1, ASAGM_1), which reacted *in vitro* with murine NK cells to deplete selectively NK activity by complement-dependent killing (23,24). To deplete the mice of NK cells <u>in vivo</u>, mice were given an iv injection of rabbit antiserum to $ASGM_1$. Corresponding to selective depletion of $ASGM_1^+$ cells and NK activity by a minute amount of 1 μl antiserum, the spleen cells showed increased number of B cells and augmented motogenic responses to B cell- but not T cell- mitogens. No changes in the percentages of $Thy1^+$, $Lyt1^+$ or $Lyt2^+$ cells, or phagocytic cells were induced in the spleen of mice injected with anti-$ASGM_1$ even at a maximum amount used, 10 μl/mouse. Treatment of spleen cells with this antiserum in the presence of complement <u>in vitro</u> also resulted in the complete removal of NK activity, without any deterioration of T cells and their helper and suppressor subsets(23,24). These NK-depleted spleen cells showed production of PWM-driven plaque forming cell (PFC) to the levels much higher than that of control spleen (Table 1). When T cells isolated from these NK-depleted spleens were also examined for their helper activity on PFC response of purified B cells to PWM, a much higher helper activity on PWM-driven PFC production as compared with NK-conataining T cells, was obtained. In addition, when the NK cells (LGL) purified from the spleen to the levels greater than 90% by Percoll-gradient centrifugation and complement-dependent killing of T cells ($Thy1^+$, $Lyt1^+$) were added into NK-depleted spleen cells, the PFC-response was suppressed depending on the number of cells added (Table 2). These results indicate that NK cells in the spleen of mice exhibit a suppressor property on B cells stimulated <u>in vitro</u> by PWM.

Table 1. Correlation between the decreased NK activity and the increased PFC response in the spleen of mice treated with $ASGM_1$ antiserum

Anti-$ASGM_1$ μl/mouse	NK activity LU/10^7	No. of PFC / 10^5 spleen cells
0	5.9	303
0.6	ND	417
1.25	2.2	504
2.5	1.1	604
5.0	0.1	1050
10.0	0.1	ND
20.0	0.1	1096

C3H/He mice were given iv 0.6 to 20 μl of anti-ASGM1 in 0.2 ml or 0.2 ml saline containing 20 % rabbit sera 24 hr before harvesting the spleen. The isolated spleen mononuclear cells were then examined for the NK activity and the number of PWM-driven PFC. The data represent the mean values of five mice.

Table 2. Suppression by NK-enriched fractions of PWM-induced PFC response

NK-depleted spleen cells	% NK-enriched cells added	PFC/10^5 spleen cells	% inhibition
+[a]	50%	164 ± 79	77
+	10%	389 ± 34	46
+	2%	639 ± 95	3
+	-	716 ± 36	-
-	50%	1 ± 1	/
Normal spleen	-	317 ± 74	/

[a]Spleen mononuclear cells from C3H/He mice that received 0.2 ml of 1/20 anti-ASGM$_1$ (10 μl) 24 hr before harvest.

Nabel et al. (25) demonstrated that a murine NK cloned line growing dependently on IL2 lysis lipopolysaccharide-activated B lymphocytes, and that when added into murine T plus B lymphocyte systems even at 1%, these cloned NK cells greatly suppress the antibody response. These cloned cells, unlike in our case, inhibit the Ig secretion by B cells induced by the preformed soluble factor, indicating that the primary cellular target of suppression is the B cell. We have also shown the the murine NK cells activated by IL2 suppressed PWM-induced B cell differentiation much more effectively than did the endogenous NK cells. These IL2-activated NK cells also inhibited the preformed soluble factor-induced B cell differentiation (Unpublished observations). Therefore, endogenous NK cells and IL2-dependent growing NK cells may have different suppressor activities on B cell differentiations, by acting through helper T cells and directly on B cells, respectively. Alternatively, it is possible that there may be at least two distinct B cell subsets as a target for endogenous NK cells and IL-2 activatd NK cells, respectively. We have shown that freshly isolated human LGL suppressed Ig production by B cells transformed by Esptein Barr (EB) virus, which needs no presence of helper T cells, suggesting the possibility that NK cells have an ability to directly suppress a subset(s) of B cells (Arai et al. in preparation).

Abruzzo and Rowley (11) reported that induction of high NK activity by poly IC in mice immunized 2 or 3 days previously promoted early termination of the ongoing primary production of IgM antibody against phoshphorylcholine coupled to SRBC. The addition of NK cell-enriched spleen cells into the cultures also suppressed PFC response. S.B. Pollack also showed that the primary anti-SRBC antibody response was enhanced 2-3 fold, compared to controls, in mice of which NK cells were depleted by injection of NK1.1 antiserum, (Another article in this chapter). All these data indicate that NK cells in human and mice may play a suppressor role in primary antibody response.

III. ACCESSORY FUNCTION IN CTL GENERATION

We have examined the role of NK cells in generation of alloimmune CTL in response to allogeneic stimulations *in vivo* and *in vitro*. Administration of asialo GM_1 antiserum into the mice completely depleted NK activity in the spleen without any deterioration of helper and suppressor T cell subsets, as mentioned above. Administration of the antiserum in mice one day before immunization of allogeneic cells completely suppressed generation of alloimmune CTL. These NK-depleted spleen cells also generated no alloimmune CTL in response to alloimmunization *in vitro*. A good correlation was found between the reduction of NK activity in the spleen and their reduced ability to generate alloimmune CTL *in vitro* (26). On the other hand, the NK-enriched fractions were purified by Percoll-density centrifugation, followed by complement-dependent lysis of $Thy1^+$ cells and enrichment of NK cells by a panning method using $ASGM_1$ antiserum, as mentioned above. When these cells with greater than 90% $ASGM_1^+$ cells and neither $Lyt1^+$ CTL helper nor $Lyt2^+$ precursor were stimulated with alloantigens *in vitro*, no CTL activity was produced. When the NK-depleted cells and NK-enriched fractions were cocultured, CTL activity, similar in magnitude to that in unfractionated spleen cells, was generated (Table 3). The NK-depleted T cell fractions deficient in CTL generation was also found to have the remarkably reduced ability to produce IL-2 and IFN

Table 3. Accessory function of NK-enriched cells for alloimmune CTL responses against C57BL/6 (vs EL-4) in the NK-depleted cells of C3H/He mice

Responder cells	Stimulator cells	% cytotoxicity[a] E/T = 10 EL-4	Meth-A
NK-depleted T	C57BL/6	6	-1
NK-depleted T+NK	C57BL/6	60	2
- NK	C57BL/6	2	1
Control T -	C57BL/6	34	1

[a] The values represent the means of triplicate culture.

Table 4. Effect of IL-2 and IFN on in vitro generation of alloimmune CTL to C57BL/6 cells (vs EL-4) in the NK-depleted responder T cells of C3H/He mice

Responder T cells[a]	IL-2 (2 U/ml)	IFN (100 U/ml)	% cytotoxicity[b] (E/T = 10) EL-4	Meth-A
NK-depleted	-	-	4	0
NK-depleted	+	-	39	0
NK-depleted	-	+	43	1
NK-depleted	+	+	64	2
Control	-	-	34	1

[a]Each responder was stimulated with X-ray irradiatead C57BL/6 spleen cells.
[b]The valuse represent the means of triplicate culture

in response to allogeneic stimulations in vitro. In conrast, purified NK cells when stimulated with allogeneic cells produced IL-2 and IFN to the levels found in control spleen cells when incubated with allogeneic cells. These results suggest the possibility that NK cells have an accessory function for CTL generation, and this function is mediated by soluble factors such as IL-2 and IFN produced by NK cells. In order to confirm this, NK-depleted T cells were stimulated by the allogeneic cells in addition of a recombinant human IL-2 (2 units/ml) and/or a recombinant IFNα A/D (100 units/ml), both of which were effective in murine as well as human cell systems. The NK-depleted T cells completely restored the ability to generate alloimmune CTL in response to alloantigens (Table 4). When the mice depleted of NK cells in vivo were once given an injection of IFNα A/D within 6 days after alloimmunization or of IL-2 by 4 to 6 days restored the ability to generate CTL in the spleen. These results indicate that NK ($ASGM_1^+$) cells may interact with allogeneic cells and produce IL-2 and IFN, and through these soluble factors, contribute to differentiation of CTL induced by allospecific antigen stimulation.

Murine and human NK cells are known to produce IFN in response to tumor cells or viral infections (18,19). We also showed that murine (20) and human (27) NK cells were responsive to IL-2 with their replication and production of IFN. Therefore, it is presumed that NK cells might be a major source for IFN which was induced in response to alloantigen stimulations. To date, it remains obscure whether NK cells exhibit the ability to produce IL2. Neverthless, recent reports infered that LGL fractions enriched highly for NK cells of human lymphoid cells and their IL-2-dependent growing cells produced IL-2 in addition to many other lymphokines in response to lectins and TPA (28). We have also found that fractions enriched for $ASGM_1^+$ cells (LGL), which contained neither macrophages nor Thy_1^+ cells, produced IL-2 in response to lectins or macrophage products (interleukin 1), in addition to allogeneic cells (Unpublished observation). These results suggest the possibilities that a subset of NK cells exhibit the ability to produce IL-2, in addition to IFN, or alternatively, cells of morphologically LGL but belonging to a distinct population of non-T cell type produce IL-2. Previously, we showed that low density fractions contained a population, if minor, of cells, which were $ASGM_1$ positive but negative for cytoplasmic granules (large agranular lymphocytes, LAL) and not involved in the NK cytotoxicity (24,29). Further investigations will be needed for characterization of cells responsible for the IL-2 and IFN productions.

Recently, Burlington et al. (30) reported that human LGLs have an accessory function in the development of CTL against virus infected cells. LGL and T cells were separated on Percoll gradients and the ability of the separated LGL and T cells to generate influenza A virus-specific CTL was measured. When stimulated by virus-infected, irradiated, adherent cells, neither LGL nor T cells cultured separately generated CTL. When they were cocultured, however, even if separated by a 0.22 μm pore size mambrance, CTL generation was readily detected in the small T cell precursors. Thus, LGL were required as accessory cells for CTL response to occur and the effect was mediated by a soluble factor(s). They demonstrated that the accessory cells in the LGL populations were OKT3, Leu-1 and $Leu-M_1$ surface antigen-negative, but possibly positive for Fc receptors. It seems therefore likely that the accessory cells may be non T, non B cells, and possibly NK cells. The nature of soluble factors involved in the assessory function remains undetermined, although it was shown that neither IFNα nor IFNγ appeared to be the LGL-produced soluble factors that mediated the assessory effect of LGL on CTL responses.

IV. CONCLUSION

The emphasis in this communication has been on the regulatory role of NK cells in immune responses as evidenced by the suppressor effects on the antibody production and the accessory effects on the CTL generation. Our data and recent reports by others have demonstrated that human NK cells (LGL) or those when stimulated with immune complex were suppressive on the polyclonal Ig production induced by PWM. In the murine systems, it has been shown by the experiments using depletion and augmentation of NK cells *in vivo* and *in vitro* that NK cells also play a suppressor role in antibody production induced by polyclonal mitogens or specific antigens. We have also shown that murine NK cells (LGL) have an accessory role in generation of alloimmune CTL in mice *in vivo* and *in vitro*, and that the effect is mediated by the LGL produced soluble factors, IL-2 and IFN. It has also been shown by others that human LGL are required as accessory cells for CTL response to occur *in vitro*. These results demonstrate that NK cells (LGL) have a regulatory cell function in a variety of immune reponses, in addition to a known killing activity to the tumor cells and virus infected cells. We are, however, only beggining to delineate some of the effects of NK cells on the level of phenomenology and have not begun to expore the mechanisms involved. Further studies may provide a valuable information on the genetic, physiologic and pathologic roles of NK cells in the regulation of immune responses.

REFERENCES

1) Herberman, R.B., Nunn, M.F., and Lavrin, D.H. (1975). Int. J. Cancer. 16, 121.
2) Kiessling, R., and Wigzell, H. (1979). Immunol. Rev. 44, 165.
3) Kiessling, R., Hochman, A.S., Halter, O., Shearer, G.R., Wigzel, H., and Cudkowicz, G. (1977). Eur. J. Immunol. 7, 655
4) Hansson, M., Kiessling, R., and Andersson, B., Karre, K., and Roder, J. (1979). Nature. 284, 662.
5) Hansson, M., Kiessling, R., and Andersson, B. (1981). Eur. J. Immunol. 11, 8.
6) Hansson, M., Beran, M., Anderson, B., and Kiessling, R. (1982). J. Immunol. 129, 126.
7) Lobo, P.I. (1981). J. Clin. Invest. 68, 431.
8) Arai, S., Yamamoto, H., Itoh, K., and Kumagai, K. (1983). J. Immunol. 131, 651.
9) Suzuki, S., Suzuki, R., Onta, T., and Kumagai, K. (1984). In: Natural Killer Activity and Its Regulation (Hoshino, T., Koren, H., Uchida, A., ed.), p.296, Excerpta Medica, Amsterdam.

10) Suzuki, S., Suzuki, R., Onta, T., and Kumagai, K. (1984). J. Immunol. Submitted for publication.
11) Abruzzo, L.V., and Rowley, D.A. (1983). Science. 222, 581.
12) Herberman, R.B., Nunn, M.E., Holden, H.T., Staal, S., and Djeu, J.Y. (1977). Int. J. Cancer. 19, 155.
13) Welsh, R.M. (1978). J. Exp. Med. 148, 163.
14) Heron, I., Berg, K., and Cantell, K. (1976). J. Immunol. 117, 1370.
15) Kumagai, K., Itoh, K., Kurane, I., and Saitoh, F. (1981). In: Self-Defense Mechanisms. Role of Macrophages. (Mizuno, D., Cohn, Z.A., Takeya, K., Ishida, N., ed.) p.183, Univ. Tokyo Press, Tokyo, Elsvier Biomedical Press, New York.
16) Bukowski, J.F., Wada, B.A., Habu, S., Okumura, K., and Welsh, R.M. (1983). J. Immunol. 131, 1531.
17) Welsh, R.M., Biron, C.A., Parker, D.C., Habu, S., Okamura, K., Haspel, M.V., and Holmes, K.V. (1983). In: Human Immunity to Virus (Ennis, F.A., ed.) P.21, Academic Press, New York.
18) Timonen, T., Saksela, E., Virtanen, I., and Cantell, K. (1980). Eur. J. Immunol. 10, 422.
19) Minato, N., Reid, L., Cantor, H., Lengyel, P., and Bloom,B. R. (1980). J. Exp. Med. 152, 124.
20) Handa, K., Suzuki, R., Matsui, H., Shimizu, Y., and Kumagai, K. (1983). J. Immunol. 130, 980.
21) Farrar, W.L., Johnson, H.M., and Farrar, J.J. (1981). J. Immunol. 126, 1120.
22) Tilden, A.B., Abo, T., and Balch, C.M. (1983). J. Immunol. 130, 1171.
23) Kasai, M., Yoneda, T., Habu, S., Maruyama, S., and Saitoh, F. (1982). J. Immunol. 129, 388.
24) Kumagai, K., Itoh, K., Suzuki, R., Hinuma, S., and Saitoh, F. (1982). J. Immunol. 129, 388.
25) Nabel, G.W., Allard, W.J., and Cantor, H. (1982). J. Exp. Med. 156, 658.
26) Suzuki, R., Suzuki, S., Ebina, N., and Kumagai, K. J. Immunol. in press.
27) Shiiba, K., Itoh, K., Shimizu, Y., and Kumagai, K. (1984). In: Natural Killer Activity and Its Regulation (Hoshino, T., Koren, H., Uchida, A., ed.), p.187, Excerpta Medica, Amsterdam.
28) Kasakura, T., Djeu, J.Y., Dougherty, S.F., and Oppenheim, J.J. (1983). J. Immunol. 131, 2379.
29) Itoh, K., Suzuki, R., Umezu, Y., Hanaumi, K., and Kumagai, K. (1982). J. Immunol. 129, 395.
30) Burlington, B., Djeu, J.Y., Wells, M.A., Kiley, S.C., and Quinnan, G.V. (1984). J. Immunol. 132, 3154.

REGULATION OF THE B CELL RESPONSE BY NK CELLS

Claire P. Robles
Peter Pereira
Pascale Wortley
Sylvia B. Pollack

Department of Biological Structure
University of Washington
Seattle, Washington 98195

Natural killer (NK) cells were first described as unprimed lymphoid cells that are cytotoxic for certain transformed cells *in vitro* (1-3). This observation has been extended convincingly to the *in vivo* situation (4-6). In general, the presence of NK cell activity has correlated with slower growth and more rapid clearance of tumor cells *in vivo* (7,8). These data support the hypothesis that NK cells provide an immunological surveillance mechanism which recognizes and destroys aberrantly differentiating and/or proliferating cells.

In addition, NK cells recognize normal cells such as syngeneic or allogeneic bone marrow cells (9,10), immature thymocytes (11), and granulocytic progenitor cells (12). NK cells have also been implicated in the rejection of bone marrow transplants in mice (13). These observations suggest that NK cells have a role in the regulation of differentiating normal cells as well as being a first line of defense against developing malignant cells. It may well be that their role in defense against malignant growth is a consequence of a more ubiquitous role in the regulation of cell differentiation.

The recognition mechanism by which NK cells are able to bind and lyse tumor targets has not been clearly defined.

[1]Supported by NIH Grants CA-37006 and CA-32553 and DOE Contract EV 10270.

ISBN 0-12-341370-2

However, it seems probable that there are multiple NK subpopulations (14) capable of interacting with more than one target recognition antigen (15). It is possible that at least some target antigens are differentiation antigens that are expressed on tumor cells as well as on normal differentiating cells. Consistent with this possibility is the demonstration that an activated, cloned NK cell line is able to bind and lyse LPS-induced B cell blasts (16).

We have shown that endogenous (unstimulated) NK cells are also capable of binding and lysing LPS-activated B cells (Table I). FITC-anti-μ labeled LPS-blasts were bound by nylon column-passed spleen cells in numbers comparable to those found with the NK cell target, YAC-1. In an agarose target-binding assay, conjugates were found in which sIg^+, FITC-labeled blasts were bound by unstained lymphocytes and identified as killed targets by their permeability to trypan blue.

TABLE I

TARGET BINDING AND LYSIS OF LPS BLASTS BY NK CELLS[a]

	Direct Binding[b]	Binding and Lysis in Agar[c]	
	% TBC	% TBC	% Lytic TBC
YAC-1	10.2	10.8	7.8
LPS-blasts (FITC)	6.7	13.7	6.3
LPS-blasts	7.6	ND	ND

[a]B6 spleen cells were cultured with 10 μg/ml LPS for 48 hr. Target binding assays were performed with nylon column-passed spleen cells as a source of target binding cells (TBC) that were mixed with equal numbers of YAC-1 or LPS-blast target cells. The LPS-blasts were labeled with FITC-anti-μ or were unlabeled.

[b]Conjugates were counted directly in a fluorescence microscope. Only sIg^+ blasts bound by unstained lymphocytes were counted.

[c]Conjugates were counted in agar after incubation for 4 hr at 37°C, 5% CO_2 and stained with trypan blue to identify targets killed by the TBC.

The ability of NK cells to recognize and lyse activated B cells, B cell leukemias (17) and some B cell lines (18) does not necessarily mean that NK cells have a regulatory role in the immune response. However, there is a growing body of evidence which suggests that NK cells do regulate the B cell response to antigenic stimulation. As few as 2 x 10^3 activated cloned NK cells added to 10^6 B cells were able to mediate 75% suppression of the mouse LPS-induced PFC response *in vitro* (19). This NK cell line was also able to suppress the PFC response in an adoptive transfer experiment. Human peripheral blood lymphocytes have been found to dramatically inhibit the mouse primary PFC response to both SRBC and to DNP-Ficoll using an effector to target cell ratio of only 0.67 to 1 (20). Brieva *et al*. found that addition of IFN to a cell fraction containing NK cell activity resulted in the suppression of immunoglobulin synthesis by human lymphoblastoid B cells (21). The suppressive effect could be cold-target inhibited with a small number of K-562 NK target cells. Arai *et al*. demonstrated that the human LGL fraction was capable of suppressing the generation of both IgM and IgG PFC but appeared to be dependent on the presence of T cells, suggesting that regulation of B cells by NK cells might occur through an interaction with T cells (22). Abruzzo and Rowley have implicated a population of accessory cells as the NK target involved in the regulation of the immune response by NK cells (23).

All of these studies which indicate that NK cells are able to suppress the PFC response have relied on the addition of NK cells to lymphocytes during antigenic stimulation. We hypothesized that manipulation of endogenous NK activity *in vivo* might lead to measurable effects on the levels of antibody forming cells during a primary antibody response. To test this hypothesis B6 mice were injected with poly I:C in order to increase endogenous levels of NK activity. Eighteen hours after poly I:C treatment the mice were immunized with SRBC. At this time the increase in NK activity due to the poly I:C treatment is at a maximum. Four days after immunization the spleens were harvested and analyzed for the presence of plaque-forming cells (PFC). The results, listed in Table II, demonstrate that there is a significant decrease in the number of PFC in the primary antibody response against SRBC when the endogenous NK activity is stimulated *in vivo*. Similar results have been reported by Abruzzo and Rowley (23).

Obviously, other effects besides an increase in NK activity could be mediated by the poly I:C treatment and, therefore, indirectly affect the PFC response. A more di-

TABLE II

DEPRESSED PRIMARY ANTIBODY RESPONSE IN MICE TREATED WITH POLY I:C

		PFC/2.5 x 10^5 cells[a]	% Decrease
Experiment 1	Control	30.7 ± 7.8	
	Poly I:C[b]	13.3 ± 2.9	57*
Experiment 2	Control	15.7 ± 1.4	
	Poly I:C	8.8 ± 2.8	44*

[a]B6 mice immunized on day 0 with 0.2 ml of 20% (v/v) SRBC. Spleens were harvested on day 4 after immunization and analyzed for PFC.
[b]200 μg poly I:C were injected intraperitoneally 18 hr prior to immunization.
*$p < 0.05$.

rect approach to test the effect of NK cells on the B cell response is to deplete animals *in vivo* of endogenous NK activity. We have developed a model system for the depletion of NK activity *in vivo* using an alloantiserum against the NK-1.1 antigen. Either intraperitoneal or intravenous administration of the anti-NK-1.1 antiserum results in the rapid depletion of NK activity (4). When this antiserum was used to deplete NK activity 4 hours prior to immunization with SRBC a two-fold increase in the number of PFC against SRBC was observed (24). There is, therefore, a reciprocal relationship between NK activity and the PFC response. Suzuki *et al.* have used antiserum to asialo-GM_1 to deplete NK cells *in vivo* and found a concomitant increase in both the mitogenic and PFC response to PWM and LPS (25).

To investigate whether T cells are involved in the regulation of B cells by NK cells, we compared the response to both T-independent and T-dependent antigens in anti-NK-1.1 treated mice (Table III). The two T-independent antigens, TNP-*Brucella abortus* (BA) and TNP-Ficoll, as well as the T-dependent antigen, TNP-SRBC, generated 60% more PFC's in animals depleted of NK activity. These data suggest that the effect of NK cells on the antibody response does not involve an interaction with T cells.

The results discussed support the hypothesis that *in vivo* manipulation of NK activity has a concomitant effect on the B cell response to antigen. Stimulation of endogenous NK activity causes a decrease in the PFC response, whereas depletion of NK activity results in an increase in the PFC response. The observed difference in the PFC responses appears to be due to a change in the magnitude of the response rather than a shift in kinetics. The PFC response to SRBC has the same kinetics, both in NMS- and anti-NK-1.1-treated mice (unpublished observations).

TABLE III

EFFECT OF *IN VIVO* NK CELL DEPLETION ON THE PRIMARY ANTIBODY RESPONSE TO T-INDEPENDENT AND T-DEPENDENT ANTIGENS

Antigen	PFC/Spleen[a] Control[b]	PFC/Spleen[a] Experimental	% Increase in PFC	*p* Value
TNP-BA	14574 ± 9219 n = 12	23659 ± 9440 n = 12	62	< 0.05
TNP-Ficoll	8322 ± 3024 n = 14	13038 ± 4905 n = 13	64	< 0.01
TNP-SRBC	22578 ± 4820 n = 7	36871 ± 23789 n = 9	63	< 0.10

[a]B6 mice (age 5-8 wk) were immunized intraperitoneally on day 0 with either 0.2 ml of TNP-BA (2 x 10^{10} organisms/ml), TNP-Ficoll (50 μg/ml), TNP-SRBC (2% solution v/v). Spleens were harvested on day 4 and analyzed for PFC using TNP-SRBC as indicator cells. The values listed in the Table represent the mean of the mice in each group (n = number of mice in each group).

[b]Mice were injected i.p. with either 0.5 ml of 1:20 dilution of normal (BALB/c x C3H) serum (control) or anti-NK-1.1 antiserum (experimental) 4 hr prior to immunization.

Another important question is the temporal relation of NK cell regulation on the B cell during the course of the antigen response. Table IV presents the results from an experiment aimed at this issue. Mice were treated with anti-NK antiserum at various times relative to immunization with SRBC. All immune spleens were harvested on day 4 after immunization. As expected, the control groups which received normal mouse serum all demonstrate similar numbers of PFC's. In the expermental groups (those receiving anti-NK-1.1 antiserum) there is a dramatic increase (310%) in the number of PFC's when the antiserum was administered 24 hours prior to immunization. This increase tapers off when antiserum is administered simultaneously with the immunizing antigen but is still significantly greater (130%) than the control. When anti-NK-1.1 antiserum is given to animals 8, 24 or 62 hours after immunization there is no significant difference between the control groups and the experimental groups. These results suggest that depletion of NK cell activity needs to occur prior to antigenic stimulation in order to have a maximal effect on the PFC response and that NK cells probably regulate an early event in the response of the B cell to antigen. Yamamoto *et al.* also found that suppression of the *in vitro* PFC response was greatest if NK cells were added to the cultures at the time of antigen addition rather than later during the culture period (20).

TABLE IV

EFFECT OF TEMPORAL RELATIONSHIP BETWEEN ADMINISTRATION OF ANTI-NK-1.1 ANTISERUM AND IMMUNIZATION ON THE PRIMARY ANTIBODY RESPONSE TO SRBC

Time[a]	PFC/Spleen[b]		% Change in PFC	p Value
	Control[c]	Experimental		
-24	26380 ± 22474	108577 ± 13827	+310	< 0.02
0	15375 ± 7474	36121 ± 14275	+130	< 0.05
+ 8	25310 ± 11958	16635 ± 9618	-35	*
+24	26023 ± 12830	16384 ± 14070	-38	*
+62	28382 ± 16978	28973 ± 19033	+2	*

[a]Time (hr) of anti-NK-1.1 antiserum administered relative to immunization.

[b]B6 mice (4 per group), age 5-6 wk, were immunized intraperitoneally on day 0 with 0.2 ml of a 2% solution (v/v) of SRBC. On day 4 after immunization spleens were harvested and analyzed for PFC using TNP-SRBC as indicator cells. The values listed in the Table represent the mean of each group.

[c]Mice received either 250 µl of a 1:10 dilution of normal (BALB/c x C3H) serum or anti-NK-1.1 antiserum intravenously.

*Not significant.

The data presented here address some of the issues surrounding NK cell regulation of B cells. The ability of NK cells to regulate the antibody response to T-independent antigens suggests that NK cells can have a direct effect on B cells and do not necessarily have to act through another cell type such as the T cell. However, this does not rule out the possibility that the B cell response might also be regulated through interactions of NK cells with T cells and/or accessory cells. Our data on the temporal relation of NK cell depletion relative to antigen exposure suggest that there is a limited window through which the NK cell is able to regulate B cells and that this occurs early on in the response to antigen. How the NK cell exerts a regulatory effect on the B cell that is undergoing a response to antigen is not known. Even though NK cells have been shown to be able to lyse LPS-activated B cells, regulation may not depend on lysis of antigen-activated B cells. Our preliminary observations suggest that NK cells may preferentially bind to LPS-blast cells at one stage of the cell cycle (G2), but it is not at all clear whether these cells are lysed. It is possible that the interaction of NK cells with some target cells can induce cytostasis that prevents further proliferation and/or differentiation. Whatever the mechanism, the consistent increase observed in the PFC response when NK cells are depleted *in vivo* suggests that NK cells are regulatory cells that act to keep B cell proliferation and differentiation within homeostatic limits.

REFERENCES

1. Herberman, R. B., Nunn, M. D., and Lavrin, D. H. (1975). Int. J. Cancer 16: 216.
2. Kiessling, R., Klein, E., and Wigzell, H. (1975). Eur. J. Immunol. 5: 112.
3. Sendo, F., Aoki, T., Boyse, E. A., and Buafo, C. K. (1975). J. Natl. Cancer Inst. 55: 603.
4. Pollack, S. B., and Hallenbeck, L. A. (1982). Int. J. Cancer 29: 203.
5. Warner, J. F., and Dennert, G. (1982). Nature (Lond.) 300: 31.
6. Habu, S., and Okumura, K. (1982). *In* "NK Cells and Other Natural Effector Cells" (R. H. Herberman, ed.) p. 1323. Academic Press, New York.
7. Hanna, H., and Burton, R. C. (1981). J. Immunol. 127: 1754.

8. Kasai, M., Leclerc, J. C., McVay-Boudreau, L., Shen, R. W., and Cantor, H. (1979). J. Exp. Med. 149: 1260.
9. Riccardi, C., Santoni, A., Barlozzari, T., and Herberman, R. B. (1981). Cell Immunol. 60: 136.
10. Nunn, M. E., Herberman, R. B., and Holden, H. T. (1977). Int. J. Cancer 20: 381.
11. Hansson, M., Kiessling, R., Anderson, B., Kärre, K., and Roder, J. (1979). Nature (Lond.) 278: 174.
12. Hansson, M., Beran, M., Andersson, B., and Kiessling, R. (1982). J. Immunol. 129: 126.
13. Lotzova, E., Savary, C. A. and Pollack, S. B. (1983). Transplantation 35: 490.
14. Jensen, P. J., and Koren, H. S. (1980). J. Immunol. 124: 395.
15. Roder, J. C., Beaumont, T. J., Kerbel, R. S., Haliotis, T., and Kozbor, D. (1981). PNAS USA 78: 6396.
16. Nabel, G., Bucalo, L. R., Allard, J., Wigzell, H., and Cantor, H. (1981). J. Exp. Med. 153: 1582.
17. Pattengale, P. K., Gidlund, M., Nilsson, K., Sundstrom, C., Sallstrom, J., Simonsson, B., and Wigzell, H. (1982). Int. J. Cancer 29: 1.
18. Jondal, M., Spina, C., and Targan, S. (1978). Nature (Lond.) 272: 62.
19. Nabel, G., Allard, W. J., and Cantor, H. (1982). J. Exp. Med. 156: 658.
20. Yamamoto, J. K., Blalock, J. E., and Johnson, H. M. (1982). Eur. J. Imunol. 12: 222.
21. Brieva, J. A., Targan, S., and Stevens, R. H. (1984). J. Immunol. 132: 611.
22. Arai, S., Yamamoto, H., Itoh, K., and Kumagai, K. (1983). J. Immunol. 131: 651.
23. Abruzzo, L. V., and Rowley, D. A. (1983). Science 222: 581.
24. Pollack, S. B. (1983). J. Reticuloendothel. Soc. 34: 184.
25. Suzuki, S., Suzuki, R., Onta, T., and Kumagai, K. (1984). *In* "Natural Killer Activity and Its Regulation" (T. Hoshino, M. S. Koren, and A. Uchida, eds.) p. 296. Excerpta Medica

DEFECTIVE NK CELL MECHANISM IN PATIENTS WITH LEUKEMIA[1]

E. Lotzová[2]
C. A. Savary[2]
M. J. Keating[3]
J. P. Hester[3]

[2]Laboratory of Immunogenetics
Department of General Surgery
[3]Department of Hematology
The University of Texas System Cancer Center
M.D. Anderson Hospital & Tumor Institute
Houston, Texas

I. INTRODUCTION

Implication of natural cytotoxicity in immune surveillance against malignancies (1-2) initiated studies on natural killer (NK) cell profile in cancer patients. Interest of our laboratory concentrated on patients with leukemia for the following reasons: Firstly, NK cells appear to be most reactive to leukemia-lymphoma target cell lines, suggesting NK cell role primarily in defense against hemopoietic and lymphoid malignancies. Secondly, patients with low NK cell *in vitro* cytotoxic potential exhibit high susceptibility to leukemias and lymphomas, and patients with leukemia and preleukemic disorders display low NK cell cytotoxicity (3-6). The latter two observations indicate that low NK cell activity could predispose to tumors of lympho-hemopoietic system. Thirdly, NK

[1]*Supported by Grant CA 31394 from the National Cancer Institute.*

ISBN 0-12-341370-2

cells are involved in regulation of normal hemopoiesis and lymphopoiesis, the function, distortion of which, may be related to leukemic disease. Lastly, the NK cell cytotoxic profile has not been explored in patients with leukemia as thoroughly as in the other groups of cancer patients. The data presented in this chapter are concerned with analysis of NK cell morphological and functional profile of patients with leukemia. Furthermore, the effect of biological modification on natural cytotoxicity of leukemic patients is discussed.

II. PROFILE OF LARGE GRANULAR LYMPHOCYTES (LGL) IN LEUKEMIC PATIENTS

We have shown earlier that patients with chronic (CML) and acute (AML) myelogenous leukemia exhibit low peripheral blood (PB) and bone marrow NK cell activity to K-562 and CEM target cell lines (4, 7-8). Quite recently we have tested also NK cell cytotoxic profile of patients with lymphoid leukemia, and in congruence with other laboratories (9-11) found that also patients with acute (ALL) and chronic (CLL) lymphoid leukemia displayed low PB NK cell cytotoxicity. These data are illustrated in Table I. It can be seen from the table that the majority of normal donors could be classified as high (48%) and medium (42%) NK cell responders, and the minority (10%) as low NK cell responders. In contrast, all of the leukemic patients exhibited low responder status, with the NK cell cytotoxicity levels below those displayed by low NK cell-responding normal donors. Interestingly, the NK cell cytotoxic potential was restored in patients in remission, indicating that the defect in NK cell cytotoxicity is acquired rather than intrinsic. However, low NK cell activity recurred in patients in relapse of leukemia, suggesting that the NK cell decline may be prognostic in early detection of leukemia relapse.

Possible NK cell importance in antileukemia defense and in prognosis of this disease initiated more detailed studies on the mechanism of NK cell impairment in leukemic patients. In compatibility with most of the other investigators, we have found that the low NK cell activity could not be attributed to dilution of NK cells by leukemic blasts, since no correlation between percent of blasts or lymphocytes and NK cell cytotoxicity was observed (4, 9-11). Furthermore, mixing of leukemic blasts with normal donor's PB mononuclear cells did not result in inhibition of the NK cell activity of the latter lymphocyte population.

TABLE I. PERIPHERAL BLOOD NK CELL CYTOTOXICITY OF PATIENTS WITH LEUKEMIA

Type of effector cells	Mean percent of cytotoxicity ± S.E.[a]
Normal donors	
High NK responders	61.8 ± 2.0
Medium NK responders	35.2 ± 1.3
Low NK responders	16.4 ± 1.7
Leukemic patients	
Preleukemic	7.2 ± 2.9
AML	7.3 ± 1.2
CML	3.3 ± 2.5
ALL	12.9 ± 2.0
CLL	6.3 ± 1.7
Remission (ALL, AML)	35.3 ± 5.4
Relapse (ALL, AML)	4.3 ± 0.6

[a]Cytotoxicity of Ficoll-Hypaqued PB against K-562 tumor was measured in a 3 hr ^{51}Cr release assay (12) at 1:50 target:effector (T:E) cell ratio. Seventy-seven normal donors and 58 patients were tested. All patients were untreated, except for those in remission or relapse (treatment-free for >3 weeks prior to NK cell test).

NK cell activity has been associated with a class of lymphoid cells, designated large granular lymphocytes (LGL), (13) which may be analyzed on May-Grünwald and Giemsa stained cytocentrifuge slides (12). Thus, it was of interest to examine the quantitative and qualitative profile of these cells in leukemic patients. In contrast to distinct differences in NK cell cytotoxicity among low, medium and high NK cell-responding normal donors, no distinction in LGL content was observed among these 3 responder classes in Ficoll-Hypaqued or nylon wool(NW)-filtered fraction of PB. Analysis of LGL content of leukemic patients demonstrated that most patients with myeloid and lymphoid leukemia (68%) had significantly lower levels of LGL than the normal donor population (Table II). This observation indicates that low LGL content may be responsible for inferior NK cell cytotoxicity of some of the leukemic patients. Nevertheless, since low NK cell cytotoxicity was displayed

TABLE II. NK CELL CYTOTOXIC POTENTIAL AND LGL CONTENT OF PB MONONUCLEAR CELLS OF NORMAL DONORS AND LEUKEMIC PATIENTS

	Ficoll-Hypaqued PB Mean (range)		Nylon wool-filtered PB Mean (range)	
Donors[a]	Cytotoxicity[b]	LGL content	Cytotoxicity	LGL content
Normal individuals				
High responders	61.5(50.0-69.8)	12.5(5.0-21.0)		
Medium responders	33.4(22.3-55.2)	15.9(2.0-38.5)	39.9(26.8-55.0)	27.0(4.5-51.5)
Low responders	17.2(13.2-19.6)	10.4(3.5-13.5)	25.1(14.9-36.4)	20.9(5.0-40.3)
Leukemic patients				
Myeloid	6.5(0.2-18.6)	1.9(<0.1-9.2)	9.2(1.3-24.5)	2.4(<0.1-8.2)
Lymphoid	10.3(3.6-17.0)	2.4(<0.1-6.0)	11.0(3.8-18.4)	7.6(1.0-18.5)

[a]*Twenty-six normal donors (7 high, 13 medium and 6 low NK cell responders) and 25 leukemic patients (14 with myeloid and 11 with lymphoid leukemia) were tested.*

[b]*Cytotoxicity against K-562 was measured in a 3 hr ^{51}Cr release assay at 1:50 T:E cell ratio.*

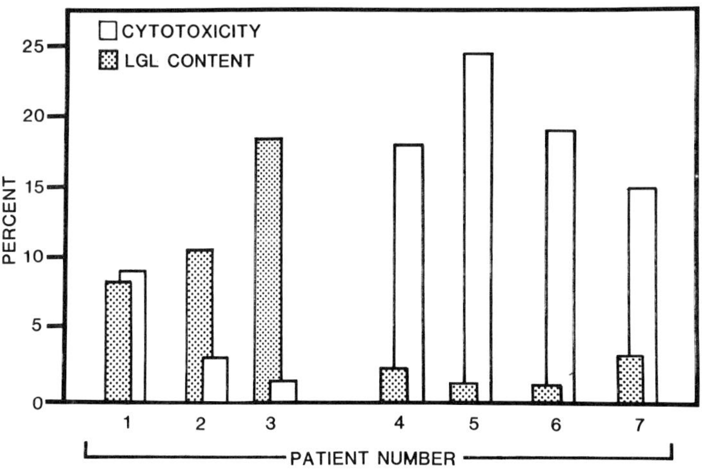

Fig. 1. Lack of correlation between LGL content and NK cell cytotoxic potential of leukemic patients. NW-filtered PB lymphocytes were tested for cytotoxicity against K-562 target in a 3 hr ^{51}Cr release assay, at 1:50 T:E cell ratio.

also by patients with LGL content comparable to medium and high NK cell-responding normal donors (Fig. 1, patients 1-3), NK cell defective activity cannot be attributed solely to low LGL content. In fact, the observation that a few patients with higher NK cell activity (comparable to NK cell low responding normal donors) did not have virtually any LGL (Fig. 1, patients 4-7), indicated that the LGL content in leukemic patients did not correlate with NK cell cytotoxicity.

To examine more precisely the content and the role of LGL in NK cell cytotoxicity of leukemic patients, NW-filtered PB lymphocytes were separated on Percoll gradient, and individual fractions analyzed for NK cell activity and LGL profiles. Percoll gradient separation was performed as described previously (13), modified by addition of two lighter concentrations of Percoll, in order to remove the majority of leukemic cells from PB samples. One ALL and two AML patients were studied, together with two normal donors. We observed two different patterns of NK cell cytotoxicity and LGL content after Percoll separation in leukemic patients.

The patient with ALL had low LGL content in NW-filtered fraction of PB (4%). However, after Percoll separation and depletion of cells forming rosettes with sheep erythrocytes, high enrichment of LGL (60%), comparable to that of normal donor, was observed in pooled fractions 3 and 4 (Fig. 2). In contrast to normal donor, whose LGL enrichment was associated

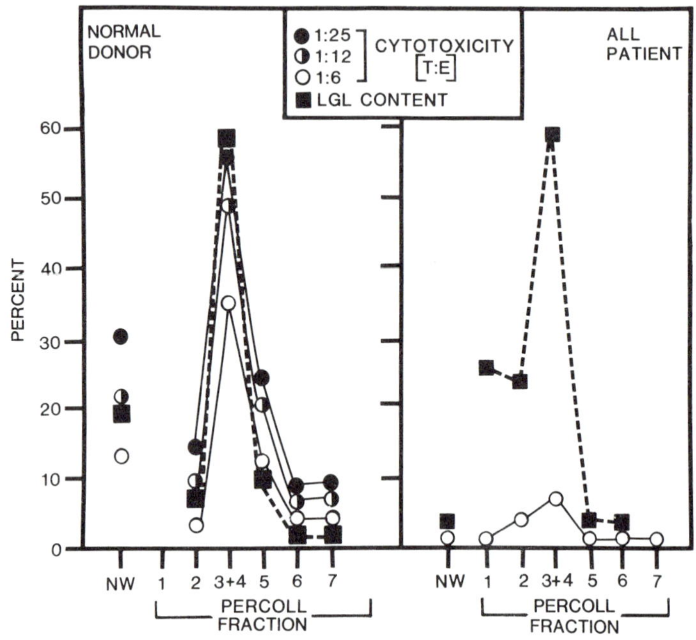

Fig. 2. Percoll density gradient separation of NW-filtered PB lymphocytes of normal donor and ALL patient. Percoll gradient was composed of the following Percoll concentrations: 35.0, 37.5, 40.0, 42.5, 45.0, 47.5 and 50.0% (fractions 1-7, respectively). Cytotoxicity was tested against K-562 target in a 3 hr ^{51}Cr release assay.

with increase in NK cell cytotoxicity (P<0.001 in pooled fractions 3 and 4 at all T:E cell ratios, Student's t-test analysis), virtually no NK cell activity was displayed by LGL-enriched fractions of ALL patient. This observation suggests that the LGL of ALL patient were defective in their cytotoxic potential, or represented less mature non-lytic LGL population.

Percoll separation of NW-filtered PB lymphocytes of AML patients resulted in a slight enrichment of LGL content in fractions 5-7 (from 3% to 8% in patient 1, and from 2% to 12% in patient 2), the fractions with low NK cell cytotoxicity. On the contrary, NK cell cytotoxicity was significantly enriched in fractions 3 and 4 (P<0.001), where virtually no LGL were present (Fig. 3). This was in contrast to normal donor PB demonstrating parallelism between enrichment of LGL and augmentation of NK cell cytotoxicity in fractions 4 and 5

(P<0.001). It was suggested previously that leukemic blasts could mediate spontaneous cytotoxicity in patients with leukemia; thus, we analyzed the individual Percoll fractions for blast content. Since only 7.5% and 3.5% of leukemic blasts were detected in NK cell-active fractions (fractions 3 and 4, respectively), and approximately 80% of blasts in fractions displaying low NK cell activity (fractions 1 and 2), the possibility that the leukemic blasts were involved in cytotoxicity phenomenon is quite unlikely. The discrepant pattern between cytotoxicity and LGL levels in patients with AML suggests that the NK cell cytotoxicity in these patients may be mediated by effector cell population other than LGL.

III. STUDIES ON NK CELL CYTOTOXIC MECHANISM OF PATIENTS WITH LEUKEMIA

The observation that some of the leukemic patients with relatively high LGL content displayed low NK cell cytotoxi-

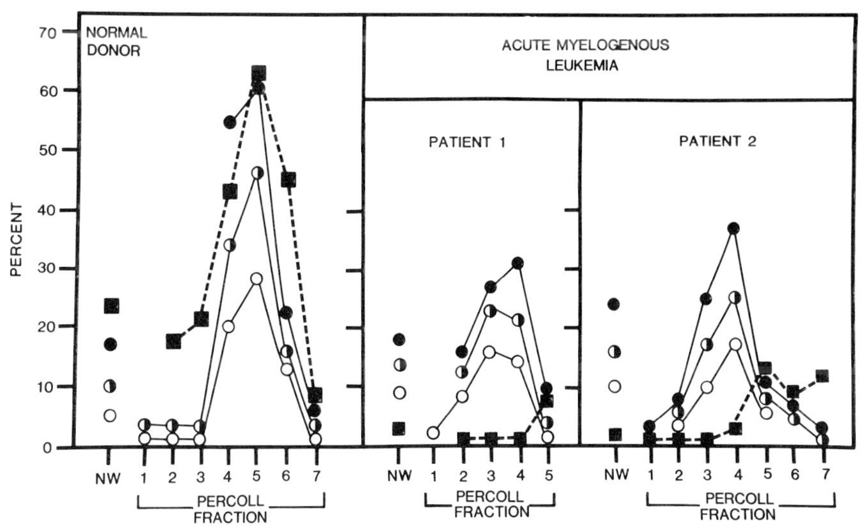

Fig. 3. Percoll density gradient separation of NW-filtered PB lymphocytes of normal donor and AML patients. Gradient was prepared as described in legend of Fig. 2. Cytotoxicity was tested against K-562 target in a 3 hr ^{51}Cr release assay, at ○ 1:12, ◑ 1:25, ● 1:50 T:E cell ratios; ■ LGL content.

city, together with the observation of the cytotoxic inefficiency of Percoll-separated LGL, led to the studies on the defective NK cell cytotoxic mechanism of leukemic patients. These investigations revealed that both, patients with myeloid and lymphoid leukemia (data pooled) experienced several defects in NK cell lytic machinery (Fig. 4). These defects were represented by low capacity of NK cells of leukemic patients to bind to and to lyse the K-562 tumor cell line in a single cell assay. Furthermore, V_{max} (reflecting rate of lysis of NK cells), the frequency of NK cells, as well as NK cell recycling capacity, were significantly impaired in leu-

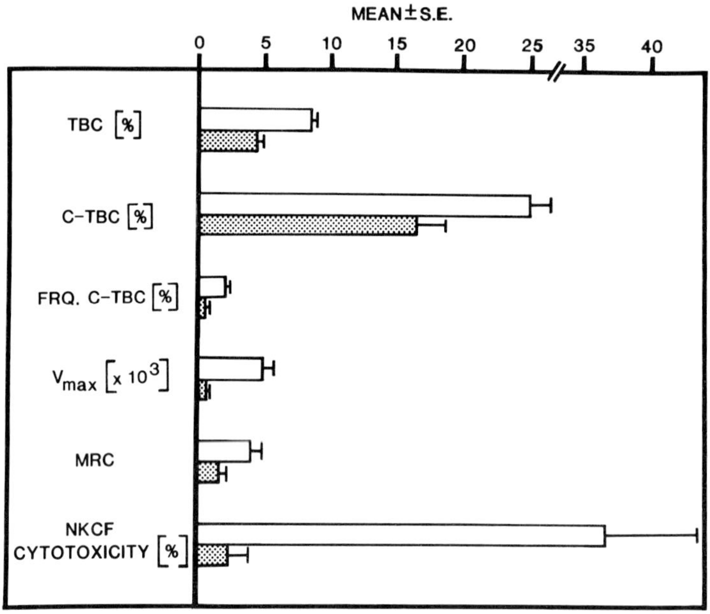

Fig. 4. NK cell functional profile of NW-filtered PB lymphocytes of normal donors and leukemic patients. Five to 10 normal individuals (□) and 5-14 leukemic patients (⁙) were tested for tumor-binding cells (TBC) and cytotoxic-TBC (C-TBC) against K-562 target in a single cell assay (14). C-TBC frequency (frq), the maximal recycling capacity (MRC) and generation and cytotoxicity of NKCF were determined as described earlier (4, 15-16). Maximum rate of tumor lysis (V_{max}) was analyzed by Michaelis-Menten kinetic model (15), using 10^5 effector cells and 3 x 10^3 to 10^5 K-562 target cells. Significant decrease in all NK cell functions was found in leukemic patients (P values ranged from <0.001 to <0.005, Student's t-test analysis).

kemic patients in comparison to normal donors. Additionally, in contrast to NK cells of normal donors, NK cells of leukemic patients did not possess the capacity to produce NK cell cytotoxic factor (NKCF).

IV. MODULATION OF NK CELL CYTOTOXICITY OF CML PATIENTS WITH INTERFERON-α

In three patients with CML we investigated the effect of exogenously administered, partially purified interferon-α

TABLE III. EFFECT OF EXOGENOUS IFN-α ON NK CELL PROFILE OF CML PATIENTS

Donor no.	Percent[b] ^{51}Cr	LGL	TBC	C-TBC	Frq. C-TBC	V_{max}[b] (x 10^4)	MRC[b]
			IFN-α treated patients[a]				
1	64.1	16.2	8.1	25.6	2.1	4.6	4.4
2	63.4	19.7	4.2	28.6	1.2	4.9	8.1
3	47.4	18.0	8.4	30.6	2.6	19.7	15.1
			Normal individuals				
4	35.7	9.5	4.7	28.3	1.3	1.3	2.0
5	13.6	13.5	6.1	13.7	0.8	1.7	4.3
6	34.3	38.5	5.5	16.2	0.9	1.1	2.5

[a]IFN-α was administered at the Department of Clinical Immunology and Biological Therapy, M. D. Anderson Hospital, in the doses of 3-9 x 10^6 U (daily for 8 months in patients 1 and 2; daily, or every other day for 24 months in patient 3). The leukopheresed PB samples were provided by Dr. J. Hester.

[b]Cytotoxicity was tested in a 3 hr ^{51}Cr release assay (1:50 T:E cell ratio); TBC in a single cell assay (2 hr). Frequency (frq.) of C-TBC and MRC were determined as indicated in Fig. 4. V_{max} was determined by the Michaelis-Menten kinetic model (15) using 5 x 10^5 effector cells and 6 x 10^3 to 10^5 target cells. K-562 target was used in all studies.

(IFN-α) on the lytic mechanism of PB NK cells. The schedule and the dose of administration of IFN-α is indicated in Table III. This table also illustrates that in contrast to untreated patients (Table I), IFN-α treated CML patients exhibited high levels of NK cell cytotoxicity in ^{51}Cr release assay, efficient tumor-binding and killing capacity in a single cell assay, and high frequency of cytotoxic NK cells. In fact, the efficiency of some of the parameters of NK cell mechanism of CML patients, for instance, V_{max} and recycling, were superior in comparison to those of normal donors, tested together with patients. Morphological analysis demonstrated that also LGL content was within the higher range of normal donors.

Separation of NW-filtered PB lymphocytes of two IFN-α treated CML patients on Percoll gradient (Fig. 5), demonstrated high enrichment of LGL in both patients in fractions 1-3, with the highest increase in LGL content in fraction 2 (from 31% to 89% in patient 1, and from 15% to 84% in patient 2). The high LGL content in Percoll fraction 1 indicates that some of the LGL of IFN-α treated patients were of low densi-

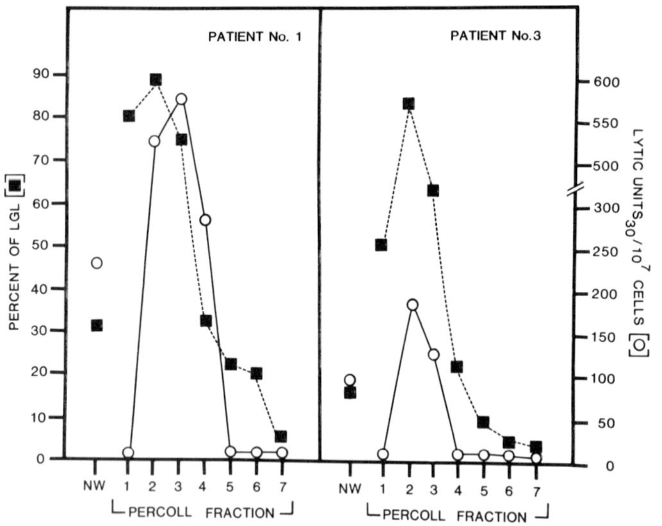

Fig. 5. Percoll density gradient separation of peripheral blood NW lymphocytes of IFN-α treated CML patients. Percoll gradients were prepared as described in legend of Fig. 2. NK cell cytotoxicity is expressed in lytic units (LU), determined from log linear plot of the data points and is reported as $LU_{30}/10^7$ lymphocytes.

ty, the characteristic shown to be associated with NK cell activation (17-18).

The correlation between LGL enrichment and augmentation of NK cell activity was detected in fraction 2 (in both patients) and fraction 3 (in patient 1), but not in fraction 1 of either patient. The lack of NK cell activity in highly LGL-enriched fraction 1 suggests that some of the LGL may be cytolytically inactive, and/or may represent NK cells with other functional activities. In fact, it has been well recognized that NK cells mediate also other functions, as represented by resistance to infections, regulation of hemolymphopoiesis, and production of various soluble factors (19-21). In this context, it is relevant to mention that IFN-α treatment resulted in a disturbance of hemopoiesis within bone marrow compartment of some of the CML patients. Specifically, bone marrow stem cells of these patients were deficient in their potential to form granulocytic colonies *in vitro* (Dr. M. Talpaz, Department of Clinical Immunology and Biological Therapy of this Institution - personal communication). Consequently, the question arises whether activated NK cells (LGL) could have been involved in such inhibition of granulopoiesis. Further research is needed to address this possibility.

V. CONCLUDING REMARKS

Our investigations demonstrated that patients with myeloid and lymphoid leukemia experience several defects in NK cell system, on both quantitative and qualitative level. In one group of patients, the low NK cell activity could be attributed to the low levels of LGL present in the peripheral blood compartment. In the other group of patients, low cytotoxic activity was exhibited in spite of the high levels of LGL, suggesting LGL defect. This was further substantiated by the lack of NK cell activity in highly enriched LGL fraction after Percoll gradient separation. In the third group of patients, the dissociation between NK cell cytotoxicity and LGL content, especially the presence of NK cell activity in Percoll fractions with no LGL content, suggested that NK cell cytotoxicity may be mediated also by effector cells, other than LGL.

Analysis of the mechanism of NK cell defective activity revealed that patients with leukemia displayed low tumor-binding and tumor-killing potential, low maximum rate of lysis, impaired recycling activity, and did not produce NK cell cytotoxic factor. We have observed, in limited number of CML patients, that NK cell cytotoxic potential was corrected by long-term *in vivo* treatment with IFN-α. This restoration of NK cell lytic mechanism was associated with increase in LGL content. The positive modulation of NK cell activity by IFN-α is encouraging; it suggests that the defect in NK cell lytic mechanism is a reversible phenomenon. However, since IFN-α treated CML patients exhibited retardation of hemopoiesis, which could be related to NK cell hyperactivity, the role of interferon in treatment of leukemia remains questionable.

REFERENCES

1. Herberman, R. B., and Holden, H. T. (1978). Adv. in Cancer Res. *27*, 305.
2. Lotzová, E., and McCredie, K. B. (1978). Cancer Immunol. Immunother. *4*, 215.
3. Klein, M., Roder, J., Haliotis, T., Korec, S., Jett, R., Herberman, R. B., Katz, P., and Fauci, A. S. (1980). J. Exp. Med. *151*, 1049.
4. Lotzová, E., Savary, C. A., and Keating, M. J. (1982). Experiment. Hematol. *10* (Suppl. 12), 83.
5. Lotzová, E. (1983). Surv. Synth. Path. Res. *2*, 41.
6. Lotzová, E. (1984). The Cancer Bull. (In press)
7. Lotzová, E., McCredie, K. B., Maroun, J. A., Dicke, K. A., and Freireich, E. J (1979). Transplant. Proc. *11*, 1390.
8. Lotzová, E., McCredie, K. B., Muesse, L., Dicke, K. A., and Freireich, E. J (1979). *In* "Experimental Hematology Today" (S. J. Baum and G. D. Ledney, eds.), p. 207, Springer-Verlag, New York.
9. Behelak, Y., Banerjee, D., and Richter, M. (1976). Cancer *38*, 2274.
10. Platsoucas, C. D., Fernandes, G., Gupta, S. L., Kempin, S., Clarkson, B., Good, R. A., and Gupta, S. (1980). J. Immunol. *125*, 1216.
11. Ziegler (Löms), H. W., Kay, N. E., and Zarling, J. M. (1981). Int. J. Cancer *27*, 321.
12. Lotzová, E., Savary, C. A., Gutterman, J. U., Quesada, J. R., and Hersh, E. M. (1983). J. Biol. Resp. Mod. *2*, 482.
13. Timonen, T., Ortaldo, J. R., and Herberman, R. B. (1981). J. Exp. Med. *153*, 569.

14. Grimm, E., and Bonavida, B. (1979). J. Immunol. *123*, 2861.
15. Ullberg, M., and Jondal, M. (1981). J. Exp. Med. *153*, 615.
16. Wright, S. C., Weitzen, M. L., Kahle, R., Granger, G. A., and Bonavida, B. (1983). J. Immunol. *130*, 2479.
17. Lotzová, E., Savary, C. A., Khan, A., and Stringfellow, D. A. (1984). J. Immunol. *132*, 2566.
18. Biron, C. A., and Welsh, R. M. (1982). J. Immunol. *129*, 2788.
19. Lotzová, E., Savary, C. A., and Pollack, S. B. (1983). Transplant. *35*, 490.
20. Herberman, R. B., and Ortaldo, J. R. (1981). Science *214*, 24.
21. Herberman, R. B. (ed.) (1982). "NK Cells and Other Natural Effector Cells", Academic Press, New York.

CHARACTERIZATION OF NK CELLS IN PATIENTS WITH CHRONIC MYELOGENOUS LEUKEMIA[1]

Yoshiaki Fujimiya
Paul K. Pattengale

Departments of Pathology and Microbiology
University of Southern California
School of Medicine
Los Angeles, California

I. INTRODUCTION

Since the vast majority of well defined, NK-susceptible targets are tissue culture-derived(1), we have deemed it important to focus on noncultured, freshly-occurring targets of NK lysis. We have further concentrated on fresh lymphohematopoietic cells as NK targets(2,3), since this approach eventually allows for clinical strategies aimed at ridding the host of NK-susceptible tumor cells. This has particular relevance to NK-susceptible bone marrow and peripheral blood based leukemia cells, which are presumably accessible to bone marrow and peripheral blood based human NK cells.

Patients with chronic myelogenous leukemia(CML) are of particular interest and relevance, since the neoplastic cell in these chronic leukemia patients is a primitive stem cell (blast) which is still capable of full differentiation. In time, however, the majority of CML patients accelerate from a chronic phase to a uniformly fatal blast crisis phase, which is most often characterized by increasing numbers of terminal

[1]This work was supported in part by NIH grant CA-29910; by ACS grant IM-387; by a fellowship grant(to Y.F.) from the Norris Cancer Center; and by a Leukemia Society of America Scholar Award(to P.K.P).

ISBN 0-12-341370-2

transferase negative, Philadelphia chromosome positive, primitive, myeloid-type blasts in the bone marrow and peripheral blood(4,5). Since a sizeable percentage of fresh myeloid-type blasts are NK-susceptible(3,6), as are cell lines derived from such blasts(e.g., K-562)(7), certain CML patients may be potentially accessible to immunotherapeutic strategies involving augmentation of autologous natural killer cell activity. For these reasons, we further deemed it important to characterize NK cells in CML patients.

II. NK ACTIVITY IN CML PATIENTS

As might be expected from the dilutional effect of large numbers of circulating neoplastic granulocytes, NK activity in the peripheral blood of CML patients was significantly depressed as compared to normals(Table I). Unexpectedly, however, we observed using the monoclonal antibodies HNK-1(8,9) and B73.1(10,11), that CML patients had normal to increased absolute numbers of large granular lymphocytes(LGLs) circulating in their peripheral blood(Table I).

In order to separate the possible extrinsic influences of contaminating neoplastic granulocytes, as well as the possibility of serum and/or lymphoid cell mediated suppressor effects on NK activity, we rigorously sorted HNK-1^+ and B73.1^+ lymphoid cells from the E^+ fraction(i.e., sheep red blood cell rosetting lymphoid cells at 4°C) of patients and normals, and then tested for NK activity. We focused on the E^+ fraction in CML patients, since sorting from the E^- fraction was logistically not feasible due to heavy contamination with neoplastic granulocytes. In contrast, positive sorting of HNK-1^+ lymphoid cells from the E^+ fraction of CML patients was technically feasible, both because of low percentages of contaminating neoplastic granulocytes, and higher starting percentages of HNK-1^+ and B73.1^+ lymphoid cells. As seen in Table II highly purified HNK-1^+ and B73.1^+ lymphoid cells from CML patients were still markedly defective in NK activity as compared to similarly prepared normal controls.

In addition to these findings, we have evidence that the NK-defective, HNK-1^+ and B73.1^+ lymphoid cells from CML patients, although not defective in their ability to bind to K-562, are unable to be augmented with alpha interferon preparations, do not function as suppressors of normal NK activity, and are not part of the neoplastic CML hematopoietic clone as evidenced by the presence of normal diploid metaphases from CML patients which are Philadelphia chromosome positive(data not shown)(12).

Table I. CML patients have normal to increased numbers of peripheral blood HNK-1$^+$ and B73.1$^+$ large granular lymphocytes(LGLs) which are not able to efficiently kill K-562

Group	n[a]	No. HNK-1$^+$ lymphoid[b] cells/mm^3 ± S.E.M.	No. B73.1$^+$ lymphoid[b] cells/mm^3 ± S.E.M.	%S.L. K-562[c]
Normals	12	510±60	450±74	54±5
Patients	9	1240±134[d]	680±43	7±2[e]

[a]Twelve normal donors(mean age = 43) and nine CML patients(mean age = 54) each tested once. S.E.M. = standard error of the mean.

[b]As determined by multiplying the absolute number of lymphocytes/mm^3 from CML patients and normals by the percentage of HNK-1$^+$ or B73.1$^+$ lymphoid cells. Fluorescence was performed on the FACS-IV on gated lymphocyte populations only.

[c]Specific lysis at an E/T ratio of 100/1. Ficoll-hypaqued band cells as effectors.

[d]p less than .05 as compared to normal donors.

[e]p less than .001 as compared to normal donors.

Table II. FACS-sorted HNK-1$^+$ and B73.1$^+$ lymphoid cells from CML patients are unable to efficiently lyse K-562 as compared to normals

		Mean % S.L. of K-562 ± S.E.M.[a]		
Group	n[b]	HNK-1$^+$ cells	n[c]	B73.1$^+$ cells
Normals	22	24±2	10	48±4
Patients	20	3±1[d]	8	4±2[d]

[a]Specific lysis at an E/T ratio of 4/1.
[b]Fourteen normal donors and twelve CML patients.
[c]Six normal donors and six CML patients.
[d]p less than .001 as compared to normals.

III. CULTURED CYTOTOXIC CELLS FROM CML PATIENTS

Since we have now demonstrated that CML patients seemingly have an intrinsic NK defect with regard to freshly isolated NK cells, we now chose to culture these defective NK cells with the intention of restoring NK activity to normal using *in vitro* methodology. To this end, we have either bulk-cultured or single-cell cloned HNK-1$^+$(Table III) or B73.1 (Table IV) lymphoid cells from CML patients for 3-8 weeks in the presence of 10% interleukin-2(IL-2) and 10^5 lethally-irradiated allogeneic feeder cells. As is evident from these *in vitro* results, we have been successful in restoring NK-like cytotoxic activity in CML patients to normal levels.

It was of further interest that successfully-restored cytotoxic clones and/or bulk cultures initiated from positively sorted HNK-1$^+$,E$^+$ lymphoid cells from either normals or patients, lost their HNK-1 positivity with time, and also failed to mark with OKT-3(Table III). In contrast, unsuccessfully restored cytotoxic clones and/or bulk cultures as measured against K-562 from some CML patients, had also lost their HNK-1 positivity with time, but did possess a high percentage of cells which marked with OKT-3. This finding may relate to the *in vivo* presence of increased numbers of OKT-3$^+$, HNK-1$^+$ lymphoid cells, which are observed in CML patients(12). It is

Table III. Clones and/or bulk cultures initiated from HNK-1$^+$ lymphoid cells from CML patients are restorable to normal NK activity[a]

Group	n[b]	%S.L. K-562[c]	%HNK-1[d]	%OKT-3[d]
Normals	15	66±6[e]	15±4	14±3
Patients	14	68±4[e]	16±1	12±8
	16	3±1[f]	14±2	92±4[f]

[a]Single cell cloning of HNK-1$^+$, E$^+$ mononuclear cells was performed using limiting dilution techniques in the presence of 5% normal, pooled human AB serum, 10% IL-2(Associated Biomedics or Electronucleonics) and 10^5 irradiated feeder cells. Bulk cultures were similarly cultured and initiated at 10^6/ml. Feeder cells are allogeneic and isolated from normal donor Ficoll-hypaque mononuclear bands and subsequently irradiated with 5000R. Cloning efficiency was 1% for CML patients as compared to 3% for normal donors.

[b]Four patients with CML and four normal donors.

[c]Specific lysis at an E/T ratio of 4/1.

[d]As measured on the FACS-IV.

[e]Also able to kill autologous myeloid-type blasts(32% specific lysis) from one patient who subsequently developed a blast crisis.

[f]p less than .01 as compared to normals.

unclear at this time, whether this cytotoxically-inactive population represents an immature population of NK cells, or an entirely NK-unrelated population of T cells bearing HNK-1. In addition, as stated in footnote [e] to Table III, one CML patient, whose clones and/or bulk cultures were available for testing as effectors, developed a myeloid-type blast crisis. His clones and/or bulk cultures, which bore a low percentage of OKT-3, were capable of killing his autologous blasts as well as K-562 as measured in short term(4 hr.) chromium release assays(average of 32% specific lysis of autologous blasts at an E/T ratio of 4/1).

Table IV. Clones and/or bulk cultures initiated from B73.1$^+$ lymphoid cells from CML patients are restorable to normal NK activity [a]

Group	n[b]	%S.L.K-562[c]	%B73.1[d]	%OKT-3[d]
Normals	10	86±4	62±6	10±6
Patients	9	82±4	68±4	75±4[e]

[a]Single cell cloning and bulk culturing of B73.1$^+$,E$^+$ lymphoid cells was performed as described in footnote a, Table III. Cloning efficiency was 1% for CML patients as compared to 3% for normal donors.

[b]Two patients with CML and two normal donors.

[c]Specific lysis at an E/T ratio of 4/1.

[d]As measured on the FACS-IV.

[e]p less than .01 as compared to normals.

In contrast to clones and/or bulk cultures initiated from HNK-1$^+$ lymphoid cells, successfully-restored cytotoxic clones and/or bulk cultures initiated from positively-sorted B73.1$^+$ lymphoid cells from either CML patients or normals, retained their B73.1 positivity with time(Table IV). Interestingly, clones and/or bulk cultures from CML patients possessed high percentages of lymphoid cells which marked positive for surface OKT-3, as compared to normal donors, which marked significantly lower(Table IV). Autologous killing has not yet been tested, since both CML patients are still in the chronic phase of their disease.

IV. CONCLUSIONS AND FUTURE DIRECTIONS

In summary, the experimental data, obtained from patients with chronic myelogenous leukemia(CML), would indicate that NK cells from these patients are intrinsically defective in their NK activity. Furthermore, this defect is able to be overcome and fully restored _in vitro_ using IL-2 and lethally-irradiated, allogeneic feeder cells. Whether this NK defect in CML patients results from inadequate IL-2 production and/or the presence of immature, cytotoxically-inactive NK cells and/ or other unknown mechanisms is unclear at the present time. It is clear, however, that high levels of restorable NK-like activity can be achieved in intermediate term cultures. The term "NK-like" should be stressed in this *in vitro* context, since these cytotoxically-restored, cultured lymphoid cells, as described in the present experiments, are also capable of killing Raji targets as well as allogeneic, more-differentiated, neoplastic CML targets(data not shown). Although somewhat controversial, others have termed these cultured, NK-like cytotoxic cells as lymphokine activated killers(LAKs)(13) and/ or anomalous killers(14). Despite these differences between cultured cytotoxic(NK-like) cells and freshly occurring NK cells, this *in vitro* approach may eventually allow for the use of these cytotoxic clones and/or bulk cultures(autologous and/or allogeneic) as an effective immunotherapeutic modality for some CML patients, particularly those who are not good candidates for bone marrow transplantation. Similar approaches using systemically-injected cytotoxic cells, especially when combined with parralel injections of recombinant IL-2, have already proven successful in significantly reducing established pulmonary metastases in sarcoma-bearing mice(15). Such experimental studies in mice are providing the basis for parralel clinical trails in man(15).

ACKNOWLEDGMENTS

The authors wish to thank Drs. Toru Abo and Giorgio Trinchieri for their valuble suggestions, and for kindly providing the monoclonal antibodies HNK-1 and B73.1.

REFERENCES

1. Pattengale, P.K., Gidlund, M., Nilsson, K., Sundstrom, C., Orn, A., and Wigzell, H.(1981). Int. J. Canc. 28, 459.
2. Pattengale, P.K., Gidlund, M., Nilsson, K., Sundstrom, C., Sallstrom, J., Simonsson, B., and Wigzell, H.(1982). Int. J. Canc. 29, 1.
3. Pattengale, P.K., Sundstrom, C., Yu, A.L., and Levine, A. Nat. Immun. Cell Growth Regul.(in press).
4. Koeffler, H.P., and Golde, D.W.(1981). N. Engl. J. Med. 304, 1201.
5. Koeffler, H.P., and Golde, D.W.(1981). N. Engl. J. Med. 304, 1269.

6. Zarling, J.M., Eskra, L., Borden, E.C., Horoszewicz, J., Carter, W.A.(1979). J. Immunol. 123, 63.
7. Lozzio, C.B., and Lozzio, B.B.(1975). Blood 45, 321.
8. Abo, T., Cooper, M.D., and Balch, C.M.(1982). J. Immunol. 129, 1752.
9. Abo, T., Miller, C.A., Gartland, G.L., and Balch, C.M. (1983). J. exp. Med 157, 273.
10. Perussia, B., Starr, S., Abraham, S., Fanning, V., and Trinchieri, G.(1983). J. Immunol. 130, 2133.
11. Perussia, B., Trinchieri, G., Jackson, A., Warner, N.L., Faust, J., Rumpold, H., Kraft, D., and Lanier, L.L.(1984). J. Immunol. 133, 180.
12. Fujimiya, Y., Bakke, A., Linker-Israeli, M., Udis, B., Horwitz, D., and Pattengale, P.K.(submitted for publication).
13. Grimm, E.A., Mazumder, A., Zhang, H.Z., and Rosenberg, S.A.(1982). J. exp. Med. 155, 1823.
14. Seeley, J.K., Mascucci, G., Poros, A., Klein, E., and Golub, S.H.(1979). J. Immunol. 123, 1303.
15. Mule, J.J., Shu, S., Schwarz, S.L., and Rosenberg, S.A. (1984). Science 225, 1487.

NATURAL KILLER CELL ACTIVITY IN GENETIC DISEASES

Hugh F. Pross[1]
Jerome A. Werkmeister[2]

Departments of Radiation Oncology
and Microbiology and Immunology
Queen's University at Kingston
Ontario, Canada

I. INTRODUCTION

The study of patients with inherited diseases of metabolism or cellular development has proven to be an extremely useful tool in defining the role of various enzymes, metabolic pathways or cell populations in normal development and cell function. This has been especially true in the case of the human immune system and much of our knowledge of the different components of the immune system, their interactions and their importance with respect to defence against various microorganisms or types of malignancy has come from the examination of patients with well-defined inherited or congenital immunodeficiency diseases. In this paper, data will be presented on the natural killer (NK) cell activity of lymphocytes from a patient with mucolipidosis III (ML III)(1), a condition caused by a defect in N-acetylglucosamine-1-phosphotransferase (2). This enzyme is responsible for the phosphorylation of mannose residues, which is necessary for the mannose-6-phosphate (man-6-P) receptor-mediated uptake and transfer of lysosomal enzymes into lysosomes (3). In addition to its potential usefulness in studying the role of lysosomal enzymes in NK

---

[1]Supported by the Ontario Cancer Treatment and Research Foundation, the National Cancer Institute (Canada) and the Medical Research Council of Canada.

[2]Present address: The Walter and Eliza Hall Institute of Medical Research, Melbourne, Australia.

ISBN 0-12-341370-2

activity, the enzyme defect is interesting because it has been found that NK cells and susceptible target cells have receptors for man-6-P on their surface and that man-6-P (and certain other phosphorylated and unphosphorylated sugars) will inhibit NK lysis of tumor target cells (4, 5). It is thus theoretically possible that man-6-P-bearing molecules act as the ligand between the killer cell and target recognition site or that the man-6-P receptor on the target acts as the acceptor site for the lytic molecule. Experiments to determine the effects of this enzyme defect on the NK activity of lymphocytes from a patient with ML III were therefore carried out to test the validity of these hypotheses. The data presented indicate that NK-mediated lysis is independent of the phosphorylation of lysosomal mannose residues or the existence of a man-6-P-dependent ligand mechanism. A brief review is presented of the effects of other genetic diseases on natural killer (NK) cell activity and the contribution which the understanding of these diseases has made to our knowledge of NK cell development, function and mechanism of action.

II. METHODS

Natural killer cell activity was assessed using lymphocytes from iron-plus-magnetism treated, Ficoll-Isopaque-purified heparinized blood in conventional 18 hour ^{51}-Chromium release assays as described in detail previously (6, 7). The target cells used were the highly NK-sensitive (6) erythroleukemic cell line, K562, the T cell line, Molt-4, the myelomonocytic line HL-60, and the murine mastocytoma line, P815 ($\pm$ rabbit anti-P815 to detect antibody-dependent cell-mediated cytotoxicity). Four to 8 effector to target cell ratios were plated for each lymphocyte preparation. The data were expressed as lytic units per million lymphocytes. One lytic unit (LU(20%)) was defined as the number of effector cells capable of lysing 20% of 5000 targets per well. The LU(20%) values, and the standard deviations of these values, were calculated using the exponential fit equation, which incorporates all of the points on the dose response curve (7). Sugar inhibition studies were done as described previously by Werkmeister *et al* (8). Lymphocyte markers and lytic conjugates were determined as described by Rubin et al (9).

The patient described in this report was born in 1955 of normal parents with no family history of similar conditions or of consanguinity. Two sisters are normal and one younger sister also has ML III. The diagnosis of ML III was made clinically and on the basis of elevated serum acid hydrolases in the absence of mucopolysacchariduria (1). The diagnosis was confirmed by Dr. V. McKusick who examined the patient and reviewed the case in 1976. At the time of NK testing the patient was healthy, with no recent infections. The affected sister did not consent to be tested.

III. RESULTS

Table I shows the NK activity of the ML III patient in comparison with that of normal controls selected on the basis of several hundred previous assays. These donors are at the upper and lower 10th percentiles of NK activity respectively. It is apparent from this table that the patient has NK activity which is markedly higher than normal, and that there is no evidence of any defect in NK activity on repeat testing. This phenomenon was not limited to the K562 target cell (see footnote to Table I) and similar relative levels of antibody-dependent cell-mediated cytotoxicity (ADCC) were observed.

To determine whether or not the enhanced NK activity of the ML III lymphocytes was subject to the same modulating agents that affect normal NK cells, NK assays were done in the presence of man-6-P, the carboxylic ionophore, monensin, and interferon. The results are shown in Table II. It can be seen from this table that ML III NK cells are inhibited or enhanced by these agents to the same extent that normal donor NK cells are. In addition to the studies shown in this table, NK assays against K562 were also done in the presence of 30% serum from normal donors or the patient. The incorporation of serum had no effect on NK lysis by the patient's or the control cells.

Marker studies done on the patient's lymphocytes showed

TABLE I. The effect of ML III on NK(K562) activity[a]

	High normal[b]	Low normal[b]	ML III
Exp. 1	82.1±9.3	12.2±0.4	110.7±9.8
Exp. 2	186.4±12.2	18.9±0.9	286.5±21.1
Exp. 3	141.1±6.9	28.1±1.3	211.9±37.5

[a]LU(20%)/10^6 lymphocytes ± standard deviation.

[b]Normal donor controls previously determined to be at the high and low extremes of normal NK (10). A similar rank order of the ML III patient's cytotoxicity *vs* the normals was seen using MOLT-4, HL-60 and antibody-coated P815 (not shown).

the following - Fc receptor (EA)-15%, ERFC-74%, T11-90%, T8-48%, T4-51%, B1-14%, B73.1-30%, HNK-1-7.3% and large granular lymphocytes - 1.5%. These percentages are within the range for normal donors although the values for T8 and B73.1 are somewhat high. The percentage of lymphocytes capable of lysing K562 in the single cell assay was 3.9±0.6, which was somewhat higher than the high normal donor value of 2.7±0.4. This suggests that the increase in NK activity seen is due to a simple increase in the frequency of active NK cells.

The high NK activity on the part of the patient's lymphocytes could be due to the abnormality in enzyme localization seen in ML III, or it could be a variant of normal, unrelated to the patient's disease. To assess this, NK assays were performed on all of the members of the immediate family, except for the youngest (affected) sister. The results of this experiment are shown in Table III. It is apparent that the patient's father is also a high NK donor, the mother has low NK activity, and the two normal sisters are intermediate but low. Although the patient's NK is significantly higher than the father's, without NK data from other ML III patients there is insufficient evidence to state unequivocally that the patient's high NK activity is due to the enzyme defect.

TABLE II. The effect of NK modulators on ML III NK [a]

Modulator	Donor	Control	+Modul.	% Control
Man-6-P (50 mM)[b]	High norm.	186.4	27.3	14.6
	Low norm.	18.9	3.6	19.0
	ML III	286.5	26.8	9.4
Man-6-P (12.5 mM)	High norm.	186.4	128.3	68.8
	Low norm.	18.9	16.3	86.2
	ML III	286.5	183.8	64.2
Monensin (10 ug/ml)	High norm.	195.3	5.3	2.7
	Low norm.	29.8	1.4	4.7
	ML III	541.6	5.0	0.9
Alpha IFN (125 U/ml)	High norm.	83.4	124.3	149.0
	Low norm.	15.7	21.1	134.4
	ML III	112.8	179.2	158.9

[a]Legend as per Table I. Standard deviations were proportionally similar to those in Table I.

[b]Lymphocytes were preincubated with monensin or alpha interferon for 1 hour at 37°. Man-6-P was in the assay (8).

II. DISCUSSION

Natural killer (NK) cell-mediated cytolysis has been divided into 4 discrete stages: (i) recognition and binding to the target, (ii) activation or triggering of the NK cell, (iii) delivery of the lytic factor, and (iv) target cell death (11, 12). The majority of published data has centered around the initial stages of interaction and activation and little information is available on the lytic phase of the response. Elucidation of the lytic mechanism involved in NK-mediated lysis would help identify the nature of the lytic molecule and would facilitate our understanding of impaired responsiveness associated with certain diseases.

Mannose-6-phosphate is known to be involved in the targeting of lysosomal enzymes from their site of synthesis in the rough endoplasmic reticulum to lysosomes (13). Selected high mannose-type oligosaccharide residues are phosphorylated by the combined action of two enzymes, UDP-N-acetylglucosamine:lysosomal enzyme N-acetylglucosaminyl-phosphotransferase and N-acetylglucosamine-1-phosphodiester-alpha-N-acetylglucosaminidase (13). The final step in the targeting process is believed to be the specific binding of the phosphorylated

TABLE III. NK activity in ML III family members [a]

Relationship to patient	NK(K562)[b] (LU(20%)/10^6	Proportional activity vs Father	Mother	Normal[c]
Patient	48.6±3.5	1.32	4.58	1.68
Father	36.7±6.9	1.00	3.46	1.26
Mother	10.6±1.0	0.29	1.00	0.37
Sister	16.8±2.1	0.46	1.58	0.58
Sister	17.3±2.4	0.47	1.63	0.60
High norm	46.5±4.2	----	----	1.61
Low norm	11.4±0.7	----	----	0.39

[a]Three similar assays were done on fresh and cryopreserved lymphocytes and showed the same rank order of cytotoxicity. The experiment shown used cryopreserved lymphocytes except for the low normal control (fresh). Cryopreservation and recovery was done as described by Pross and Maroun (14).

[b]Data are shown ± standard deviations.

[c]Calculated by dividing test LU by each parental LU or the mean LU value of the two normal controls.

enzymes to the man-6-P receptor (2). It is possible, in view of recent results from sugar inhibition studies (4, 8), that such a receptor could represent a means of recognition between NK cells and target cells with respect to the acceptance and uptake of a soluble lytic factor. As stated above, in ML III, the basic defect has been shown to be a deficiency in phosphotransferase enzyme (2) which results in the extracellular accumulation of non-phosphorylated enzymes lacking the man-6-P recognition marker (3). Thus, if man-6-P receptors play a role in NK cell-mediated cytotoxicity, lymphocytes from patients with ML III should be NK-defective.

The results reported in this paper show that NK (and ADCC) activity is not in any way impaired, but in fact lies beyond the highest range of activity based on our previous characterization of normal NK activity from selected donors. Furthermore, the response of the ML III patient's NK cells to *in vitro* modulators was similar to that of normal cells. Normal NK function has also been observed in a patient with ML I (E. Saksela, personal communication) and two patients with MLII (H.-D. Haubeck, personal communication). ML I and ML II involve a similar enzyme defect to ML III, although the clinical picture differs and the conditions are more severe (1).

The increased level of endogenous NK cell activity in the patient reported here is probably due to a higher than normal frequency of NK cells in the circulation, since the proportion of lytic conjugates was slightly higher than that of the high normal control. The actual level of NK activity observed is probably unrelated to the enzyme defect in ML III since the patient's father also high had NK activity. As it happened the patient's mother had low normal NK activity, effectively ruling out an association between high NK activity and the carrier state for ML III. These data suggest that man-6-P recognition, either by the "lytic moiety acceptor site" or as part of NK-target cell binding is *not* a necessary event in NK-mediated cytotoxicity. It is not possible, however, to draw such a firm conclusion concerning the role of lysosomal enzymes in NK cytotoxicity. Varki *et al* (15) have shown that while phosphotransferase activity is markedly decreased in the leukocytes of ML III patients, extracts of both the mononuclear and granulocytic fractions of these cells had normal levels of lysosomal enzymes. These enzymes lacked the phosphomannosyl recognition marker, however, indicating that some other, as yet undefined, mechanism exists for targeting the enzymes to lysosomes, and, possibly, to NK-sensitive targets.

Thus the mucolipidoses I-III can be added to the list of genetic diseases which have contributed to our knowledge of the nature of human natural killer cells. The role of genetic factors in establishing "endogenous" normal human NK activity is undoubtedly very complex and and has not been worked out.

TABLE IV. Genetic diseases and NK activity in man

Condition	Effect on NK[a] Decreased	Normal[b]	References
X-linked agamma-globulinemia	-	15	(18-23)
Common variable hypogamma.	8	57	(20-25)
Selective IgA immunodef.	-	16	(22,24)
Severe combined immunodef.(B?SCID)	1	4	(22,26)
Severe combined immunodef.(B+SCID)	11	1	(18,19,21,27)
Severe combined immunodef.(B-SCID)	4	3	(18,19,23,27,28)
Immunodef. with hyper IgM	-	1	(21)
Ataxia telangiectasia	5	8	(18,19,22)
Wiskott-Aldrich syndrome	1	5	(22,23)
Immunodef. with thymoma	1	1	(20,25)
Nezelof's syndrome	-	1	(29)
Di George's syndrome	-	4	(18,19,27)
X-linked lympho-prolif. syndrome	7	3	(30,31)
Chediak-Higashi syndrome	18	-	(32-40)
Chronic granulo-matous disease	-	7	(41-43)
Defective interferon production	4	-	(23)
Down's syndrome (trisomy-21)	-	11	(44)
Thallassemia major	-	25	(45)
Mucolipidosis III	-	1	This report.

[a]Number of patients reported.
[b]Includes patients with "increased" NK.

Although human NK cells are influenced by a number of stimuli (16), basal NK levels are probably genetically determined. It is well established that an individual donor's NK activity relative to the normal mean is consistent over many years (10) although in the normal population as a whole there is wide variation in NK activity. At birth, NK levels are about 50% of adult levels, which are reached in adolescence. The mean level for adult males is slightly higher than that of females and this is maintained into old age, in contrast to the age-related changes in NK activity seen in mice (10, 11). HLA-identical twins are identical in NK activity (10, 11) and it has been shown by Trinchieri et al (17) that low NK activity is associated with the HLA-A3, B7 haplotype, at least in males.

The effect of various disease states on NK activity has been studied by many investigators in recent years and a summary of some of those studies is shown in Table IV. In this Table, a number of genetic diseases have been listed according to their effect on NK activity and the number of patients studied. These diseases can be classified into several overlapping categories according to the information they can give about NK cells - those in which well-defined cell lineages are affected (Bruton's aggammaglobulinemia, Di George syndrome, severe combined immunodeficiency syndrome), those in which a discrete mechanism of action or enzyme function is affected (agammaglobulinemias, ML III), those in which the chromosomal abnormality is established (trisomy-21), and those in which a "selective" defect in NK function itself may indicate the biological relevance of NK cells (Chediak-Higashi syndrome). Some genetic diseases are not informative in this regard, either because the exact nature of the defect is not known (Nezelof's syndrome, ataxia telangiectasia) or because the disease affects many components of the immune system besides NK cells (X-linked lymphoproliferative syndrome).

A number of facts have emerged from the study of these conditions (for references see Table IV). First, severe combined immunodeficiency can affect B, T, and NK cells, leaving monocyte function intact. Although this places NK cells in the lymphocyte lineage, it is apparent that NK cells are distinct from both B and T cells, since normal NK activity has been described in patients completely deficient in either T or B cell development, and conversely, T and B cell function is normal in NK-deficient Chediak-Higashi patients. The fact that NK cell function is normal in patients with chronic granulomatous disease also suggests that the lineage of these cells is independent of monocytes and macrophages. Although the evidence is compatible with NK cells being a separate lineage, it is possible that they are a pre-T cell for the following reasons: patients with E rosette-positive SCID often have functioning NK cells (Table IV), cultured NK cells develop into cytotoxic "NK-like" cells with T cell markers (46), and the NK cells of patients with NK-leukocytosis or leukemia are usually T gamma cells (36, 47, 48).

Genetic and developmental diseases have also given useful information about the mechanism of cytotoxicity mediated by NK cells. Lymphocytes from many of the agammaglobulinemics referred to in Table IV were completely unable to synthesize immunoglobulin, either in vivo or in vitro, and yet were still capable of normal NK function in serum-free medium. These observations rule out an antibody-dependent cell-mediated cytotoxicity mechanism as a necessary mode of action of these cells and show that the nature of the NK cell receptor-target structure binding mechanism is not an Fc receptor-antibody interaction. As indicated in this report, it is not mediated by man-6-P receptors, either.

As discussed above, post NK-binding events leading to target cell lysis are still poorly worked out. Patients with Chediak-Higashi syndrome (CHS) have large abnormal granules in the cytoplasm, suggesting that such lysosomal granules are important in lysis. In CHS, NK cells are apparently present in normal numbers, but require prolonged contact before the target cell is lysed (40). It was proposed by Roder et al (49) that one of the early post-binding events leading to lysis was the release of oxygen radicals by the NK cells. This hypothesis is "testable" using mononuclear cells from patients with chronic granulomatous disease (CGD). Granulocytes and monocytes from these patients are unable to generate oxygen radicals and the patients suffer from frequent infections because of this defect. Several authors (Table IV) have now shown that patients with CGD have normal NK activity in the absence of any detectable radical formation, essentially ruling out this mechanism as being involved in NK cell-mediated lysis.

Some of the conditions listed in Table IV apparently affect the overall levels of NK activity by influencing the interferon-NK regulatory circuit (16). Lipinski et al (23) have described four patients with defective interferon production who were low in NK activity. This suggests that basal NK levels are dependent on constant interferon stimulation. The converse was observed in a group of trisomy-21 patients, in whom it was found that, while basal NK function was normal, NK responsiveness to interferon was increased and stimulated peripheral blood mononuclear cells produced more interferon in culture than normal. However, there was no difference in these parameters between children who had repeated infections and those who did not. Augmented levels of NK activity were also observed in patients with thalassemia major, while carriers of the condition were normal (45). It was noted that the levels appeared to correlate with the age of the patient and the number of transfusions which the patient had received. The possibility that repeated antigenic stimulation causes an elevation in NK should be kept in mind when assessing NK function in patients who suffer from repeated infections or who require regular transfusions since these factors could obviously alter NK cell activity independently of any effects caused by the underlying disease itself.

Perhaps the most interesting application of our knowledge of genetic diseases is the attempt to determine the biological significance of various cell types. A great deal has been written about the increased incidence of malignancy in patients with immunodeficiency diseases (50), but the majority of the diseases which were studied were not solely NK deficiency diseases and many were undoubtedly associated with normal NK function. The presence of normal NK activity in many T and B cell defects has been used to explain the fact that these patients do not have the incidence and types of tumors that might be expected if T cell immune surveillance were a fundamental form of host defence against malignant disease (20). Two of the genetic diseases listed in Table IV, Chediak Higashi syndrome and X-linked lymphoproliferative syndrome (XLP), are of interest because they are associated with a high incidence of malignant lymphoproliferative disorders. In XLP, a number of immunological defects exist (31), but in the Chediak-Higashi syndrome the non-granulocyte defect is primarily of NK cells (34). This implies that the NK cell system is responsable for surveillance against tumors of lymphoid origin.

In conclusion, the study of genetic and developmental diseases of man has led to many valuable contributions to our understanding of natural killer cells. The information which these studies have yielded has complemented that derived from experiments done with the blood of normal donors and it is only a matter of time before we fully understand what NK cells are, how they work and what their biological significance is.

ACKNOWLEDGMENTS

We thank Mrs. Mabel Chau and Mrs. Pamela Bandy-Dafoe for excellent technical assistance, Miss Helen Roughton for drawing bloods, Mrs. Christine Jackson for typing the manuscript. We also thank Dr. M. Partington for referring the patient to us and Drs. J. Roder and T. Oeltmann for helpful disussion. We would like to express our appreciation to the ML III patient's family for their cooperation, and to the patient himself whose cheerful and cooperative nature made working with him a pleasure.

REFERENCES

1. Kelly, T. E., Thomas, G. H., Taylor, H. A., McKusick, V. A., Sly, W. S., et al. (1975). Johns Hopkins Med. J. 137, 156.
2. Sly, W.S., and Fischer, H. (1982). J.Cell.Biochem. 18, 67.
3. Hasilik, A., and Neufeld, E. (1980). J.Biol.Chem. 255, 4946.
4. Laughton, C. (1984). Analyt. Biochem. 140, 47.
5. Forbes, J. T., Bretthauer, R. K., and Oeltmann, T. N.(1981). Proc. Natl. Acad. Sci. USA 78, 5797.
6. Jondal, M., and Pross, H. (1975). Int. J. Cancer 15, 596.
7. Pross, H. F., Baines, M. G., Rubin, P., et al. (1981). J. Clinical Immunol. 1, 51.
8. Werkmeister, J., Roder, J., Curry, C., and Pross, H. (1983). Cellular Immunol. 80, 172.
9. Rubin, P., Pross, H. F., and Roder, J. C. (1982). J. Immunol. 128, 2553.
10. Pross, H., and Baines, M. (1982). Int. J. Cancer 29, 383.
11. Roder, J., and Pross, H. F. (1982). J. Clin. Immunol. 2, 249.
12. Roder, J. C., Karre, K., and Kiessling, R. (1981). Prog. in Allergy 28, 66.
13. Waheed, A., Hasilik, A., and Von Figura, K. (1981). J. Biol. Chem. 256, 5717.
14. Pross, H. F., and Maroun, J. A. (1984). J. Immunol. Meth. 68, 235.
15. Varki, A., Reitman, M., Vannier, A., Kornfeld, S., et al. (1982). Am. J. Hum. Genet. 34, 717.
16. Herberman, R. B. (1982). Molecular Immunology 19, 1313.
17. Santoli, D., Trinchieri, G., Zmijewski, C. M., and Koprowski, H. (1976). J. Immunol. 117, 765.
18. Peter, H.H. (1983). Blut 46, 239.
19. Peter, H., Friedrich, W., Dopfer, R., Muller, W., et al. (1983). J. Immunol. 131, 2332.
20. Pross, H., Gupta, S., Good, R., and Baines, M. (1978). Cell. Immunol. 43, 160.
21. Koren, H., Amos, D., and Buckley, R. (1978). J. Immunol. 120, 796.
22. Nelson, D. (1980). In, Primary Immunodeficiencies, M. Seligmann and W.H. Hitzig (eds.), Elsevier/North-Holland Biomedical Press pp. 141.
23. Lipinski, M., Virelizier, J., Tursz, T., and Griscelli, C. (1980).Eur. J. Immunol. 10, 246.
24. Strannegard, O., Bjorkander, J., Hanson, L., and Hermodsson, S. (1982). Clin. Immunol. Immunopathol. 25;, 325.

25. Huwyler, T., Morell, A., Hirt, A.,et al. (1984). Clin. exp. Immunol. 56, 486.
26. Lopez, C., Kirkpatrick, D., Sorell, M., et al. (1979). Lancet ii, 1103.
27. Sirianni, M., Businco, L., Fiore, L., et al. (1983). Birth Defects 19, 107.
28. Buckley, R., Gard, S., Haynes, B., Sindel, L., Davis, K., Sampson, H., Ruff, M., and Koren, H. (1983). Birth Defects 19, 101.
29. Flomenberg, N., Welte, K., Mertelsmann, R., et al. (1983). J. Immunol. 130, 2644.
30. Sullivan, J., Byron, K., Brewster, F., and Purtilo, D. (1980) Science 210, 543.
31. Harada, S., Bechtold, T., Seeley, J., and Putilo, D. (1982). Int. J. Cancer 30, 739.
32. Roder, J., Haliotis, T., Laing, L., Kozbor, D., et al. (1982) Immunology 46, 555.
33. Haliotis, T., Roder, J., Klein, M., Ortaldo, J., Fauci, A., and Herberman, R. (1980). J. Exp. Med. 151, 1039.
34. Klein, M., Roder, J., Haliotis, T., Korec, S., et al. (1980). J. Exp. Med. 151, 1049.
35. Abo. T., Roder, J., Abo, W., Cooper, M., and Balch, C. (1982). J. Clin. Invest. 70, 193.
36. Van De Griend, R.J., Astaldi, A., De Bruin, H.G., and Weening, R. S. (1982). New Engl. J. Med. 306, 1369.
37. Brahmi, Z. (1983). Human Immunol. 6, 45.
38. Merino, F., Klein, G., Henle, W., et al. (1983). Clin. Immunol. Immunopathol. 27, 326.
39. Katz, P., Zaytoun, A., Lee, J. J., and Fauci, A. (1984). J. Immunol. 132, 571.
40. Targan, S., and Oseas, R. (1983). J. Immunol. 130, 2671.
41. Kay, H., Smith, D., Sullivan, G., Mandell, G., and Donowitz, G. (1983). J. Immunol. 131, 1784.
42. El-Hag, A., and Clark, R. (1984). J. Immunol. 132, 569.
43. Lopez, C., Kirkpatrick, D., Fitzgerald, P. A., and Et Al. (1982). J. Immunol. 129, 824.
44. Funa, K., Anneren, G., Alm, G., and Bjorksten, B. (1984). Clin. exp. Immunol. 56, 493.
45. Neri, A., Brugiatelli, M., Iacopino, P., Callea, V., and Ronco, F. (1984). Acta haemat. 71, 263.
46. Kedar, E., Ikejiri, B., Timonen, T., Bonnard, G., et al. (1983). Eur. J. Cancer Clin. Oncol. 19, 757.
47. Pross, H., Pater, J., Dwosh, I., Giles, A., et al. (1982). J. Clin. Immunol. 2, 126.
48. Ferrarini, M., Romagnani, S., Montesoro, E., et al. (1983). J. Clin. Immunol. 3, 30.
49. Roder, J. C., Helfand, S. L., Werkmeister, J. A., McGarry, R., Beaumont, T., and Duwe, A. (1982). Nature 298, 569.
50. Gatti, R., and Good, R. (1971). Cancer 28, 89.

Part V.

Cultured Cells with NK-Like Activity: Models for Analysis of NK Activity

NK CLONES: ISOLATION AND CHARACTERIZATIONS

Reinder L.H. Bolhuis

Department of Immunology
Rotterdam Radio-Therapeutic Institute, Rotterdam;
and Radiobiological Institute TNO, Rijswijk, The Netherlands

INTRODUCTION

Cells originating of separate lineages such as monocytes, macrophages, lymphocytes and natural killer (NK) cells can mediate various *in vitro* cytotoxic activities and they have been studied as to their occurrence, generation, lytic mechanisms and target cell specificity range. NK cells distinguish themselves from other lymphoid effector cells in that they need no prior purposeful stimulation, and their lytic function is not restricted by the major histocompatibility (MHC) complex. In contrast to cytotoxic T lymphocytes, "fresh" or resting NK cells are $T3^-$. The target structures which direct nonspecific lytic activities are, unlike allospecific CTL histocompatibility target cell structures, poorly defined.

Descriptions of the phenotypes of "fresh" or resting NK cells provided evidence of their heterogeneity. Moreover, they were shown not to mediate classic $T3^+$ cell functions. Later, it was shown that $T3^+$ lymphocytes could also mediate nonspecific cytotoxicity after allostimulation and, in contrast to the allospecific lysis, the nonspecific MHC nonrestricted cytolysis was not inhibited by anti-T3 or anti-T8 monoclonal antibodies (MAbs). All these findings added to the confusion as evident from the literature which attempts to describe the relationship of these various populations of cells capable of mediating nonspecific cytotoxicity.

With the introduction of the hormone interleukin 2 (IL-2) the possibility of cloning and expanding T cells became a reality. Moreover, the capacity to separate cells on the basis of their cell surface phenotype and to establish clones from these purified populations allows the analysis of the heterogeneity of cells that are or can be induced *de novo* to mediate nonspecific cytolysis, to study the role of the cell lineage specific structures and other

ISBN 0-12-341370-2

membrane antigens in their lytic functions, either specific, nonspecific or both, and to explore the various factors that regulate these functions. Thus, cloning of lymphoid cells allows further dissection of the immune regulation network and, in particular, to investigate the origins of nonspecific lytic effector cells and their mechanisms and specifities of target cell interactions.
It has to be realized that the process of cloning and expansion involves activation of the lymphoid cells and hence the triggering and regulating stimuli may differ from those acting upon resting (NK) cells. Moreover, prolonged culturing of cloned cells may ultimately be associated with changes in functions and chromosomal aberrations representing a "transformative process" and thus not always reflect normal physiological activities. Therefore, direct comparison with fresh (resting) or short term cultures is mandatory. These factors, however, can be studied and taken into account before drawing general conclusions.

From the summary of the presentations of this session of "The Second International Workshop On NATURAL KILLER CELLS", it will become evident that virtually all T cells can or can be induced to exert nonspecific, MHC nonrestricted cytolysis.

It will be shown that, in contrast to recent reports, cloned allospecific and thus MHC restricted $T3^+$ CTL can simultaneously mediate MHC nonrestricted cytolysis. The implications of the various finding regarding the lineages or subpopulations of a lineage to which the (cloned) effector cells belong and the involvement of various immune structures in the lytic processes will be discussed.

G. TRINCHIERI

Trinchieri reported on the purification of B73.1 positive cells by fluorescent activated cell sorting (FACS). He had obtained an 85 to 100% pure $B73.1^+$ population. T cells were then bulk cultured for up to one week and the cells were tested for their capacity to respond to PHA, Con-A or in a mixed lymphocyte culture reaction (MLR). Special care was taken to look for possible changes in the kinetics of the response or the dose requirements. In spite of this, no mitogen response of $B73.1^+$ cells was observed. To determine whether stimulation could be observed under conditions which were optimal for proliferation of cells as described by Golub, Bolhuis and others, $B73.1^+$ cells were cultured for 6 days. Use was made of a mixture of irradiated B-LCL feeder cells and lymphocytes which give synergetic growth promoting activity. At day 6 a strong induction of T8 antigens on the $B73.1^+$ cell derived cultures was observed. Moreover, Trinchieri showed that OKT10 and HLA-Dr and three T-cell activation antigens i.e. 5E9 to define the transferrin receptor, DR antigens on monocytes and activated lymphocytes and anti-Tac monoclonal antibody identifying the IL-2 receptor were expressed on 15-30% of control cells. Full expression of the same antigens was also detected in conditioned medium stimulated B73.1$^-$ T cells. Eighty to 90% of the cells in the B73.1$^+$ fraction constantly expressed the IgG-Fc receptor. About 60% of the cells expressed the activation antigens detected by anti-Tac, 4F2 and 5E9 monoclonal antibodies. No cells that coexpressed IgG-FcR and the Leu4 antigen were detected. About 45% of the FCR positive cells also expressed the DR 46% Tac (31%); all expressed the OKT10 antigen. Thus, the appearance of the activation antigens appeared to precede the maximum proliferation of cells in culture by about 24 hours. Trinchieri then looked into whether the activation antigens detected on the B73.1$^+$ cells after a six-day culture in the presence of Daudi cells and IL-2 containing conditioned medium were expressed on the actively cycling NK cells. Two-wavelength immunofluorescence experiments in which cell cycle analysis was done by using the propidium iodate staining technique combined with indirect immunofluorescence were performed. The results obtained showed that 90% of the cells expressing the 3G8$^+$ antigen were in the S/G2/M phase of the cell cycle and were HLA-Dr$^+$ and Tac$^+$. Fifty to 70% of these cells also expressed the antigen identified by 5E9. Antigens were expressed at lower density on cells in the G0/G1 phase. Fifteen to 20% of the total cells appeared to be in the S/G2/M phase of the cell cycle. A slightly higher proportion of cycling cells was observed among 3G8$^+$ cells, confirming that the 3H/TDR incorporation observed in the B73.1$^+$ cells was due to proliferation of NK cells rather than contaminating T cells.

Trinchieri and co-workers stated that transitions of B73.1$^-$ cells to B73.1$^+$ and vice versa were never observed nor induction of T3 antigens on B73.1$^+$ cells.

Discussion

The question of whether one looked here at activated killer cells (since Daudi stimulates the NK cells) or at NK cells raised and effectively addressed by stating that they represented "activated natural killer cells".

Ortaldo asked Trinchieri why he did not see any response to PHA, whereas such responses, although low, were seen and reported by him and others. Trinchieri favoured the view that the response to PHA was due to contaminating responder T cells.

H. KOREN

Koren reported on the work in his laboratory with Travis. The purpose of his study was to investigate the proliferative capacity of MAb purified subpopulations of NK cells. To this end, mononuclear cells after depletion of monocytes were isolated on Percoll followed by a panning technique using MAb to determine whether the proliferation in the NK cell fraction was confined to a particular subpopulation of NK cells. Other NK/ADCC specific

MAbs, OKM1 and the Leu-11b, which have a similar reactivity pattern as B73.1 MAb, were also used. The Leu-11b separated cells contained almost all of the NK cytotoxic activity. When the proliferative capacity in response to phorbol esters such as PMA, the lectin PHA and the lymphokine IL-2 were studied, they found that the Leu-11b$^+$ cells, i.e., the NK cells, did not proliferate. Leu-11b$^-$ cells even showed significantly more proliferation than did the unseparated cells. According to Koren, this may indicate a suppressor cell population in the Leu-11b$^+$ Percoll low density cell fraction. When using the OKM1 MAb for separation of cells, similar results were obtained. Some proliferative capacity was still observed in the OKM1$^+$ fraction, but this could most likely be ascribed to contamination of the OKM1$^+$ cells with Leu-4$^+$ or OKT3$^+$ cells and needs to be investigated by double-labelling studies. Koren suggested that the the lack of proliferative capacity may serve as a useful parameter for checking the purity of separated NK cells.

J. ORTALDO

Ortaldo reported on the generation of clones derived from large granular lymphocytes (LGL), their phenotypes, functional activities in terms of lysing of tumor cells and both NK-sensitive and NK-nonsensitive target cells and on lymphokine production by these clones. One of the general characteristics noted by most investigators is the limited life-span of human cloned lines. It would be of interest to immortalize cytotoxic clones so that the limitation of the growth (25-30 generations) can be overcome. The limited number of generations may not be an intrinsic property of the cells, i.e. it may be due to the culture conditions, since cloned mouse T-cell lines can be propagated for years. He also reported on the cloning of cells by either direct cloning from the LGL fraction or after further purification of the LGL fraction. Cells were plated at 0.5 cells per well and cultured in the presence of a low concentration of PHA and autologous feeder cells. A number of striking differences between CTL clones and those derived from NK cells was found. The most obvious difference was that the NK cells did not show MHC restriction. The cell surface expression of antigens on various clones derived from the LGL fraction was T3$^+$ and T8$^+$ as do CTL clones, but their lytic activity was not inhibited by anti-T3 and anti-T8 monoclonal antibodies (MAbs). It is known that allospecific cytolysis of human CTL clones is inhibited by anti-T3 and/or anti-T8. The major phenotype found was T3+, T8+ and/or B73.1+. Thus, a major difference is the expression of the IgG-Fc receptor recognized by the B73.1 MAb, as described by Perussia and Trinchieri. The authors also found some B73.1$^-$ LGL derived clones. In other cloning experiments, LGLs were selected for a positive reaction with a particular MAb by flow cytometry and grown in bulk culture for about 1 week and subsequently cloned. The overall plating efficiency appeared to be

15%. The capacity of the cells to lyse NK targets varied considerably when tested against the panel of 5 target cells. Those clones which were derived from B73.1$^+$ purified cells always showed the broadest target cell spectrum and thus exhibited almost unrestricted cytolysis, as reported earlier by Bolhuis and van de Griend. It was generally found that about 22% of the clones showed lysis against at least 3 of the 5 target cells tested. At this point, Ortaldo suggested the possibility that the low percentage of cytotoxic clones observed may be due to the fact that these cells were tested against the "wrong" target cell (thus, extending the panel of target cells may increase the percentage of clones showing lytic activity). Another possibility is that these cells had not yet differentiated due to a lack of factors necessary to induce cytotoxicity, as is known to be the case for CTL. The analysis of the specificity of cytotoxicity exerted by the LGL derived clones showed that, again, there was a wide variety in lytic patterns: some clones had a broad reactivity against many different target cells, and others a more restricted one.

In summary, of the 22% of cytotoxic clones, about 60% of them showed lytic activity against all of the target cells and 40% showed a restricted pattern of specificity, reacting against only 1 or 2 of the target cells tested. All of the B73.1$^+$ clones always simultaneously expressed antibody dependent cellular cytolysis (ADCC) as reported by Bolhuis et al. The lytic capacity of the clones appeared to be constant and there was no loss of lytic activity after a number of generations. Of importance is that cytotoxicity was determined after resting the cells to avoid lytic reactions due to activation by medium components. Only with a few clones did Ortaldo observe a change in the reactivity, i.e., the capacity to lyse a particular target cell was lost. The opposite was never observed; i.e., no gain of lytic activities occurred. One of the suggestions made by him was that clones which lost their lytic activity were about to die. Thus, there was a great heterogeneity in terms of both the target cell spectrum which could be lysed by a particular clone as well as the phenotype. No correlation was found between a particular phenotype and the lytic pattern of clones.

Since a number of investigators had already reported that lymphokines such as interferon (IF) can be produced by LGLs, lymphokine production by clones derived from the LGLs was also studied. A possible correlation between the type of lymphokine produced and the phenotype of clones was also investigated. No such correlation was found. Most of the clones which were tested produced gamma inter-feron in response to either target cells or lectins. IL-1 production was observed in response to LGLs and LPS lectin. Production of lectin, IL-2 and BCGF was also induced by PHA lectin. One clone was shown to produce more than one lymphokine irrespective of lytic activity. Gamma IF was even produced without stimulation, although the level of production was increased after stimulation. Production of both IL-1 and IL-2 by one particular clone was also observed. This needs to be verified, since these data are quite preliminary. As far as correlations with phenotype could be made, it appeared that the cells which expressed T3 antigens did not produce IL-1 but did produce IF and IL-2.

Discussion

Discussing his own data, he referred back to the question raised by Herberman that the receptors on NK cells may or may not be clonally distributed and whether these clones have specific, genetically stable capabilities in terms of producing lymphokines. Thus, it could be anticipated that if a cell goes through various differentiation states, it may change both its lytic pattern and its lymphokine production. The question of whether a particular clone changed in its capacity to produce a particular lymphokine as a function of time or even produced one lymphokine at one time and another later was then raised. This could indeed be the case: clones can "flip-flop" and thus produce IL-1, stop producing it and start to produce IL-2. At the same time, gamma IF could continuously be produced. This observation was made in a few clones only.

The small percentage of clones which exerted cytolytic activity was also discussed and the possibility that this could be due to the 4-6 hour cytolytic assays was considered. Especially Saksela had had the experience that, when the assay was extended up to sixteen hours, lysis could be observed at this time point, whereas almost no lysis was seen after 4 hours of effector cell/target cell incubation in a ^{51}CR assay. There was also concern about the low plating efficiency and the possibility that other than NK cells were cloned in this assay was mentioned. Ortaldo refuted the argument, saying that these starting populations were about 99.5% pure. He also stated that, when subpopulations of LGLs were isolated, different plating efficiencies could be found, i.e., as high as 25%. Thus, although Ortaldo shared the concern, he believed that it was not "non-NK" cells which they had cloned. Brooks suggested that the possibility that selective outgrowth of non-NK cells took place, even in a highly purified population as used by Ortaldo, should still be considered. Ortaldo replied that micro manipulation of effector/target cell conjugates and subsequent cloning of the target cell binder yielded the same types of clones in the hands of Zagury.

One could argue, however, that binding cells are not all killer cells. The point of heterogeneity in lytic activity and phenotype remained a subject of discussion throughout the session.

T. TIMONEN

Timonen reported on the results obtained with clones derived from healthy donors. OKM1 positive or NK 9 positive cells were isolated by FACS and deposited by the cell sorter at 1 cell per well. The cloning efficiency appeared to be in the range of 10 to 20%. Thirty-percent of the clones obtained appeared to be cytolytic in a 16-hour 51-Cr release assay at high effector/target cell ratios. Quite unexpectedly, rather strong cytotoxicity against anchorage dependent target cells was also observed. Thirty-percent of the cytotoxic clones studied appeared to exert ADCC. LGLs of tumor patients were also cloned. In a particular patient with an ovarian carcinoma, clones were derived and a variety of cytolytic activities were observed, including lysis of the autologous tumor cells. One of the obvious aims of cloning cytotoxic T cells, especially those with anti-autologous tumor reactivity, is to use these cells as an immunotherapy treatment modality by passively transferring these clones into the patient. Therefore, cytotoxic clones were tested on either carcinoma cells obtained from the plural effusion of cancer patients or on K-562 target cells. He observed a tremendous loss (about 70%) in the lytic capacity of these effector cells after overnight incubation against both the autologous target cells and NK-sensitive target cells. Even when a clone was reactive against K-562 but not the autologous tumor, the autologous tumor cells could still inactivate the effector cells with respect to reacting against the K-562 target cells. It was concluded that activation of the lytic machinery was not necessary to have abrogation of the lytic capacity against another type of target cell but that cell to cell contact is necessary for the inactivation process. In a search for suppression of effectors, 50% of the 3-day culture supernatant of tumor cells which were obtained from the ascites and cultured in medium was used. A slight inhibition was observed after overnight incubation of effector cells in the 50% tumor cell supernatant. The ascitic fluid could inactivate other cytolytic clones, irrespective of their target cell specificity. Only partial restoration of the cytotoxic capacity could be achieved by using IF. The strong inhibition of cytolytic activity was effected by proteins with molecular weights in the range of 20 - 30,000 daltons and the inhibition of lytic capacity occurred for all types of effector cells, including fresh LGLs and autologous cytotoxic clones, irrespective of their target cell specificity.

To investigate whether the inactivation process was a pre- or post-target cell binding event, fresh LGLs from a normal donor were used as effector cells. The effector/target cell binding percentage decreased from 42% to about 12%, which coincided with a drastic decrease in the lytic capacity. Timonen drew special attention to the fact that there was also a decrease in the percentage of purified LGLs after overnight incubation with as little as 5% of the tumor cell culture supernatant. Nevertheless, no toxic effects were apparent.

Discussion

From the discussion, it became clear that others also found inhibitory substances in the about 20,000 molecular weight fraction in human serum and this appeared to be associated more with carcinomas and sarcomas rather than with lymphomas. The question of whether or not the ascitic fluid would affect the expression of IL-2 receptors was raised. Timonen found that the ascitic fluid actually had good growth promoting activity but did not maintain cytolytic activity. The effector cells became immobile as a result of the inactivation procedure and therefore this represented a true inactivation procedure.

The possibility was raised by Herberman that it could be the shedding of tumor target cell structures by the tumor cells. Although he was not excluding this possibility, Timonen considered it rather surprising that this would then inactivate the (lytic activity of) cells against a range of other tumor cells which were lysed by the effector cell, and one had to assume that different target cells all shared that structure.

DR. GAIL

Gail presented data in relation to the lysis of fresh lymphoreticular tumor cells. His group studied patients with lymphoma, especially centrocytic lymphomas, and acute lymphoblastic leukemia. Their main interest was to determine whether the lymphoblasts were susceptible to lysis by fresh NK cells. He also presented data on patients with chronic myeloid leukemia (CML) and chronic granulocytic leukemia (CGL). Philadelphia chromosome positive CML blasts were the most NK susceptible tumor target cells of hemapoietic origin. In CML patients in myeloid blast crisis, there was a very strong correlation between the number of blast cells and the percentage of lysis observed. If they purified the blast cells, the percentage lysis increased; thus, one apparently purifies the NK susceptible target cells. After determining the susceptibility of blast cells to NK cell lysis, it became of interest to investigate the NK capacity of cells of these patients in view of its possible relationship to the course of the disease or the control of it. The low lytic activity observed could be explained on the basis of the dilution effect of the granulocytes present in the blood of these patients. NK cells were thus purified by making use of the Leu-7 (HNK-1) MAb. At the time of their studies, this was the best MAb available for the identification and isolation of NK cells. Peripheral blood lymphocytes of CML patients were purified over Ficoll and the nonadherent cells were then isolated and divided into the E-rosette positive ($E\text{-}RFC^{+}$) and $E\text{-}RFC^{-}$ subsets. A good correlation between the percentage of $HNK\text{-}1^{+}$ cells and the lysis of K-562 was found in healthy donors. The percentage of $HNK\text{-}1^{+}$ cells was greatly increased in the $E\text{-}RFC^{+}$ fraction of the CGL patients and there was a very low degree of contamination with granulocytes in the E-RFC$^+$, HNK-1$^+$ cells. The E$^+$, HNK-1$^+$ cells were purified by using the FACS-IV cell sorter to avoid possible inhibition of NK cell function by (neoplastic) granulocytes as reported by Kay et al. The HNK-1$^+$, E-RFC$^+$ lymphocytes of the CML patients were still very defective in their capacity to kill K-562 cells in comparison with those from normal donors. Cells from the HNK-1$^+$ fraction were cloned by limiting dilution, using irradiated lymphocytes as feeder cells.

Four healthy donors were used and clones were obtained from the HNK-1$^+$ cell fraction. Thirty-six clones were obtained (cloning efficiency, 3%) and it was remarkable that in about 3 weeks of culture the HNK-1 marker was lost from the surface of these cells. A very efficient lysis (up to 70%) at low E/T ratios was observed. Thirty-four clones were derived from the patients (cloning efficiency, 1%) and they could be divided into three groups. Cells of clones in the first group were low in OKT-3 and in HNK-1 and were also effective killers of K562. The second group of clones contained cells that showed intermediate kill and increased T3 Ag expression. Cells of the third group were not capable of lysing K-562 and showed high T3 Ag expression and were also strongly HNK-1$^+$. One patient went into blast crisis after the clones were established. It was found that the autologous and allogeneic clones as well as fresh NK cells from CML patients could lyse the blast cells. A correlation was found between the per cent lysis of K-562 and that of blast cells; i.e., if K-562 was not lysed, the blast cells were not lysed either.

Discussion

An interesting point raised for the discussion by Gail was that the Leu7$^+$ cells which also express T3 antigen on their surface were not capable of lysing. Thus, it might well be that these were not NK cells at all. One hypothesis could be that the Leu-7$^+$ cells represent immature NK cells. Gail asked Saksela whether he thought that immature T3$^+$ Leu-7$^+$ NK cells would fit into the differentiation scheme. Saksela replied that it might well be that the effector cells were inactivated as a result of the presence of neoplasm, since lymphoma patients have reasonable numbers of LGL in their peripheral blood but low NK activity. Although Leu-7$^+$ cells in the bone marrow are found among the NK-9$^+$ cells, they represent only 20% of this cell fraction. However, they could not conclude at this point whether these were Leu-7$^+$ or Leu-7$^-$, NK9$^+$ cells. The argument could also be turned the other way around, however, i.e., Leu-7$^+$ cells may rather represent a more mature population: the Leu-7$^+$ and Leu-11$^-$ cells are not capable of lysis and cannot be induced to lyse K-562 cells by using IL-2 or IF.

Bolhuis then pointed out that shortly after bone marrow transplantation high numbers of Leu-7$^+$ cells (up to 60%) are found in the peripheral blood of bone marrow recipients. The discussion led to the conclusion that these cells may be of interest, but that at present no function can be ascribed to them.

R.L.H. BOLHUIS

Bolhuis summarized the data on the clonal expansion of human B73.1$^+$ NK cells which exert strong ADCC cytotoxicity and occasionally lectin dependent cytotoxicity (LDCC). He emphasized that in spite of the tremendous attention which is received by the NK cells, they still continue to be difficult to define. The fact that NK cells are multifunctional, i.e., they lyse a wide variety of tumor target cells, exert operationally defined distinct lytic mechanisms, lyse virus infected cells and may have regulatory activities in hemopoiesis as well as for the production of various kinds of lymphokines. This suggests their general importance and justifies the continued attention given to these cells.

To clone NK cells, Bolhuis and co-workers used the procedure of taking Ficoll-isolated lymphocytes which were depleted of monocytes by adherence techniques and separated the cells by using the B73.1 MAb, provided by Trinchieri, for purification of the B73.1$^+$ and B73.1$^-$ cells by FACS. Clones were then derived by limiting dilution of the B73.1$^+$ and B73.1$^-$ cell fractions. Use was made of a culture system which was especially developed for the rapid expansion of either cloned or noncloned lymphocytes of T-cell origin or related lineages. As specified earlier by Trinchieri, the system consisted of a mixture of either autologous or allogeneic irradiated lymphocytes and B-LCLs. It appeared that the clones which were derived from the T3$^+$ B73.1$^-$ fraction were more heterogeneous with respect to their target cell repertoire. Most of the clones derived from this fraction showed absolutely no lytic activity, especially when cultured in the presence of lectins, although some did clearly show cytotoxicity against NK susceptible target cells and NK resistant B-LCLs. Some of these clones were also capable of LDCC. The phenotype of the clones derived from the B73.1$^-$ fraction appeared to be T3$^+$, T4, or T8$^+$, HNK-1$^-$ T11$^+$, IA$^+$ E$^+$ and EA-γ-RFC$^-$. The T3$^+$, B73.1$^-$ cell derived clones also had azurophilic granules in the cytoplasm, although less abundantly than clones derived from the T3$^-$, B73.1$^+$ fraction. The phenotype of the clones which were derived from the B73.1$^+$ fraction was also extensively analyzed. Most of the clones derived from this fraction were B73.1$^+$, T3$^-$, T4$^-$, T8$^-$, T11$^+$, OKM1$^-$, HNK1$^-$, E$^+$, EAγ$^+$ as determined at 4 to 5 weeks after establishment of the clones. One was B73.1$^+$, T3$^-$, T4$^-$, T8$^-$, HNK1$^-$, T11$^-$, E$^-$, EAγ, RFC$^+$. The target cell spectrum was also broad. Both types of clones were derived from the B73.1$^+$ fraction. Another cytotoxic clone derived from the B73.1$^-$ fraction was B73.1$^-$, T3$^+$, T4$^-$, T8$^-$, HNK1$^-$, T11$^+$, E$^+$, EAγ, RFC$^-$.

Bolhuis then emphasized that special care should be taken in using the term LGL clones. First of all, because of the presence of azurophilic granules in both B73.1$^+$ and B73.1$^-$ cell derived clones, independent of shared lytic characteristics, and, secondly, because no HNK1 or OKM1 antigens were found on these cloned cells, two surface markers which are generally present on "fresh" LGL. Thus, the B73.1$^+$ clones exhibited multiple lytic activities, a stable phenotype and showed a broad target cell spectrum, including freshly biopsied tumor cells.

Data concerning the nonspecific cytolytic effect of cloned T3$^+$ cells which were derived from mixed lymphocyte culture (MLC) responder cells were also presented. Extensive analysis of the clones derived from the MLC responder cells were performed. Cloning and expansion were done as for the T3$^-$ B73.1$^+$ cells, with the exception that PHA was added to the medium. The data were summarized because these clones were used only to illustrate how heterogenous the effector cells are which can lyse NK-sensitive and NK-nonsensitive tumor target cells. Three OKT3$^+$, OKT8$^+$, E-RFC$^+$ EA-γ RFC$^-$ clones with cytotoxic activity against the stimulator cell and one noncytolytic clone were expanded for over 50 generations. One clone, G9, showed allospecific lysis of Cw3$^+$ lymphocytes and a Cw3+ B-LCL. Three cytolytic clones (G9, D11 and A3) exerted cytotoxicity against the stimulator B-LCL, human plasma cell leukemia derived cells (LICR-LON) and short-term cultured, freshly biopsied melanoma cells (O-Mel).

Interestingly, four other EBV transformed B-LCLs unrelated to the stimulator B-LCL were not lysed. The clones also effectively lysed NK sensitive target cells such as K-562. Other NK sensitive target cells such as Daudi were killed only when they were pretreated with PHA; thus, these effector cells showed LDCC. The cytolytic activity against B-LCL, the LICR-LON and O-Mel but not against the NK sensitive K562 or PHA treated target cells (LDCC) was inhibited by MAbs against Class I antigens. The clones showed differential susceptibilities to inhibition of lysis by various OKT3 and anti-MHC Class I and II Mabs (data elsewhere in this issue).

Therefore, the investigators concluded that human CTL cell clones derived from MLCs can exert multiple lytic activities in the sense that the operationally defined lytic mechanisms differ at least at certain stages of the lytic cycle and that they, like the T3$^-$ B73.1$^+$ derived clones, can also exert nonspecific cytolysis against NK sensitive and nonsensitive target cells, albeit with a narrower target cell specificity range and lower lytic activity. This, however, does not imply that the mechanisms of induction of lysis are identical. The unequivocal demonstration of true allospecific CTL clones being capable of lysing NK susceptible targets again illustrates that the term "NK" or "NK-like" clone is not a valid designation for these clones.

Another observation of interest was that the relatively narrow target cell spectrum of CTLs can be broadened and cytolytic activity induced or tremendously augmented by a relatively short incubation of the T3$^+$ CTLs with either the lectin PHA or the anti-T3 MAb. Together with other observations made in his laboratory, it was concluded that T3 and PHA act through the T3 antigen receptor complex and the T3 antigen is the site which delivers their activation signals.

Cells derived from three patients with Tγ lymphocytosis were also cloned. It is known that patients with Tγ lymphocytosis represent a heterogeneous group within the group of chronic T-proliferative disorders and proven malignancies. Three patients showed an expansion of phenotypically and functionally different types of Tγ cells: one had $T3^{+}$, $B73.1^{+}$, $T4^{-}$, $T8^{+}$, $OKM1^{+}$ Tγ cells exerting ADCC and NK cell cytotoxicity; another had $T3^{+}$, $B73.1^{-}$, $T4^{-}$, $T8^{+}$, $OKM1^{-}$, $ADCC^{+}$, NK^{-} and the third $T3^{-}$, $B73.1^{+}$, $T4^{-}$, $T8^{-}$, $OKM1^{+}$, $ADCC^{+}$, NK^{+} cells.
Two categories of Tγ patients have been reported, namely, the Tγ killer and the Tγ K/NK type with suppressor cell function, K cell function (ADCC) and both killer and natural killer selectivity. Data on one patient with Tγ killer and two with the Tγ K/NK type were presented.
Although the proliferative response to polyclonal mitogens was almost nonexistent in lymphocytes from patients with the Tγ lymphocytosis (as reported earlier by them and which is in agreement with the reactivity of normal Tγ cells), induction of the expression of IL-2 receptors and proliferation of the lymphocytes of each of the patients was achieved by using the culture system already described for the expansion of clones from $T3^{-}$, $B73.1^{+}$ and $T3^{+}$, $B73.1^{-}$ MLC responder cell populations. Thus, the absence of a proliferative response to polyclonal mitogens in PBL does not imply that the lymphocytes derived from these patients cannot be expanded in vitro. Morphologically, most (32-80%) of the lymphocytes of these patients had the morphological appearance of LGL. Twelve clones (eleven were noncytotoxic) were obtained from patient 1 and the phenotypes were $T3^{+}$ $B73.1^{-}$, in spite of the fact that the majority of such lymphocytes in fresh peripheral blood of this patient were T3+ B73.1+.

Also 12 clones were derived from patient 3 and the phenotypes of these clones were identical to that of the fresh peripheral blood lymphocytes of these patients: $T3^{-}$, $B73.1^{+}$ or $T3^{-}$, $B73.1^{-}$. All of these clones exerted activated killer cell activity against a wide variety of tumor lines and freshly biopsied melanoma tumor cells and showed ADCC. Although all three patients showed increased numbers of LGL as compared with normal donors, activated killer clones were easily generated from patient 3 but not from patient 1, although 80% of the lymphocyte population of this patient consisted of LGLs. The twelve $T3^{-}$ cytotoxic clones of lymphocytes from patient 3 showed quantitative differences regarding the binding to the B73.1 MAb as determined by cytofluorometry. No such variations in the percentages of EA-gamma-RFC or the lytic activities against an array of freshly biopsied or tissue cultured tumor cells was seen and ADCC activity was also found in all clones. The existence of patients with expansion of various types of NK cells thus provides information on the heterogeneity of NK cells and their capacity to proliferate in vitro and additional data on the characteristics and origin of lymphocytes in patients with Tγ lymphocytosis.

Discussion

After the presentation of Bolhuis, the question arose of how Ortaldo reconciled his data with those just presented. Ortaldo stated that when they sorted B73.1$^+$ cells they could reproduce the findings reported by Bolhuis. When they also started with B73.1$^+$ cells, T3$^+$ clones were indeed very rare. In the clones derived from B73.1$^-$ LGLs, more heterogeneity was seen. One could argue that either the LGL is not an absolute criterion for NK cells or that NK cells are very heterogeneous in phenotype.

Ortaldo stressed the point that if clones are derived from B73.1$^-$ LGL cells which resemble the "null" cell, i.e. T3$^-$, T4$^-$, T8-, these LGL may be rather immature cells and thus only a small percentage of the LGL cell population.

Herberman raised the possibility that, since the group of Bolhuis used direct FACS sorting, the majority of the cells in the B73.1$^-$ fraction were T cells, and the chance of obtaining clones with the characteristics described by Ortaldo would be very low indeed. This, however, appeared not to explain their findings, because in that case, as Bolhuis replied, these types of clones would have emerged from the B73.1$^-$ fraction. Moreover, as Ortaldo stated, the percentage of cytotoxic cells was very low indeed, so one would have to again assume a much greater plating and cloning efficiency in comparison with the B73.1$^-$ LGLs to explain the difference in types of clones obtained. Bolhuis et al. observed that both T3$^+$ and T3$^-$ NK cells have similar plating efficiencies (on the average, 5%); however, if PHA was added, the cloning efficiency of the T3$^+$ cells but not T3$^-$ was dramatically increased by up to 30-40%. The question of why no HNK-1 was observed in these clones, this in view of the fact that about 60% of the B73.1$^+$ cells express the HNK-1 antigen, was then raised. However, it has been observed that the HNK-1 marker is rather unstable in tissue culture of HNK-1$^+$ cells and that it disappears from the surface after a number of weeks in culture.

Saksela asked whether a variation was seen in the phenotype between the clones of one patient. Such variation was seen: T3$^-$, B73.1$^+$ clones were relatively the easiest to clone. Cells of one of the 12 B73.1$^+$, T3$^-$ clones from patient 3 expressed T8 on the surface, whereas the other 11 clones were T4$^-$ T8$^-$ and T3$^-$. It was reemphasized that there was a great heterogeneity in the density of the B73.1 antigen on the surface of these clones, i.e., varying from 0% to 92%, and in varying intensities. Saksela then asked for a comment on this: in view of the malignancy, one would expect the malignant cells of the peripheral blood to be "cloned" in vivo. So, if one would then, so to speak, clone from this "cloned" population what one then obtains are the subclones which should be identical. Bolhuis replied that it was still under debate whether Tγ lympho-cytosis is a true malignant disease or perhaps relatively benign. The follow-up period of these types of patients, however, has not yet been long enough to make conclusive statements. Therefore, this remains to be established. Moreover, the possibility that residual normal $T3^-$, $B73.1^+$ cells were cloned could not be ruled out. A detailed review of these types of patients by Reynolds and Foon will appear in the December issue of Blood (1984).

Pross asked the question of whether granulocytes of the patient after incubation in autologous serum were lysed by the clones or by the fresh blast cells of these patients, this in view of the recurrent infections and granulocytopenia that he also observed in his patients with Tγ lymphocytosis. This type of experiment had been considered, but had not yet been performed due to the limited number of cells available.

Koren came back to the question of whether these Tγ lymphocytes are malignant cells and suggested chromosome analysis might answer this question. Chromosome analysis was performed in some cases including the patient of which cytotoxic clones were derived, but no aberrations were found. If so, it would also have answered the question of whether the clones were derived from a normal subpopulation of lymphocytes still present or from the Tγ cells involved in the disease. Bolhuis stated that it will be of interest to investigate identical chromosomal aberrations which can be identified in different clones of an individual patient, to solve the problem of malignancy.

C. BROOKS

In the final presentation, Brooks opened by stating that four years ago cytotoxic T cells were viewed as functionally differentiated end cells. Now, we know that cytotoxic T cells will undergo changes during culture, some of these being reversible and some irreversible. He showed that antigen specific T cells can be driven by high concentrations of IL-2 through a whole pathway, i.e., they can acquire nonspecific cytotoxicity against only NK sensitive targets and, at a later stage, become promiscuous killers, i.e., any type of target cell, including (autologous) PWM blasts and NK resistant tumor cells, were lysed. The phenotype of the cells appeared to be $Thy1^+$, $Ly2^+$ but also negative cells had been observed.

The change from an antigen specific cytolytic stage to one where both the antigen specific and NK susceptible target cells were lysed appeared to be reversible upon removal of the IL-2. Thus, it appeared that, under the influence of high concentrations of IL-2, these cells became multifunctional and, if further driven by IL-2, they even lost the antigen specific cytolytic activity and antigen induced proliferation. This latter change was irreversible.

The regulation of the transition into these various stages was also considered. The progress from the antigen specific to the NK specific state and subsequently into promiscuous killer cell state occurred very quickly and the cells entered into a stage in which they were radically different biochemical entities from the starting population, i.e., new antigens appeared on their surfaces.

The answer to the question of whether the promiscuous "NK-like" cells could be brought back to the resting antigen specific stage was "no", even when antigen specific stimulator cells were used. The cells express the unusual asialo GM2 antigen which is different from the GM1 antigens present on the cell surface at the beginning of the differentiation pathway. Thus, he reported that new surface antigens of high molecular weight proteins appeared and these changes were rather dramatic. The transition from one stage to the other appeared to be a general phenomenon for the mouse CTL clones. In spite of the fact that all clones acquired the capacity to kill nonspecifically, the kinetics for each clone were very different and varied from clone to clone in terms of days and for others weeks before they acquire nonspecific cytotoxicity.

Brooks pointed to the fact that the variation in the specificities observed between clones was due to an intrinsic variation in sensitivity to activation for nonspecific cytolytic capacity. Actually, he said, "flip-flopping" between a number of activation stages was observed. At the early stages of the pathway of the activation process, the lytic activity can be boosted by natural IFs. Bèta IF was an order of magnitude more potent than alpha IF but gamma IF had no effect. The induction of NK specificity on the antigen specific CTL clones did not affect the antigen specific cytolysis. The NK lysis by these antigen specific clones could also be induced by using crude as well as highly purified IL2 and the induction by IL2 was not affected by adding anti-IF Abs. The specific stimulator antigen could not be used for the acquisition of NK cell lysis and third party cells had no effect.

Discussion

The possibility that if one would clone end terminal cells at low concentrations of IL-2, the subclones would display the same nonspecific cytolytic activity or, alternatively, only a minority of the cells would be carried further was discussed. Brooks found that the clones required this high concentration of IL-2 in order to be expanded or because another growth factor which was limiting was necessary. The subcloning at relatively low concentration of IL-2 was consequently not performed.

Ortaldo asked whether, if these cells had reached the stage of promiscuous killing, the changes continued upon repeated testing. The group had not looked at further changes in specificity going back from the promiscuous stage, although this would indeed be very interesting. They did investigate whether the cells could repeatedly go back and forth through these cycles and vice versa, i.e., from the antigen specific resting state to the NK specific lytic state; and this was indeed found. It could even be done several times.

Brooks raised a question concerning the inhibition of lysis with MAbs against LY2 in relation to the work presented by Bolhuis. They had not done such experiments except in the early stages of their research where they were studying the promiscuous killer activity and found that cells that carried the $Ly2^+$ were inhibited by anti-Ly2 Ab. However, other investigators have found that $Ly2^+$ clones could acquire NK activity, but "NK-like" activity was not inhibited by anti-Ly2 Abs. The interpretation of the data, however, was complicated by the fact that there were $Ly2^+$ clones which could not be inhibited by the anti-Ly2 Abs, not even when using the antigen specific target cells. Thus, there apparently also exists a heterogeneity in the susceptibilities to inhibition of lysis of mouse cloned T cells. No change in Ly2 expression within a particular clone was observed.

Saksela asked for information on the chromosome pattern in the clones. Brooks had the impression that the irreversible change to promiscuous killer cell activity was associated with chromosomal aberrations.

Herberman was interested in the time course of the events in the case of interferon and IL-2 boosting and inductional cytolysis, in view of the fact that a decrease in cytotoxic capacity has been observed on resting the cells in medium without the IF or IL-2 activation signals; i.e., a rapid decrease in lytic activity was seen after a short incubation in normal medium. Their observations were that indeed over a period of 12 hours there is a decrease in cytolytic activity. It is important that no trace amounts of IL2 are left in the medium used for "resting", because trace amounts of IL-2, although not inducing lytic activity, could maintain induced activity. The fact of whether induced activity on antigen specific clones was a stable feature of these cells remained to be established.

Ortaldo stated that, in the few CTL clones he had been looking at, the nonspecific killer activity appeared to decrease if they rested the cells in medium without an activation signal. This might be an unstable feature therefore.

Eva Klein asked whether a good criterion could be provided for the stage where there is a transition from the antigen specific resting state to the antigen specific splenic NK pattern.

Brooks specified that there were two elements to the promiscuous stage. There is a point from where they can go back to the antigen specific memory CTL and one where the transition is irreversible. During the promiscuous stage of cytotoxicity, prior to the irreversible stage, he observed that autologous LPS^+ blasts but not third party LPS^+ blasts were lysed. After the irreversible transition, however, all types of LPS^+ blasts were lysed.

Brooks also mentioned that it could be that antigens are capable of inducing NK type of killing in the cells. One would really have to ensure that after antigen stimulation the responder cells are "rested", because IL-2 production may be induced as a result of antigen stimulation and be involved in the regulation of the induction of nonspecific killing.

Herberman asked whether the changes in the lytic pattern of these effector cells could be correlated with studies on binding target cells. Thus, a loss of a certain target cell specificity might involve the loss of the binding capacity to that particular target cell.

The binding studies, although of particular interest, were not performed because of the already stickiness of the effector cells and this may have yielded spurious results. They did, however, investigate the inhibition of the antigen specific target at the antigen specific stage using NK sensitive targets and no inhibition of the antigen specific target cell lysis was observed. They then concluded that the acquisition of "NK-like cytotoxicity" is associated with the acquisition of a NK cell receptor on these cells.

In reply to the question concerning the NK-1 antigen expression asked by Dennert, Brooks stated that they had published on the NK-1 and, depending on the genetics, NK2 is also expressed on CTL. Four antisera, three from Burton and one from Pollack, had been used. Dennert stated that the antiserum NK-1 from Pollack is much more specific. They found that these represent T-killer clones and this was contrasted with the presence of the NK-1 antigen on the NK clones generated in the laboratory of Dennert. Using the antiserum obtained from Burton, reactivity with both types of clones (T clones and NK clones) was seen. Brooks replied that the antigen expression was indeed very low.

Dennert discussed the possibility that they were indeed looking at different types of cloned cells. The question of whether the clones studied by Brooks can exert the ADCC like the mouse NK clones reported by Dennert and human NK clones by Bolhuis was raised. This was reminiscent of the data presented during this session that in the human system the NK cell derived clones without exeption exerted ADCC, whereas the CTL derived clones, although capable of nonspecific killing, did not show ADCC. This aspect was not studied in detail by Brooks.

As to the relationship between NK cells and other types of effector cells, Brooks commented that there was no doubt regarding the fact that the cells differentiated from CTL. However, he believed his system offered the possibility of studying the regulation of NK cell activity and suggested a close relationship between T cells and NK cells.

SUMMARY AND CONCLUSIONS

The studies reported by Trinchieri and Koren clearly demonstrate that NK cells not only phenotypically but also functionally differ from "classic" T cells ($T3^+$) in terms of their spontaneous lytic activity but also mitogen unresponsiveness. Although they do not express the T3 antigen which serves as the site for transmission of activation signals (antigens; PHA; anti-T3 etc.), they can be induced to express T-cell activation antigens such as the Tac antigen, transferrine receptor and DR antigen.

The fact that no cells were ever found which expressed Leu-4, T3 or IgG-FcR illustrates their separate origin. The appearance of the activation antigens preceded the maximum proliferative phase of the NK cells.

The finding that the OKT8 antigen is expressed on a proportion of the $B73.1^+$ "fresh" NK cells already indicates the phenotypic heterogeneity of these cells. This becomes even more obvious from the reported results on cloning of lymphocytes. Results were presented showing that MHC nonrestricted cytolysis was observed by clones derived from LGL which differed, both in phenotype as well as in functional activities. No correlation between phenotype and functions were observed. These clones showed heterogeneous cytotoxicity patterns, i.e. some clones lysed only one or a few NK susceptible target cells whereas others lysed a wide range of target cells including NK resistant target cells and the effector cells could either be $T3^+$ as well as $T3^-$. When NK cells were isolated using the B73.1 McAb, recognizing the IgG-FcR, the lytic pattern was rather homogeneous, involving a broad range of target cells. All clones also exerted ADCC. This of course raises questions about the clonal distribution of putative NK cell receptors or interactions sites, i.e., whether ubiquitous or nonubiquitous target structures are recognized. In order to answer that question it is necessary to decide whether one, for instance, considers LGL derived $T3^+$ clones as NK cell derived. This question becomes even more compelling since cloned allospecific T cells ($T3^+$, T8 [and sometimes $T4^+$]) thus exerting MHC restricted target cell lysis, can _also_ exert MHC nonrestricted cytolysis.

Thus, the capacity of effector cell to lyse K562 cells does not allow the conclusion that this cell is an NK-cell or NK derived cell. Unless one assumes the existence of $B73.1^-$ LGL with NK activity that may change their lytic pattern during the process of prolonged clonal expansion, this cell may rather be a classic T-cell derived and not NK-cell derived.

This, however, does not affect the importance of the findings reported that lymphoid cells from various lineages, including $T3^-$ and $T3^+$ cells, are capable of (acquiring) nonspecific lytic activity, both in the human and mouse system. These cells may have important biological functions if occurring _in vivo._

Coming back to the question of whether or not the receptors on effector cells for nonspecific lysis of (tumor) target cells are clonally distributed, it is clear that the cold target cells inhibition studies support that idea. Whether the receptor on B73.1$^+$, T3$^-$, ADCC$^+$ derived NK-cell clones are also clonally distributed is at present under investigation. The fact that the 13.1 MAb described by Newman directed against the T200 antigen inhibits the lysis of K562 but not other target cells favours the involvement of different interaction sites or, alternatively, of different epitopes on one particular interaction site in the recognition and/or lysis of different types of target cells. Moreover, the 13.1 MAb can inhibit the lysis of K-562 by both resting T3$^-$ NK cells, cloned T3$^-$ NK cells as well as cloned T3$^+$ cells and this suggests that similar or identical structures or epitopes are expressed on effector cells of different cell lineages which are involved in the lysis of a particular target cell for these different types of effector cells. The wide range of tumor target cells lysed by B73.1$^+$, T3$^-$ clones versus the narrower range of target cells exerted by T3$^+$ clones would then suggest either the presence of different types of interaction sites on these cells or different numbers of interaction sites in the membrane leading to different conformational presentations thus determining the target cell specificity range of these effector cells.

The T3 MAb and lectin PHA induce various nonspecific cytolysis by T3$^+$ cloned CTL. Moreover, human and mouse "aged clones" can acquire nonspecific cytolytic capacity (Pawelec et al., elsewhere in this issue) and antigen specific CTL can be induced to exert nonspecific cytolysis either by a T3 MAb, lectin (PHA) or by the IL-2 hormone treatment. Taken these observations together with the stable lytic capacity of B73.1$^+$, T3$^-$ human and mouse NK cell derived clones (Dennert et al.) showing a wide target cell specificity range and ADCC one might conclude that: 1) the latter indeed represent a separate cell lineage with limited phenotypic and target cell heterogeneity; 2) the nonspecific lytic activity of activated T3$^+$ cells is related to their differentiation stage. Important are the findings that fresh (autologous) tumor cells of hemopoietic origin or carcinomas and sarcomas can also be lysed by the cloned effector cells. Fresh LGLs from these patients also lyse the autologous tumor cells. The applicability of such effector cells for immunotherapy remains a matter of serious investigation since it was shown that such effector cells were inactivated after prolonged incubation of these effectors on tumor target cells.

It remains to be investigated whether this will occur *in vivo*. Animal studies are therefore required to investigate these important issues. An important issue for future investigations will be whether certain stages of the lytic cycle and/or lytic substances produced by the effector cells, as well as whether interaction sites on (cloned) T3$^+$ and T3$^-$ effector cells are identical in relation to their nonspecific lytic function.

The ultimate and most important study will be to unequivocally establish their *in vivo* tolerance and to investigate their applicability to positive interference with and correction of immune system inbalancies and control of residual disease in cancer patients.

K562 TUMOR CELLS PREFERENTIALLY ACTIVATE LEU 11+ HUMAN LARGE GRANULAR LYMPHOCYTES IN A MIXED LYMPHOCYTE RESPONSE CULTURE

Joseph H. Phillips
Lewis L. Lanier

Becton Dickinson Monoclonal Center, Inc., Mountain View, CA 94043

I. INTRODUCTION

Natural killer cells are a heterogenous population of large granular lymphoid cells that recognize and kill a variety of tumor cell lines and virus-infected target cells (reviewed in 1-3). Unlike antigen specific cytotoxic T cells, NK cells do not require deliberate presensitization and kill susceptible target cells by a process that is not MHC (major histocompatibility complex) restricted.

MHC restricted cytotoxic T cells are usually generated *in vitro* by co-culturing peripheral blood mononuclear cells with mitomycin-C treated lymphocytes (MLR) or tumor cells (MLTR). It has been reported by numerous investigators that non-MHC restricted killer cells are also generated during these MLR/MLTR cultures (4-11). These non-specific killer cells have been termed "NK -like cells","anomalous killer cells", and "lymphokine activated killer cells" primarily because they do not express the typical pan T cell antigens and kill NK-sensitive targets (4-11).

Recently, by co-culturing peripheral blood mononuclear cells (PBMC) with irradiated allogeneic B lymphoblastoid cells for 5 days, we observed a small population of lymphoblasts (approximately 10%) that expressed the NK cell associated Leu 11 antigen (CD16, Fc receptor expressed by NK cells and neutrophils) (12). These cells demonstrated potent cytotoxic activity against the NK-sensitive tumor cell line K562, and did not express the T cell associated

ISBN 0-12-341370-2

markers Leu 1 (CD5) and Leu 4 (CD3). Elimination of these Leu 11+ lymphoblasts from MLTR effector cell suspensions removed the majority of cytotoxic activity against K562, while having no effect on the specific MHC-restricted cytolytic T cells (12). These MLTR-activated Leu 11+ cells differed from freshly isolated Leu 11+ NK cells in that a significant proportion of the MLTR-activated Leu 11+ cells co-expressed class II MHC antigens and surface transferrin receptor (12).

Since the vast majority of the cytotoxic cells (approximately 90%) in the MLTR generated against the B lymphoblastoid were antigen specific cytotoxic T cells, it was unclear whether the Leu 11+ cells proliferated in the MLTR as a direct consequence of recognition of the DR+ B lymphoblastoid cell, or simply due to the indirect activity of various factors secreted by the antigen-specific T lymphoblasts. In an attempt to study directly the activation of NK cells, we have co-cultured human PBMC or Percoll-enriched large granular lymphocytes (LGL) with mitomycin-c treated NK-sensitive tumor cell line, K562. Since K562 lacks serologically detectable class I and II MHC antigens, it was hoped that a MLTR using K562 as the stimulator would not generate a classic T cell directed MLR ,but would instead allow for the preferential expansion of activated Leu 11+ effectors.

II. MATERIALS AND METHODS

A. Preparation of Human Peripheral Blood Mononuclear Leukocytes

The white cell fraction of plasmaphoresed human peripheral blood from normal donors was obtained form the American Red Cross, San Jose, CA. The mononuclear leukocytes were isolated by standard Ficoll/Paque gradient centrifugation (Pharmacia Fine Chemicals, Piscataway,NJ), were washed 5-10 times in Hank's balanced salt solution (HBSS) at low centrifugal speed, and were resuspended in RPMI 1640 medium (GIBCO) containing 5% heat-inactivated horse serum (Sterile Systems, Logan, Utah), 2mM glutamine (GIBCO) ,and antibiotics (referred to as "culture medium"). In certain experiments, the cell populations were enriched for LGL by the method of Timonen and Saksela (13), with slight modifications (14,15).

B. Reagents

All monoclonal antibodies used in these studies were prepared at Becton Dickinson Monoclonal Center, Inc., with exception of the anti- Tac (IL-2 receptor) monoclonal antibody, that was generously provided by Dr. T. Waldmann, NIH.

C. Two-color Immunofluorescence Staining

Immunofluorescent staining was performed as described in detail previously (16). Isotype matched FITC or PE conjugated myeloma proteins were used to control for nonspecific Fc binding.

D. Two Color Flow Cytometry

Two-color flow cytometry was performed using a FACS 440 system (Becton Dickinson FACS Division, Sunnyvale, Ca.) equipped with an argon ion laser (Spectra Physics, Mountain View, CA), as described elsewhere (16).

E. Cytotoxicity Assay

Human tumor cell lines, obtained from the American Type Culture Collection, Rockville, MD, were labelled with ^{51}-Cr and were used as targets in a 4 hour radioisotope release cytotoxicity assay as described previously (16).

F. Allogeneic Mixed Leukocyte Tumor Reaction

Ten to 20 x 10^6 allogeneic K562 tumor cells were incubated with mitomycin-C (80 ug/ml; Sigma Chemicals, St. Louis,MO.) in serum-free RPMI 1640 medium for 1 hour at 37°C. After incubation, the tumor cells were washed three times in HBSS and resuspended at a concentration of 1 x 10^6 in culture medium. PBMC or Percoll-enriched LGL were suspended in culture medium at a concentration of 1 x 10^6 cells/ml. Responder cells (10 x 10^6) were mixed with mitomycin-C treated K562 (10×10^6) in a total volume of 10 ml culture medium in T-25 tissue culture flasks (Falcon Plastics,Oxnard, CA).. The flasks were incubated in an upright position in a humidified 5% CO_2 atmosphere at 37°C for 6 days. On the sixth day, the cells were harvested, washed in HBSS, and the viable cells isolated by centrifugation on Ficoll/Hypaque.

G. Complement-mediated Cytotoxic Depletions

Complement-mediated cytotoxic depletion was performed as described previously (12), using a pre-selected batch of young rabbit complement (Pel-frez, Rogers, AR).

H. Proliferation Assay

Cellular proliferation in the MLTR cultures was determined by ^{3}H-thymidine incorporation. Responder cells (3×10^5) and mitomycin-C treated K562 (3×10^5) were mixed in 0.2 ml volume in round bottom microtiter plates (Linbro, Flow Labs, McLean, VA). Responder cells and K562 were cultured alone as controls. On the fifth day in culture 1 uCi of ^{3}H-thymidine was added to each well and the plates incubated for an additional 24 hours at 37°C. The proliferation assay was terminated on the sixth day by harvesting the cultures using a Cell Harvester (Cambridge Technology, Inc.) and determining incorporated radioactivity using a Beckman scintillation counter. The results are presented as mean counts per minute of triplicate wells ($\pm$ standard deviation) and as a stimulation index (S.I.). The stimulation index is equal to the cpm of the MLTR divided by the cpm of the control cultures.

I. Light Microscopic Morphology

Aliquots of cell suspension were cytocentrifuged onto ethanol cleaned glass slides. The slides were fixed in absolute methanol for 10 minutes and rapidly air dried. The fixed cells were then stained in a 15% aqueous solution of Giemsa for 10 minutes and washed in distilled water. Light microscopic analysis was performed by oil immersion microscopy according to standard morphological criteria.

III. RESULTS

A. Proliferative Response of Peripheral Blood Mononuclear Cells to K562

Peripheral blood mononuclear cells were co-cultured for 6 days with or without mitomycin-C treated NK-sensitive tumor cell line, K562. The viable cells were recovered using Ficoll/Hypaque and the low forward angle light scatter was determined using flow cytometry. Previous studies have shown that low forward angle light scatter is essentially proportional to cell size for lymphoid cells (17). A significant proportion of the cells isolated form the K562-stimulated MLTR (K562-MLTR) cultures were lymphoblasts (approximately 30%) as determined by flow cytometric analysis and phase-contrast microscopy. In contrast, the PBMC from the same individual cultured for 6 days without K562 stimulators but in the same media demonstrated very few lymphoblasts (approximately 7%). The percentage of lymphoblasts in K562-MLTR cultures from 4 individuals varied from 25-48% as compared to control cultures without K562 which showed very few lymphoblasts (3-10% ; Table I).

Blastogenesis, indicative of proliferation, was also confirmed by the incorporation of ^{3}H-thymidine by cells isolated from K562-MLTR cultures (Table I). PBMC co-cultured with K562 for six days showed significant stimulation indices ranging in magnitude from 7 to 20 (Table I). Although incorporation of ^{3}H-thymidine substantiated the existence of blastogenesis revealed by flow cytometric analysis, the magnitude of blastogenesis in K562-MLTR cultures did not necessarily correlate with the magnitude of the stimulation index (Table I).

B. Phenotypic Analysis of Cells in K562-MLTR and Control Cultures

The expression of cell surface antigens was examined on cells in both the large (blast) and small (nonblast) cell fraction of K562-MLTR and control cultures (Figs. 1 & 2).

Table I. ^{3}H-Thymidine and blastogenesis of control and K562-MLTR cultures using peripheral blood mononuclear cells as responders

Experiment (Individual)	^{3}H-Thymidine (cpm + S.D.)		S.I.	% Blasts	
	Control	MLTR		Control	MLTR
1	650 ± 150	4560 ± 680	7	8	50
2	200 ± 96	4130 ± 320	20	10	36
3	750 ± 310	11250 ± 910	15	4	25
4	360 ± 125	430 ± 180	12	11	33

S.I. stimulation index (MLTR/Control). % Blasts refers to the percentage of viable cells with lymphoblastoid characteristics based on light scatter measurements.

Phenotypic analysis of PBMC cultured for 6 days without K562 showed that approximately 7-10% of these cells expressed the NK cell associated antigen, Leu 11. However, when these same PBMC were co-cultured with K562 for six days, 20-30% of the small cells and 65-75% of the blasted cells clearly expressed the Leu 11 antigen (Fig. 1).

More extensive phenotypic analysis of these cells by two-color immunofluorescence revealed that a significant proportion of the Leu 11+ cells isolated from the K562-MLTR culture co-expressed the Leu 2a (cytotoxic/suppressor antigen CD8), and Leu 15 (CRIII) antigens (Fig. 1). Leu 11+ cells isolated from control cultures also expressed these antigens, however, the proportion of Leu 11+ cells co-expressing Leu 15 was significantly higher in the control cultures (not shown). A significant proportion (10%) of the lymphoblasts isolated from the K562-MLTR culture co-expressed Leu 11 and the pan T cell antigen, Leu 4 (CD3). Leu 11+ cells isolated from control cultures or fresh peripheral blood have <u>not</u> been observed to co-express Leu 4 (12,16). Leu 11+ cells from K562-MLTR cultures did not express the monocyte specific antigen, Leu M3 (Fig. 1).

With respect to expression of activation antigens, a significant proportion of the Leu 11+ cells isolated from K562-MLTR , but not control cultures co-expressed the HLA-DR antigen (Fig.1). Approximately 30-40% of the small Leu 11+ and 50-80% of the Leu 11+ lymphoblasts from the K562-MLTR cultures were phenotypically Leu 11+,DR+ (Fig 1 & 2). A smaller percentage of the Leu 11+ cells from the K562-MLTR cultures also expressed low density Leu 10 (HLA-DQ), which was not expressed on control culture cells (Fig. 1). Tac antigen (IL-2 receptor) and transferrin receptor also were selectively expressed on a significant proportion of the Leu 11+ cells from the K562-MLTR cultures (not shown).

Phenotypic analysis for another NK cell associated antigen, Leu 7 revealed that there was a significantly higher proportion of Leu 11+ cells than Leu 7+ in the K562-MLTR cultures (Fig. 1). Two-color immunofluorescence clearly indicated that the majority of the Leu 7+ cells were found in the small cell population. Most small Leu 7+ cells did not co-express the activation antigens HLA-DR (Fig.2) or IL-2 receptor. The proportion of these small Leu 7+ from the K562-MLTR cultures that co-expressed the Leu 11 antigen was variable from individual to individual. Only a small percentage of the blasts expressed Leu 7+ (Fig. 1). These Leu 7+ blasts, however, did co-express Leu 11 antigen as well as the activation antigens HLA-DR and Tac (data not shown).

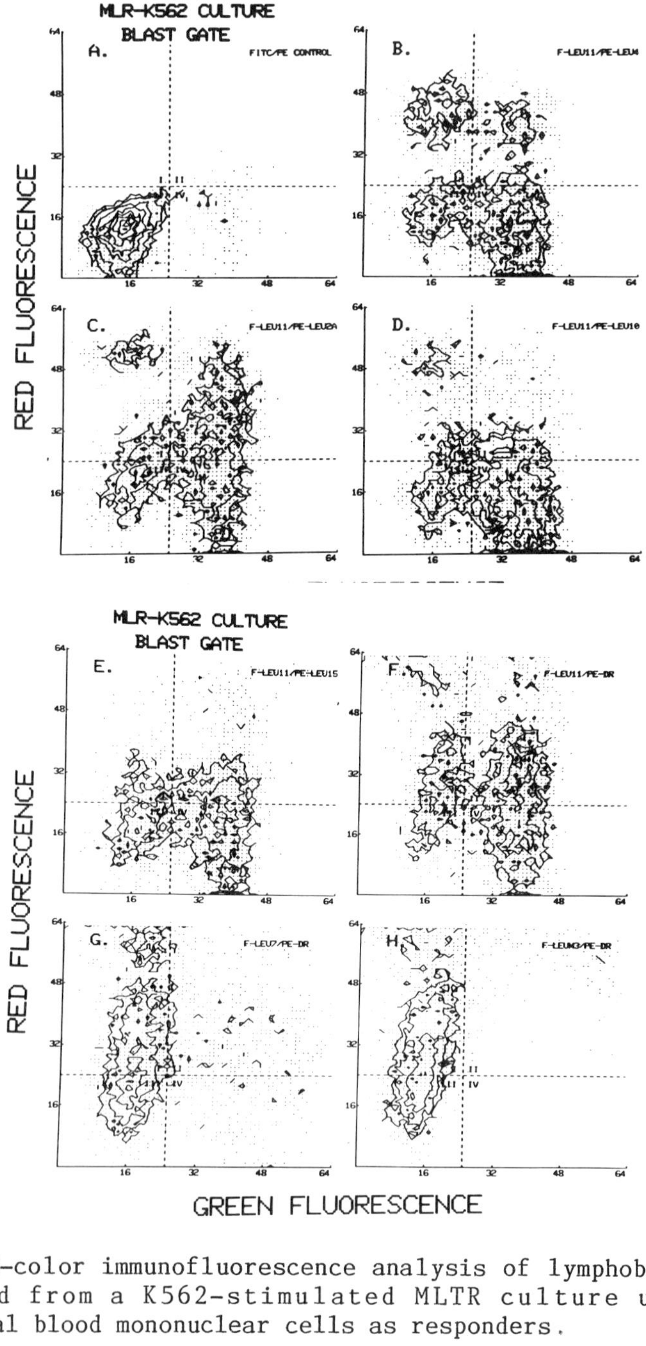

Fig. 1. 2-color immunofluorescence analysis of lymphoblasts obtained from a K562-stimulated MLTR culture using peripheral blood mononuclear cells as responders.

Fig. 1 legend (continued)

Viable cells isolated from a K562-MLTR culture were stained with: FITC -Ig and PE- Ig control myeloma proteins (panel A); FITC anti-Leu 11a and PE anti-Leu 4 (B); FITC anti-Leu 11a and PE anti-Leu 2a (C); FITC anti-Leu 11 and PE anti-DQ/Leu 10 (D); FITC anti-Leu 11a and PE anti-Leu 15 (E); FITC anti-Leu 11a and PE anti-HLA-DR (F); FITC anti-Leu 7 and PE anti-HLA-DR (G); and FITC anti-Leu M3 and PE anti-DR (H). Quantitation of the immunofluorescence was achieved using a FACS 440 system. Data are presented as contours of green (x-axis,log scale) and red (y-axis, log scale) fluorescence. The plots were divided into quadrants, based on the negative control to indicate areas which contained cells stained only with PE (quadrant I); cells stained with FITC and PE (quadrant II); unstained cells (quadrant III); and cells stained only with FITC (quadrant IV).
The low forward light scatter profile of the K562-activated MLTR cells was used to discriminate large cells (i.e. lymphoblasts) from small lymphocytes (i.e. nonblasts). A light scatter gate was set on the lymphoblasts and the fluorescence of the cells in the blast gate is presented in the displayed contours. Quantitation of the results revealed:
panel A: <5% of cells stained with either the FITC-Ig or PE-Ig control.
panel B: 15 % Leu 4+,11-; 10 % Leu 4+,11+; and 56% Leu 4-,11+.
panel C: 16 % Leu 2+,11-; 34 % Leu 2+,11+; and 31% Leu 2-,11+.
panel D: 6 % DQ+,Leu 11-; 8 % DQ+,Leu 11+;and 58% DQ-, Leu 11+.
panel E: 45 % Leu 11+,15-;20 % Leu 11+,15+; and 7% Leu 11-,15+.
panel F: 20 % DR+,Leu 11-;35 % DR+,Leu 11+; and 33% DR-, Leu 11+.
panel G: 48 % DR+,Leu 7-; 10 % DR+, Leu 7+; and 6% DR-, Leu 7+.
panel H: 54 % DR+,LeuM3-; <5 % DR+, Leu M3+; and <1% DR-, Leu M3+.

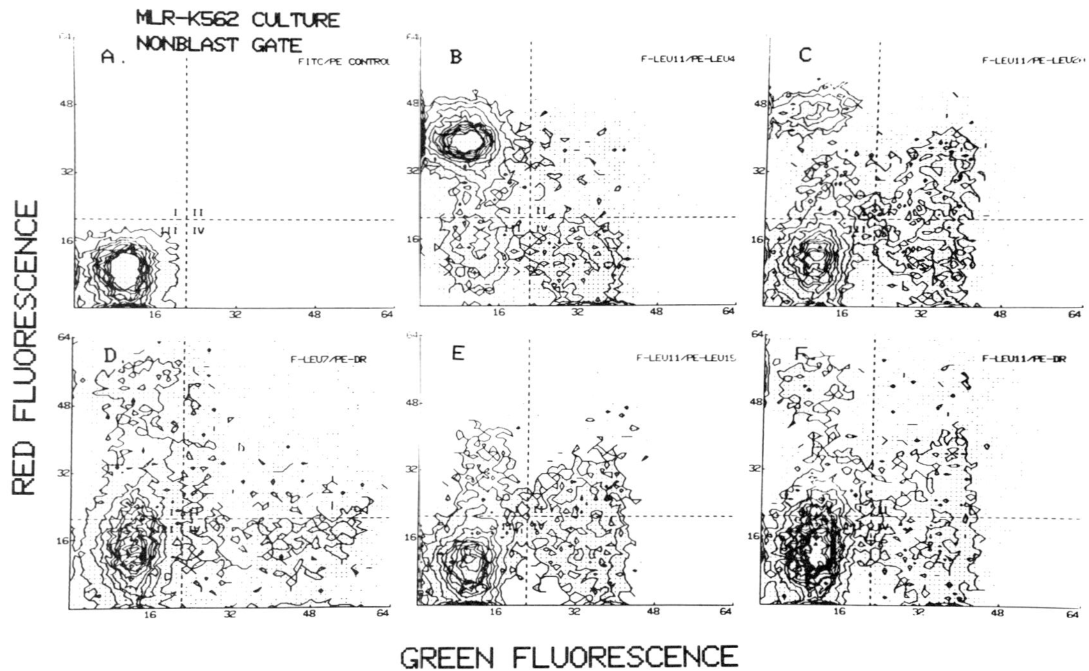

Fig. 2. 2-color immunofluorescence analysis of lymphocytes obtained from a K562-activated MLTR culture using peripheral blood mononuclear cells as responders.

Fig. 2 legend (continued)

Viable cells obtained from a K562-MLTR culture were stained with FITC-Ig and PE-Ig control myeloma proteins (panel A); FITC anti-Leu 11a and PE anti-Leu 4 (B); FITC anti-Leu 11a and PE anti-Leu 2a (C); FITC anti-Leu 7 and PE anti-HLA-DR (D); FITC anti-Leu 11a and PE anti-DQ/Leu 10 (E); or FITC anti-Leu 11a and PE anti-HLA-DR (F). By setting a gate on the small cells and reprocessing the list mode data, it was possible to determine the immunofluorescence profile of the small, nonblast cells and compare it with the profile of blast cells in the same sample. Above are presented the data for the "nonblast" (small) gated cells. Quantitation of the results revealed:

panel A: <5% cells stained with either FITC-Ig or PE-Ig controls.

panel B: 73% Leu 4+,11-; <5% Leu 4+,11+; and 14 % Leu 4-,11+.

panel C: 16% Leu 2+,11-; 10 % Leu 2+,11+; and 8 % Leu 2-,11+.

panel D: 7 % Leu 7+,DR-; 5 % Leu 7+,DR+; and 20 % Leu 7-,DR+.

panel E: 13% Leu 11+,15-; 5 % Leu 11+,15+; and 7 % Leu 11-,15+.

panel F: 13% Leu 11+,DR-; 5 % Leu 11+,DR+; and 16 % Leu 11-,DR+.

C. Cytotoxic Activity of Lymphoid Cells Isolated from Control and K562-MLTR Cultures

Phenotypic studies indicated that a large proportion of the cells isolated from a MLTR using K562 as the stimulator expressed the NK cell-associated Leu 11 antigen. Functional analysis was undertaken to assess the cytotoxic activity of these effector cells. As shown in Fig.3, cells isolated from the K562-MLTR cultures demonstrated substantially augmented cell-mediated cytotoxicity against the NK-sensitive tumor target, K562, compared to cells from control cultures. Significant levels of cytotoxicity against K562 were consistently observed at effector to target (E:T) ratios as low as 0.75 to 1 (Fig. 3). Enhanced cytotoxicity mediated by MLTR-generated effectors also was observed against another relatively sensitive NK cell target, HL-60 (Fig.3). In addition to demonstrating augmented cytotoxicity against NK-sensitive tumor targets, effector cells from the K562-MLTR cultures, but not control cultures,s were able to lyse a broad range of relatively NK-resistant tumor targets, such as CEM and CCRF-SB (Fig.3).

Depletion experiments with anti-Leu 11b monoclonal antibody and complement conclusively demonstrated that the cytotoxic activity against K562 (as well as against HL-60, CEM ,and SB) present in the K562-MLTR population was mediated by Leu 11+ effector cells (Fig. 4). In contrast, depletion of the K562-MLTR activated cells with anti-Leu 1 (pan T cell antigen CD5), or anti- Leu 7 did not substantially reduce the cytotoxic activity against K562 (Fig. 4).

In order to assess the relative cytotoxic capacity of the Leu 11+ cells in control and MLTR cultures, Leu 11+ cells were isolated to high purity by cell sorting using a FACS system. In addition, cells from the K562-MLTR culture were sorted into Leu 11+ small (non-blast) and Leu 11+ large (blast) cell populations and were analyzed for cytotoxic activity against K562. Figure 5 clearly demonstrates that both the small and large Leu 11+ cell populations sorted from K562-MLTR cultures possessed equivalent cytotoxic function. Furthermore, both populations of effectors were substantially more cytotoxic than the Leu 11+ cells sorted from control cultures.

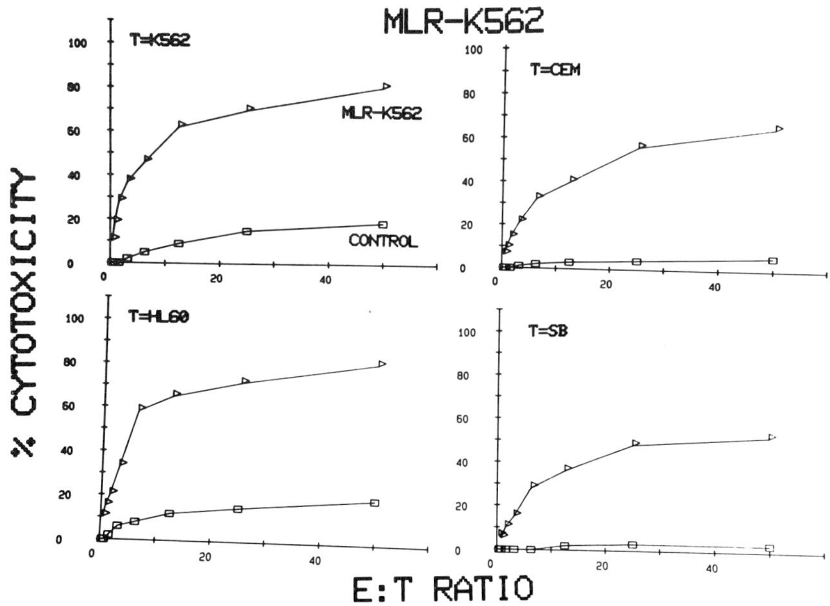

Fig. 3. Cytotoxic activity of cells obtained from a K562-activated MLTR culture

Viable lymphocytes from K562-MLTR (▷) or control cultures (□) were tested for cytotoxic activity against ^{51}Cr labelled K562, CEM, HL60, and SB cells in a 4 hr radioisotope release assay.

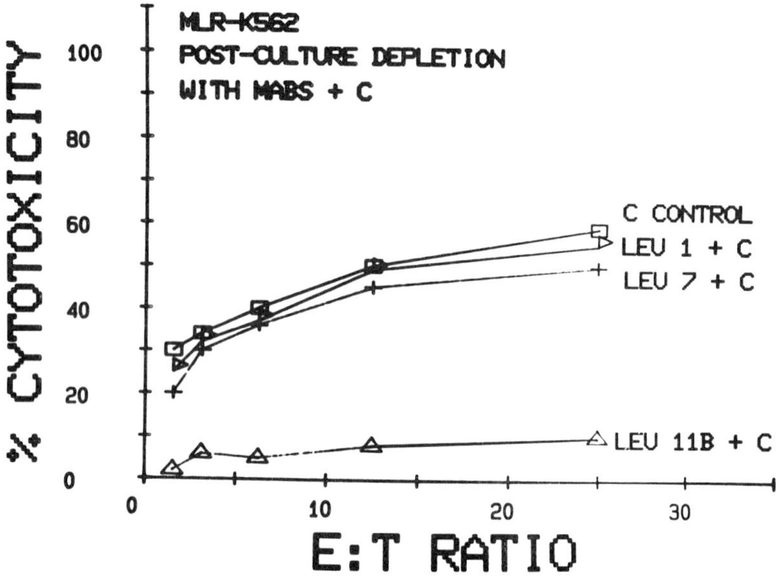

Fig. 4. Cytotoxic depletion of K562-MLTR activated effector cells using monoclonal antibodies and complement.

Viable lymphocytes from K562-MLTR cultures were treated with anti-Leu 7 + C', anti-Leu 1 + C; anti-Leu 11b + C'; or C' only (control). These cells were then tested for cytotoxic activity against ^{51}Cr labelled K562 targets in a 4 hr radioisotope release assay. Untreated cells demonstrated the same cytotoxic profile as C' control cells (not shown).

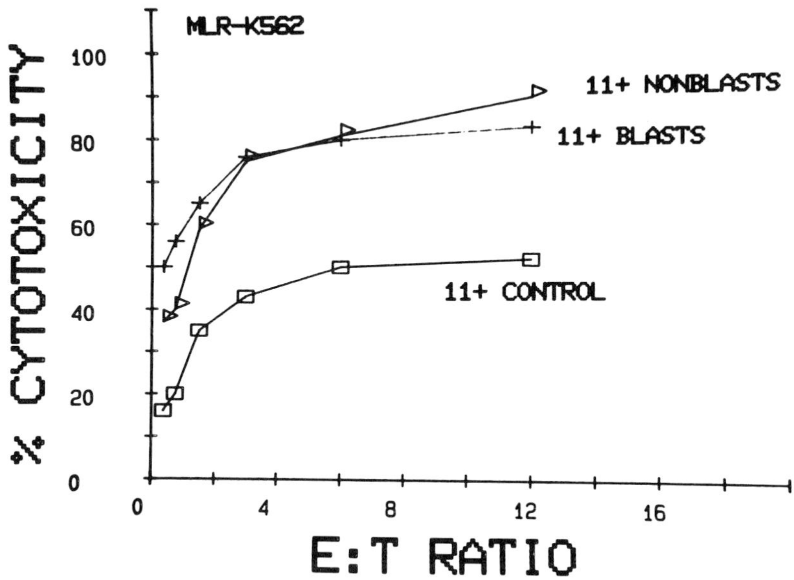

Fig. 5. Cytotoxic activity of Leu 11+ Lymphocytes Isolated from K562-MLTR or Control Cultures.

Viable lymphocytes from control or K562-MLTR were stained with FITC anti-Leu 11a and small (nonblast) Leu 11+ cells, Leu 11+ blasts, or unstimulated Leu 11+ cells from control cultures were isolated using a FACS system. The isolated subpopulations (>90% pure) were tested for cytotoxic activity against K562 target cells in a 4 hr radioisotope release assay.

Examination of Giemsa-stained cells described in Fig. 5 indicated that the Leu 11 + cells from control cultures and the small Leu 11+ cells from K562-MLTR cultures were a relatively homogeneous population of LGL with prominent azurophilic granules. The Leu 11+ blasts isolated from the K562-MLTR culture , however, were extremely large cell with abundant cytoplasm, euchromatic nuclei, prominent azurophilic granules and varied numbers of cytoplasmic vacuoles. Approximately 10-20% of the Leu 11+ blasts isolated from MLTR culture demonstrated typical mitotic configurations.

D. Proliferative, Phenotypic, and Functional Characterization of LGL Co-cultured with K562

Experiments were designed to question whether or not the response of Leu 11+ cells to K562 in a MLTR required the presence of accessory monocytes or B cells. Monocytes and B cells were removed from peripheral blood mononuclear cells by sequential plastic adherence and passage over nylon wool. The LGL in these nonadherent nylon wool passed suspensions were enriched by centrifugation on modified Percoll cushions, which allowed for the maximal recovery of Leu 11+ and Leu 7+ subpopulations, although the magnitude of the LGL enrichment was not a great as achieved with standard discontinuous Percoll gradients (13,14). Cells not stained with either anti-Leu 11 or anti-Leu 7 were typical Leu 4+ T lymphocytes. LGL enriched suspensions were then co-cultured with mitomycin-C treated K562. Flow cytometric analysis of low forward angle light scatter of these K562-activated MLTR generated with LGL (designated LGL/K562-MLTR) revealed a greater percentage of lymphoblasts (40-60%) than was observed in K562 activated -MLTR using PBMC (designated PBMC/K562-MLTR) as responders (Table II). Likewise, the proliferative stimulation indices of the LGL/K562-MLTR were significantly greater (S.I.=30-50) than those observed in the PBMC/ K562-MLTR (Table I).

Table II. ^{3}H-Thymidine and blastogenesis of control and K562-MLTR cultures using large granular lymphocytes as responders

Experiment (Individual)	^{3}H-Thymidine (cpm + S.D.)		S.I.	% Blasts	
	Control	MLTR		Control	MLTR
1	800 ± 210	24,800 ± 830	31	12	41
2	620 ± 190	27,900 ± 911	45	11	65
3	1,000 ± 670	50,120 ± 1229	50	15	58

S.I. = stimulation index (MLTR/Control).

Phenotypic analysis of the cells isolated from a LGL/K562-MLTR culture is shown in Figure 6. Control cultures of LGL incubated without K562 showed phenotypic profiles identical to freshly isolated LGL (14). The phenotype of the cells from a LGL/K562-MLTR culture was essentially similar to that observed for the cells isolated from PBMC/K562-MLTR cultures. In general, however, a greater proportion of the Leu 11+ cells from the LGL/K562-MLTR cultures co-expressed the activation associated antigens, HLA-DR, Tac, and transferrin receptor than was observed in the PBMC/K562- MLTR (Fig. 6). Similar to the PBMC/K562-MLTR, the effector cells generated in LGL-K562 MLTR demonstrated extremely augmented levels of cell-mediated cytotoxicity against K562,HL-60, CEM, and SB, and were completely depleted when effector cell suspensions were treated with anti-Leu 11b plus complement. Since these LGL-enriched suspensions were completely devoid of monocytes (<1%), the results of these experiments clearly indicate that accessory monocytes and/or monocyte-derived factors were not require to generate the proliferative response of Leu 11+ cells to K562.

E. Analysis of the Peripheral Blood Precursor of the K562-MLTR-Activated Leu 11+ Killer Cells

Studies to determine the potential precursor of the Leu 11+ cells generated in the K562-MLTR were performed by complement depletion of PBMC of either Leu 7+ or Leu 11+ cells before the initiation of the MLTR. The results shown in Figure 7 indicate that treatment of PBMC with anti-Leu 7 plus complement had essentially no effect on the subsequent generation of K562-activated killer cells. Depletion of Leu 11b+ cells from the responder PBMC prior to MLTR with K562 ,however, severely inhibited the subsequent generation of Leu 11+ cells, as well as cytotoxic activity against K562. The results of these depletion studies indicate that the peripheral blood precursor for the K562-activated NK cell population is derived from a population of Leu 11+ cells that lack the expression of detectable Leu 7 antigen.

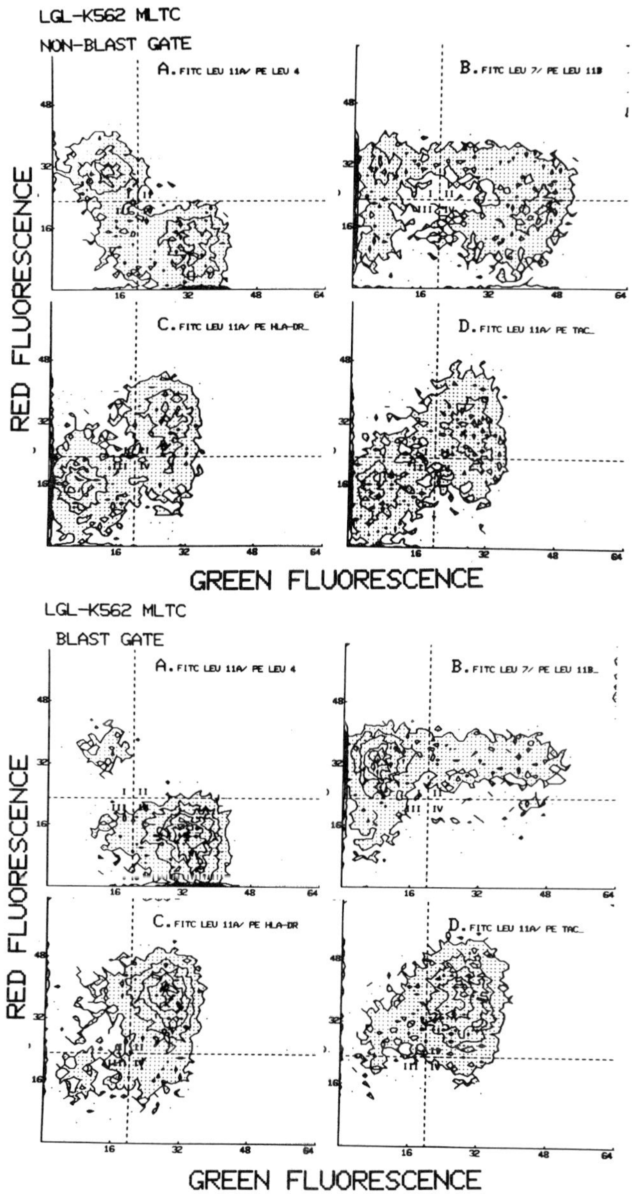

Fig. 6. Phenotype of Lymphocytes Isolated from a K562-MLTR Culture using Large Granular Lymphocytes as Responders.

Fig. 6 legend (continued)

Viable lymphocytes isolated from LGL/K562-MLTR or control cultures were stained with: FITC anti-Leu 11a and PE anti-Leu 4 (panel A); FITC anti-Leu 7 and PE anti-Leu 11b (panel B); FITC anti-Leu 11a and PE anti-HLA-DR (panel C); and FITC anti-Leu 11a and PE anti-TAC (IL-2 receptor) (panel D). Data are displayed for the small (non-blast) and large (blast) lymphocytes, as determined by light scatter measurements. Data are presented as described in Fig. 1. Quantative analysis indicated:
Non-Blast Gate:
panel A: Leu 4-,11+, 52%; Leu 4+,11+, 2%; Leu 4+,11- 32%.
panel B: Leu 7-,11+, 25%; Leu 7+,11+, 27%;Leu 7+,11- 30%.
panel C:Leu 11+,DR-,12%;Leu 11+,DR+, 41%;Leu 11-,DR+,17%.
panel D:Leu 11+,Tac- 11%;Leu 11+,Tac+ 40%;Leu 11-,Tac+ 10%.
Blast Gate:
panel A: Leu 4-,11+, 82%; Leu 4+,11+, 3%; Leu 11-,4+, 7%.
panel B: Leu 7-,11+, 54%; Leu 7+,11+, 27%; Leu 7+,11- 6%.
panel C:Leu 11+,DR- 10%;Leu 11+,DR+ 69%; Leu 11-,DR+ 15%.
panel D:Leu 11+,Tac- 5%;Leu 11+,Tac+, 79%; Leu 11-,Tac+,6%.
In the Ig-isotype control sample >95% of lymphocytes were presented in quadrant III (unstained) in both the blast and non-blast gate (not shown).

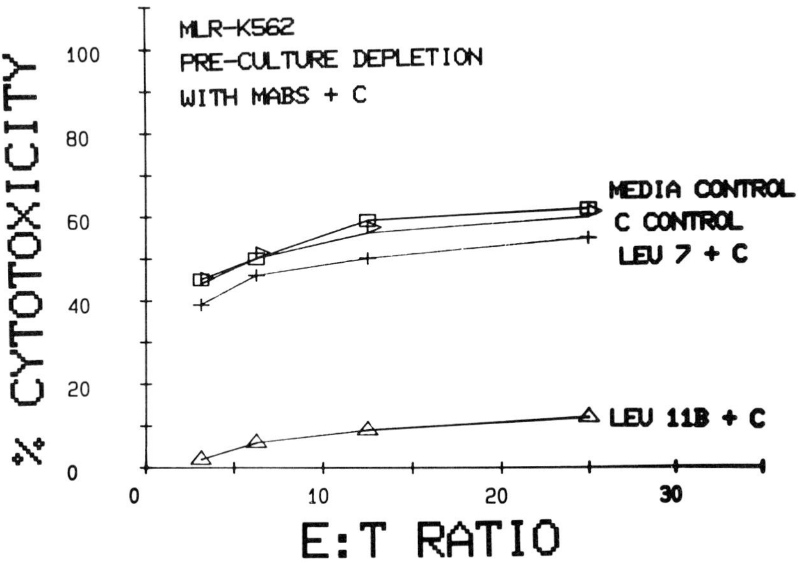

Fig. 7. Precursor analysis of the responsive cells in K562-stimulated MLTR culture.

Peripheral blood mononuclear cells were treated with anti-Leu 7+ C'; anti-Leu 11b + C'; C' only (control) or were untreated (media control). These cells were then co-cultured for 6 days with mitomycin-C treated K562 cells. Viable cells were isolated using Ficoll/Hypaque, then assayed for cytotoxic activity against 51Cr labelled K562 target cells in a 4 hr radioisotope release assay.

IV. DISCUSSION

The mixed lymphocyte response is a valuable model to study the activation of lymphoid cells in response to autologous and allogeneic cellular antigens. In many situations, this response involves antigen-specific proliferation of an amplifier T lymphocyte population that in turn promotes the generation of antigen-specific cytotoxic T lymphocytes (18-20). Often in this assay, T lymphocytes respond to class I or class II major histocompatibility antigens and the ensuing cytotoxic response is MHC restricted. From a clinical standpoint, these mechanisms may be important *in vivo* in tissue transplantation (i.e. graft rejection) and bone marrow transplantation (i.e. graft versus host disease).

In this "conventional" mixed lymphocyte response, typically generated using relatively NK-resistant cells (i.e. allogeneic DR+ B cells and B lymphoblastoid cell lines, or autologous DR+ B cells) as the stimulator, the predominant cytotoxic effector is an antigen-specific, MHC-restricted cytotoxic T lymphocyte. However, non-MHC restricted cytotoxicity against NK-sensitive targets also has been observed in these MLR cultures (4-10). This activity has been ascribed to "NK-like" cells, lymphokine-activated killers, anomolous killers, and other "non-specific" cytotoxic cells (4-10). Recently, we have definitively identified this non-MHC restricted, activated killer cell population in an MLR using a B lymphoblastoid cell line as stimulator. The MLR-activated killer cells which lysed NK-sensitive NK targets expressed the NK cell-associated antigen Leu 11 (12). Furthermore, the MLR-activated Leu 11+ killer population could be distinguished from control, unactivated cells by the presence of class II MHC antigens on a significant proportion of the MLR-activated cells.

Since T cells respond preferentially when relatively NK-resistant B cell lines are used as stimulators in a MLR, we questioned whether or not co-culture with a NK-sensitive stimulator similarly would result in activation of NK cells. Early studies by Poros and Klein (11) suggested that stimulation of peripheral blood mononuclear cells with K562 resulted in expansion of Fc+,E- cells. In our present studies we have confirmed this observation using a monoclonal antibody, anti-Leu 11, directed against the Fc-IgG receptor present on human NK cells (16,21).

These studies have conclusively demonstrated that Leu 11+ large granular lymphocytes have the capacity to proliferate in response to direct stimulation with a NK-sensitve tumor cell line. Significant proliferation and blastogenesis in the Leu 11+ subset of lymphocytes were achieved by using either unfractionated peripheral blood mononuclear cells or with Percoll-gradient enriched large granular lymphocyte preparations. Additionally, it was clear that the vast majority of cytotoxic effector cells generated in the K562-activated MLTR cultures were contained within the Leu 11+ population and by morphological criteria were large granular lymphocytes. We also have shown that the level of cytotoxicity mediated by the K652-MLTR activated Leu 11+ cells was significantly higher than of Leu 11+ cells isolated from control cultures. This argues not only for an increased total number of effectors, but also more potent cytotoxic potential by the Leu 11+ lymphocytes. Single cell cytotoxicity assays will be required to address the question of whether or not there is an expansion of the total proportion of cytotoxic cells within the Leu 11+ population, or alternatively enhanced killing potential mediated only by a proportion of the Leu 11+ cells. Similarly, single cell cytotoxicity assays and cold target inhibition assays will be necessary to determine whether the apparent increase in target specificity of the K562-MLTR activated Leu 11+ cells results from a qualitative or quantitative effect.

In a conventional mixed lymphocyte response cultures using a B lymphoblastoid cell as stimulator (designated B-MLR), we demonstrated that treatment of peripheral blood mononuclear cells with anti-Leu 11b and complement prior to culture prevented the generation of Leu 11+ -activated killer cells (12). Anti-Leu 7 and complement pre-treatment of the responder cell population neither diminished nor enhanced the generation of Leu 11+ -activated killer cells in the B-MLR culture, thus suggesting that the peripheral blood precursor of the Leu 11+ MLR-activated killer expressed the phenotype Leu 7-,11+. Similar results were obtained in these studies using K562 as stimulator. Although a significant proportion of Leu 7+ cells were detected in the K562-activated MLTR cultures, they appeared to contribute little to the total cytotoxic capacity of the population, as assessed by cytotoxic depletion with anti-Leu 7 and complement. Furthermore, Leu 7+ cells were not necessary to generate potent cytotoxic effectors in the K562-activated MLTR, since elimination of this population with antibody and complement prior to co-culture with K562 did not greatly diminish the cytotoxic potential of the effector cells obtained after 6 days. Again, these experiments would argue that peripheral blood derived Leu 7+ cells do not serve as precursors for the Leu 11+ population. These observations are inconsistent with the hypothesis proposed by Abo *et al.* (22,23) that the Leu 7 antigen is a marker of immature NK cells in peripheral blood. Rather, the alternative possibility, i.e. that the Leu 7 antigen is expressed late during the differentiation of peripheral blood NK cells, is not excluded based on our observations. This latter hypothesis can be tested by determining whether or not Leu 7-,11+ can be induced to differentiation *in vitro* and subsequently express the Leu 7 antigen.

Finally, results of these studies suggest an approach to select and expand NK cell population for long term culture. In the past, it has been exceedingly difficult to establish long term lines of cells with typical NK phenotype, morphology, and function. The majority of cloned "NK" cell lines have possessed characteristics more akin to T lymphocytes than freshly isolated NK cells or the Leu 11+ activated-killer cells that we have described in mixed lymphocyte cultures (24-27). Our preliminary attempts to grow these K562-activated Leu 11+ cells in culture have been quite promising, with cultures now thriving after 2 months of growth. These lines would provide an excellent resource for generation of new monoclonal antibodies, the production and characterization of NK-cell derived factors, and a biochemical study of the recognition and lytic mechanism of unique cells.

REFERENCES

1. Herberman, R.B. *et. al.* (1979). Immunol. Rev. 44,43.
2. Roder, J.C., and Pross, H.F. (1982). J. Clin. Immunol. 2,249.
3. Oldham, R.K. (1982). J. Biol. Response Modif. 1,217.
4. Strassman, G., Bach, F.H., and Zarling, J.M. (1983). J. Immunol. 130,1556.
5. Goto, M., and Zvaifler, N.J. (1983). J. Exp. Med. 157,1309.
6. Lopez-Botet, M., Silva, A., Rodriguez, J., and deLandazuri, M.O. (1982). J. Immunol. 129,1109.
7. Santoli, D., Perussia, B., and Trinchieri, G. (1981). *In* "Mediation of Cellular Immunity in Cancer by Immune Modifiers" (M.A. Chirigos et al.,eds.) p. 143. Raven Press, New York.
8. Seeley, J.K., Masussi, G., Poros, A., Klein, E., and Golub, S.H. (1979). J. Immunol. 123,1303.

9. Svedmyr, E., Wigzell, H., and Jondal, M. (1974). Scand. J. Immunol. 3,499.
10. Jondal, M., and Targan, S. (1978). J. Exp. Med. 147,1621.
11. Poros, A., and Klein, E. (1979). Cell. Immunol. 46,57.
12. Phillips, J.H., Le, A.M., and Lanier, L.L. (1984). J. Exp. Med. 159,993.
13. Timonen, T., and Saksela, E. (1980). J. Immunol. Meth. 36,285.
14. Phillips, J.H., Warner, N.L., and Lanier, L.L. (1984). Natl. Immun. Cell Growth Regul. 3:73.
15. Lanier, L.L., Benike, C.J., Phillips, J.H., and Engleman, E.G. (1985). J. Immunol., In press.
16. Lanier, L.L., Le, A.M., Phillips, J.H., Warner, N.L., and Babcock, G.F. (1983). J. Immunol. 133,180.
17. Salzman, G.C. (1982). In "Cell analysis" (N. Catsimpoolas, ed.), vol.1, p. 111. Plenum Press, New York.
18. Lanier, L.L., Engleman, E.G., Gatenby, P., Babcock, G.F., Warner, N.L., and Herzenberg, L.A. (1983). Immunol. Rev. 74,143.
19. Thomas, Y., Rogozinski, L., and Chess, L. (1983).
20. Reinherz, E.L., and Schlossman, S.F. (1980). Cell 19,821.
21. Perussia, B., Trinchieri, G., Jackson, A., Warner, N.L., Faust, J., Rumpold, H., Kraft, D., and Lanier, L.L. (1984). J. Immunol. 133,180.
22. Abo, T., Cooper, M.D., and Balch, C.D. (1982). J. Immunol. 129,1752.
23. Abo, T., Miller, C.A., Gartland, G.L., and Balch, C.D. (1983). J. Exp. Med. 157,273.
24. van de Griend, R.J., van Krimpen, B.A., Ronteltap, P.M., and Bolhuis, R.L. (1984). J. Immunol. 132,3185.
25. Hercend, T., Reinherz, E.L., Meuer, S., Scholssman, S.F.,and Ritz, J. (1983). Nature 301,158.
26. Kornbluth, J., Flomenberg, N., and Dupont, B. (1982). J. Immunol. 129,2831.
27. Allavana, P., and Ortaldo, J.R. (1984). J. Immunol. 132,2363.

DISCRIMINATION BETWEEN ANOMALOUS KILLER CELLS AND NK-LIKE CELLS[1]

Gordon F. Burns
Tony Triglia
Jerome A. Werkmeister[2]

Lions Clinical Cancer Research Laboratory, Clinical Research Unit of The Walter and Eliza Hall Institute of Medical Research, Melbourne, Victoria, Australia.

I. INTRODUCTION

Attempts to grow tumor-specific CTL from lymphocytes within melanoma tissue (1) or from the peripheral blood of patients with melanoma (2) or lymphoma (3) have resulted in the isolation of cytotoxic lymphocytes able to kill not only the patient's own tumor cells but, in addition, a range of allogeneic tumor cell targets. Similar cytotoxic cells can be grown from the blood of normal subjects following mixed lymphocyte culture (MLC) with allogeneic lymphocytes or with autologous or allogeneic EBV-infected B lymphoblasts. Recent experimental data suggest that MLC generates the production of different kinds of effector cells, with the NK-like cells which are able to kill K562 targets being readily distinguishable from the CTL which kill the stimulating allogeneic target cells (4,5). It is not clear, however, what is the nature and origin of the cells from the same cultures (anomalous killer, AK cells) which are able to kill, without prior sensitization or apparent MHC restriction, tumor cell targets which are resistant to NK cell killing. We have addressed this question by the use of monoclonal antibodies (MAb) and sugars which can inhibit the generation of one or other cell type, and by the use of clones. We have also examined the cytolytic mechanisms used by NK and NK-like cells and by AK cells by producing MAb which block target cell killing at the effector phase (6-8). Such experiments lead us to conclude that NK-like cells which preferentially kill K562 target cells are derived from NK cells and that AK cell function is a property acquired by CTL *in vitro* (5).

II. DIFFERENT PRECURSORS FOR AK AND NK-LIKE CELLS

We have used mixed lymphocyte culture between blood mononuclear cells and irradiated autologous or allogeneic

---

[1]This work was supported by the Lions Cancer Research Fund of Victoria.
[2]Supported by a fellowship from the National Health and Medical Research Council of Australia.

ISBN 0-12-341370-2

B lymphoblasts under conditions of limiting dilution to determine the frequency of the precursors of CTL, AK cells and NK-like cells. Each effector cell type was defined by the lysis of different target cells as follows: CTL, by lysis of the stimulating B lymphoblast; AK cells, by lysis of a melanoma cell line resistant to killing by fresh NK cells; NK-like cells, by the lysis of K562. Among mononuclear cells from several donors the frequency of precursors of allospecific CTL ranged between 1/900 and 1/7,500, that of AK cells between 1/250 and 1/750 (ref. 9) and that of NK-like cells between 1/50 and 1/1,200 (Werkmeister et al, submitted for publication). Several agents were found to alter the frequencies of each effector cell type induced in culture, and in a manner which suggested that the precursors of AK cells were the same or similar to those of CTL but different from NK-like cell precursors (Table I). Thus, the addition of PHA to the MLC at the beginning of culture resulted in multiphasic cell dose-response curves for the precursors of CTL and AK cells, rather than the linear single hit responses seen for NK-like cell precursors under these conditions, perhaps suggesting a regulatory role for suppressor cells in the induction of AK cells and CTL (9). The monosaccharide N-acetyl-D-galactosamine (NADG) when present from the beginning of culture reduced the frequency of NK-like cells but not of AK cells or CTL (Table I and Werkmeister et al, submitted for publication) whereas a MAb named LeoA1, with specificity for an activation antigen on T cells (Burns et al, submitted for publication), greatly reduced the frequency of CTL and AK cells generated, but did not affect the frequency of NK-like cells (Table I).

TABLE I Agents affecting the generation of CTL, AK cells and NK-like cells from their precursors[a]

Agent added	Reciprocal frequency of precursors[b] of					
	CTL		AK cells		NK-like cells	
	-	+	-	+	-	+
PHA	7,000	NR[c]	650	NR[c]	700	700
NADG	5,500	6,000	550	600	450	1,400
LeoA1	900	1,750	400	1,800	350	350

[a]Results from limit dilution analysis of 1 donor for each agent; other donors gave similar results in each case. The agents (PHA, 1%; NADG, 20 mM; LeoA1, 1/10 hybridoma supernatant) were added at the start of a 10 day culture.
[b]Calculated according to the zero order term of the Poisson probability distribution; each point was determined from 48 replicate cultures.
[c]NR: not recordable (see text).

The mechanism of inhibition by each of the reagents mentioned remains to be elucidated, but it should be noted that LeoA1 detects quite a different antigen from the T3 antigen of T cells and does not inhibit cytotoxicity at the effector cell level, although it was found previously in bulk culture studies that the presence of anti-T3 antibodies from the beginning of MLC abrogated the induction of CTL and AK cells but had no effect on, or even increased, the generation of NK-like cells (5). The above MAb (LeoA1 and anti-T3) inhibited CTL and AK effector cell generation in the absence of complement (C), presumably by interfering with some essential differentiation signal. We have also begun to investigate the progenitors of NK-like and AK cells by removing defined cell types from blood mononuclear cells with MAb + C and then testing for any reduction in the progenitor frequency of the effector cells which are generated subsequently in MLC. One MAb which has proved of interest in this regard is termed 25 E11. The antibody binds to large granular lymphocytes, platelets, monocytes and weakly to granulocytes in peripheral blood and bone marrow, and immunoprecipitates a protein of 110 kilodaltons (Burns and López, unpublished data); it does not bind to small lymphocytes or erythrocytes. When peripheral blood mononuclear cells were pretreated with 25 E11 and C prior to analysis in the limiting dilution assays described above, the frequency of precursors of NK-like cells was reduced by 70%. The effect of 25 E11 + C upon the progenitor frequency of CTL and AK cells has not been tested, but the data obtained are suggestive of an NK cell progenitor for NK-like cells.

Further evidence suggesting discrimination between the precursors of AK and NK-like cells was found by examining the blood of patients with Hairy cell leukemia (Werkmeister et al, submitted for publication). Mononuclear cells from 4 patients with decreased NK cell function were unable to generate the NK-like cells in MLC whereas normal levels of AK effector cells (able to kill melanoma target cells) were generated in the same cultures by cells from 3 of the patients.

III. CLONED CTL DISPLAYING AK CELL FUNCTION

We have demonstrated previously by use of the split clone technique that the great majority of CTL clones able to kill the sensitizing B lymphoblasts were unable to kill K562 target cells, this latter property being a function of NK-like cells (1,5). Others have come to the same conclusion by counting double cell conjugates in the single cell binding assay (4). In our studies, however, many of the CTL clones did lyse melanoma cells which is a function of AK cells. It was possible that this result was attributable to inadequate cloning, therefore experiments were performed in which blood mononuclear cells were stimulated with autologous or allogeneic B lymphoblasts and then cloned on irradiated autologous filler cells with IL-2 at one cell per well. From these cultures was selected a number of primary clones which appeared to be specific CTL since they were effectively cytolytic towards the stimulating B lymphoblasts but did not kill melanoma or K562 target cells. The primary clones were re-seeded at 1 cell per well on filler cells and with restimulation with the original stimulating B lymphoblast. After expansion, some 60 subclones from each of the primary clones were retested for cytotoxicity against the stimulating B lymphoblast and melanoma cells. The results (Table II) revealed that almost all of the CTL subclones could kill melanoma cells with or without loss of their original CTL function making it clear that in man, as in the mouse (10), CTL clones can acquire AK cell function. Whether this activity is an *in vitro* artefact or the result of further differentiation of CTL cells awaits further study.

IV. DISSECTION OF THE CYTOLYTIC MECHANISM OF NK AND NK-LIKE CELLS

A mouse MAb raised against NK cells and termed 9.1C3 has proved of value in attempts to probe the cascade of events which culminate in target cell death (6). The antibody blocked the lysis of K562 target cells by NK and NK-like cells at a discrete step after the Ca^{++}-dependent programming for lysis; it did not block conjugate formation between effector and target cells and, since it does not bind to K562 cells, it functions at the effector cell level (8). Nevertheless the

TABLE II Summary of target cell specificity of CTL clones[a]

Expt.	Stimulating B blast	# Primary clones selected	Total # clones tested	# Clones killing B blasts	# Clones killing Melanoma	# Clones killing Both
1 & 2	Allogeneic	7	384	0	0	308
2	Autologous	3	180	13	93	29

[a]Primary clones were selected from cells seeded at 1 cell per well by their ability to kill the stimulating B lymphoblast but not melanoma cells or K562. The primary clones were then recloned at 1 cell per well and subsequently retested for cytolytic activity against B lymphoblasts and melanoma cells.

inhibitory effect of 9.1C3 was found to be relatively specific for K562 target cells, and it did not at all inhibit, for example, melanoma cell lysis by AK cells generated in MLC. To identify the target cell structure involved in this late event in the lytic cascade, anti-idiotype antibodies (anti-id) to 9.1C3 were raised in rabbits (7). The anti-id were purified by affinity chromatography and demonstrated to bind specifically to 9.1C3 MAb. Of more importance it was found that the anti-id bound to K562 cells but not to a range of unrelated cell lines and that $F(ab')_2$ fragments of anti-id were able to inhibit the lysis of K562 by NK and NK-like cells by binding to the target cell (7). The anti-id did not block conjugate formation and it was postulated that these antibodies were binding to a (secondary) receptor, or cytotoxic acceptor site, on the K562 cell membrane (7).

To investigate this, K562 cells were tested with the mutagen MMS and variants which bound low amounts of the anti-id were selected on the fluorescence activated cell sorter (FACS). By repeated cloning and FACS selection with the antibodies, clones of K562 cells were obtained which expressed little of the antigen detected by anti-id. These clones formed normal numbers of conjugates with purified NK cells but were 6- to 7-fold less sensitive to lysis than the parent cell line. Moreover effects such as the kinetics of lysis and the effects of interferon were excluded, and the resistance to lysis of these clones was demonstrated to be inversely related to the binding of anti-id (Table III and Werkmeister et al, submitted for publication).

TABLE III Characteristics of resistant mutants of K562 cells selected with anti-idiotype antibodies

K562 cell clone	Binding of anti-id[a]	% CFC[b]	LU[c]/10^7 Untreated	LU[c]/10^7 IFN activated[d]
Parent	105	26 ± 2	211	309
Resistant mutant	7	25 ± 1	42	64

[a]Mean fluorescent intensity on an arbitrary scale from the FACS after indirect immunofluorescent staining with anti-id.
[b]% conjugate-forming cells with Percoll-purified NK cells.
[c]Lytic units (LU) causing 20% lysis.
[d]NK cells pretreated with 500u of recombinant human leukocyte interferon for 1 h at 37°C.

These findings substantiate the possibility of a stimulus secretion model for NK cell killing (11) which is currently being investigated directly with 9.1C3 MAb (R.B. Herberman; personal communication), and also point to the relationship between NK and NK-like cells at the effector cell level.

Of considerable interest in relation to the in vivo role of NK cells is our recent finding that, among the cells in normal bone marrow, the anti-id appeared to bind to the progenitors of erythroid cells. It is unclear whether the antigen involved is essential for some process during erythropoiesis but it has been found that the NK-resistant mutants of K562 described above were never completely negative for binding of anti-id and that these cells consistently reverted to the parental state (of anti-id antigen expression and high sensitivity) after culture of between 3 and 4 months.

V. THE STRUCTURE ON MELANOMA CELLS RECOGNIZED BY AK CELLS

AK cells, like CTL, express a T cell phenotype, usually $T3^+T8^+$. In contrast to the MHC-restricted function of CTL, however, AK cell killing cannot be blocked by MAb to the T3 or T8 antigens when tested at the population (6,9) or the cloned cell level (12 and Table IV). To investigate whether AK cells expressed a receptor for tumor cells which was distinct from the classical T cell receptor we raised a series of MAb against AK cells and screened them for their ability to inhibit the killing of melanoma cells. Of several hundred such MAb tested,

TABLE IV Antibodies to T3 and T8 inhibit CTL but not AK cells

Expt.[a]	Blocking antibody	$LU^b/10^6$ cells against: B lymphoblasts	Melanoma
1	Nil	41	132
	Anti-T3	10[c]	129
	Anti-T8	15[c]	133
	LeoLFA-1	12[c]	42[c]
2	Nil	40	19
	Anti-T3	18[c]	21

[a]Expt. 1 is representative of 10 independent experiments with cells from different donors. Expt. 2 is the results from a cloned AK cell line and shows the % lysis at a 4:1 effector to target ratio.
[b]Lytic units (LU) causing 20% lysis.
[c]$p<0.001$ by Student's t test.

one antibody was selected for its ability to block consistently this AK cell function. Further testing of this antibody (LeoLFA-1) revealed that it also inhibited killing mediated by CTL (Table IV) and NK cells, and the cellular distribution and biochemical characteristics of the antigen bound by LeoLFA-1 suggested that it had the same specificity and function as other MAb to LFA-1 (T. Triglia, unpublished data).

It was then decided to approach this problem by identifying antigenic structures on the melanoma target cells which might be recognized by AK cells. Mice were immunized with a melanoma cell line and the hybridoma antibodies were screened for their ability to block AK cell-mediated killing of the same cell line. Concurrently the antibodies were tested for their ability to bind to melanoma cells and to fetal fibroblasts in an RIA. Over 2000 hybridoma supernatants from 3 fusions were tested in this way and some 50% bound to the melanoma cells. Of these, 3 antibodies inhibited AK cell-mediated lysis. One antibody, termed LeoMel 3, was selected for detailed analysis (Werkmeister et al, in preparation). Surface labeling of melanoma cells with ^{125}I followed by immunoprecipitation and SDS-PAGE analysis failed to reveal a protein band, but it was seen that most of the labeled material ran ahead of the dye front suggesting the presence of a glycolipid. Hence the lipids from melanoma cells and (control) T cells were extracted, separated by thin layer chromatography, and probed with MAb. The results demonstrated that LeoMel 3 was binding to an acidic glycolipid on melanoma cells, probably to both GD2 and GD3. These data suggested that the AK cell receptor was functioning as a lectin-type receptor for carbohydrate.

To challenge this notion we requested a large number of documented melanoma 'specific' antibodies from several groups. The response was generous and we were able to test over 25 such antibodies for blocking of AK-mediated cytotoxicity: moreover, since LeoMel 3 was an IgMκ antibody, it was important also to test a number of antibodies of the same isotype. The results of blocking studies carried out with a number of these antibodies are shown in Table V and, in summary, the only other antibody tested which was able to inhibit AK cell-mediated lysis was R24 (13) (kindly provided by Dr. L. Old) which is known also to be directed against an acidic glycolipid, GD3. Anti-melanoma protein antibodies of IgM isotype did not inhibit lysis. In single cell assays it was demonstrated that LeoMel 3 inhibited conjugate formation between AK and melanoma cells (mean % inhibition 70% in 2 assays) suggesting that LeoMel 3 was blocking a primary recognition structure on the melanoma cells.

TABLE V The blocking effect of LeoMel 3 and other anti-melanoma monoclonal antibodies on the effector stage of AK cell function

Antibody added	LU[a]/10^6 cells
None	210
LeoMel 3	68[b]
Nμ 4B	209
$J_2$4014	231
140.240	245
SC.8128	252
256-1	234
R24	105[b]

[a] Lytic units (LU) causing 20% lysis.
[b] $p<0.001$ by Student's t test.

Further analysis of the Mel 3 antigen revealed that it was relatively specific for melanoma cells since LeoMel 3 binding was not observed on normal melanocytes or on a large selection of tumor cell lines. These data suggest that AK cells identify and bind to melanoma cells via a lectin-type receptor for carbohydrate, and that, perhaps, AK cells must possess a number of such receptors in order to recognize different tumor cell types.

VI. CONCLUSIONS

Considerable effort has been directed towards identifying the non-specific effector cells which are generated during the course of MLC with the conclusions reached differing according to the assay used. Advocates for both an NK cell origin and a T cell origin for these cells can be found. We believe that this apparent conflict is resolved by the identification of the two major populations of non-specific effector cells described in this report: the NK-like cells which kill predominantly NK sensitive target cells by a stimulus secretion mechanism similar to that used by 'fresh' NK cells and which probably come from the same progenitors as NK cells; and AK cells which are the direct progeny of cytotoxic T lymphocytes and whose precursors are subject to the same regulatory influences as CTL (1,5,9). AK cells, however, appear to employ a different recognition structure from the classical T cell receptor (9) which may be similar to that used by NK cells (Werkmeister et al; in preparation).

The *in vivo* role of these cells requires further elucidation but preliminary evidence reported here suggests that NK and NK-like cells may play a part in the regulation of erythropoiesis.

REFERENCES

1. Burns, G.F., Good, M.F., Riglar, C., Bartlett, P.F., Crapper, R.M. and Mackay, I.R. (1984) Clin.Exp.Immunol. 57:487.
2. Knuth, A., Danowski, B., Oettgen, H.F. and Old, L.J. (1984) Proc.Natl.Acad.Sci.USA 81:3511.
3. Yssel, H., Spits, H. and de Vries, J.E. (1984) J.Exp.Med. 160:239.
4. D'Amore, P.J., Golightly, M.G. and Golub, S.H. (1984) In: NK Cells and Other Natural Effector Cells. Ed. R.B. Herberman. Academic Press Inc., p.785.
5. Burns, G.F., Triglia, T. and Werkmeister, J.A. (1984) J.Immunol. 133:1656.
6. Burns, G.F., Triglia, T., Bartlett, P.F. and Mackay, I.R. (1983) Proc.Natl.Acad.Sci.USA 80:7606.
7. Werkmeister, J.A., Burns, G.F. and Triglia, T. (1984) J.Immunol. 133:1385.
8. Burns, G.F., Werkmeister, J.A. and Triglia, T. (1984) J.Immunol. 133:1391.
9. Burns, G.F., Werkmeister, J.A. and Triglia, T. (1984) Cell.Immunol. In press.
10. Brooks, C.G., Urdal, D.L. and Henney, C.S. (1983) Immunol.Rev. 72:43.
11. Roder, J.C. and Werkmeister, J.A. (1983) In: Cancer: Etiology and Prevention. Ed. R.G. Crisper. Elsevier Science Publishing Co., p.225.
12. Yssel, H., Spits, H. and de Vries, J.E. (1984) J.Exp.Med. 160:239.
13. Pukel, C.S., Lloyd, K.O., Trevassos, L.R. et al (1982) J.Exp.Med. 115:1133.

DIFFERENTIATION OF LARGE GRANULAR LYMPHOCYTES DURING INDUCTION OF ACTIVATED KILLER CELLS IN VITRO

Tuomo Timonen

Department of Pathology
University of Helsinki
Finland

I. INTRODUCTION

Natural killer activity of freshly isolated peripheral blood lymphocytes is principally mediated by large granular lymphocytes (Timonen *et al.*, 1979; Timonen *et al.*, 1981). Lymphoid cells activated in culture (activated killer cells, AK cells) have cytotoxic activity which partially resembles that of NK cells (Seeley *et al.*; Grimm *et al.*, 1982). However, AK cells are phenotypically T cells (T3+, M1-) and thus differ from majority, but not all LGL which are mostly T3-, M1+ (Ortaldo *et al.*, 1981). Morphological characteristics of AK cells have mainly been inspected from proliferating cultures. While some common features between AK cells and LGL have been described (Hercend *et al.*, 1983), the exact relationship between LGL and AK cells has remained obscure.

In this study we have investigated morphological and phenotypic characteristics of AK cells induced to cytotoxic activity by culturing human peripheral blood mononuclear cells with the Burkitt lymphoma cell line Raji.

II. MATERIALS AND METHODS

The growth medium used throughout the experiments was RPMI 1640 (Biofluids, Rockville, Maryland) supplemented with 10 % heat-inactivated newborn calf serum (Flow Laboratories, Irvine, UK), 0.29 mg/ml of glutamine and antibiotics.

ISBN 0-12-341370-2

Peripheral blood mononuclear cells were isolated by Ficoll-Isopaque gradient centrifugation, and nonadherent cells were further enriched by sequential incubation of the cells in glass flasks and nylon wool columns. Nonadherent cells were fractionated to seven fractions by discontunious density gradient centrifugation on Percoll (Timonen _et al_., 1982). Surface phenotypic analyses were performed with various monoclonal antibodies at 1:50 dilution and with the NK-9 antibody (Nieminen and Saksela, 1984) at 1:1000 dilution of ascites. Fluorescence activated cell sorter (Becton Dickinson FACS IV) was used for both the phenotypic analyses and cell sorting.

Surface adherent macrophages were isolated by incubating Ficoll-Isopaque-purified mononuclear cells in 24-well tissue culture plates (Nunc, Roskilde, Denmark), $3x10^6$ cells/well. After two hours, nonadherent cells were thoroughly washed away and the remaining adherent cells were further incubated for and additional two hours, after which the plates were washed again. Adherent cells were > 95 % monocytes as judged by morphological criteria, and virtually no LGL were detected among these cells.

Cytotoxicity was measured by the standard 16 hour $chromium^{51}$-release assay. Morphology was analyzed from methanol-fixed, Giemsa-stained cytocentrifuged slides.

III. RESULTS

A. Induction of AK Cells

In Figure 1, the seven fractions obtained from the Percoll gradient have been incubated for one day with either medium only, K562 cells or Raji cells (at 1:40 tumor cell:lymphocyte ratio). The incubation with Raji has resulted in the appearance of cytotoxic activity against K562 in the intermediate density fractions containing medium-sized lymphoycytes. Co-culturing lymphocytes with K562 resulted in an increased lytic actitivy only in the low density, LGL-enriched fractions.

B. Morphological and Phenotypic Analyses of Raji-Adherent Lymphocytes

We next studied whether the killer cells induced by Raji were also adherent to Raji. The Raji-adherent cells from medium-sized lymphocytes were adsorbed and eluted from conjugates by

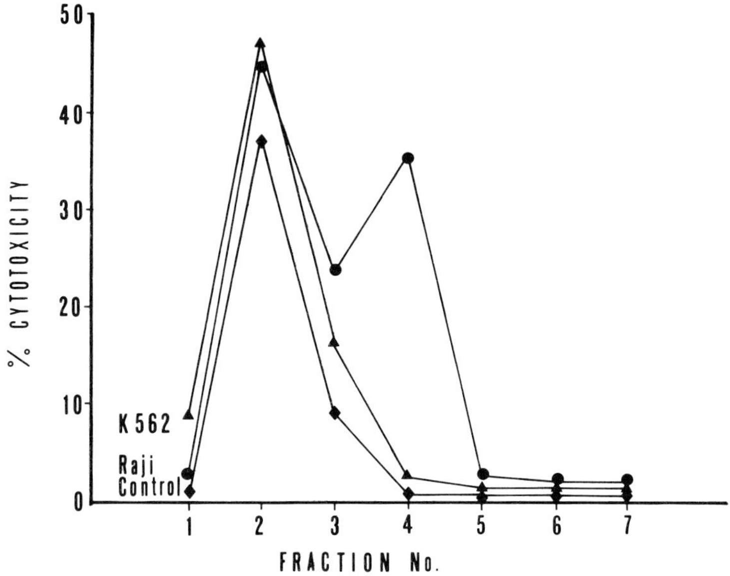

Fig. 1. Stimulation of cytotoxic activity in peripheral blood lymphoycytes fractionated by discontinuous density gradient centrifugation on Percoll. The cells were cultured with medium only (control), K562 or Raji.

TABLE I. Morphological and surface marker characteristics of Raji-adherent medium-sized lymphocytes

Morphology[a]	Per cent of total[b]	Per cent reactivity with antibodies			
		MI	T3	T8	T4
AL	86 (69-92)	10(1-15)	92(80-96)	66(50-75)	7(2-15)

[a]AL: agranular lymphocytes. The rest of the cells were LGL.

[b] range from four experiments.

the density gradient centrifugation method described previously (Timonen *et al.*, 1979). The cytotoxic activity of Raji-adherent cells was clearly stronger than that of Raji non-adherent cells or unfractionated cells (Figure 2).

Morphologically, the Raji-adherent cells were > 85 % agranular lymphocyte with a surface phenotype of T cells (Table I).

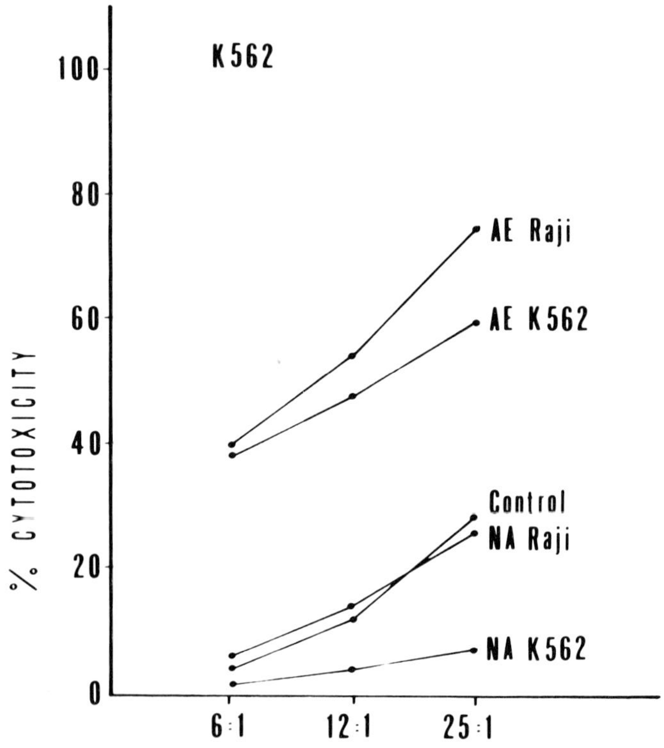

Fig. 2. Cytotoxic activity of peripheral blood lymphocytes absorbed and eluted (AE) from Raji or K562. The cells have been cultured overnight after the fractionation. NA: cells nonadherent of tumor cells. Control: unfractionated mononuclear cells.

TABLE II. Morphological characteristics of Percoll fractions cultured in various conditions

Culture condition	Fraction[a]						
	1	2	3	4	5	6	7
A control	60	81	48	21	9	3	1
B + Raji	58	77	50	20	9	5	2
C + Raji + macrophages	55	80	58	44	17	8	5
D + macrophages	50	75	44	15	5	1	1
C - A	< 1	< 1	10	23	8	5	4

C. Differentiation of Medium-Sized Lymphocytes to LGL in Culture

As the data described above suggested that AK cells induced by Raji were agranular lymphocytes, it was of interest to study whether they would include progenitors of LGL. Various immunostimulants and differentiation inducers have been tested, but the only treatment which has caused an increase in the frequency of LGL among medium-sized lymphocytes has been incubation of Raji-stimulated cells on monolayers of autologous macrophages (Table II). The total amount of cells did not increase during the incubation and no blasts or mitotic figures were detected in morphological analyses. This indicated that an expansion of the pre-existing LGL-pool did not take place in these experiments.

We next fractionated the medium sized lymphoycytes to various subpopulations by monoclonal antibodies and FACS. As shown in Table III, the LGL-morphology could be induced to heterogeneous lymphocyte subsets, the only constant features being an almost complete lack of LGL-progenitors in NK-9- and T4+ populations.

When used alone, M1+ cells did not transform to LGL under these culture conditions. However, when unfractionated cells were stimulated with Raji and then cultured on macrophages, the frequency of LGL among M1+ cells increased without a change in the total amount of M1+ cells. This suggests that M1- auxilliary cells are required in the differentiation of M1+ medium-sized lymphocytes to LGL (data not shown).

TABLE III. Differentiation of agranular lymphocytes to LGL in various subpopulations of medium-sized lymphocytes

Marker	% LGL	
	Control	+Raji+macrophages
T3 +	10	34
−	38	47
MI +	50	45
−	6	23
T8 +	13	63
−	5	24
NK-9 +	16	40
−	< 1	8
T4 +	2	4
−	24	45
HNK-1 +	50	52
−	4	20

IV. DISCUSSION

Natural killer cells originate in the bone marrow, yet there are only few LGL and very little cytotoxic activity in this organ, at least in the rat (Reynolds *et al.*, 1982). Probably the functional and morphological maturation takes place in the periphery. As cytotoxic NK cells do not belong to the recirculating lymphocytes (Zöller *et al.*, 1982) and most of the lytic activity is found in blood and spleen, it is quite likely that the terminal diferentiation of NK cells takes place in these organs. Splenectomy does not influence NK activity (Haller *et al.*, 1977). Therefore it could be assumed that the differentiation of NK cells takes place in blood, either inherently without a requirement of maturation factors, or somehow through the control of hematopoietic and reticuloendothelial systems. So far there is no data on whether the typical LGL-morphology and cytotoxic capacity appear in NK cells simultaneously.

The results of this study show that morphologically agranular mononuclear cells containing activated killer cells are partially transformed to LGL when cultured with macrophages. It could be hypothesized that this process may reflect the differentiation of LGL also in vivo. The present data also suggests that the cytotoxic activity and LGL morphology do not appear in NK cells simultaneously, the cytotoxic progenitors being agranular.

However, important questions remain to be answered before the role of macrophages in the facilitation of LGL differentiation from agranular precursors can be established. The major concern is whether the LGL which differentiate in culture really represent the LGL encountered in vivo. AK cells are phenotypically T cells whereas LGL for the most part are not. It is also so in this case: the Raji-adherent cells are mostly T3+ and M1-, and LGL-differentiation can be induced to T3+, M1- agranular lymphocytes. To explain the T cell phenotype of culture-induced progenitors of LGL, two possibilities remain. Either the expression of the T3 determinant is down-regulated during the contact of AK cells with macrophages, or the phenomena described here concern mainly T3+ LGL, which exist as a minor subpopulation of NK cells also in vivo (see Table III). In the latter case, the cytolytic AK cells would represent only a part of LGL-progenitors. So far we have no evidence on an alteration of T3-expression on AK cells during culturing with macrophages. However, it is known that the T3 determinant rapidly appears on LGL when they are cultured in the presence of T cell growth factor (Timonen _et al._, 1982). It could be hypothesized that a shift to the opposite direction may in some situations also take place.

Another issue in regard to the relationship of AK cells and LGL is the clear difference in their target cell specificity (Grimm _et al._, 1982). We have not yet studied whether the pattern of cytotoxicity against NK-insensitive target cells would change as a result of culturing AK cells with macrophages.

Although conclusions in regard to the biological significance of in vitro differentiation of LGL would be premature, the present approach will probably open possibilities to study the interrelationship of LGL and cytotoxic T cells at a precursor cell level: if precursors of LGL are exhausted from lymphoid cells by the methods described in this paper, will the precursors of cytotoxic T cells still remain in the non-LGL pool? Such studies are presently under way.

ACKNOWLEDGEMENTS

Supported by Grant CA-23809 from the National Cancer Institute, National Institutes of Health, Bethesda, MD. The technical assistance of Miss Pirkko Kalliomäki is gratefully acknowledged.

REFERENCES

Grimm, E.A., Mazumnder, A., Zhang, H.Z., and Rosenberg, S.A. (1982). J. Exp. Med. 155, 1823.

Haller, O., Kiessling, R., Örn, A., and Wigzell, H. (1977). J. Exp. Med. 145, 1411.

Hercend, T., Reinherz, E.L., Meuer, S., Schlossman, S.F., and Ritz, J. (1983). Nature 301, 158.

Nieminen, P., and Saksela, E. (1984). J. Immunol. 132, 702.

Ortaldo, J.R., Sharrow, S.O., Timonen, T., and Herberman, R.B. (1981). J. Immunol. 127, 2401.

Reynolds, C.W., Timonen, T., Holden, H.T., Hansen, C.T., and Herberman, R.B. (1982). Eur. J. Immunol. 12, 577.

Seeley, J.K., Masucci, A., Poros, E., Klein, E., and Golub, S.H. (1979). J. Immunol. 123, 1303.

Timonen, T., Saksela, E., Ranki, A., and Häyry, P. (1979). Cell. Immunol. 48, 133.

Timonen, T., Ortaldo, J.R., and Herberman, R.B. (1981). J. Exp. Med. 153, 569.

Timonen, T., Ortaldo, J.R., Stadler, B.M., Bonnard, G.D., and Herberman, R.B. (1982). Cell. Immunol. 72, 178.

Timonen, T., Reynolds, C.W., Ortaldo, J.R., and Herberman, R.B. (1982). J. Immunol. Methods 51, 269.

Zöller, M., Bellgrau, D., Axberg, I., and Wigzell, H. (1982). Scand. J. Immunol. 15, 159.

PATTERNS OF PROLIFERATION IN HUMAN NK CELLS: PRACTICAL AND THEORETICAL IMPLICATIONS[1]

William W. Travis[2]
Hillel S. Koren

Division of Immunology
Duke University Medical Center
Durham, North Carolina

I. INTRODUCTION

The proliferative capacity of natural killer cells has been the subject of numerous investigations in recent years. The practical applications of this research, such as the generation of NK clones for analytical and clinical use and the consequences for our theoretical constructs of the NK system, underscore the importance of understanding this aspect of NK cell biology.

Here we will examine recent progress in dissecting the phenomena surrounding NK cell expansion, especially as it concerns the identification of NK cells and their progeny. We will also look at the heterogeneity of form and function within the NK system and the implications for an NK lineage.

---

[1]Supported by National Institutes of Health Grants CA23354 and CA09058.

[2]Present address: Department of Microbiology and Immunology, Baylor College of Medicine, Houston, Texas.

ISBN 0-12-341370-2

II. NK PROLIFERATION: ANALYSIS OF PUTATIVE PROGENY

When IL2-dependent NK clones were first described by Dennert in 1980 (1), the endogenous NK cell was still poorly defined and difficult to isolate in large quantities free from non-NK contaminants. It was presumed that clones with NK cytolytic function would mirror the NK cell found in vivo and that these clones represented the product of an NK lineage rather than another lymphocyte subclass. This section is a survey of research on proliferating NK-like effectors, in which we will analyze their usefulness for the study of NK cell growth characteristics. The particular focus will be on our ability to identify the origin of clones based only on their NK-like attributes.

A. Cytolytic Activity

The primary criterion for the identification of an NK cell is its characteristic, selective pattern of cytolysis on certain tumor cell lines in the absence of MHC restriction (2). A number of investigators have been successful in establishing long term lines and clones capable of killing NK sensitive targets, suggesting a kinship with the in vivo NK cell (3-12). Yet several lines of experimental evidence challenge that correlation. Although not tested in all cases, many clones will lyse a broader spectrum of targets than will endogenous NK cells. It is not clear whether this is a difference in kind or degree, and hence, such cytolytic activity is often referred to as NK-like or as activated killing. Second, NK-like activity can be induced in cells with no known relationship to NK cells. Allospecific cytotoxic T lymphocytes (CTL) under appropriate culture conditions also exhibit NK-like killing potential (13), as is similarly seen in the lymphokine activated killer (LAK) (14) and anomalous killer (AK) (15) phenomena. MHC-unrestricted killing is clearly not limited to the freshly isolated NK effector. These various killer systems may potentially represent a fundamental process in the generation of NK cells in vivo. However, because of their superficial likeness to fresh NK, they render doubtful the identification of NK progeny by cytolytic function alone.

B. Morphology

Timonen, et al. demonstrated the close association of fresh NK activity with a morphologically distinct class of lymphocytes which they termed large granular lymphocytes (LGL) (16). By light microscopy, Giemsa-stained LGL exhibit an eccentric, often slightly indented nucleus with abundant cytoplasm containing azurophilic granules. By electron microscopy another characteristic feature, a cytoplasmic inclusion termed the parallel tubular arrays (PTA), is found predominantly, if not exclusively, in LGL (17). LGL-like morphology has also been used to identify NK clones and appears on most clones with NK-like cytolytic function. However, many investigators find that cloned CTL (18) as well as noncytolytic clones (12) also resemble LGL. Conversely, clones lacking LGL morphology have also been shown to express NK activity (19). Consequently, morphology of cultured lymphocytes may not be a good indicator of their origin nor adequate for defining a true clone of an NK cell.

C. Cell Surface Markers

Freshly explanted effectors of NK possess a distinctive array of cell surface markers as shown in Table I. They lack the pan T cell antigen T3 (26) but can be recognized by the NK specific monoclonal antibodies (mAb) B73.1 (23), OKM1 (28), and Leu-7 (29). Among NK-like cytolytic clones there seems to be no preferred antigenic phenotype. Timonen, et al. (5) have demonstrated that LGL, upon culture with PHA and IL2-containing media, tend to gain reactivity with OKT3. Other authors corroborate this finding with T3-bearing NK-like clones (4, 6, 9, 11, 12). Yet there are many examples of clones lacking this antigen (7-12) and therefore these cells may more closely resemble the endogenous NK effector. With regard to the expression of NK-associated markers on cloned lines, most authors find the antigens recognized by OKM1 and Leu-7 to be absent or at low levels. In some cases the presence of Fc receptors and B73.1 antigen has been reported (11, 12). A recent communication from van de Griend, et al. (12) details an interesting negative correlation between the expression of the antigens defined by the mAbs B73.1 and OKT3; in no instance were these antigens coexpressed on the

TABLE I. Monoclonal antibodies used to identify NK cells

Antibody	Cell Specificity	Reference
OKT3	pan T, non-NK	(20)
OKT11	pan T, NK, (ER)	(21)
OKT8	supp. T, NK	(22)
B73.1/Leu-11	NK, PMN, (FcR)	(23)
OKM1	NK, monocyte, PMN, (LFA-1)	(24)
HNK-1/Leu-7	LGL	(25)

same clone. This is more reminiscent of the phenotype of fresh NK cells and may indicate that B73.1 clones represent the true progeny of endogenous NK effectors.

D. Can NK Cell Lineage Be Inferred From NK-Like Characteristics Alone?

The obvious difficulty with NK-like clones is determining whether they are derived from an endogenous NK cell or from non-NK cells. NK-like, MHC-unrestricted cytolytic function and LGL-like morphology are not confined to freshly isolated NK cells, although their appearance in cultured lymphocytes may reflect the *in vivo* pathway for the generation of NK cells. To date no one has produced a clone with all of the characteristics of the endogenous NK effector. Theoretically, there is no reason to suppose that the progeny of an NK cell would resemble the forebear in either form or function. The use of NK-like clones as models of NK function may be valid, but their relationship to the lineage of the endogenous NK cell is at present ill-defined.

III. NK PROLIFERATION: ANALYSIS OF PROGENITORS

Our enhanced ability to identify and isolate the endogenous NK effector, resulting in large part from recent advances in monoclonal antibody technology, now permits the study of growth responses in antigenically well-defined NK-enriched populations. The advantage of this approach lies in the certain knowledge of the identity of the progenitor and the ability to make direct correlations between marker and function. This approach may however be prejudiced against detecting the growth of ontogenically closely related preor post-NK cells, should they exist, which may differ antigenically from the NK effector. Fine definition and positive selection of the starting population has produced results which should be directly applicable to our knowledge of NK proliferation.

A. Short-Term Bulk Cultures

The mitogenic responses of NK enriched populations in short term cultures are poor compared with those of T cells. T-gamma cells, now known to include a large proportion of NK effectors, are deficient in their proliferative response to Con A and PHA (29). Abo and Balch have similarly demonstrated that Leu-7 bearing cells exhibit little if any detectable growth in the presence of PHA, ConA, pokeweed mitogen (PWM), or allogeneic cells (30). Our laboratory has examined the mitogen responsiveness of NK cells enriched by panning with the Leu-11, B73.1, or OKM1 monoclonal antibodies and found that all three reagents select for lymphocytes giving poor responses to PHA or tetradecanoyl phorbol acetate (TPA)(manuscript in preparation). In particular, Leu-11 and B73.1 positive cells have a profound deficiency in their growth capacity which could not be reconstituted with exogenously added IL2-containing medium (Table II). The low level proliferation consistently observed with OKM1 positive cells was found to be associated with a Leu-4(OKT3) subpopulation (31), whereas most of the remaining cells in the OKM1 population express the B73.1 antigen which indicates mitogen unresponsiveness in our hands. Analogously, the only proliferation detected in Leu-7 positive cells was found in a minor subpopulation coexpressing the OKT3 antigen. The Leu4/OKT3 antigen is strongly associated with T cells, but has never been detected on endogenous NK cells, as discussed

TABLE II. Proliferative response of B73.1 or OKM1 selected cells to mitogens or IL2[a]

Expt.	Cell	Treatment	Day of Harvest d.3	d.5	d.7
1	B73.1+	Medium	0.0[b]	0.0	0.0
		PHA(1ug/ml)	1.2	1.5	1.4
		IL2-CM(10%)[c]	0.5	0.8	1.0
		PHA+IL2-CM	1.2	1.5	2.1
	Unsep.	Medium	0.3	1.2	1.9
		PHA	21.5	54.4	50.0
		IL2-CM	8.5	56.5	46.0
		PHA+IL2-CM	15.5	43.6	81.4
2	OKM1+	Medium	0.6		
		PHA (5ug/ml)	9.7		
		Leu-4 (2ug/ml)	13.0		
	Unsep.	Medium	3.4		
		PHA	38.0		
		Leu-4	69.1		

[a]Selected cells cultured at 10^5 per well with 10^4 irradiated autologous PBMC in microtiter wells.

[b]Tritiated thymidine incorporation measured in counts per minute x 10^{-3}.

[c]IL2-containing medium (IL2-CM) prepared by pulsing PBMC with PHA and TPA, and recovering supernatant 48 hr later after culture in mitogen-free medium. This preparation may contain residual mitogens.

earlier. There are indications, however, that NK cells may pass through an obligatory Leu-4/OKT3 expressing stage in ontogeny (32), which these cells may represent.

B. Long-Term Clonal Expansion

In contrast to short-term bulk cultures, at least two groups have reported successful attempts to clonally expand B73.1 positive NK cells in the presence of PHA and IL2-containing medium (11, 12). Not only did almost all B73.1 derived clones express NK-like cytolytic function, but they also retained expression of the marker. No clones derived from B73.1 negative progenitors exhibited this marker. This strong linkage of B73.1 expression with progenitor and progeny and the tight correlation of this antigen with NK activity in fresh cells provide convincing evidence that freshly explanted NK cells can indeed be expanded in culture while retaining at least one important marker as well as cytolytic function, albeit altered to a broader target spectrum.

C. Heterogeneity of Growth Capacity in NK Cells

The apparent discrepancy between the efficiency of NK associated proliferation in short term bulk culture and long term clonal expansion may simply result from differences in the sensitivity of the readout system. If a very small percentage of NK cells are capable of growth, they may escape detection in the bulk culture yet be prominent in a limiting dilution system; in the experiments just described, the cloning efficiency of B73.1 cells was on the order of five percent. Alternatively, NK cells may mediate a form of self regulation such that when cultured at high density they are maximally inhibitory, but when cultured at low cell numbers they are free from blocking. A final consideration might be the source of IL2 used for expansion. Impurities in the preparation may have adverse effects on the ability of NK cells to respond to growth stimuli. TPA, for example, is a common additive used to elicit high titers of IL2. A recent communication from Trinchieri, et al. (33), however, states that phorbol esters are potent modulators of the Fc receptor (B73.1 antigen) on NK cells, leaving open the possibility of other functional alterations. Although TPA as found in our IL2 preparations seems to have no deleterious effects on T cell responses or on the maintenance of T cell lines, these results cannot be extrapolated to NK cells. In a timely report, Trinchieri, et al. (34) have shown that at least a portion of the NK population, as defined by B73.1 antigen expression, can give a strong proliferative response to pure recombinant IL2 in short term culture.

IV. SUMMARY AND CONCLUSIONS

The role and importance of proliferation by NK cells in the mediation of their effector function is at present unclear. Biron and Welsh have suggested that dividing NK or NK-like cells can be found in virally infected mice (35). NK cells have also been implicated in tumor surveillance (2), regulation of hemapoietic (36) and immune (37) processes, and anti-microbial defense (38). In analogy to T and B lymphocyte responses to specific antigens and mitogens, NK cells may respond to specific environmental pressures by expansion and thus enhance their effector function. Study of NK cell responses along this vein will certainly give us more information with respect to the ontogeny and fate of the NK cell, whether it is an end stage cell or if the NK effector found in the blood is only an immature form capable of further maturation or differention. Practical knowledge of the regulation of NK cell expansion will also be useful for the growth and maintenance of long term lines for possible therapeutic or research uses.

One way to assess NK cell proliferation is through its progeny. T and B lymphocyte progeny can be readily identified via distinct markers of function or functional maturity, such as T3 antigen on T cells or immunoglobulin production in B cells. Our knowledge of NK cell ontogeny has been limited (39) and requires more direct evidence from ontological studies correlating markers and function with the state of differentiation or maturation. Other criteria, such as LGL-like morphology and MHC-unrestricted cytotoxic function, have often been invoked as markers of NK lineage in cultured populations, but these seem to also be associated with the progeny of T cells or other non-NK cells. Such parameters have probably not been fully exploited, and may still produce useful techniques for the identification of NK progeny. The widespread availability of recombinant DNA technology suggests a different route to the identification of NK cells or cells of an NK lineage. In both T and B cells, genomic material is rearranged to produce functional genes coding for the respective lineage specific antigen receptor (40, 41). Should these rearrangements not occur in the fresh NK cell, this would provide an elegant marker and possibly a definition for NK cells, both _in vivo_ and _in vitro_. Although NK-like clones have been described which bear the putative T cell antigen receptor (42), they also coexpressed, perhaps in an obligatory manner, the T3 antigen which is not found on fresh NK cells or on many other NK-like clones.

To date the most useful information regarding NK related proliferation has come from studies where the NK cell progenitor has been well defined prior to culture. NK cells as defined by B73.1 or Leu-11 mAbs do seem to expand in culture, in response to T cell mitogens and growth factors, although evidence would suggest that they do so with low efficiency. Optimization of the growth conditions may significantly enhance the proliferative efficiency. In no case do these cells respond to lectins in the absence of IL2. IL2 may be necessary and sufficient for their growth. This should not be taken as an indication of T cell character in view of recent reports describing the direct dependence of B cell growth on this hormone (43). The low efficiency of growth of NK cells may define a heterogeneity within the population of B73.1 positive cells which should be amenable to further analysis. Alternatively factors in the growth medium, such as PHA or TPA, may have unsuspected harmful effects on the proliferative activity of NK cells. Further experimentation with pure growth factors, such as recombinant IL2 could resolve the question of NK cell growth regulation.

There had been early indications that NK cells behaved toward lectins and IL2 in a manner similar to T cells with respect to growth in culture. However, the present picture is considerably more complex. There is good reason to believe that endogenous NK cells can proliferate in response to IL2 alone, and that there is heterogeneity within the NK population in this respect. This has strong implications for NK cells responding to situations of T cell expansion and may speak to a possible feedback regulation mechanism (39). Moreover, the ability of NK cells to express Class II antigens suggests a more complex interaction with other cells of the immune system (44).

REFERENCES

1. Dennert, G. (1980). Nature 287, 47.
2. Herberman, R. B., and Holden, H. T. (1978). Adv. Can. Res. 27, 305.
3. Sheehy, M. J., Quintieri, F. B., Leung, D. Y. M., Geha, R. S., Dubey, D. P., Linner, C. E., and Yunis, E. J. (1983). J. Immunol. 130, 524.
4. Pawelec, G. P., Hadam, M. R., Schneider, E. M., and Wernet, P. (1982). J. Immunol. 129, 2271.
5. Timonen, T., Ortaldo, J. R., Stadler, B. M., Bonnard, G. D., Sharrow, S. O., and Herberman, R. B. (1982). Cell. Immunol. 72, 178.
6. Abo, T., and Balch, C. M. (1983). Eur. J. Immunol. 13, 383.

7. Krensky, A. M., Ault, K. A., Reiss, C. S., Strominger, J. L., and Burakoff, S. J. (1982). J. Immunol. 129, 1748.
8. Kornbluth, J., Flomenberg, N., and Dupont, B. (1982). J. Immunol. 129, 2831.
9. Hercend, T., Reinherz, E. L., Meuer, S., Schlossman, S. F., and Ritz, J. (1983). Nature 301, 158.
10. Hercend, T., Meuer, S., Reinherz, E. L., Schlossman, S. F., and Ritz, J. (1982). J. Immunol. 129, 1299.
11. Allavena, P., and Ortaldo, J. R. (1984). J. Immunol. 132, 2363.
12. van de Griend, R. J., van Krimpen, B. A., Ronteltap, C. P. M., and Bolhuis, R. L. H. (1984). J. Immunol. 132, 3185.
13. Brooks, C. G. (1983). Nature 305, 157.
14. Grimm, E. A., Ramsey, K. M., Mazumder, A., Wilson, D. J., Djeu, J. Y., and Rosenberg, S. A. (1983). J. Exp. Med. 157, 884.
15. Seeley, J. K., Masucci, G., Poros, A., Klein, E., and Golub, S. H. (1979). J. Immunol. 123, 1303.
16. Timonen, T., Ortaldo, J. R., and Herberman, R. B. (1981). J. Exp. Med. 153, 569.
17. Babcock, G. F., and Phillips, J. H. (1983). Surv. Immunol. Res. 2, 88.
18. Sheehy, M. J., Yunis, E. J., Agostini, R. M., Jr., Quintieri, F. B., Leung, D. Y. M., Geha, R. S., and Yunis, E. J. (1983). Lab. Invest. 48, 549.
19. Schneider, E. M., Pawelec, G. P., Liangru, S., and Wernet, P. (1984). J. Immunol. 133, 173.
20. Chang, T. W., Kung, P. C., Gingres, S. P., and Goldstein, G. (1980). Proc. Nat. Acad. Sci. USA 78, 1805.
21. Fast, L. D., Hansen, J. A., and Newman, W. (1981). J. Immunol. 127, 448.
22. Perussia, B., Fanning, V., and Trinchieri, G. (1983). J. Immunol. 131, 223.
23. Perussia, B., Starr, S., Abraham, S., Fanning, V., and Trinchieri, G. (1983). J. Immunol. 130, 2133.
24. Sanchez-Madrid, F., Nagy, J. A., Robbins, E., Simon, P., and Springer, T. A. (1983). J. Exp. Med. 158, 1785.
25. Abo, T., and Balch, C. M. (1981). J. Immunol. 127, 1024.
26. Zarling, J. M., and Kung, P. C. (1980). Nature 288, 394.
27. Breard, J., Reinherz, E. L., O'Brien, C., and Schlossman, S. F. (1981). Clin. Immunol. Immunopathol. 18, 145.
28. Abo, T., Cooper, M. D., Balch, C. M. (1982). J. Immunol. 129,1752.
29. Moretta, L., Ferrarini, M., Mingari, M. C., Moretta, A., and Webb, S. R. (1976). J. Immunol. 117, 2171.
30. Abo, T., and Balch, C. M. (1982). J. Immunol. 129, 1758.
31. van de Griend, R., ten Berge, I., Tanke, H. J., Roos, D., Schellekens, P. Th. A., Melief, C. J. M., Zeijlemaker, W. P., and Astaldi, A. (1982). J. Immunol. 128, 1979.

32. Abo, T., Miller, C. A., Balch, C. M., and Cooper, M. D. (1983). J. Immunol. 131, 1822.
33. Trinchieri, G., O'Brien, T., Shade, M., and Perussia, B. (1984). J. Immunol. 133, 1869.
34. Trinchieri, G., Matsumoto-Kobayashi, M., Clark, S. C., Seehra, J., London, L., and Perussia, B. (1984). J. Exp. Med. 160, 1147.
35. Biron, C. A., and Welsh, R. M. (1982). J. Immunol. 129, 2788.
36. Hansson, M., Beran, M., Andersson, B., Kiessling, R. (1982). J. Immunol. 129, 126.
37. Tilden, A. B., Abo, T., Balch, C. M. (1983). J. Immunol. 130, 1171.
38. Weston, P. A., Jensen, P. J., Levy, N. L., and Koren, H. S. (1981). J. Immunol. 126, 1220.
39. Abo, T., Miller, C. A., and Balch, C. M. (1984). Eur. J. Immunol. 14, 616.
40. Hedrick, S. M., Cohen, D. I., Nielsen, E. A., and Davis, M. M. (1984). Nature 308, 149.
41. Brack, C., Hirama, M., Lenhard-Schuller, R., and Tonegawa, S. (1978). Cell 15, 1.
42. Hercend, T., Meuer, S., Brennan, A., Edson, M. A., Acuto, O., Reinherz, E. L., Schlossman, S. F., and Ritz, J. (1983). J. Exp. Med. 158, 1547.
43. Zubler, R. H., Lowenthal, J. W., Erard, F., Hashimoto, N., Devos, R., Macdonald, H. R. (1984). J. Exp. Med. 160, 1170.
44. Phillips, J. H., Le, A. M., and Lanier, L. C. (1984). J. Exp. Med. 159, 993.

CULTURE OF NK CELLS: SPECIFICITY, PHENOTYPES, FUNCTIONS AND COMPARISON WITH ACTIVATED T CELLS

John R. Ortaldo
Paola Allavena
Llewellyn Mason
Ronald B. Herberman

Biological Therapeutics Branch
Biological Response Modifiers Program, DCT
NCI-Frederick Cancer Research Facility
Frederick, Maryland

I. INTRODUCTION

Previous studies have indicated that natural killer (NK) cells are a heterogeneous population in regard to their tumoricidal activity (1-3). In humans and rodents, NK cells are morphologically identifiable with large granular lymphocytes (LGL) (4-7). It has been suggested that each subset of NK cells recognizes one or a limited number of target cells. One means to more directly define this heterogeneity and their specificity is the testing of many NK cell-derived clones. LGL can be cultured and expanded in vitro in interleukin 2 (IL-2)-containing medium, with retention of their morphological characteristics (8,9). One of the more complex questions presently being addressed by a number of investigators is whether cytotoxic cells growing in IL-2 are derived from, or directly related to NK cells or T cells. A number of investigators have identified Lyt 2^+ MLC-activated or lymphokine-activated mouse lymphocytes that demonstrate tumoricidal activity similar to that of fresh NK cells (10,11). Similar results have been obtained with lectin-activated cultures of human T cells, which can demonstrate activated killer effects which resemble fresh natural killer cell activity. Recently a number of investigators (e.g. 12,13,14, 15) have demonstrated that clones of cytotoxic T lymphocytes (CTL), which demonstrate MHC or allorestricted killing,

ISBN 0-12-341370-2

can also demonstrate additional cytotoxic reactivity, mediated via separate receptors, which resembles that of the NK cell. Therefore, a critical question which must be addressed regarding either cultures or clones of mouse or human T lymphocytes can undergo alterations of specificity such that they mimic NK cells. Perhaps the final resolution of this issue will be possible when the structures recognized on the target cells by both natural killer cell cultures and activated T-cell cultures are identified (i.e., are NK recognition receptors distinct from T cell receptors). At present, however, a number of interesting differences in the mouse and the human have been examined and will be discussed regarding the phenotype and characteristics of the cells from both human and mouse when cultured in the presence of IL-2. In addition, the heterogeneity of human LGLs, clones and subsets of LGL recognized by monoclonal antibody and sorted by flow cytometry will be discussed.

II. RESULTS

Percoll-derived LGL and T cells (5) were cloned by limiting dilutions at 0.5 cell/well (1). Table I shows a summary of the data from all the clones tested. Only 22.4% of the clones (44 of 196) were significantly cytotoxic ($p<0.05$ Student's t-test) against a panel of tumor targets. Most of the lytic clones lysed all the targets tested, but a considerable proportion, 18 of 44 (41%), preferentially lysed only one or two targets.

Table II shows the actual level of cytotoxicity from 10 representative clones (9), some with a pattern of broad reactivity and others with more restricted reactivity. It is worth noting that these results were obtained both with clones derived from whole preparations of LGL and from those derived from MoAb-selected subsets of LGL. To demonstrate that the restriction in specificity of some of the clones was not simply due to the low number of effectors used; we increased the effector to target ratio (E:T). Even under these conditions there were clones able to lyse some, but not all the targets tested (data not shown), suggesting that the lack of reactivity was due to the absence of recognition receptors for that particular target.

LGL clones were also tested for the expression of surface markers. Table III presents a summary of the phenotype of LGL clones and also of T cell clones for comparison. All the LGL clones presented in this table had significant NK or ADCC activity. LGL clones reacted with some of the following

TABLE I. Summary of the cytotoxicity of LGL clones[a]

	No. of clones
Tested for cytotoxicity[b]	196
Cytotoxic[c]	44
" with broad reactivity	26
" with restricted activity	18
" " " " --only K562	6
" " " " --only Daudi	2
" " " " --only Molt-4	1
" " " " --only in ADCC	3
" " " " --only K562 + Molt-4	5
" " " " --only in ADCC + Alab	1
Cytotoxic against other NK targets but no ADCC	14

[a]LGL were obtained from peripheral blood lymphocytes by centrifugation on discontinuous gradients of Percoll (5). Contaminant T lymphocytes were removed by rosetting at 29°C with sheep red blood cells. The preparations were 80-90% LGL, >60% OKM1$^+$, and <5% OKT3$^+$. In some experiments LGL were treated with MoAb and positive cells were sorted by flow cytometry. LGL were cloned by limiting dilutions at 0.5 cells/well in medium RPMI 1640, supplemented with 10% human AB serum and IL-2 and autologous irradiated (3000 R) mononuclear cells as feeders as described in (9).

[b]Clones were tested in an 8-hr ^{51}Cr-release assay against a panel of NK targets (K562, Molt-4, Alab, Daudi) and on the ADCC target (RL♂1 + anti-Thy 1).

[c]Significant activity against at least one of the panel of 5 targets.

markers: OKT3, OKT4, OKT8, OKT10, OKM1, and B.73.1. All clones derived from B73.1$^+$ cells retained this marker during culture. The three clones derived from the unfractionated LGL were also positive for B73.1. The OKT8 antigen was expressed in half of the clones derived from unselected LGL, while none of the OKM1$^+$ LGL clones reacted with OKT8, but most were OKT4$^+$.

Because of the heterogeneous phenotype of fresh cytotoxic LGL, it was of interest to correlate the cytotoxic activity and the surface phenotype of the LGL-derived clones. However, the same pattern of cytotoxicity was often represented by clones with different phenotypes and some clones had the same

TABLE II. Target specificity of representative cytotoxic LGL-derived clones[a]

Characteristics of original population	Clone	E:T	% specific lysis K562	ADCC	Molt-4	Alab
whole LGL	H.50	3:1	66	49		13
B73.1^{+}-LGL	L.7	3:1	80	70	84	28
OKM1^{+}-LGL	C.7.1	3:1	66	20	62	
whole LGL	R.F.3	3:1	62		56	33
whole LGL	P.D8	3:1	27	0		0
B73.1^{+}-LGL	L.A1	20:1	100	0	0	
whole LGL	H.49	3:1	0	22		13
OKM1^{+}-LGL	C.8.2	3:1	0	38	0	
OKM1^{+}-LGL	C.6.8	3:1	25	0	15	
OKT10^{+}-LGL	C.4.3	3:1	32	0	22	
LGL uncultured[b]		20:1	69	70	68	55
LGL cultured[c]		20:1	65	60	55	49

[a]Data from (9).
All values >6% are significantly positive ($p \leqslant 0.05$) by Student's t-test. Specific lysis <6% is listed as 0. Blank spaces mean the clone was not tested against that target. Subsets of LGL were fractionated by flow cytometry.

[b]Whole LGL population tested immediately after the separation from the blood.

[c]Whole LGL population cultured *in vitro* for 30 days in IL-2-containing medium.

surface antigens but divergent patterns of cytotoxicity (data not shown). No clear correlations were seen either for clones of different donors or for clones from the same donor, between phenotype and cytotoxic function.

LGL-derived clones were concomitantly tested for their (a) production of cytokines, (b) cytotoxicity against tumor targets, and (c) surface markers. Some clones had NK activity as well as cytokine-producing ability, whereas others secreted cytokines but had no detectable NK or ADCC.

Table IV provides an overall summary of the cytokine-producing capability of all the LGL-derived clones tested (16, 17). All of the first 12 clones tested spontaneously produced

TABLE III. Surface phenotype of cytotoxic LGL and noncytotoxic T-cell clones[a]

	Characteristics of original population prior to culture			
MoAb[b]	Whole LGL	B73.1.$^+$ LGL	OKM1$^+$ LGL	T cells[c]
OKT3	8/12[d] (66)[e]	2/4 (50)	3/6 (50)	7/7 (100)
OKT4	NT[f]	2/5 (40)	5/6 (83)	NT
OKT8	7/12 (58)	2/5 (40)	0/6 (0)	3/7 (43)
OKT10	4/9 (55)	NT	2/6 (66)	2/6 (33)
OKM1	10/12 (83)	NT	3/6 (50)	2/6 (33)
B73.1	3/3 (100)	5/5 (100)	NT	NT

[a]Data summarized from (9).

[b]Reactivity of cloned cells with MoAb was determined by the *Staphylococcus aureus* protein A assay. A clone was considered positive when at least 30% of the cells were coated with ⩾ 10 bacteria.

[c]T lymphocytes were obtained from the high-density fraction of Percoll gradient.

[d]Number of positive clones out of total number tested.

[e]Percentage of positive clones in parentheses.

[f]NT, not tested.

IFN$_\gamma$; when the number of clones studied was enlarged to 40, 5 clones among the noncytotoxic ones did not produce any detetable IFN even when stimulated with PHA, while the other clones spontaneously produced IFN$_\gamma$ and even higher quantities when stimulated with PHA.

The incidence of cytotoxic clones secreting IL-1 was higher than that of clones without detectable cytotoxic reactivity. However, more clones will need to be tested in order to determine to what extent these two functions are really associated. In contrast, the frequency of production of IL-2 was quite similar for the cytotoxic and noncytotoxic clones. In regard to BCGF, some of the noncytolytic and cytolytic clones were positive, but the number of clones tested will have to be enlarged before conclusions might be drawn regarding the possible connection between these two functions.

The possible association of cytokine production with a particular surface phenotype also was examined. In fresh LGL, IL-1-producing cells have been found to comprise a

Table IV. Cytokine production by clones of human LGL[a]

Cytokine[b]	Cytotoxicity[c]	No. positive/ No. tested	% +	Range of units/ml
IFN_{γ} sp.[d]	-	7/7	100	10-120
	+	5/5	100	25-125
IFN_{γ}[e]	-	32/37	86	10-625
	+	3/3	100	10-250
IL-1	-	7/48	14	20-85
	+	2/6	33	50-60
IL-2	-	14/41	34	3-165
	+	3/8	37	4-120
BCGF	-	5/24	21	51-166
	+	1/1	100	90

[a]Clones were obtained by limiting dilution at 0.5 cells/well and expanded in medium + IL-2. Data summarized from (16).

[b]Assayed in culture fluids from cells incubated for 48 hr with PHA or LPS but in absence of added IL-2.

[c]Cytotoxic activity was determined in a 6-8-hr chromium 51 release assay. A clone was considered positive when a significant level of reactivity (*P* $\leqslant$ 0.5, Student's *t*-test) was observed with at least one of a panel of three to five target cells.

[d]Spontaneous production of IFN was determined in the culture supernatant of clones incubated for 48 hr in the absence of PHA or LPS. Culture medium + IL-2 alone contained <5 U/ml.

[e]IFN_{γ} levels indicated are above the spontaneous levels in unstimulated control.

distinct subpopulation (B73.1^{+}, OKM1^{+}, Ia^{+} cells) (16). It has been more difficult to identify a particular LGL subset that is responsible for production of IFN_{γ} and IL-2, although the LGL producing IL-2 appear to be predominantly OKT11^{+} (17).

A summary of the phenotype of the cytokine-producing clones is presented in Table V. In agreement with the phenotype of the subset of LGL that is predominantly responsible for production of IL-1, all the clones producing IL-1 were OKM1^{+}, and three clones were also B73.1^{+}. However, six B73.1^{-} clones also were found to produce detectable levels of IL-1 activity. Most of the clones, as already reported (1), were OKT3^{+}, a T-cell marker that is expressed by most LGL-

Table V. Monoclonal antibody reactivity of cytokine producing LGL-derived clones

Cytokine produced	Reactivity with monoclonal antibody of clone producing the particular cytokine[a]			
	OKT3	OKT8	OKM1	B73.1
IL-1	3/9[b]	4/8	9/9	3/9
IL-2	8/9	3/6	8/9	1/9
IFN_{γ}	10/14	3/12	13/14	6/14

[a]Reactivity with MoAb was assessed by the *Staphylococcus aureus* protein A assay. A clone was considered positive when ≥30% of the cells were coated with 10 or more bacteria. Data summarized from (16).

[b]Number of positive clones out of total clones tested.

derived cultures (1,8), but is not detectable on LGL freshly isolated from the peripheral blood. The OKT8 antigen was also expressed on 58% of the cytokine-producing LGL-derived clones. A previous report (20) using fresh LGL indicated that uncultured $OKT8^+$ cells might have a suppressive effect on IL-1 production by LGL; however, 4 of the 8 clones that produced IL-1 were $OKT8^+$.

The above results (16) are disparate from our results of cytokine production by phenotypically distinct subsets of fresh LGL (17). This may be explained by the fact that the phenotype of cultured cells appears to differ significantly from that of the cell of origin, with some surface markers being lost in culture and others, which are undetectable on the surface of fresh LGL. Moreover, studies performed on cloned cells have shown that expression of some surface markers is not consistent, with considerable variation during culture (1,8).

Data from mouse studies regarding identification of effector cells lysing NK targets as being LGL-derived or T-cell derived have been confusing. Although a number of cytotoxic mouse clones have been discussed elsewhere (11,18,19) the derivation of such clones and identification of characteristics of their progenitors as well as their phenotype after culture needs further analysis. Most of these cultures are derived from unfractionated spleen cells or at most nylon wool nonadherent cells. We have therefore begun studies to

TABLE VI. Cytotoxicity and phenotype of cultured mouse LGL and T cells

Source of cells and pretreatment	LU 30% against YAC-1	% Positive with antibody to		
		Thy 1.2	Lyt 2.1	$asGM_1$
Day 0				
LGL				
+C	16.1	13[a]	4	48
+α$asGM_1$ + C	<0.1			
+αLy 2.1 + C	15.1			
T				
+C	<0.1	83	29	21
+α$asGM_1$ + C	<0.1			
+αLy 2.1 + C	<0.1			
Day 8[b]				
LGL	1317	74	10	99
T	202	99	58	69
Day 14[b]				
LGL	676	75	1	94
T	149	99	72	92

[a]Weak reactivity based on mean channel fluorescence.
[b]Cells were initiated from LGL- or T-enriched population shown above (complement control).

examine the nature of the progenitors as well as the nature of the effectors of IL-2-derived cultured and of clones which mediate NK-like activated killer cell activity against NK-susceptible targets such as YAC. Table VI represents a summary of the phenotype and the anti-YAC-1 cytotoxic reactivity of cultures derived from LGLs or T cells purified on discontinuous Percoll gradients in combination with treatment by monoclonal antibodies and complement. As can be seen, comparison of the phenotype at Day 0, LGL and T cells demonstrate considerable differences. The LGLs were very weakly Thy 1^+ compared to T cells, both regarding the percentage of Thy 1^+ cells and the mean channel fluorescence. Both fractions contained asialo GM_1^+ cells, whereas only some T cells exhibit significant amounts of the Lyt2. If one analyzes the effector function present in these cells (by antibody and complement lysis), the natural killer cells were asialo GM_1^+, Lyt2.1$^-$ effectors. This type of effector cell has been described previously for NK cells in a variety of strains (1,2). Upon culture, however, the observation that at both Day 8 and Day 14 significant levels of YAC killer cells can be culture-derived from both LGL-enriched populations (~50% LGL) and T cells. Although LGL-derived populations consistently demonstrated more activity than T cells, high levels of killing were demonstrated in the T-lymphocyte-derived population. Considerable phenotypic differences were observed in the populations which are derived from LGLs and/or T cells with regard to the expression of the Lyt2. Both LGL and T cells exhibited high levels of Thy 1 expression (data not shown). However, consistent in a number of experiments was the high level of expression of the Lyt2 which progressively increased with time of culture of the T-cell fraction but in contrast was generally not present in the cultures of LGL-enriched cells. Studies previously performed with Lyt2$^+$ mouse clones have demonstrated that activated killers, in addition to allospecific reactivity, have developed a broad pattern of cytotoxic activity in culture (10,11). Although present studies are being performed to elucidate the phenotype of the effector cells, the data to date (not shown) indicate the major phenotype of the LGL-derived activated killers with reactivity against NK-susceptible targets was Lyt2$^-$ in contrast to Lyt2$^+$ in the cultures derived from T cells. Whether the Lyt2 marker will reliably discriminate LGL-derived and T cell-derived effector cells remains to be elucidated.

III. DISCUSSION

The major purpose of the analysis of human LGL-derived clones was to clarify whether the cytotoxic specificity of NK cells and their cytokine-producing capabilities are clonally distributed characteristics. A substantial number of clones were tested for their cytotoxicity against a panel of 3-5 tumor cell lines for their phenotype as well as their cytokine production. Overall, the results demonstrated considerable heterogeneity among the clones with regard to the restrictive pattern of specificity, phenotype, and cytokine production. Some clones lysed only certain NK-susceptible target cells, with no detectable reactivity against the other target cells. Cytokine production also varied considerably among the clones, but with cytotoxic as wall as noncytotoxic clones exhibiting this functional capacity to similar degrees. Neither the cytokine-producing function nor the cytotoxic activity of these effector cells was correlated with a specific phenotype, with several different antigenic patterns being represented. Although considerable heterogeneity was seen in both the specificity as well as the cytokine production of LGLs, it should be pointed out that most of the clones were broadly reactive, similar to fresh NK cells, against the majority of the NK target panel, and these also produced at least some cytokines. In fact, a large number of clones had to be tested to accurately assess more restricted cytotoxic patterns. These result indicate that the heterogeneity of the entire NK population may be attributable, at least in part, to a mixture of clones with variability in their target specificity, phenotype, and function.

A central question regarding the observed heterogeneity among the clones derived from LGL-enriched populations is whether some of the clones might have been derived from contaminating T cells rather than LGL. In the present studies, the conditions used for both the human and the mouse studies were designed to optimally select enrich for LGLs and to minimize the number of contaminating T-cells. Human populations were highly purified for LGL and then depleted of T cells by E rosetting at 29°C and by treatment with monoclonal antibodies (OKT3 or anti-Leu-1). Similarly, in the mouse studies, cells which were Ly 1^+ or $Lyt2^+$ were depleted prior to culture. Further, some human clones were derived from a variety of positively selected subsets of LGLs, the $B73^+$, the $OKM1^+$ as well as the $OKT11^+$ subsets. Heterogeneity was seen in all LGL subsets, which would indicate that the heterogeneity is intrinsic to the LGL population and not due to T-cell contaminants. The cultures derived from mouse LGL-enriched populations had little or no expression of Lyt2, in striking contrast to the high percentage of $Lyt2^+$ cells in the cultures derived from T cells. Some investigators (e.g. 11) have contended that the majority of cells which given in culture in the presence of IL-2 are in fact derived from contaminating T cell populations. We agree that one cannot conclusively state that all of the cultured cells derived from the human or mouse LGL-enriched populations were all derived from LGL and not from a small subset of contaminating T cells. Admittedly, the frequency of cells responding in

LGL populations, based on progenitor limiting dilution assays, (9) was only 5-20%, whereas 10-50% of T cells demonstrated proliferative responses. However, despite these potential problems, we believe that data are much more consistant with a derivation of most of the cytotoxic cultured cells from LGL rather than T cells. The frequency of contamination by T cells in the LGL-enriched population was very small, usually 1% or less. Also, T cells which exhibit activated killer activity against NK-susceptible targets to the presence of high concentrations of IL-2 lose this cytotoxicity rapidly in the absence of IL-2 (8). In contrast, LGL-mediated activity has been shown to be stable and could be maintained for several days in the absence of IL-2. In addition, single-cell manipulation studies, which have been performed by Zagury, et al. (21), support the derivation of cytotoxic clones from NK cells. These studies have indicated that individual cells selected for a phenotypic marker of LGL or for NK activity have demonstrated similar heterogeneity with regard to specificity of cytotoxicity as well as to phenotypic markers after cutlure in the presence of IL-2.

Therefore, we conclude that most cultures of LGL-enriched populations are indeed derived from LGLs and not from contaminating T cells and that most of the observed heterogeneity is due to the intrinsic nature of the NK cell population. The heterogeneity within the cultured populations actually resembles the heterogeneity that exists in fresh NK cells, both with regard to cytolytic and cytokine-producing functions and phenotypic markers. The activated killing against NK-susceptible target cells by CTL clones or the loss of specificity of CTL clones in the presence of high concentrations of IL-2 may be a separate or unrelated phenomenon. Certainly one cannot term any cell which can kill YAC-1 in the mouse, or K562 in the human, as being a natural killer cell. Our results conclude that cultures initiated from purified T cells, depleted of detectable LGL or NK activity, could develop considerable cytotoxicity versus NK-susceptable targets after culture in the presence of IL-2. One should consider the phenotype, the functional stability, morphology, and other characteristics of these cells before attempting to attribute them to either being T cells or NK cells. However, the presence of an NK receptor on the surface of an activated cytotoxic T cell certainly has been supported by clones which have been analyzed by Binz et al. (12) and Bolhuis (13), demonstrating multiple effector receptors for MHC-restricted as well as NK-susceptible targets. Elucidation of these points may require the determination of whether recognition of targets by NK cells and by restricted T killers are through the same recognition receptors. We must determine whether clones of cytotoxic cells derived from LGL or T cells can acquire NK-reactivity and whether the recognition of NK targets by activated T cells is mediated via distinct or identical target surface molecules from those employed by fresh or cloned NK cells.

REFERENCES

1. Ortaldo, J. R., and Herberman, R. B. (1982). In "Annual Review of Immunology" (W. Paul, C. Fathman, H. Metzger, eds.), p. 359. Annual Reviews Inc., Palo Alto.
2. Minato, N., Reid, L., and Bloom, R. B. (1981). J. Exp. Med. 154, 750.
3. Stutman, O., Figarella, E. F., and Wisum, R. (1980). In "Natural Cell-Mediated Immunity Against Tumors" (R. B. Herberman, ed.), p. 1073. Academic Press, New York.
4. Timonen, T., and Saksela, E. (1980). J. Immunol. Methods 36, 285.
5. Timonen, T., Ortaldo, J. R., and Herberman, R. B. (1981). J. Exp. Med. 153, 569.
6. Taglibue, A., Luini, W., Soldaleschi, D., and Boraschi, B. (1981). Eur. J. Immunol. 11, 919.
7. Kumagai, K., Itoh, K., Suzuki, R., Hinuma, S., and Saitoh, F. (1981). J. Immunol. 127, 282.
8. Timonen, T., Ortaldo, J. R., Stadler, B. M., Bonnard, G. D., Sharrow, S., and Herberman, R. B. (1982). Cell. Immunol. 72, 198.
9. Allavena, P., and Ortaldo, J. (1984). J. Immunol. 132, 2363.
10. Stutman, O., Paige, C. J., and Figarella, E. F. (1978). J. Immunol. 121, 1819.
11. Brooks, C. G., Kuribayashi, K., Sale, G. E., and Henney, C. (1982). J. Immunol. 128, 2326.
12. Binz, H., Fenner, M., Frei, D., and Wigzell, H. (1983). J. Exp. Med. 157, 1252.
13. Bolhuis, R. H., Griend, R. J., and van de Ronteltep, C. P. M. (1984). Natural Immunity and Cell Growth Regulation 3(2), 61.
14. Hercend, T., Reinherz, E. L., Meuer, S., Schlossman, S. F., and Ritz, J. (1983). Nature (England) 301, 158.
15. Hercend, T., Meuer, S., Reinherz, E. L., Schlossman, S. F., and Ritz, J. (1983). J. Immunol. 129, 1299.
16. Allavena et al., submitted

17. Herberman, R. B., Allavena, P., Scala, G., Kasahara, T., Domzig, W., Procopio, A., Blanca, I. R., Ortaldo, J. R., and Oppenheim, J. J. (1983). In "Natural Killer Activity and Its Regulation" (T. Hoshino, H. Koren, A. Uchida, eds.) p. 409. Exerpta Medica, Tokyo.
18. Dennert, G., Yogeeswaran, G., and Yamagata, S. (1981). J. Exp. Med. 153, 545.
19. Kedar, E., Ikejiri, B. L., Sredni, B., Bonavida, B., and Herberman, R. B. (1982). Cell Immunol. 69, 305.
20. Arai, S., Yamamoto, H., Ihoh, K., and Kumagai, K. (1983). J. Immunol. 131, 651.
21. Zagury, D., Morgan, D. A., Fouchard, M., Bernard, J., Feldman, M., and Herberman, R. B. (1984). JNCI, in press.

ANALYSIS AND FUNCTIONAL CHARACTERIZATION OF T3$^-$, NK CELL DERIVED AND T3$^+$, CTL PRECURSOR CELL DERIVED CLONES

R.J. van de Griend
R.L.H. Bolhuis

Rotterdam Radio-Therapeutic Institute, Rotterdam; and Radiobiological Institute TNO, Rijswijk, The Netherlands

INTRODUCTION

Two major groups of cytotoxic effector cells are known: 1) Natural Killer (NK) cells, lymphocytes from, e.g., peripheral blood, endowed spontaneous with nonspecific and MHC' nonrestricted cytolytic activity against NK susceptible target cells such as K562 without deliberate prior antigenic stimulation. Most of these cells also show antibody-dependent cellular cytolysis (ADCC). They generally have the phenotype T3$^-$, B73.1$^+$, T4$^-$, T8$^{-/+}$, T11$^+$, OKM1$^+$ and the morphological appearance of large granular lymphocytes (LGL). Minor subpopulations with partially different characteristics may also exist; 2) cytotoxic T lymphocytes (CTL) having MHC restricted cytolytic activity after prior antigenic stimulation, e.g., after mixed lymphocyte culture (MLC). Their cytolytic activity is directed against target cells presenting the relevant stimulator antigens. Nonspecific cytolytic activity of MLC responder cells against, e.g., K562 but no ADCC, had been reported earlier (1) and has been defined as activated killer (AK) cell activity. They have the phenotype T3$^+$, B73.1$^-$ and usually T8$^+$, T11$^+$, OKM1$^-$. Our knowledge about target cell recognition and lytic mechanism of cytotoxic lymphocytes has been improved after cloned T3$^+$ effector cells expanded by using IL2, had come available. Until recently the expansion of NK cells up to large cell numbers sufficient to allow extensive analysis of NK cells at the clonal level, was difficult (2, 3).

We have developed a culture system for cloning and rapid expansion of both T3$^-$ and T3$^+$ clones (4) which enabled us to compare the phenotypic and functional characteristics of these

ISBN 0-12-341370-2

two major categories of cytotoxic effector cells. The culture system involves the use of a mixture of feeder cells, i.e., irradiated lymphocytes and cells from an EBV transformed B-cell line (B-LCL) (5). By using this procedure, which allows rapid expansion of $T3^+$ noncytotoxic effector cells, we have cloned and expanded $T3^-$ and $T3^+$ effector cells from several normal individuals up to 10^9 cells in about 30-40 days. Our results show that the clones derived from $B73.1^+$ and $B73.1^-$ cells, obtained by fluorescence activated cell sorting and from MLC responder cells differ markedly in their cytolytic activities, cell surface phenotypes and their response to activation signals. Moreover, in contrast to $T3^+$ clones, $T3^-$ clones do not respond to certain activation signals such as PHA. The functional and phenotypic heterogeneity within the group of $T3^-$ and $T3^+$ cloned effector cell types will be discussed.

MATERIALS AND METHODS

Generation and expansion of cytotoxic clones

$T3^-$ clones were obtained from either total peripheral blood lymphocytes (PBL) or cells enriched for $T3^-$, $B73.1^+$ cells by fluorescence activated cell sorting (FACS) or depletion of high affinity rosette forming cells at 29°C ($E29°^-$ cells). Clones were obtained by limiting dilution in the presence of two types of irradiated feeder cells, e.g., allogeneic PBL and B-LCL cells as described in detail elsewhere (4, 6). A particular set of rapidly growing $T3^-$ cytotoxic effector clones was derived from PBL from an individual with so-called Tγ lymphocytosis (7). Lymphocytes of this patient had the following phenotype: 8% $T3^+$, 5% $T4^+$, 4% $T8^+$, >95% $T11^+$, 79% $WT1^+$, 94% $OKIa^+$, 37% $B73.1^+$, 90% $OKM1^+$ and 35% $Leu7^+$. Morphologically most of its lymphocytes resembled large granular lymphocytes (LGL). This donor had 26 x 10^9/l lymphocytes among 30 x 10^9/l leukocytes, but did not show other clinical symptoms.

$T3^+$ clone n-4 was generated from a $T3^+$ cell fraction cultured in the same way as the $T3^-$ clones. Clone D11 was derived from PBL from another donor after priming in MLC (8, 9) against the B-LCL BSM and then expanded with feeder cells. Clone G9 was derived from a third donor, also after priming against BSM. Clone R1 was derived from donor 1 by priming against APD. All clones were subsequently expanded in 96-well round bottom microtiter plates in a system that contained irradiated (40 Gy) allogeneic PBL (2 x 10^4/well) and irradiated (20 Gy) BSM + APD feeder cells (10^4/well) (for details, see refs. 3 and 4). The HLA phenotypes of the various B-LCL and tumour target cells were: BSM: A2 Bw62 Cw3 Dw4,R4; APD: A1 Bw60,40.1 C- Dw6,Rw6; WVB: A2 Bw38 C- Dw6mRw6; WEN: A2,w19 B12,40 Cw3,w5 DR4; LICR-LON: A2,3 Bw35,7 Cw3,w4 DR4,R6; O-Mel: A2,11 B8,w62 Cw3 DR3,4.

Phenotyping with monoclonal antibodies (MABs)

Cell membrane phenotyping of cloned T cells was carried out by indirect immunofluorescence. Monoclonal antibodies (MAbs) OKT3,4,8,11 and OKM1 were obtained from Ortho Pharmaceutical Corp. (Raritan, NJ, U.S.A.). B73.1, which recognizes virtually all NK cells (10) was a gift from Dr. G. Trinchieri (Wistar Institute, Philadelphia, Pa., U.S.A.). MAb V D2, directed against Fc receptors from granulocytes, but also reactive with Tγ cells, was obtained from Dr. A.E.G.Kr. von dem Borne (Central Laboratory of Red Cross Blood Transfusion Service, Amsterdam, The Netherlands). As a second label, we used FITC-labeled goat-anti-mouse IgG (Nordic, Tilburg, The Netherlands). After washing, the cells were analyzed on a FACS II, Becton Dickinson, Sunnyvale, CA., U.S.A.).

Inhibition or induction of cytolytic activity with MAbs

Cytolytic activity was determined by using ^{51}Cr-labelled target cells (10^3 per well) in a 3 hr assay. Inhibition by MAbs was determined by adding the MAbs to the effector cells and incubating this mixture for at least 30 min at 37°C before adding the target cells. The following dilutions were used: anti-class I (w6/32, Mas 032c), 1:100; anti-class II (HLA DR 7.5,10.1 from Dr. F. Koning, Dept. of Immuno-Haematology, Academic Hospital, Leiden, The Netherlands), 1:100; anti-T3 (OKT3), 1:100; anti-T8 (OKT8 or FK-18; from Dr. F. Koning, see above), 1:100 or 1:1000, respectively; and anti-T4 (OKT4), 1:100.

Induction of cytolytic activity was done by preincubation of effector cells for 30 min at room temperature (continuous shaking) with OKT3 (25 μg/ml/10^6 cells), SPV-T3c (a gift from Dr. H. Spits, Dept. of Immunology of the Netherlands Cancer Institute, Amsterdam, The Netherlands) (10 μl culture supernatant/10^6 cells) or PHA (10 μg/ml/5.10^5 cells). The PHA but not anti-T3 was then washed away and effector cells were seeded for cytotoxicity assay.

RESULTS

T3$^-$ clones

T3$^-$ cytotoxic effector cells were cloned from lymphocytes enriched for T3$^-$ B73.1$^+$ cells by direct culture under limiting dilution conditions with feeder cells as previously described (11). All T3$^-$ clones exerted strong cytolytic activity against a variety of tumour target cells and, without exception, also ADCC activity (14/14) clones tested). The tumour cells lysed by a single clone included NK sensitive target cells (K562 and T-cell lines Molt4 and HSB); B-LCL cells (APD and BSM); class I nega-

Table 1

CYTOLYTIC ACTIVITY OF T3$^-$ ACTIVATED KILLER CELL CLONES

clone no.	target cells							
	K562	Molt-4[1]	P815-IgG[2]	APD	BSM[3]	Daudi	T24	O-Mel
- 1	71	74	63	37	53	68		
- 3	63	62	55	15	16	71		
- 4	67	77	61	27	35	87		
- 5	72	69	72	57	43	72	52	
- 6	75	57	71	45	43	71	28	
- 7	71	53	69	51	25	68	36	
+ 1	66	63	44	24	36	69		
+ 3	67	68	47	46	50	90		
+ 6	64	64	40	1	2	39		
+ 9	44		14	37		45		
+10	75	45	32	30	35	70	67	65
+15	66	44	37	8	34	58	75	84
+16	64	55	37	11	26	36	83	68
+17	47	25	20	7	27	49	34	59

Results expressed as % ^{51}Cr-release in a 3 hr assay.
Effector/target cell ratio 9:1.

[1] Similar results were obtained with the T cell line HSB.

[2] No cytotoxicity was observed against nontreated P815 cells.

[3] Other B-LCL cells such as WVB as well as the plasma cell leukemia derived cell line LICR-LON were also lysed by most of these clones.

tive B cells (Daudi); anchorage dependent target cells (T24 bladder carcinoma cells) and short-term cultured freshly biopsied melanomal cells (O-Mel) as well as LICR-LON human plasmacytoma cells (Table 1).

All except one of these clones showed the phenotype T3$^-$, T4$^-$, T8$^-$, T11$^+$, B73.1$^+$ and formed rosettes with sheep red blood cells (E_{sh} rosettes), confirming the binding of OKT11 MAb, and formed rosettes with IgG coated ox erythrocytes confirming the binding of B73.1 MAb and ADCC. One clone (p+16) of 14 tested from this individual did not express T11 antigens and did not form rosettes with E_{sh} and thus had the phenotype of null cells. T3$^-$ clones expressed IL2 receptors as determined with the anti-Tac MAb (data not shown). Also one clone showing the general phenotype of the T3$^-$ clones additionally expressed T8 antigens on about 50% of the cells. Another set of T3$^-$ clones with cytolytic activity was derived from an individual with a lymphocytosis of T3$^-$Tγ cells. As shown in Table 2, again all clones

Table 2

CYTOLYTIC ACTIVITY OF T3$^-$ CLONES DERIVED FROM AN INDIVIDUAL WITH Tγ LYMPHOCYTOSIS

clone no.	code in fig.1	target cells K562	P815-IgG	Daudi	T24	O-Mel	EAγ RFC %
RG71	h	29	86	46	4	8	26
RG72	f	30	84	50			13
RG73	g	65	77	74	16	11	14
RG74	c	30	95	45			36
RG75	b	54	75	72	18	14	61
RG76	d	69	79	77	17	14	32
RG77	a	75	77	85	53	46	62
RG78	i	57	68	56			11
RG79	k	50	75	55	26	13	20
RG80	e	27	82	45			17
RG81	j	42	76	52	8	16	18
RG82	l	42	80	43			15

Results expressed as % ^{51}Cr-release assay in a 3 hr assay. Effector/target cell ratio 9:1.

analysed, 12 out of 12 showed cytolytic activity, including ADCC. The level of ADCC activity was similar but that against other target cells varied from clone to clone. These clones were all T3$^-$ but, at variance with clones obtained from the sorting experiment, the expression of the antigen recognised by the B73.1 MAb varied from clone to clone. The fluorescence intensity varied from 0% in clone RG79 up to 92% in clone RG77 as shown in Fig. 1. In addition, the number of EAγ rosette forming cells also varied from 11-62% in these twelve clones (Table 2). However, two clones (RG77 and RG76) when tested with V D2 MAb, directed against Fc receptors from granulocytes, reacted equally well, i.e., all cells were positive. One of the clones (RG73) expressed the T8 antigen on 38% of the cells. It should be noted that only T3$^-$Tγ cells were expanded because neither from normal individuals nor from patients with a T3$^+$ B73.1$^+$ or T3$^+$ B73.1$^-$ Tγ lymphocytosis, clones with EAγ rosette forming properties were obtained (12). To estimate the frequency of T3$^-$ cytotoxic T cells that give rise to cytotoxic clones we have searched for these cells in 12 different individuals now.

Several T3$^-$ cytotoxic T-cell clones with broad target cell specificity and various lytic activities were generated from 7 out of 12 individuals tested as summarized in Table 3. Cells from two donors (1 and 2) were already cultured for more than 30

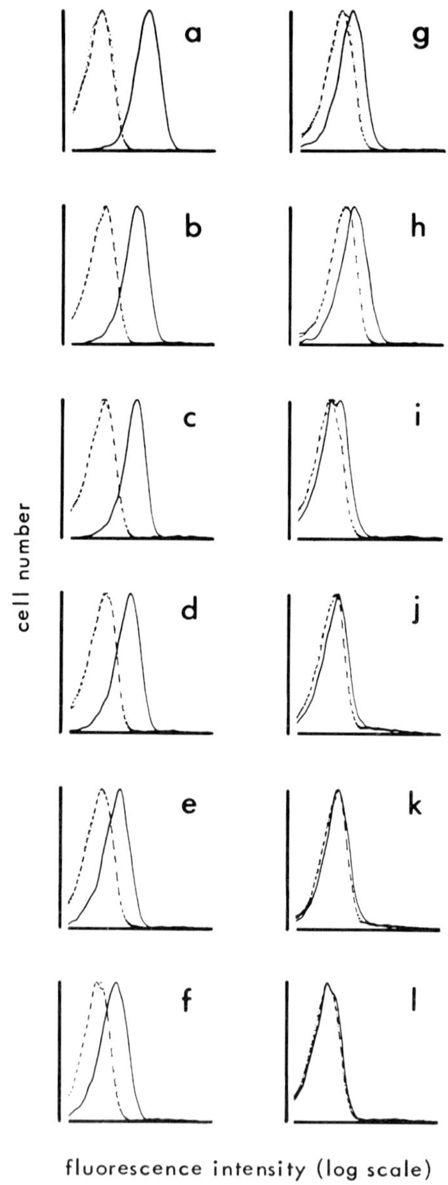

Figure 1.
Fluorescence histogram after staining with B73.1 of twelve clones derived from a patient with T3⁻ Tγ lymphocytosis.
..... = GAM/FITC control; ----- = OKT3; ——— = B73.1.

Table 3

FREQUENCY OF T3$^-$ CYTOTOXIC CLONES GENERATED FROM PBL OF NORMAL DONORS OR PATIENTS WITH T3$^-$ Tγ LYMPHOCYTOSIS

donor	pre-en-riched by	phenotype T3$^+$ (%)	phenotype B73.1$^+$ (%)	fresh/ frozen	cloning effi-ciency (%)	no. of clones tested total	T3$^-$CTX$^+$[1]
1	FACS[2]	nt	84	fresh	5	7	7
2[3]	not	8	37	frozen	5	12	12
3	not	68	nt	frozen	<1	10	5
4	not	67	14	frozen	<1	9	0
5	not	67	nt	frozen	<1	8	1
6	not	78	nt	frozen	10	7	1
7	E29°$^-$	21	45	fresh	1	9	3
8	E29°$^-$	19	48	fresh	<1	3	0
9	E29°$^-$	16	30	fresh	1	6	5
10	E29°$^-$	45	11	fresh	<1	9	7
11	E29°$^-$	31	19	fresh	<1	4	0
12	E29°$^-$	12	39	fresh	2	8	8

[1]CTX$^+$ means: cytolytic activity against K562, P815-IgG as well as Daudi cells is present.
[2]B73.1$^+$ cells were obtained by sorting and both B73.1$^+$ and B73.1$^-$ fractions were cultured with or without leucoagglutinin (1 µg/ml).
[3]T3$^-$Tγ lymphocytosis patient (12).

generations with a calculated cell yield of more than 10^9 in about 40 days. The cloning efficiency (5%) of the T3$^-$ cells, in contrast to that of T3$^+$ clones (5), was not enhanced by addition of leucoagglutinin plus feeder cells to the culture medium. The feasibility of large scale expansion of T3$^-$ cloned lymphocytes from a panel of randomly choosen blood donors is under study.

T3$^+$ clones

Clones derived from the B73.1$^-$ cell fraction and cloned and expanded identically as the T3$^-$, B73.1$^+$ cells showed the phenotype T3$^+$, B73.1$^-$, T4$^+$ or T8$^+$, OKM1$^-$, T11$^+$. Only one clone out of 20 (clone n-4) showed appreciable cytolytic activity against various tumour target cells. At E/T ratio of 20:1 the following lytic activities (% ^{51}Cr-release) were obtained: 49, 15, 32, 77, 26 and 58 for BSM, APD, LICR, Daudi, K562 and Molt-4, respectively. Although this clone did not express the B73.1 antigen and hardly formed (2%) rosettes with IgG coated ox erythrocytes, it exerted significant ADCC activity. Some clones exerted low levels of lectin dependent cellular cytolysis (LDCC). $T3^+$ cytotoxic clones were also derived from MLC responder cells, using as stimulator cells $Cw3^+$ and $DRW6^+$ B-LCLs (BSM and APD, respectively). The cells were then cloned and expanded and three of them were studied in detail for immune specific and nonspecific cytolytic capacity, phenotype and response to MAbs and lectin (D11, G9 and R1). The clones showed the phenotype $T3^+$, $B73.1^-$, $T4^+$ or $T8^+$, $T11^+$, $OKM1^-$.

The CTL clones lysed tumour target cells (K562), but higher effector to target cell ratios were required than with $T3^-$ clones and lysis against K562 was not blocked by either anti-T3, anti-T8 or anti-class I and II MAbs. The tumour target cell specificities are presented in Table 4. No ADCC was observed.

Clone D11 (Table 4) lysed the stimulator B-LCL BSM, but not lymphocytes expressing identical MHC class I or II antigens. Lysis was not blocked by anti-T3 or anti-T8 MAb, but surprisingly, anti-class-I MAb but nit class II MAb blocked lysis of BSM as well as LICR-LON and O-Mel.

$T3^+$,$T8^+$ clone G9 was also primed against BSM cells and showed a similar tumour target cell specificity and level of lytic activities. No ADCC activity was detected. Clone G9 also showed Cw3 antigen specific cytolysis of all $Cw3^+$ B-LCLs and lymphocytes. The lysis of $Cw3^+$ B-LCLs and lymphocytes was inhibited by anti-class I as well as anti-T3 and T8 MAbs.

Table 4

CYTOLYTIC ACTIVITY, SPECIFICITY AND SENSITIVITY FOR MAbs IN TWO $T3^+T4^-T8^+$ CLONES (D11 AND G9) AND ONE $T3^+T4^+T8^-$ CLONE (R1)

target cell	E/T ratio[1]	clone D11 CTX[2]	clone D11 inhibition[3] ClI/T8/T3	clone G9 CTX[2]	clone G9 inhibition[3] ClI/T8/T3	clone R1 CTX[2]	clone R1 inhibition[3] ClII/T4/T3
BSM	5	42	79/24/10	44	98/100/34	0	
APD	80	0		0		50	72/32/30
WEN	80	0		60		0	
WDV	80	0		0		30	
LICR	10	49	94/14	50	88/98	0	
Daudi	80	0		0		0	
O-Mel	20	38	89/-10/13	59	86/78	5	
$Cw3^+$/ $DRw6^+$ Ly	20	0		40		40	
K562	40	37	-3/-11/-22	22	14/73/-36	12	/-25
Molt-4	80	0		0		0	
HSB	80	0		5		0	

[1]Effector to target cell ratio.
[2]Results as % ^{51}Cr-release in a 3 hr assay.
[3]Inhibition (percentage) by adding various MAbs in the cytotoxic assay.

T3$^+$,T4$^+$ clone R1 showed only a low lytic activity against K562 target cells, but no other tumour target cells were lysed. No ADCC was observed. This clone also showed class II DRW6$^+$ specific lysis of B-LCLs and lymphocytes which could be inhibited by anti-class-II and anti-T3 and anti-T4 MAbs.

Responsiveness of T3$^-$ and T3$^+$ clones to activation signals by PHS and anti-T3.

Lectins (e.g. PHA) enhance the proliferation rate of T3$^+$ lymphocytes but not of T3$^-$ lymphocytes. In addition, the plating efficiency of T3$^+$ cells increased five-fold or more in the presence of lectin (leucoagglutinin) whereas the plating efficiency of T3$^-$B73.1 cells was not enhanced by lectins (6). As we demonstrate here, PHA (and anti-T3 MAbs) can also induce or augment cytolytic activity of some T3$^+$ cytotoxic clones against some target cells. The lytic activity of clones D11, G9 and R1 against K562 was enhanced by PHA and, to a lesser extend by anti-T3 MAb whereas anti-T3 MAb had a stronger effect on the cytolytic activity of clones D11 and G9 on Daudi cells (Table 5). Although anti-T3 MAbs can induce non-specific cytolytic activity, the allospecific cytolytic activity of, e.g., clone G9

Table 5

ANTI-T3 MAb AND PHA CAN INDUCE NONSPECIFIC CYTOLYTIC ACTIVITY IN SOME T3$^+$ BUT NOT IN T3$^-$ CLONES AND FRESH NK CELLS

clone	phenotype	additions	target cells		
			K562	Daudi	BSM
D11	T3$^+$T8$^+$	none	17[c]	0[d]	32[a]
		+ PHA	37	21	32
		+ anti-T3	20	23	29
G9	T3$^+$T8$^+$	none	30[c]	4[d]	55[a]
		+ PHA	58	24	46
		+ anti-T3	41	41	34
R1	T3$^+$T4$^+$	none	8[c]	4[d]	33[a]
		+ PHA	19	13	35
		+ anti-T3	6	4	25
					ADCC
RG75	T3$^-$T8$^-$	none	17[b]	34[a]	42[a]
		+ PHA	19	36	39
		+ anti-T3	17	37	35

E/T ratio a) 3:1; b) 9:1; c) 27:1; d) 81:1.

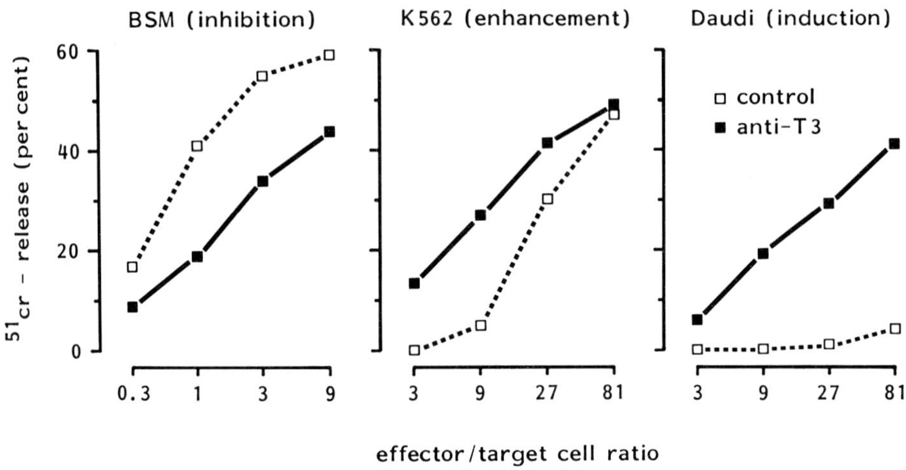

Figure 2.
Effect of anti-T3 MAb on the cytolytic activity of clone G9 depends on the target cell.

against BSM is inhibited. This differential effect that anti-T3 MAbs can have on cytolytic activity against different target cells is illustrated in Fig. 2. As also shown in Table 5 the lytic activity of $T3^-$ clone RG75 is not enhanced by either PHA or anti-T3 MAb. Thus, lytic activity of $T3^+$ cells can be augmented and induced by both PHA and anti-T3, but not anti-T8 MAbs in a target cell dependent manner. These results suggest that the T3 antigen is involved in the transmission of the activation signal by PHA.

DISCUSSION

We have compared the lytic activities of $T3^-$ and $T3^+$ cloned T cells and their sensitivity for MAbs and proliferation inducing signals. The $T3^-$ clones were derived from $B73.1^+$ FACS isolated lymphocytes (6, 10). Most were $T3^-$, $B73.1^+$, $T4^-$, $T8^-$, T11, $OKM1^-$ and thus phenotypically resemble Tγ-lymphocytes (11). One clone did not express T11 and therefore resembled null cells (11). Two clones derived from the peripheral blood of a patient with Tγ lymphocytosis were $T3^-$, $B73.1^-$ but showed normal cytolytic activity including ADCC activity. All $T3^-$, $B73.1^+$ NK cell derived clones without exception derived either from normal donors or from patients with Tγ lymphocytosis exerted strong cytolytic activity including ADCC (6, 12). Thus, the lack of the

T3 antigen, the capacity for ADCC and the high MHC non-restricted lytic activity and wide tumour target cell specificity range clearly distinguishes NK cell derived clones from $T3^+$ CTL with both MHC restricted nonspecific cytolysis. We have demonstrated here that, although $T3^+$ cloned CTL can lyse K562 and other tumour cells, their lytic activity is lower and the target cell specificity range in narrower in comparison with $T3^-$ clones as recently described in detail (13). Moreover, $T3^+$ cells can exert non-specific cytolysis without concomitant allospecific cytolysis (e.g., $T3^+$ $T4^-$ $T8^-$ clone, n-4). In addition, PHA and anti-T3 MAb do not influence the plating efficiency or cytolytic

Table 6

DIFFERENCES BETWEEN $T3^-$ AND $T3^+$ ACTIVATED KILLER CLONES

	$T3^-$	$T3^+$
phenotype	$B73.1^+$ $T4^-$ $T8^-$ $T11^+$ $B73.1^+$ $T4^-$ $T8^-$ $T11^-$ $B73.1^+$ $T4^-$ $T8^+$ $T11^+$ $B73.1^-$ $T4^-$ $T8^-$ $T11^+$	$T4^+$ $T8^-$ $T4^-$ $T8^+$ $T4^-$ $T8^-$
EAγ RFC	yes	no
lytic activity	$ADCC^+$	$ADCC^-$
target cell specificity broad	broad	narrow
induction of proliferation by PHA and/or anti-T3	no	yes
induction of nonspecific lysis by PHA or anti-T3	no	yes
level of non-specific lytic activity	high	low
specific lytic activity	no	yes
all clones have cytolytic activity	yes	no

activity of cloned T3$^-$ cells, whereas they increase the plating efficiency of T3$^+$ cells and augment and induce the nonspecific, non-MHC restricted lytic activity of cloned T3$^+$ CTL and inhibit the concomitant allospecific lysis of these clones. The anti-T8 MAb may inhibit allo-specific cytolysis but does not induce cytolysis by these clones. Therefore, these data also suggest that the T3 antigen is involved in the induction of IL-2 receptors, in proliferation and in cytolytic activity in T3$^+$ cells.

In conclusion, T3$^-$ cells derived clones activated during cloning and expansion can be distinguished from T3 CTL activated during MLC followed by cloning and expansion as summarized in Table 6.

ACKNOWLEDGEMENTS

We thank Dr. G. Trinchieri for providing the MAb B73.1 and M.J. Giphart for providing the various HLA-typed B-LCL cells. We thank Mrs. B.A. van Krimpen, Mr. C.P.M. Ronteltap and Mr. F. Smekens for technical assistance and Mrs. M. van der Sman for typing the manuscript.

REFERENCES

1. Bolhuis, R.L.H. and Schellekens, H. (1981), Scand. J. Immunol. 13, 401.
2. Hercend, T., Reinherz, E.L., Meuer, S., Schlossmann, S.F. and Ritz, J. (1983). Nature 301, 158.
3. Allavena, P. and Ortaldo, J.R. (1984). J. Immunol. 132, 2363.
4. Griend, R.J. van de and Bolhuis, R.L.H. (1984). Transplantation 38, no. 4.
5. Griend, R.J. van de, Krimpen, B.A. van, Bol, S.J.L., Thompson, A. and Bolhuis, R.L.H. (1984). J. Imm. Methods 66, 285.
6. Bolhuis, R.L.H., Griend, R.J. van de and Ronteltap, C.P.M. (1984). Nat. Imm. and Cell Growth Regulation 3, no. 2, 61.
7. Rümke, H.C., Miedema, F., ten Berge, R.J.M., Terpstra, F., van de Reyden, H.J., van de Griend, R.J., de Bruin, H.G., von dem Borne, A.E.G. Kr., Smit, J.W., Zeylemaker, W.P. and Melief, C.J.M. (1982). J. Immunol. 129, 419.
8. Spits, H., Vries, J.E. de and Terhorst, C. (1981). Cell. Immunol. 59, 435.
9. Tanaka, Y., Sugamura, K. and Hinuma, Y. (1981). J. Immunol. 128, 1241.
10. Perussia, B., Starr, S., Abraham, S., Fanning, V. and Tinchieri, G. (1983). J. Immunol. 130, 2133.

11. Griend, R.J. van de, Krimpen, B.A. van, Ronteltap, C.P.M. and Bolhuis, R.L.H. (1984). J. Immunol. 132, 3185.
12. Griend, R.J. van de and Bolhuis, R.L.H. (1985). Blood 65, in press.
13. Griend, R.J. van de, Giphart, M.J., Krimpen, B.A. van and Bolhuis, R.L.H. (1984). J. Immunol. 133, 1222.

NK CELLS AND T CELL DIFFERENTIATION

Colin G. Brooks

Basic Immunology,
Fred Hutchinson Cancer Research Center,
1124 Columbia Street,
Seattle, WA 98104

Whether mature NK cells are capable of sustained proliferation *in vitro* (or *in vivo*) is still unclear. Nonetheless, a large literature exists describing cloned lines of "NK-like" cells. While certain characteristics of these clones are indeed very similar to those of NK cells, there are also substantial discrepancies, most commonly in target cell specificity and in surface phenotype, the latter often being reminiscent of activated T cells. For example, mouse cell clones, composed of large granular cells, with potent lytic activity towards YAC-1, and with the general phenotype Ig^-, Thy-1^+, Ly-1^-, Ly-2^-, Ly-5^+, Qa-5^+, asialo-GM_1^+, NK alloantigen$^+$ (the same as NK cells *in vivo*), can readily be generated (1). However, these clones also lyse many NK-resistant targets and sometimes express Ly-2, a marker highly specific for cytotoxic/suppressor/class-I-reactive T cells, and absent from splenic NK cells. This latter observation led us to directly examine the relationship between NK cells and various T cell subsets. In this chapter I shall review our published (2-5) and unpublished work in this area and discuss how these findings have contributed new ideas to our understanding of NK cells. The general area of NK clones and NK regulation by IL-2 has been reviewed elsewhere (6, 7).

NK ACTIVITY IN CLASSICAL CTL

Antigen-specific clones were produced by standard methodology (8), and in several cases were recloned to ensure monoclonality. *All* clones generated in these initial experiments were Thy-1^+, Ly-2^+, Ig^-. At intervals of 2-6 weeks, the cells were stimulated with irradi-

ISBN 0-12-341370-2

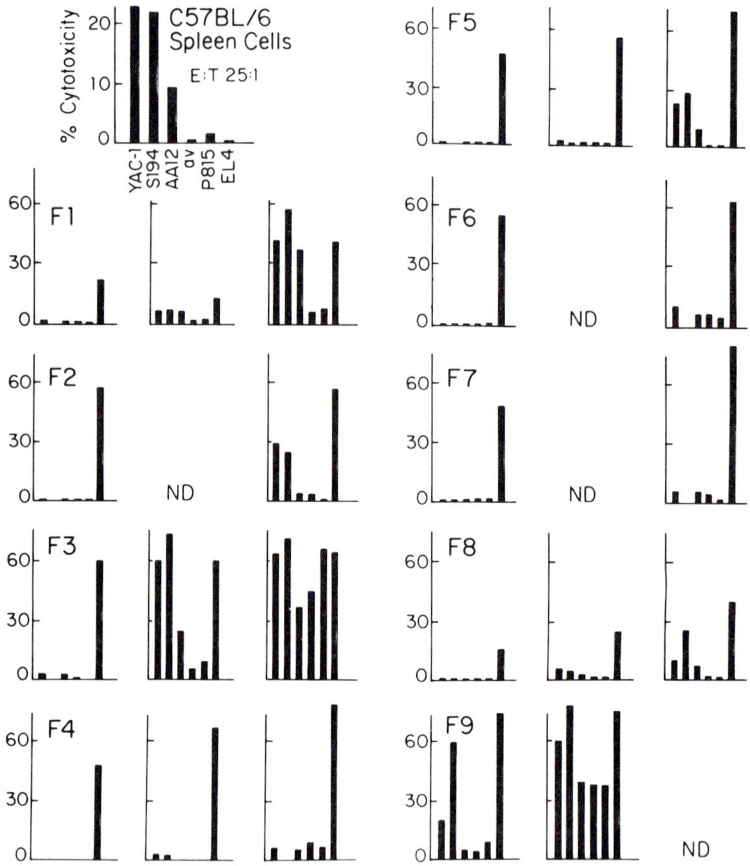

Fig.1. Specificity of bm1 anti-b CTL clones at 1:1 E:T ratio after 0, 3, or 7 d. culture in medium with 30% CS.

ated allogeneic spleen cells in medium containing 10% rat Con-A-induced spleen cell supernatant (CS), and after 1 week were returned to basal medium containing 2% CS, whereupon the cells reentered the quiescent state. Most of our work has utilized bml anti-b clones, where antigen-specific reactivity is determined by the mutation of 3 amino acids in the H-2K^b molecule (9).

Contrary to previous belief, it soon became clear that reactivity of the CTL clones could be readily manipulated by exposing the cells to suitable culture conditions and to appropriate lymphokines. The essence of our original findings is exemplified by the data in Fig. 1, which shows the cytolytic specificity of all 9 clones obtained from plating tertiary MLR cells at 1 cell/well. The cloning frequency was such that the probability of monoclonality was >95%. No selection was made other than that the cells show sufficient proliferative activity through cloning and subsequent expansion to provide adequate numbers of cells for testing. Thus, these results are fully representative of CTL clones generated in this manner, and essentially identical results have been obtained in many other independent experiments.

The specificity of the clones was tested at day 0 (left panel), day 3 (middle panel), and day 7 (right panel) at a 1:1 E:T ratio on a collection of mycoplasma-free target cells of which YAC-1 and S194 are highly NK sensitive, AA12 has intermediate NK sensitivity and av, P815, and EL4 are NK resistant (the lysis of these targets by fresh spleen cells is shown at top left). At day 0, 8/9 clones showed significant lysis only of the $H-2^b$-bearing target EL4 (antigen-specific killing). Clone F9, however, also lysed the highly NK sensitive targets YAC-1 and S194. This clone, in essence, displayed a dual specificity: antigen-specific lytic activity with a superimposed NK lytic activity. Lines of this type are not composed of two separate clones, one with antigen-specific lytic activity and the other with NK specificity, because recloning at limiting dilution produces daughter clones which also kill both YAC-1 and EL4, and in cold target competition assays there is cross inhibition between the NK and antigen-specific targets.

The existence of such dual reactive clones led to two important conclusions. Firstly, cells of indisputable T cell origin can express NK activity, providing confirmation at the clonal level of a hypothesis advanced earlier by Klein (10). Secondly, CTL are not restricted to the expression of an exquisite antigen-specific lytic activity, as previously supposed, but can display other, antigen non-specific and NK-related specifities.

The possibility that the supplementary lytic activity of these dualistic CTL clones was simply due to recognition of cross reacting antigens on certain target cells was ruled out by several observations. 1. The same NK-related pattern of reactivity was seen in clones raised in other strain combinations, including the reverse combination (b-anti-bml), and in clones derived from A/Sn responder cells, syngeneic with YAC-1 tumor cells. 2. NK reactivity in CTL is MHC-unrestricted because a partially NK-sensitive mutant lymphoma line, RI.E, which lacks both class I and class II MHC antigens, was lysed as well as its MHC-bearing parent line. 3. NK reactivity was inducible in CTL under conditions where there was no change in antigen-specific reactivity (see below). 4. Elegant experiments with similar CTL clones in rats showed that their antigen-specific reactivity, but not their NK-like reactivity, was inhibited by specific anti-idiotype serum (11), strongly suggesting that the two forms of lytic activity are mediated through distinct receptor systems. (Contradictory findings have been reported by Hercend et al. (12): a clone-specific antibody, apparently directed against a T-cell receptor (T_i)-like structure, inhibited lysis by a human NK-like clone, implying that T_i, or a related surface structure, can play a crucial role in non-antigen-specific cytolysis.)

INDUCIBILITY OF NK ACTIVITY IN CTL

A second major finding to emerge from experiments of the type described in Fig. 1, was that the NK state was inducible. The rate of induction varied markedly between clones. F3 showed the development of strong NK activity within 3 days, F1, F2, F5 and F8 showed varying degrees of induction by 7 days, and F4, F6 and F7 showed no inducibility by day 7, although after even longer exposure to high concentrations of CS some NK activity did eventually develop. We have now studied approximately 30 CTL clones in various strain combinations and all are eventually inducible for NK activity, although in one case 30 days exposure to the inducing conditions was necessary. We assume, without formal proof, that clones such as F9 are also inducible but are simply so exquisitely sensitive to the inducing factors that the 2% CS in the basal medium is adequate. Preliminary evidence indicates that the sensitivity to inducing factors is itself controlled by minor variations in the maintenance conditions for CTL clones. For example, early after antigenic stimulation, there is a dramatic augmentation in the magnitude of NK activity inducible by purified lymphokines, and increasing the amount of CS in the basal medium to 3% or 4% can lead to a marked increase in the subsequent kinetics of induction of NK activity.

We have also recently derived "helper-independent" CTL clones, which not only display antigen-specific lytic activity, but also secrete IL-2. Such clones can be induced to express NK activity, demonstrating that at least some T cells are multipotential, with the capacity to express helper function, classical CTL function, and NK function. Indeed, there is growing evidence that multipotentiality in post-thymic T cells may be the rule rather than the exception. Multifunctionality in human T cell clones has been extensively documented by Pawelec et al. (13). In the mouse, T cell clones displaying both antigen-specific suppressor activity and NK activity have been obtained (14), and recent studies have suggested that perhaps all CTL are derived from precursors which secrete IL-2 (15). Direct support for the conclusion that CTL are generally capable of expressing NK-like reactivities has come from the detailed limiting dilution studies of Shortman et al. (16). These studies are particularly important because they demonstrate that the induction of NK activity in established CTL clones is neither a peculiarity of strongly clonogenic CTL nor an idiosyncrasy associated with immortality *in vitro*. Very similar results have been reported for limiting dilution cultures of human CTL (17).

It should be emphasized that when new lytic reactivities first appear following the exposure of CTL to high concentrations of CS the specificity of these new reactivities is essentially the same as that of splenic NK cells (see in Fig. 1., F1 on day 7, F2 on day 7, F3 on day 3, F5 on day 7, F8 on day 7). This relationship holds true even when larger panels of target cells are used. Some minor variations, most likely attributable to non-specific factors which influence the affinity of effector-target interaction, are occasionally seen, such as a relatively higher lysis of S194 by the cloned lines than by spleen cells; but overall the correlation between the inducible lytic specificity in CTL and the specificity of splenic NK cells is always very high. These findings are in agreement with the earlier results of Dennert (18), and provide compelling evidence that NK specificity does not vary significantly at the clonal level. This is not to say that subpopulations of natural effector cells having quite distinct specificity do not exist (eg. NC cells), but rather that NK specificity is determined in a fundamentally different manner than is the clonotypic antigen specificity of B and T cells.

If the exposure of CTL to high concentrations of CS is prolonged, further dramatic changes can occur, the first of which is the acquisition of a second new lytic specificity which permits the lysis of a very wide range of both NK sensitive and NK resistant targets. This is well illustrated in Fig. 1. by the data for F3 at day 7 and F9 at day 3. This type of specificity has often been referred to as "NK-like" despite the fact that it bears no resemblence to the specificity of NK cells from normal individuals. The term promiscuous is more appropriate. The CTL have by now enlarged, have prominent cytoplasmic granules, and are proliferating. Up to this point, the alterations are entirely reversible: upon returning the cells to basal medium they cease to proliferate, have reduced size and granularity, and express only antigen-specific lytic activity. However, if the cells are kept in high CS medium, a series of irreversible biochemical changes occurs, resulting (sometimes after a crisis period) in cells which proliferate continuously in high CS medium, have lost all antigen-specific recognition function (both proliferative and cytotoxic), have gross alterations in surface glycoprotein display, and constitutively express promiscuous lytic activity. Optimal growth of these cells requires both IL-2 and other, as yet undefined, factors present in CS. Fig. 2 summarizes these changes.

The large granular cells that are the end products of this "differentiation" are extremely similar in ultrastructure, surface markers, and function to the "NK-like" clones we (1) and others (eg. 18, 19)

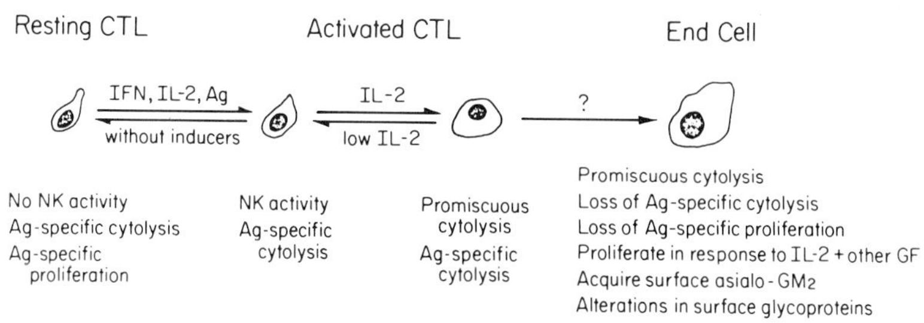

Fig.2. Differentiation of CTL in vitro.

derived by placing mouse spleen cells directly into high concentrations of CS. It now seems highly likely that many, perhaps all, of these clones arose from mature T cells. T cells proliferate so rapidly in high CS medium that they soon become predominant in bulk cultures, even when the starting populations are extensively depleted of T cells prior to culture. Our recent studies have clearly shown that Ly-2$^-$ T cells, like Ly-2$^+$ T cells, can differentiate into antigen-independent promiscuous killer cells, probably accounting for the Ly-2$^-$ NK-like cells described in many reports. The propensity of T cells to proliferate rapidly in response to IL-2, particularly if already activated *in vivo* or by traces of mitogen in the medium, has also plagued most attempts to study human NK cells in long-term culture. The abrupt change of phenotype, from that of NK cells to that of activated T cells, during culture of partially purified human NK cells (20), is most likely due to the outgrowth of T cells. Most clones obtained from such bulk cultures also have T cell characteristics (21, 22).

The possibility that NK cells can be grown in long-term culture is of course not excluded by such data. However, success will clearly depend on starting with extremely pure populations of NK cells, cloning directly at 1 cell/well or lower, and developing highly favorable growth conditions. Some encouraging results in this direction have recently been reported (23; Bolhuis, this volume). However, the specificity of the clones obtained so far differs markedly from that of blood NK cells, leaving some doubt as to their origin.

LYMPHOKINES WHICH REGULATE NK EXPRESSION IN CTL

Our recent experiments have focussed on the initial step in this transition, because it provides a unique opportunity to study the regulation of NK activity at the clonal level. Fractionation of CS revealed quite clearly that IL-2 was a major stimulator of NK activity in CTL. This was further confirmed using IL-2 purified to homogeneity from mouse LBRM-33 tumor cells or from human Jurkat tumor cells and with highly purified recombinant human IL-2. Unit-for-unit, however, such IL-2 preparations were less effective than crude CS, suggesting the presence of other factors in CS which either themselves can regulate NK activity or which promote responsiveness to IL-2.

It also became apparent that IFN was a potent regulator of NK activity in CTL. βIFN was markedly more effective than αIFN while, surprisingly, both natural and recombinant γIFN were virtually devoid of activity. Further experiments revealed that γIFN were also an extremely poor inducer of NK activity in fresh spleen cells. Induction of NK activity in CTL by βIFN and IL-2 differed in several important ways. Firstly, IFN acted entirely selectively, inducing

NK activity without affecting specific cytolysis, whereas IL-2 also induced some increase in specific cytolysis. Secondly, IFN induced NK activity very rapidly (eg. within 2 h), but IL-2 required >24 h. Thirdly, IFN caused no morphological change nor any alteration in net levels of protein, RNA, or DNA synthesis (Wayner and Brooks, unpublished), whereas IL-2 caused marked enlargement and metabolic activation of the cells. The possibility that Con A present in CS played a role in NK induction has been formally excluded by demonstrating that CTL grown for one month in medium containing lectin-free supernatant from the MLA-144 cell line or molecularly homogeneous IL-2 displayed no diminution in NK-inducibility by either IFN or IL-2.

Specific stimulator spleen cells will also induce NK activity in resting CTL, although whether this is due to a direct interaction with alloantigen or to indirect effects has not yet been determined. The most dramatic NK induction occurs, however, using a combination of stimuli. IL-2 and βIFN (but not γIFN) synergize strongly, as had previously been observed for the induction of NK activity in fresh spleen cells (24). Using an antibody putatively directed against the mouse IL-2 receptor (25), we have found that βIFN enhances IL-2 receptor expression on CTL, possibly explaining the synergistic relationship. Both βIFN and IL-2 show a synergistic interaction with antigen.

RELATIONSHIP OF NK CELLS TO T CELLS

The discovery that classical CTL can be induced to express NK specificity, and that this transition is controlled primarily by the two major regulators of NK activity *in vivo*, IFN and IL-2, demands that serious attention be given to the possibility that NK cells are in general derived from T cells. In the particular model system which we have studied, it is clear that the cells which express NK activity cannot be equated with NK cells *in vivo* if only for the reason that they usually express the $Ly\text{-}2^+$ antigen. However, the ease and frequency with which this induction occurs in CTL clones suggests either that it is part of a differentiation pathway which we are as yet unable to take to completion *in vitro*, or that it is a general property of some or all members of the T cell lineage, including pre-thymic T cells, and that it is these latter cells which are the predominant precursors of NK activity *in vivo*.

Fig. 3 illustrates some of the possible relationships of NK cells and T cells. Pluripotential stem cells (CFU-S) give rise to precursor cells for all lymphocytes and these in turn seed off committed T cell precursors which may pass through several pre-thymic differentiation states before entering the thymus and eventually becoming mature post-thymic T cells. Many fundamental questions regarding

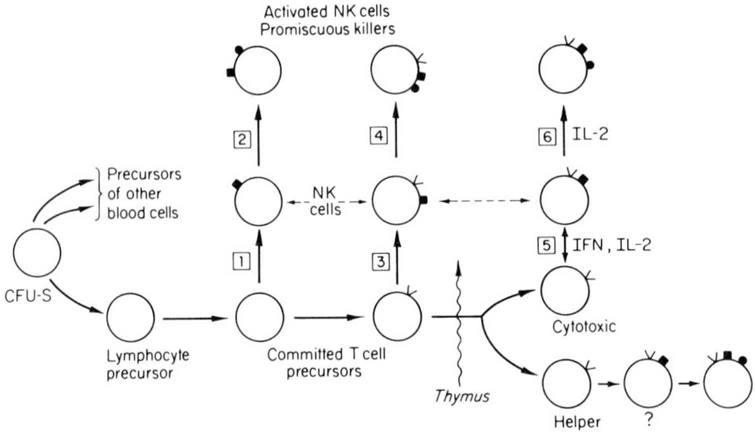

Fig.3. Possible interrelationships of NK cells and T cells. V,■,●:receptors for antigen, NK, and promiscuous killing, respectively.

this pathway are still unanswered. For example, we do not know whether commitment to helper/cytotoxic/suppressor functions (or to MHC class I or class II reactivity) occurs pre-thymically, intra-thymically or post-thymically: in Fig. 3 it has arbitrarily been shown at the post-thymic stage. Whether the thymus acts primarily to provide maturation signals or repertoire censorship or both is also unclear. However, the presence of CTL precursors in congenitally athymic mice (26) or in irradiated, thymectomized, bone-marrow reconstituted mice (27), provides clear evidence that substantial maturation of cells of CTL lineage, including the expression of functional antigen-specific receptors (T_i), can occur in the absence of the thymus.

In Fig. 3, the possible derivation of NK cells from three levels of T cell maturation - pre-thymic cells lacking T_i, pre-thymic cells expressing T_i, and post-thymic T cells - is shown. The development of NK and promiscuous recognition is illustrated by the acquisition of new receptor structures. Whether these recognition structures would necessarily be acquired sequentially and, indeed, whether classical receptor-ligand interactions account for these recognition patterns, is unclear. Development of NK or promiscuous lytic cells from precursor cells at different stages of T cell maturation would account for the heterogeneity of NK cells *in vivo* (eg Thy-1^+ and Thy-1^- subsets) and the even greater heterogeneity seen after *in vitro* culture. For example, the LAK cells described by Grimm *et al.* (28) might arise via pathway 1 + 2. Anomalous killer (AK) cells could arise in a variety of ways, eg. via acquisition of a promiscuous recognition pattern by relatively mature NK cells (29, 30; pathways 2 + 4) or by polyclonal activation of immature (31; pathways 3 + 4) or mature (32, 33; pathways 5 + 6) CTL. Because of quantitative and qualitative deficiencies in signal presentation in vitro, cells having combinations of phenotypic characteristics not readily detected on cells in vivo may arise. The possibility that cells acquiring NK status can inter-differentiate, by acquiring or losing T_i (horizontal broken arrows) or revert to conventional CTL precursors (reversal of steps 1, 3, and 5) is intriguing. It is worth noting that, in man, an antibody recognizing only NK cells and precursor CTL has been described (34; Saksela, this volume), and, in the mouse, NK cells and precursor CTL share the asialo-GM_1 marker (35). The absence of the thymus presumably creates a block or diversion in T cell maturation, which may account for the high NK activity and unusual surface phenotype (high Thy-1) of NK cells in athymic mice. Various experiments have demonstrated that the thymus, directly or indirectly, influences the maturation of NK cells (36, 37).

The unravelling of the intricate relationship of NK cells and the various T cell subsets and precursors presents a formidable but exciting challenge to immunologists, cell biologists and molecular biologists. The timely availability of methods for probing T_i gene rearrangements and gene expression, the continuing development of new and more useful monoclonal antibodies, and the increasing availability of highly purified lymphokines and growth factors, make the precise delineation of the pathways of T cell and NK cell differentiation an exciting and realistic scenario in the next few years.

REFERENCES

1. Brooks, C.G., Kuribayashi, K., Sale, G.E. and Henney, C.S. (1982). J. Immunol. 128, 2326.
2. Brooks, C.G., Urdal, D., and Henney, C.S. (1983). Immunol. Rev. 72, 43.
3. Brooks, C.G. (1983). Nature 305, 155.
4. Brooks, C.G., Burton, R.C., Pollack, S.B. and Henney, C.S. (1983). J. Immunol. 131, 1391.
5. Olabuenaga, S., Brooks, C.G., Gillis, S., and Henney, C.S. (1983). J. Immunol. 131, 2386.
6. Brooks, C.G. and Henney, C.S. (1984). In "The Interleukins. Contemporary Topics in Molecular Immunology" (S. Gillis and F.P. Inman, eds.). Plenum Press, New York.
7. Brooks, C.G. (1985). In "Immunobiology of Natural Killer Cells" (E. Lotzova and R.B. Herberman, eds.) CRC Press, Boca Raton.
8. Glasebrook, A.L. and Fitch, F.W. (1980). J. Exp. Med. 151,876.
9. Pease, L.R., Schulze, D.H., Pfaffenbach, G.M., and Nathenson, S.G. (1983). Proc. Natl. Acad. Sci. 80, 242.
10. Klein, E. (1980). Immunol. Today 1, iv.
11. Binz, H., Fenner, M., Frei, D., and Wigzell, H. (1983). J. Exp. Med. 157, 1252.
12. Hercend, T., Meuer, S., Brennan, A., Edson, M.A., Acuto, O.,

Reinhertz, E., Schlossman, S.F., and Ritz, J. (1983). J. Exp. Med. 158, 1547.
13. Pawelec, G., Schneider, E.M., and Wernet, P. (1983). Immunol. Today 4, 275.
14. Heuer, J., Opalka, B., Rassat, J., Themann, H., and Kolsch, E. (1983). Eur. J. Immunol. 13, 551.
15. von Boehmer, H., Kisielow, P., Leiserson, W., and Haas, W. (1984). J. Immunol. 133, 59.
16. Shortman, K., Wilson, A., and Scollay, R. (1984). J. Immunol. 132, 584.
17. Moretta, A., Pantaleo, G., Mingari, M.C., Melioti, G., Moretta, L., and Cerottini, J-C. (1984). Eur. J. Immunol. 14, 121.
18. Dennert, G., (1980). Nature 287, 47.
19. Nabel, G., Bucalo, L.R., Allard, J., Wigzell, H., and Cantor, H. (1982). J. Exp. Med. 153, 1582.
20. Timonen, T.T., Ortaldo, J.R., Stadler, B.M., Bonnard, G.D., Sharrow, S.O., and Herberman, R.B. (1982). Cell. Immunol. 72, 178.
21. Abo, T., and Balch, C.M. (1983). Eur. J. Immunol. 13, 383.
22. Allavena, P., and Ortaldo, J.R. (1984). J. Immunol. 132, 2363.
23. van de Griend, R.J., van Krimpen, B.A., Ronteltap, C.P.M., and Bolhius, R.L.H. (1984). J. Immunol. 132, 3185.
24. Kuribayashi, K., Gillis, S., Kern, D.E., and Henney, C.S. (1981). J. Immunol. 126, 2321.
25. Malek, T., Robb, R., and Shevach, E. (1983). Proc. Natl. Acad. Sci. 80, 5694.
26. Gillis, S., Union, N.A., Baker, P.E., and Smith, K.A. (1979). J. Exp. Med. 149, 1460.
27. Duprez, V., Maziarz, R., Weinberger, O., and Burakoff, S.J. (1984). J. Immunol. 132, 2185.
28. Grimm, E.A., Ramsey, K.M., Mazumder, A., Wilson, D.J., Syen, J.Y., and Rosenberg, S.A. (1983). J. Exp. Med. 157, 884.
29. Strassman, G., Bach, F.H., and Zarling, J.M. (1983). J. Immunol. 130, 1556.
30. Karre, K., Seeley, J.K., Eriksson, E., Burton, R.C., and Kiessling, R. (1983). J. Exp. Med. 157, 285.
31. Torten, M., Sidell, N., and Golub, S.H. (1983). J. Exp. Med. 156, 1545.
32. Bartlett, S.P., and Burton, R.C. (1982). J. Immunol. 128, 1070.
33. Burns, G.F., Triglia, T., and Werkmeister, J.A. (1984). J. Immunol. 133, 1656.
34. Nieminen, P., and Saksela, E. (1984). J. Immunol. 133, 702.
35. Beck, B.N., Gillis, S., and Henney, C.S. (1982). Transplantation. 33, 118.
36. Herberman, R.B., and Holden, H.T. (1978). Adv.Cancer Res. 27, 305.
37. Shivola, M., and Hurme, M. (1984). Immunology. 53, 17.

RECOGNITIVE HETEROGENEITY OF NK CELLS[1]

Graham Pawelec
Peter Wernet

Immunology Laboratory, Medizinische Klinik,
D-7400 Tübingen, West Germany

Walter Newman

Department of Molecular Immunology,
Ortho Pharmaceutical Corporation,
Raritan, New Jersey

Freshly isolated human NK cells, present without prior purposeful priming mostly in the large granular lymphocyte (LGL) fraction (1) of peripheral blood mononuclear cells from healthy donors, spontaneously lyse a variety of tumour and virus-infected, as well as certain normal, target cells. Because of the multitude of different cells lysed, one long-standing question regarding the specificity of the NK system has been whether NK cells recognise common target determinants shared between different susceptible cells, or whether they recognise diverse determinants via discrete receptors heterogeneously expressed on different NK cell subsets.

Some evidence for recognitive heterogeneity of NK cells has accrued from results of experiments where effector cells were absorbed on monolayers of a particular target cell, followed by assay of unabsorbed cells on different targets (2-6). Data consistent with heterogeneity of recognition have also been derived from interaction analysis of large panel

[1]Supported in part by the DFG through SFB 120 (A1).

ISBN 0-12-341370-2

studies, as well as from cold target cell or isolated target structure cross-competition studies (6-8). Further intimations of heterogeneous recognitive and lytic activities of NK cells were derived from experiments in which effector cells were separated into subsets on the basis of their different expression of cell surface structures (9,10). However, the ultimate selection process to aid in a definitive answer to the question of NK specificity would be to study effector activities of a single isolated cell. This could be approached in single cell cytotoxicity (CTX) assays using mixtures of target cells, but involves significant practical difficulties when studying more than two different targets at one time. Nonetheless, results of conjugate formation assays were also consistent with a limited recognitive heterogeneity of NK cells (11), although not all cells binding to targets necessarily lyse them.

The application of long-term culture and cloning techniques to provide homogeneous populations of effectors derived from a single cell would have the advantage of providing large amounts of material for extended analyses. Recently, propagation and cloning of cells which mediate patterns of CTX very similar to those of freshly isolated NK cells has become possible by using growth factors to maintain cultures for extended periods. Most NK and NK-like cells from various sources appear to require the T cell growth factor interleukin 2 (IL 2), as shown by their propagation in recombinant IL 2 (12,13). Although freshly isolated human LGL lack the majority of T cell markers, many express receptors for IL 2, and respond to this growth factor without further stimulation (14). T cell markers are then rapidly expressed by the cultured cells and they may be maintained for some time in culture (14). This enables, for the first time, an extensive analysis of the recognitive specificities of cloned populations. Clones have been derived from LGL isolated with monoclonal antibodies (moAb) or by physical methods, some of which were found to mediate heterogeneous CTX patterns (15,16). Certain clones lysed only one particular NK-susceptible target cell, for example only K562 (15), or only K562, or Daudi, or MOLT 4 (16). These results are clearly consistent with clonal heterogeneity amongst NK-active cells expressing receptor structures for non-ubiquitous target determinants on different NK-susceptible cells. However, other clones from the same experiments retained broad CTX patterns on numerous target cells. Furthermore, within certain isolated subpopulations of effectors, there appears to be little or no evidence for recognitive heterogeneity following cloning. Thus, a subset of LGL characterised by reactivity with the moAb B73-1 gave rise only to clones displaying broad NK CTX patterns on the range of target cells tested (17). Similarly, in mouse, CTX patterns of clones in most reports thus far would argue against recognitive heterogeneity (for example, ref. 18), but again this may be related to subset preselection processes, since some indications for restricted NK CTX activity of murine clones have been forthcoming (19).

A major dilemma in interpreting the results from such experiments stems from the possibility that cultured NK-active cells may not be representative of spontaneously CTX freshly isolated NK cells. At present, therefore, it remains prudent to designate the CTX activity of cultured cells as "NK-like", particularly since a variety of cells, differing with regard to their origins and surface marker phenotypes, appear able to lyse NK-susceptible target cells. Thus, NK-like CTX is generated in cells stimulated in mixed lymphocyte cultures (MLC), even after prior depletion of NK activity from the responding cell population (20). We have found that up to 20% of IL 2-dependent clones derived from limiting dilution of MLC-sensitised lymphocytes can mediate NK-like CTX (21-24). NK-like activity was detectable in these clones from the earliest time point after cloning at which CTX could be assayed, suggesting that they may have been derived from previously NK-active cells, or had very rapidly acquired NK-like CTX during culture. Such lytic clones were characterised as NK-like for the following reasons: 1) CTX was limited to cell line targets which were susceptible to lysis by fresh NK cells, and was not directed against resistant targets such as normal lymphocytes, activated B or T cells, most B cell lines, or PHA-pretreated normal lymphocytes. They did not manifest "anomalous" killing of NK-resistant autologous or allogeneic B cell lines; 2) clones propagated in the absence of lectins, or (short-term) in the absence of IL 2, retained NK-like CTX activity; 3) cold target cross-competition experiments suggested that the target structures recognised by the clones were not expressed by normal lymphoid cells; 4) CTX activity was boosted by gamma interferon (IFN); and 5) morphology was usually typically LGL, although with a greater range of T cell markers than fresh, but not short-term cultured, LGL (24). Thus, in terms of CTX patterns, positive regulation by IFN, and morphology, the clones were strictly NK-like. The CTX activities of representative clones are illustrated in Table I, and previously reported in detail using a greater variety of target cells (23). Some of these MLC-derived NK-like clones manifested heterogeneous patterns of lytic activity on NK-susceptible targets, consistent with a clonal distribution of receptors for a distinct variety of target determinants. Certain clones lysed only one target cell, whereas others lysed a limited variety of targets. Again, cold target competition

TABLE I Restricted cytotoxic activities of MLC-derived NK-like T cell clones

	Target cells[a]					
Effector	K562	HSB2	B-LCL	T-LCL	T-LCL + PHA	K562 + IFN
12-1	422[b]	361	0.1	0.1	nt[c]	656
12-9	323	754	0.1	0.1	nt	nt
29-15	216	201	0.1	0.7	1	423
29-16	153	225	0.1	0.1	0.1	186
29-21	266	327	0.1	0.1	nt	728
29-23	433	684	0.4	0.1	0.1	487
29-30	528	347	0.1	0.8	5	756
29-31	2	263	0.1	0.1	3	4
38-37	128	1	2	nt	nt	nt
38-27[d]	351	175	198	365	nt	nt

[a] Target cells were ^{51}Cr-labelled K562, HSB2, EBV-transformed B cell line (B-LCL) from the donor of the MLC stimulating cells, IL 2-dependent T cell line (T-LCL) from the same donor, T-LCL pretreated with 1% PHA for 1 hour, or K562 in the presence of 100 u/ml IFN.

[b] CTX activity presented as lytic units (LU) per 10^7 effector cells. One LU was defined as the number of effector cells giving 25% specific isotope release from 2 x 10^3 target cells in a 4 hour assay.

[c] nt = not tested.

[d] Clone 38-27 expresses allospecific as well as NK-like CTX.

studies suggested that for clones lysing only one cell, unique target determinants were being recognised. One exception to the exclusively NK-like CTX patterns of these clones is exemplified by clone 38-27 in Table I. This effector mediated HLA-restricted allospecific CTX against normal target cells, as well as NK-like CTX. For this clone, normal target cells and NK-susceptible target cells failed to cross-compete, suggesting the utilisation of different receptor structures for the recognition of the different target cells.

If it could be demonstrated that NK-like effectors and freshly isolated NK cells employed the same receptor structures to recognise their target cells, one of the difficulties of interpreting NK-like CTX mediated by such clones would be reduced. In an attempt to clarify this question, we have employed the moAb 13.1 to inhibit target cell recognition and lysis. This moAb was produced by immunising mice with human LGL and was selected in screening assays based on the blockade of NK lysis by freshly prepared NK cells. It is directed to an epitope of 200 kD "T200" molecules furthest removed from the cell membrane (25), and was found to block, at the effector cell level, lysis by fresh NK cells of erythro/myeloid, but not T leukaemic, target cells (26). It interferes with a post-binding, pre-Ca^{2+}-dependent, prelethal hit stage of lysis, termed "triggering" (27). It was suggested (reviewed in ref. 28) that T200 might be, or form part of, a receptor structure on NK cells. MoAb 13.1 was accordingly used to investigate further the recognitive specificities of MLC-derived NK-like clones. It was found that the pattern of lytic blockade on these NK-like cells precisely corresponded to that observed on freshly isolated NK cells. Thus, kill of K562, but not of HSB2, target cells was inhibited in a dose dependent manner (13). Representative results using saturating amounts of moAb 13.1 are shown in Table II. A similar finding with one NK-like clone was also recently reported by others (29). These findings suggest that identical receptor structures are involved in recognition of NK target determinants by NK-like T cell clones as well as by freshly isolated NK cells. Such results further imply that the T cell receptor for antigen present on helper and CTX T cells, which is known to be a 20 kD T3-associated 49/43 kD heterodimer and not a monomeric 200 kD molecule, is uninvolved in the recognition of NK target determinants by these NK-like clones.

However, that structures similar to T cell receptors may be at least sometimes involved in target recognition by certain NK-like clones is implied by a report that some such human clones expressed unique "clonotypic" determinants carried on T3-linked 49/43 kD "Ti" complexes. MoAb against these determinants either inhibited CTX of the clone against all target cells tested (30), or only against a particular type of target cell (31). The utilisation of T cell receptor complexes in NK-like target recognition was further implied by the ability of anti-T3 moAb to block lysis (13,30), although inhibition was weaker and more variable than that caused by anti-T200 (13) or anti-clonotypic (30) antibodies. However, allowing the moAb to modulate T3, and presumably also any Ti structures, from the cell surface, failed completely to prevent NK-like lysis of target cells. This

TABLE II Inhibition by moAb 13.1 of NK-like CTX by cloned effector cells

	Target cell			
	-13.1		+13.1[a]	
Effector	K562	HSB2	K562	HSB2
NAL[b]	46[c]	25	18	27
12-1	41	78	13	70
12-9	49	53	13	51
29-15	38	25	6	26
29-16	43	27	9	28
29-21	38	57	6	60
29-23	29	33	8	39
29-30	41	29	4	27
29-31	2	48	2	53
38-37	25	3	3	4
38-27	49	29[d]	11	30[d]

[a] 10μg/ml 13.1 was present for the duration of the assay.

[b] Non-adherent lymphocytes.

[c] Percent specific isotope release from 2 x 10^3 ^{51}Cr-labelled target cells in a 4 hour assay.

[d] For dual allospecific and NK-like clone 38-27 this target is the specific priming cell rather than HSB2.

would suggest no absolute requirement for T3-Ti complexes in NK-like recognition. Moreover, certain T cell clones thus treated have been reported to acquire, not lose, NK-like CTX (32). Further arguments against the involvement of T cell receptors in NK-like lysis derive from moAb blocking experiments with certain clones possessing dual CTX for NK-susceptible, and for normal allogeneic, target cells. Thus, lysis only of normal targets by such a clone was blocked by anti-Leu 2 moAb, which detects a structure thought to form part of the recognition complex specific for HLA class I antigens

(33). Conversely, lysis only of NK targets by clone 38-27 (Table II) was blocked by moAb 13.1. Furthermore, as mentioned above, normal and NK-susceptible target cells failed to cross-compete for lysis by this clone. Taken together, the above considerations would seem to reduce the likelihood of a T3-linked 49/43 kD T cell receptor structure being critically involved in NK target recognition by at least the majority of NK-like clones. The same is of course true for fresh NK activity.

The application of moAb 13.1 to block CTX also allowed an approach to be made to a further question concerning the specificity of recognition by NK-like clones. Although the isolation of a minority of NK-like clones lysing only a single target cell argued for clonal distribution of receptors for different target determinants (23), the majority lysed a broader variety of cells. The latter would be compatible with the idea either that single clones carry multiple receptors for different target determinants, or recognition structures for common determinants widely distributed between different target cells. If the former were the case, moAb against such receptors might be expected to block recognition and lysis of some but not all target cells. whereas if only one receptor was involved, lysis of all target cells should be blocked by a relevant moAb. As can be seen from Table II, the former is the case. Thus, clones lysing only K562 (for example, clone 38-37) were inhibited by moAb 13.1, whereas clones lysing only HSB2 were not. However, clones lysing both types of targets were nonetheless inhibited only in their killing of K562, suggesting the presence of multiple receptor structures on the same cloned cell line. A continuous distribution of different receptors on NK active clones might thus be envisaged, ranging from one extreme of only a single receptor, to the other extreme of a large number of, or all possible, receptors. However, most clones would fall between these two extremes, and express varying numbers of different receptors. This would account for the generally low level of recognitive heterogeneity observed in selected subsets of NK cells. In this case, arguments for clonally restricted and unrestricted lytic activities would both be applicable to oligoclonal effector populations.

Recently there have been a number of reports of NK-like CTX acquired by cloned T cells initially lacking this lytic activity, as well as of clones mediating dual allospecific CTX and NK-like CTX. These results may suggest a rather widespread distribution of NK-like activity amongst functionally different lymphocytes. For example, NK-like activity may be elicited in allospecific CTX, and even in non-CTX, clones after treatment with moAb against T3 surface structures (32). Proliferative "helper" clones maintained for extended periods in culture lose antigen specificity and may acquire NK-like CTX (34). The question again arises as to the relationship of such NK-like activities to spontaneous NK CTX. Once more, at present a definitive answer is lacking, but at least in the latter case, aged NK-like clones have been shown to have exactly the same pattern of lytic blockade by moAb 13.1 as fresh NK cells and MLC-derived NK-like clones. This again suggests that all these NK active cells utilise the same receptor structures. It seems most probable that a variety of different effector cells capable of mediating NK-like CTX would contribute to lytic activity collectively designated as NK. Some of these may express stable NK characteristics upon culture, whereas others may mediate NK-like CTX only under certain conditions. Thus far, examples of the former are rare. One report concerns phenotypic (non-T cell) and functional (NK CTX) stability after cloning and expansion of a preselected subset of LGL (17): such cells may indeed belong to a separate NK lineage, with their expression of NK CTX not being dependent on a particular differentiation phase within that lineage. In contrast, the reversible induction of NK-like CTX by high levels of IL 2 in mouse clones (35), and irreversible age-associated acquisition of NK-like CTX by human T cell clones (34), could be examples of NK-like activity associated only with a particular differentiation state of the cells. Such findings might be borne in mind when considering the origins of NK cells in vivo. Thus, it may be that at least some NK cells could be derived from aged, previously immunologically specific, T cells. The proposed regulatory and the CTX activities of NK cells in vivo would be equally applicable regardless of the origins of the effectors, whether from T cells, or completely independent lineages.

A simple minimal unifying model for the specificity of NK-like cells would thus propose the existence of widespread binding capacity of effectors to targets, which would not itself be sufficient to program lysis (36). Following initial binding, specific recognition of the target cells via receptors of limited distribution involving T200 molecules would take place. Only this recognitive event would result in the delivery of the lethal hit by cells possessing the appropriate lytic machinery. These would be predominantly LGL, but could also be experimentally manipulated T cells (by culture, antibody treatment or lymphokine treatment) which appear to depend on the same T200 molecules for recognition and lysis. There seems to be a limited recognitive heterogeneity amongst at least the latter NK-active cells, which is explicable by the possession by most NK-like clones of multiple receptors for different target determinants on susceptible cells. However, the existence of clones with very limited receptor "repertoires" suggests a clonal distribution of recognition structures for discrete target determinants.

REFERENCES

1. Timonen, T., Ortaldo, J.R., and Herberman, R.B. (1981). J. Exp. Med. 153, 569.
2. Philips, W.H., Ortaldo, J.R., and Herberman, R.B. (1980). J. Immunol. 125, 2322.
3. Saksela, E., Timonen, T., Ranki, A., and Häyry, P. (1979). Immunol. Rev. 44, 71.
4. Jensen, P.J., Weston, P.A., and Koren, H.S. (1982). In "NK Cells and Other Natural Effector Cells" (R.B. Herberman, ed.), p. 699, Academic Press, New York.
5. Ortaldo, J.R., Oldham, R.K., Cannor, G.B., and Herberman, R.B. (1977). J. Natl. Cancer Inst. 59, 77.
6. Takasugi, M., and Mickey, M.R. (1976). J. Natl. Cancer Inst. 57, 255.
7. Roder, J.C., Karre, K., and Kiessling, R. (1981). Progr. Allergy 28, 66.
8. Tai, A.S., Safilian, B., and Warner, N.L. (1982). Human Immunol. 4, 123.
9. Klein, E., Masucci, M.G., Masucci, G., and Vanky, F. (1982). Molec. Immunol. 19, 1323.
10. Fitzgerald, P.A., Kirkpatrick, D., and Lopez, C. (1982). In "NK Cells and Other Natural Effector Cells" (R.B. Herberman, ed.), p. 73, Academic Press, New York.
11. Timonen, T., Ortaldo, J.R., and Herberman, R.B. (1982). J. Immunol. 128, 2514.
12. Ortaldo, J.R., Mason, A.T., Gerard, J.P., Henderson, L.E., and Farrar, W. (1984). Immunobiol. 167, 191.
13. Pawelec, G., Newman, W., Schwulera, U., and Wernet, P. (1984). Cell. Immunol., in press.
14. Ortaldo, J.R., Timonen, T.T., Vose, B.M., and Alvarez, J.A. (1982). In "The Potential Role of T Cells in Cancer Therapy" (A. Fefer and A. Goldstein, eds.), p. 191. Raven Press, New York.
15. Hercend, T., Reinherz, E.L., Meuer, S., Schlossman, S.F., and Ritz, J. (1983) Nature 301. 158.
16. Allavena, P., and Ortaldo, J.R. (1984). J. Immunol. 132, 2363.
17. Van de Griend, R.J., Van Krimpen, B.A., Ronteltap, C.P.M., and Bolhuis, R.L.H. (1984).

J. Immunol. 132, 3185.
18. Dennert, G., Yogeeswaran, G., and Yamagate, S. (1981). J. Exp. Med. 153, 545.
19. Kedar, E., and Herberman, R.B. (1982). In "NK Cells and Other Natural Effector Cells" (R.B. Herberman, ed.), p. 859, Academic Press, New York.
20. Seeley, J.K., Masucci, G., Poros, A., Klein, E., and Golub, S. (1979). J. Immunol. 123, 1303.
21. Pawelec, G., Rehbein, A., Kahle, P., Britzelmeier, C., Ziegler, A., Müller, C., and Wernet, P. (1981). In "Leukaemia Markers" (W. Knapp, ed.), p. 355, Academic Press, London.
22. Pawelec, G., Hadam, M., Ziegler, A., Lohmeyer, J., Rehbein, A., Kumbier, I., and Wernet, P. (1982). J. Immunol. 128, 1892.
23. Pawelec, G., Hadam, M., Schneider, E.M., and Wernet, P. (1982). J. Immunol. 129, 2271.
24. Schneider, E.M., Pawelec, G., Shi, L., and Wernet, P. (1984). J. Immunol. 133, 173.
25. Newman, W., Fast, R.D., and Rose, L.M. (1983). J. Immunol. 131, 1742.
26. Newman, W. (1982). Proc. Natl. Acad. Sci. (USA). 79, 3858.
27. Targan, S., and Newman, W. (1983). J. Immunol. 131, 1149.
28. Newman, W., Targan, S.R., and Fast, L.D. (1984). Molec. Immunol., in press.
29. Van de Griend, R.J., Giphart, M.J., Van Krimpen, B.A., and Bolhuis, R.L.H. (1984). J. Immunol. 133, 1222.
30. Hercend, T., Meuer, S., Brennan, A., Edson, M.A., Acuto, O., Reinherz, E.L., Schlossman, S.F., and Ritz, J. (1983). J. Exp. Med. 158, 1547.
31. Hercend, T., Meuer, S., Brennan, A., Edson, M.A., Acuto, O., Reinherz, E.L., Schlossman, S.F., and Ritz, J. (1984). Cell. Immunol. 86, 381.
32. Spits, H., Yssel, H., Leeuwenberg, J., and De Vries, J.E. (1984). Eur. J. Immunol., in press.
33. Moretta, A., Panteleo, G., Mingari, M.C., Melioli, G., Moretta, L., and Cerottini, J.-C. (1984). Eur. J. Immunol. 14, 121.
34. Pawelec, G., Schneider, E.M., and Wernet, P. (1984). J. Immunol., in press.
35. Brooks, C.G. (1983). Nature 305, 155.
36. Piontek, G.E., Kiessling, R., Grönberg, A., and Ährlund-Richter, L. (1982). In "NK Cells and Other Natural Effector Cells" (R.B. Herberman, ed.), p. 705, Academic Press, New York.

INDEX